ACS SYMPOSIUM SERIES 677

Computational Thermochemistry

Prediction and Estimation of Molecular Thermodynamics

Karl K. Irikura, EDITOR
National Institute of Standards and Technology

David J. Frurip, EDITOR
Dow Chemical Company

Developed from a symposium sponsored by the Division
of Computers in Chemistry at the 212th National Meeting
of the American Chemical Society,
Orlando, Florida,
August 25–29, 1996

American Chemical Society, Washington, DC

Library of Congress Cataloging-in-Publication Data

Computational thermochemistry: prediction and estimation of molecular thermodynamics / Karl K. Irikura, editor, David J. Frurip, editor.

    p.   cm.—(ACS symposium series, ISSN 0097–6156; 677)

    "Developed from a symposium sponsored by the Division of Computers in Chemistry at the 212th National Meeting of the American Chemical Society, Orlando, Florida, August 25–29, 1996."

    Includes bibliographical references and indexes.

    ISBN 0–8412–3533–3

    1. Thermochemistry—Mathematical models—Congresses.

    I. Irikura, Karl K., 1963–   . II. Frurip, David J. III. American Chemical Society. Division of Computers in Chemistry. IV. American Chemical Society. Meeting (212th : 1996: Orlando, Fla.) V. Series.

QD510.C66   1998
541.3'6'011319—dc21   97–52708
                                                   CIP

This book is printed on acid-free, recycled paper.

Foreword

THE ACS SYMPOSIUM SERIES was first published in 1974 to provide a mechanism for publishing symposia quickly in book form. The purpose of this series is to publish comprehensive books developed from symposia, which are usually "snapshots in time" of the current research being done on a topic, plus some review material on the topic. For this reason, it is necessary that the papers be published as quickly as possible.

Before a symposium-based book is put under contract, the proposed table of contents is reviewed for appropriateness to the topic and for comprehensiveness of the collection. Some papers are excluded at this point, and others are added to round out the scope of the volume. In addition, a draft of each paper is peer-reviewed prior to final acceptance or rejection. This anonymous review process is supervised by the organizer(s) of the symposium, who become the editor(s) of the book. The authors then revise their papers according to the recommendations of both the reviewers and the editors, prepare camera-ready copy, and submit the final papers to the editors, who check that all necessary revisions have been made.

As a rule, only original research papers and original review papers are included in the volumes. Verbatim reproductions of previously published papers are not accepted.

ACS BOOKS DEPARTMENT

Contents

METHODS BASED ON MOLECULAR-ORBITAL OR DENSITY-FUNCTIONAL THEORY

APPLICATIONS

APPENDICES

Preface

MOLECULAR THERMOCHEMISTRY is the study of the quantitative thermodynamic stability of molecules. It is central to chemistry and critical for many industries. Although experimental data are available for thousands of molecules, there are innumerably more molecules that are important and interesting but for which no data exist. Consequently, many methods have been developed for estimating or predicting thermochemical data for new molecules, ranging from purely empirical schemes to state-of-the-art, ab initio, quantum-mechanical methods. Nearly all the methods rely on computers in some way, some very intensively. The cost of computational power is decreasing far faster than that of any experimental technique. Furthermore, rapid advances in both quantum theory and algorithm development are greatly amplifying the power of available computers. It is clear that computational methods will soon dominate the field of thermochemistry.

This volume is the first book on the topic of computational thermochemistry. Our ambitious goal is to describe all the major methods used for estimating or predicting molecular thermochemistry. The book serves as an introduction and textbook for newcomers to the field and also as a reference for current practitioners. Anyone who uses chemical thermodynamic data will benefit from this book including students, chemists, and chemical engineers, from occasional users of estimation methods to researchers developing new methods. Appendices direct the reader to relevant software and databases provide worked examples of actual calculations.

This volume was developed from a symposium titled "Computational Thermochemistry", presented at the 212th National Meeting of the American Chemical Society, sponsored by the ACS Division of Computers in Chemistry, in Orlando, Florida, August 25–29, 1996. To the best of our knowledge, this was the first symposium on the topic of computational thermochemistry. It was organized by the editors of this book to encourage mixing between the various computational and theoretical research communities whose work bears on molecular thermochemistry. A two-hour panel discussion was incorporated in the symposium to identify specific needs for thermochemical data and to encourage communication between the users and suppliers of data. The major conclusions from the panel discussion are included in the overview to this book (Chapter 1).

Acknowledgments

We would like to thank the contributing authors and the staff at ACS Books for their time, expertise, and enthusiasm. We would also like to acknowledge Dr. Tyler B. Thompson for leading us to work together on this and other projects. Finally, we are pleased to acknowledge generous financial support for the symposium from the following sponsors: National Institute of Standards and Technology (Physical and Chemical Properties Division); International Business Machines (Scientific and Technical Systems and Solutions Group); Donors of the Petroleum Research Fund (administered by the ACS); Gaussian, Inc.; and Cray Research, Inc.

KARL K. IRIKURA
Physical and Chemical Properties Division
National Institute of Standards and Technology
Gaithersburg, MD 20899

DAVID J. FRURIP
Dow Chemical Company
Analytical Sciences, Building 1897F
Midland, MI 48667

August 12, 1997

OVERVIEW

Chapter 1

Computational Thermochemistry

Karl K. Irikura[1] and David J. Frurip[2]

[1]**Physical and Chemical Properties Division, National Institute of Standards and Technology, Gaithersburg, MD 20899**
[2]**Analyical Sciences Laboratory, Dow Chemical Company, Building 1897F, Midland, MI 48667**

Demands for thermochemical data far exceed the current capabilities for experimental measurements. Fortunately, there are many methods for estimating gas-phase molecular thermochemistry. Current techniques are summarized and compared on the basis of range of applicability, cost, and reliability.

By *computational thermochemistry* we mean any method that involves using a computer to make thermochemical predictions. This ranges from empirical methods that require only an instant on a slow personal computer to high-level quantum calculations that require days on the most powerful supercomputers.

The thermochemistry that we address in this chapter is more restricted, emphasizing gas-phase enthalpies of formation. We recognize that most practical work is not in the gas phase but in the condensed phase, such as in solution and at surfaces. In many cases, however, the thermochemistry of a chemical reaction is not very sensitive to the solvent or other environment and gas-phase values are adequate. Furthermore, there are methods for estimating enthalpies of vaporization and of solution. In this book, Domalski describes a means for estimating aqueous enthalpies, Giesen et al. describe a method for estimating solvation enthalpies, and Chickos et al. describe a procedure for estimating enthalpies and entropies of phase changes. Emerging methods for predicting the energetics of heterogeneous processes (*1,2*) are beyond the scope of this book. A recent handbook describes estimation methods for several properties relevant to environmental chemistry (*3*). In addition, the classic reference by Reid et al. (*4*) includes methods for estimating a wide variety of thermochemical and thermophysical properties.

Scope and Overview

The purpose of this chapter is to summarize all the available methods for predictive thermochemistry. We do not expect to have achieved this ambitious goal. We hope that readers will alert us to any important omissions so that we can keep our list up-to-date.

More pragmatically, we have attempted to assemble enough information to allow a chemist or engineer to identify methods that suit his or her requirements and resources. There are generally three principal considerations. (1) The *range of applicability* determines whether a method can be applied to a given molecule. (2) The *cost* of a method includes not only the price of software (see Appendix A) but also the computational requirements and the amount of training needed to use each method. Cost can also affect the range of applicability, typically by limiting the size of molecules that can be handled. (3) *Reliability* determines whether the method produces useful results, and may be described in terms of accuracy (i.e., mean errors) and precision (i.e., maximum errors). With these meanings, "accuracy" indicates the error in a typical prediction and "precision" indicates the worst error that might reasonably be expected. Table I provides a qualitative summary, based primarily upon our own experience and opinions.

For many molecules, experimental thermochemical data are available and no estimations are necessary. Reliable experimental data are also critical for developing the parameters of empirical and semiempirical predictive methods. Some of the relevant databases and associated software are described briefly in Appendix A. But for less-common functional groups, inorganic or organometallic compounds, reactive intermediates (radicals and ions), or metastable molecules (transition states) one is seldom lucky enough to find good experimental data, and estimation methods are essential (although less reliable than for common molecules).

For ordinary, stable organic molecules containing only common functional groups, purely empirical estimation methods are very inexpensive and the most reliable. Such methods are heavily-parametrized schemes for interpolating between values in experimental databases. Most of these methods were developed before desktop computers were commonplace and do not require computers. Nonetheless, software packages have been developed to make the methods much easier to use and we recommend them. Molecular mechanics is the one empirical method that requires a computer, since it involves optimizing a large number of molecular parameters. Like simpler empirical methods, it relies completely upon experimental data for prototypical molecules and is therefore restricted to the better-studied functional groups.

When the standard empirical methods fail, it is usually because one or more parameters are not available. With enough chemical intuition, one can often make estimations anyway, as exemplified in the chapter by Afeefy and Liebman. This is equivalent to developing a new "method" tailored to the molecule at hand. Typically, however, one turns to standard methods that are based upon quantum mechanics. All such methods require a computer.

In making a thermochemical prediction based upon quantum chemistry, one must generally make a compromise between high accuracy and low cost. This compromise is most troubling for exotic molecules or for compounds that contain transition metals;

Table I. Some Common Methods of Computational Thermochemistry[a]

Method	Applicability		Cost			Reliability[b]	
	Type	Size	Software	Computer	Personnel	Accuracy[c]	Precision[d]
empirical (e.g., Benson groups)	common organic	any	$	¢	$	A-	B+
molecular mechanics (e.g., MM3)	common organic	100,000 atoms	$	¢¢¢	$$	A-	B
semiempirical MO theory (e.g., MNDO/d)	organic, some inorganic	500 atoms	$	$	$$	C+	C-
density functional theory (e.g., B3LYP)	all[e]	50 atoms[f]	$$$	$$	$$$	B-	C+
CBS-4 (MO theory)	all[e]	20 atoms[f]	$$$	$$$	$$$	B	B-
BAC-MP4 (corr. MO theory)	organic, some inorganic	20 atoms[f]	$$$	$$$	$$$$	B+	B
PCI-80 (corr. MO theory)	all[e]	20 atoms[f]	$$$	$$$	$$$$	B+	B-
G2 (MO theory)	all[e]	6 heavies[f]	$$$	$$$$	$$$	A-	B
CCSD(T) with basis set extrapolation (MO theory)	all[e]	3 heavies[f]	$$$	$$$$$$	$$$$	A	A-

[a] The opinions in this Table are those of the authors and do not necessarily represent those of the authors of other chapters in this book.

[b] DFT methods are more robust than MO methods for molecules that contain transition metals, although both are less accurate for transition-metal species than for organics.

[c] Typical performance for a large set of molecules (i.e., mean error).

[d] Worst-case performance among a large set of molecules (i.e., largest error).

[e] Subject to availability of basis sets.

[f] Light atoms add less than heavy atoms to the computational expense; "heavy" means lithium or heavier.

sometimes high accuracy is unavailable at any reasonable cost. In most cases, the accuracy can be enhanced by judicious application of chemical intuition.

As can be seen from Table I, the least expensive (and least reliable) quantum chemical methods are designated "semiempirical." These methods involve replacing many of the smaller quantum interactions with empirical parameters. As with any parametrized technique, semiempirical molecular orbital theory is limited in scope by the available experimental database. Unlike purely empirical schemes, however, semiempirical quantum methods can be applied with success to exotic or unusual molecules. This may be attributed to the use of quantum theory, which is valid for any molecule. In general, methods with lots of empirical parameters are less expensive than methods with few empirical parameters. Since they are essentially sophisticated extrapolation schemes, however, heavily-parametrized methods are more limited in the types of molecule (i.e., functional groups) that they can accommodate. Methods with few empirical parameters rely upon quantum mechanics for their accuracy and generality. The *ab initio* methods range from approximate calculations that can be done on a desktop computer to very accurate calculations that require powerful computers. Because of their computational requirements, the most accurate methods are severely limited in the size of molecule that they can accommodate.

Comparison of Methods: Table I. A selection of several predictive methods are compared in Table I on the basis of applicability, cost, and reliability. "Applicability" refers to the type and size of molecule that a method can accommodate. "Cost" includes the price of software (without academic or other discounts), the cost of the computer time needed to do the calculation, and the cost of training someone to perform the calculations with confidence. "Reliability" refers to the size of typical errors ("accuracy") and of the largest errors ("precision").

All these factors depend upon other considerations. For example, cost depends upon how often estimations are used, local personnel costs, and any discounts available. The limit on molecule size may be higher for symmetrical molecules. Even the rankings in the "accuracy" and "reliability" columns depend upon the type of molecule for which the predictions are needed. The methods complement each other and are all useful; the semi-quantitative comparisons in Table I reflect our subjective judgments for what we consider "typical" applications.

Empirical Estimation Methods

If the molecule of interest is not in experimental databases such as that by Pedley (*5*) or those available from NIST (see Appendix A), the next step is usually to apply an empirical estimation scheme. These are very inexpensive to use. If the necessary parameters are available, the more sophisticated schemes are generally very reliable. Of course, the quality of the results depends upon the quality of the parameter values; a parameter value derived from a single experimental measurement is probably more uncertain than one based upon measurements of several different compounds.

The simplest useful method for predicting reaction enthalpies is probably *bond additivity*, which is based upon the crude approximation that all bonds of the same type are equally strong (*6*). For example, all C-C bonds would contribute equally to the enthalpy of formation. Of course, only rough results can be expected from such a simple method. Typical bonds strengths may be found, for example, in the Handbook of Chemistry and Physics (7). Bond additivity can be cast in a form that leads to enthalpies of formation; see the chapter by Cohen and Benson for a description.

Group Additivity. Much better results can be obtained using *group additivity* methods based upon functional groups. Two simple examples are the methods due to Joback and to Cardozo (4). To use such methods one must identify and count the appropriate groups within the molecule and sum their respective contributions. This is an easy exercise and involves no special notation or nomenclature.

More sophisticated, second-order group methods include the effects of the chemical environment (second-nearest neighbors) on a group's contribution (6,8). These are more complicated and more accurate than the first-order methods and require some effort to learn. The most popular one is that developed by Benson and co-workers (4,6,9-11). Various corrections account for interactions such as steric crowding and ring strain. Benson's method has the major practical advantage that it has been incorporated into computer programs (two are available from NIST), thus sparing the user from doing the arithmetic or even identifying the groups (see Appendix A). Recent developments are described in the following chapter by Benson and Cohen.

Despite the predominance of Benson's approach, other second-order group methods do exist! In their classic book, Cox and Pilcher pointed out that the methods developed by Benson, by Laidler, and by Allen are mathematically equivalent, although they employ different chemical interpretations of the interactions (6). Parameter values and worked examples for these three methods (as well as extensive tables of evaluated thermochemical data) can be found in Cox and Pilcher's book (6). For calculations by hand, Reid et al. found the method by Yoneda (4,12) to be comparable to Benson's in accuracy, and noted that it did not require the calculation of symmetry numbers for entropy estimations (4). More recently, Pedley developed an estimation method to aid in his evaluation of experimental data in the literature (5).

Sometimes, not all the group parameters are available for the molecule of interest. The most expedient solution is often to estimate the value for the missing group by comparison with similar groups. For example, if a group with an acetylenic substituent (C_t) is missing, one could use the value for the corresponding group with a vinyl substituent (C_d). As with any departure from a prescribed method, the quality of the results depends upon the judgment (i.e., chemical intuition) of the user. One may also abandon "methods" altogether and make an estimation based upon the thermochemistry of several related compounds (see below).

Limitations of Group Additivity. As indicated in Table I, group additivity schemes are restricted to common organic molecules. There are many other types of molecule in which one might be interested! For example, the thermochemistry of gas-phase ions is not well-estimated using group additivity schemes (13). Other empirical schemes for estimating cation thermochemistry, such as the one developed by Holmes and Lossing (14), are summarized in the introduction to the GIANT (Gas-Phase Ion and Neutral Thermochemistry) tables (13). These methods are generally based upon trends in ionization energies or proton affinities within homologous series of molecules.

The thermochemistry of many inorganic and organometallic compounds, and especially transition-metal compounds, still can't be estimated by standard group-based methods because the group values are unavailable. A few estimation schemes are available (*15-17*); often the thermochemistry of related organic compounds can be used to estimate enthalpies of formation of organometallics. Drago's electrostatic-covalent model, described in this book in the chapter by Drago and Cundari, is applicable to a wide variety of compounds and has been parametrized for thermochemical quantities such as the strengths of dative and covalent bonds, solvation energies, and the energetics of gas-phase ion-molecule reactions. Estimation methods based upon trends in bond energies, bond-energy/bond-length correlations, and Laidler's estimation method have been summarized in a review (*18*). A fundamental problem with organometallic thermochemistry is the frequent difficulty of experimental measurements, and consequently the likelihood of large errors in reported values (*19*). Unfortunately, the methods for estimating the thermochemistry of organometallic compounds are not sufficiently robust to avoid the need for good user judgment.

Molecular Mechanics. A molecule can be modeled as a mechanical assembly of balls (atoms) and springs (chemical bonds), with a few other forces added to simulate electrostatics, dispersion, etc. Such models are collectively described as *molecular mechanics* (MM) and find widespread use in biological and pharmaceutical chemistry. Of the organic MM methods, only those from the Allinger group (e.g., MM3 and MM4) are parametrized for enthalpies of formation (*20,21*). MM methods require a computer, since all the geometric parameters of the molecule must be varied to minimize the molecule's computed energy. However, only a desktop, personal computer is needed for calculations on organic molecules of typical size (say, fewer than 500 atoms). These models are further described by Rogers elsewhere in this book.

Chemical Reasoning. In many cases, one cannot use the systematic methods described above because one or more essential parameters are lacking. One effective strategy is then to construct a hypothetical reaction that satisfies two conditions: (1) it is approximately thermoneutral (i.e., $\Delta_r H° = 0$), and (2) thermochemical data are available for all the compounds involved except the one of interest. The quality of the prediction depends mostly upon the correctness of judgment used to meet the first requirement. This type of unsystematic approach is illustrated by reaction 6, below, and in the chapter by Afeefy and Liebman. Note that chemical reasoning is also useful for guessing the values of missing parameters; some examples are given in the chapter by Benson and Cohen.

Quantum Mechanical Models

When experimental data are lacking and empirical estimation methods fail, one generally turns to some type of quantum mechanical method. This is often the case for free radicals or for transition states, which are essential for calculating reaction rates (see the chapters by Durant, by Petersson, and by Berry, Schwartz, and Marshall). Such techniques model a molecule as an assembly of electrons and atomic nuclei. This is extremely general, which is a good feature. However, quantum calculations are much more intensive computationally than are empirical models. There is a compromise between the cost of a calculation and the reliability of the resulting prediction, but experimental results can often be exploited to correct a theoretical prediction. Such empirical corrections may be incorporated within the theory (semiempirical methods) or applied to the final results.

There are three common types of quantum chemical model: semiempirical molecular orbital (MO) theory, *ab initio* (non-empirical) MO theory, and density functional theory (DFT). All three aim to solve the non-relativistic, electronic Schrödinger equation using various approximations to minimize the total energy of the molecule. The MO theories are based upon the concept of orbitals that vary in energy, with the lower-energy orbitals containing electrons and with the higher-energy orbitals empty. DFT is based upon a theorem that the electronic energy depends only upon the electron density distribution within the molecule. In practice, DFT calculations often use molecular orbitals to compute the electron density. Until about five years ago, MO theory completely dominated the field of quantum chemistry. Now DFT calculations are also very popular for chemical applications.

Semiempirical MO Theory. Even a small molecule contains many electrons and therefore many interactions. Decades ago, computers were much weaker and *ab initio* calculations were impossible for molecules containing more than a few atoms. To make progress, "semiempirical" methods were developed in which many of the interaction terms were neglected or replaced by empirical parameters, greatly reducing the computational cost. Some purely empirical forces were added to aid in reproducing experimental results. Other features, such as the use of an atomic-orbital basis set and the possibility of including electron correlation explicitly, are the same as in *ab initio* methods (see below). A discussion of semiempirical theory and of recent methodological advances is provided by Thiel elsewhere in this book.

Semiempirical methods were dominant for many years. Today, computers are powerful and the more rigorous *ab initio* methods have become practical, but semiempirical methods remain important for studying large molecules or large numbers of molecules. One practical difference is that semiempirical calculations produce enthalpies of formation directly (generally at 298 K), whereas *ab initio* results require both quantum and thermal corrections. The quality of the thermochemical predictions obtained using popular semiempirical methods (MNDO, AM1 and PM3) is described in Thiel's chapter in this book.

Ab Initio **MO Theory.** The relatively high cost of *ab initio* calculations limits their use. In the simplest theories, termed Hartree-Fock (HF) or self-consistent-field (SCF), each electron moves in the average electric field generated by the other electrons. The shapes of the molecular orbitals are described mathematically as a linear combination of simpler functions, which resemble atomic orbitals and constitute the *basis set*. Increasing the size of a basis sets (i.e., adding more basis functions) permits a better description of the detailed shapes of the orbitals. Consequently, bigger basis sets yield more reliable results. Of course, they also cost more; the computational expense of an HF calculation typically increases as $N^{2.7}$, where N is the number of basis functions (which is approximately proportional to the number of atoms) (*22*). Reducing this strong dependence is an active area of research (*23-28*).

More sophisticated theories begin with an HF calculation and then attempt to include the energy due to the instantaneous repulsion between electrons, termed *electron correlation energy* or simply *correlation energy*. Correlation energy is usually included either by perturbation theory (e.g., MP2, MP4), which is essentially a theory of extrapolation, by configuration interaction (CI) theory (e.g., CISD), or by coupled-cluster theory (e.g., QCISD, CCSD). Just as HF theory uses a basis set of atomic orbitals to describe each two-electron (or one-electron) molecular orbital, correlated theories use a basis set of electron configurations (i.e., individual choices of which MOs are occupied and which are empty) to describe the many-electron wavefunction. Correlated calculations are much more accurate than uncorrelated HF calculations but also increase in cost much faster (up to about N^7 for feasible calculations). Furthermore, correlated calculations generally require larger basis sets (larger N) than HF calculations. Methods for extrapolating feasible correlated calculations to the unfeasible infinite-basis limit are described in the chapters by Petersson and by Martin.

Most correlation energy is associated with electron pairs, so electron correlation is essential in quantitative calculations of any process that makes or breaks electron pairs. Since most chemical bonds involve an electron pair, electron correlation is important in the theoretical description of most chemical reactions. Correlation energy may be included explicitly within the theory, as described above, or it may be included implicitly, by means of empirical corrections or chemical strategies, as described in following sections.

To obtain accurate results, both large basis sets and highly-correlated theories are needed. However, when computational chemists discuss "high" and "low" levels of *ab initio* theory, they are referring primarily to the sophistication of the treatment of correlation energy and secondarily to the "quality" (i.e., size) of the basis set. For example, HF theory is uncorrelated and therefore the lowest level of theory. MP2 theory (second-order perturbation theory) is the lowest correlated level. MP4 (fourth-order) is obviously higher-level than MP2. But CCD theory (coupled-cluster theory including only the double-excitation operator) is also higher than MP2. CCSD(T) (CCD plus the single-excitation operator and an approximate treatment of the triple-excitation operator) is the highest level of theory in general use today (early 1997). For basis sets described with similar names, bigger basis sets typically have designations that are longer or include larger numbers. For example, the 6-311++G(2df,2pd) basis set is much bigger than the more common 6-31G(d) basis, and cc-pV5Z is far larger than aug-cc-pVDZ. In many acronyms S, D, T, and Q stand for single, double, triple and quadruple. A glossary of major terms is provided in Appendix D of this book and elsewhere by Brown et al. (*29*).

Ab initio calculations do not directly produce enthalpies of formation or any other thermochemical quantities. They produce total molecular energies expressed in the atomic unit of energy, the hartree (1 hartree/molecule = 627.51 kcal/mol = 2625.5 kJ/mol). The values are negative and represent the energy change upon assembling the molecule from its component nuclei and electrons. Entropies, heat capacities, and other quantities are derived from the computed molecular partition function using textbook methods of statistical mechanics (see Appendix B). Enthalpies of formation, however, can only be derived by computing the energy change for some chemical reaction.

Although total atomization is a common choice of reaction for this purpose, it is seldom optimal; alternative reaction schemes are discussed in a later section of this chapter.

The reliability of thermochemical predictions based upon *ab initio* MO calculations is discussed in several other chapters in this book. The popular book by Hehre, Radom, Schleyer, and Pople (*30*), although slightly dated, is a very useful text emphasizing practical applications. The more mathematical text by Szabo and Ostlund (*31*) is often recommended for learning the underlying theory.

Multireference MO Theories. The commonly-used *ab initio* MO methods are all based upon Hartree-Fock theory, which describes the molecule using a single electron configuration. In some cases, more than one electron configuration is needed for a correct qualitative description of the bonding. Such molecules are best described using a multiconfiguration SCF, or MCSCF calculation, possibly followed by perturbation theory (e.g., CASPT2) or configuration interaction (e.g., MR-SDCI) to include additional electron correlation. Unfortunately, these powerful methods are strictly for the specialist. Many decisions must be made when applying these techniques and no reliable algorithm has been developed for making those decisions. Expert judgment is the best guide, and even the greatest experts occasionally make bad choices and obtain poor results.

Density Functional Theory. The fundamental theorem of DFT states that there exists a functional (viz., a function of a function) of the electron density that yields the exact electronic energy. Unfortunately, the theorem does not provide any instructions for obtaining the correct functional, and further significant progress in DFT thermochemistry appears to depend upon the development of better functionals; there are no obvious extensions to "higher" order as there are for MO theory. Nonetheless, DFT is already very useful for thermochemistry, as described in the chapters in this book by Curtiss and Raghavachari, by Politzer and Seminario, by Blomberg and Siegbahn, and by Ziegler.

DFT has commonly-cited advantages over MO theory. It includes correlation energy comparable to highly-correlated theories such as QCISD, but for a cost between those for HF and MP2 calculations. Adequately converged results are obtained using smaller basis sets than are needed for correlated MO methods. DFT methods are more robust for difficult systems, such as molecules with multiconfiguration character, free radicals with spin contamination (i.e., undesired mixing of high-spin excited states with the ground state), or molecules containing transition metal atoms. The major disadvantage of DFT for thermochemistry appears to be its lack of extensibility, which means that one can not do a series of calculations to demonstrate convergence. In other words, there is no such thing as a "high-level" DFT calculation, although "nonlocal" functionals are fancier and more reliable than "local" functionals. Another current shortcoming of DFT is its general inability to cope with electronically excited states of molecules.

The practical difference between DFT and MO theory may be summarized by saying that DFT is less expensive computationally and provides thermochemistry of greater precision (fewer outliers) but lower accuracy (larger typical errors). Some comparisons are given in the chapters by Curtiss and Raghavachari and by Blomberg and Siegbahn. A popular text on DFT in chemistry is that by Parr and Yang (*32*).

Quantum Monte Carlo. There are a few specialists in quantum Monte Carlo (QMC) methods for *ab initio* calculations. These techniques involve numerical solution of the Schrödinger equation and achieve precisions proportional to the square root of the computational expense. A unique feature of QMC is that most of the uncertainty of the result is defined by sampling statistics. For example, recent calculations on hydrogen fluoride yielded a (non-relativistic) bond strength D_e = 592.2 ± 1.7 kJ/mol, only 0.12 kJ/mol greater than the corresponding experimental value (*33*). Unfortunately, QMC calculations are so expensive that hydrogen fluoride, at 10 electrons, is the biggest molecule currently feasible. One can use various approximations to extend the range of QMC calculations, but this adds uncertainty to the results and compromises the advantage of the technique. Thus, QMC methods are not yet practical for typical thermochemical applications.

Nearly *ab initio* Methods: Empirical Corrections

A consensus is emerging that users of thermochemical data will rely increasingly upon *ab initio* calculations. But there is also a growing consensus that pure *ab initio* theory is not accurate enough to satisfy the demands for thermochemical data. To resolve this mismatch, empirical corrections must supply the needed accuracy. Of course, empirical corrections are feasible only to the extent that (1) errors in the theory are systematic and predictable and (2) there are enough experimental data to implement the corrections.

Parametrized Methods. Many empirically parametrized methods exist and more are being developed. Some of the oldest involve atom "equivalents," which are surrogates for atomic energies that are combined with the total molecular energy to obtain an enthalpy of formation (*34-39*). These methods are easy to use, inexpensive, and accurate (usually to about 10 kJ/mol). At present, their utility is somewhat limited because they are best applied to rather common molecules, for which one would normally use an empirical group method. As the methods and underlying theories advance, however, we expect these methods to become increasingly useful.

At the opposite, expensive end of the spectrum are the Gaussian-2 (or G2) method and its variants, which are very popular among physical chemists. See the chapter by Curtiss and Raghavachari for details. These are composite methods, in which the results of several calculations are combined to estimate the result that would have been obtained with a much more expensive calculation. The few empirical parameters do not depend upon the chemical composition of the molecule under study, resulting in a general method. These techniques are easy to use and usually accurate (typically to 10 kJ/mol or so). High cost, however, limits their usefulness to small molecules. Furthermore, these and other high-level composite methods occasionally give poor results, as exemplified in the chapter by Berry, Schwartz, and Marshall.

As stated above, semiempirical MO theories include empirical parameters at the quantum mechanical level, not as simple arithmetic corrections. Some nominally *ab initio* approaches do the same, although with far fewer adjustable parameters. The most popular of these is probably the B3LYP method (*40*), which is a hybrid, parametrized mix of local DFT, nonlocal DFT, and HF theories (*41*). It is only a little more expensive than HF theory and as easy to use but is much more reliable, since it includes correlation energy self-consistently. Its efficacy is discussed in the chapters by Curtiss and Raghavachari and by Blomberg and Siegbahn.

The BAC-MP4 method combines moderately high-level calculations with several empirical parameters based primarily upon the types and lengths of chemical bonds, as described in the chapter by Zachariah and Melius. It is one of the least expensive of the established methods and is unique in providing an estimate of the uncertainty of each prediction. However, since parameter development requires benchmark (experimental) data for each bond type, predictions for unusual types of molecule are unreliable. This method would probably be much more popular if software were readily available to automate the tedious arithmetic involved.

Some methods involve simple corrections applied to *ab initio* energetics. For example, one may examine the effect of including some correlation energy and extrapolate to the fully-correlated limit (*42-44*). PCI-*X* is such a method and is described in the chapter by Blomberg and Siegbahn. Since there may be as few as one extrapolating parameter, these methods are general. Empirical corrections may be based upon other concepts as well, such as bond symmetry (σ, π, etc.), as exemplified in a method described in Martin's chapter.

Chemists' Strategies: Reaction Schemes

The most popular empirical corrections are unsystematic, in that the procedure is not uniquely determined by the molecule under study. They are based upon the notion that theoretical calculations on chemically similar molecules will have similar systematic errors. The trick, then, is to identify a molecule that is "similar" to the one of interest and for which reliable thermochemistry is already available. A well-chosen reaction will give excellent results using a much smaller basis set and a much lower level of theory than would otherwise be necessary, as illustrated below. One test of a reaction is to compare its energy change calculated at the uncorrelated HF level with that calculated at a correlated level such as MP2. If the reaction energy is unaffected by explicitly including correlation energy, then the correlation effects in the reaction are probably well-balanced (*45*).

As mentioned earlier, electron correlation is very important in the energetics of reactions that involve a change in the number of electron pairs, or equivalently, in the spin multiplicity. Unfortunately, no practical calculation can recover all the correlation energy, so *ab initio* predictions suffer systematic errors for such processes.

Bond-breaking is such a process. For example, all the electrons in chlorobenzene are paired but one pair breaks to form $C_6H_5 + Cl$. Thus, one would not expect to get a reliable bond energy simply by calculating the energy for reaction 1. Indeed, simple

$$C_6H_5Cl \rightarrow C_6H_5 + Cl \qquad (1)$$

unrestricted Hartree-Fock (HF) calculations using the popular, small 6-31G* basis set predict an enthalpy change $\Delta_r H°_{298} = 230$ kJ/mol. The HF result compares miserably with the experimental value of 396 ± 8 kJ/mol (*46*); the error is -166 $\pm$ 8 kJ/mol. (Energies calculated using the hybrid B3LYP density functional method yield $\Delta_r H°_{298} = 379$ kJ/mol, for a much smaller error of -17 $\pm$ 8 kJ/mol.)

Isogyric Reactions. An *isogyric reaction* is one in which the spin multiplicity (i.e., the number of electron pairs) does not change (*47*). Such a reaction can be constructed by adding hydrogen atoms and molecules to balance the spin multiplicity, as in reaction 2.

$$C_6H_5Cl + H + H \rightarrow C_6H_5 + Cl + H_2 \tag{2}$$

The HF prediction for the isogyric reaction is $\Delta_r H°_{298}$ = -98 kJ/mol, as compared with the experimental value of -40 ± 8 kJ/mol. Now the error is reduced to -58 ± 8 kJ/mol. (B3LYP energies yield $\Delta_r H°_{298}$ = -67 kJ/mol.) Of course, one may choose any species to balance the spin. Reaction 3 is also isogyric and more concise. For reaction 3 the HF

$$C_6H_5Cl + H \rightarrow C_6H_5 + HCl \tag{3}$$

prediction of $\Delta_r H°_{298}$ = -63 kJ/mol compares reasonably well with the experimental value of -36 ± 8 kJ/mol; the error is -27 ± 8 kJ/mol. (B3LYP energies yield $\Delta_r H°_{298}$ = -32 kJ/mol.) Finally, reaction 4 is isogyric and also avoids explicit calculations on free

$$C_6H_5Cl + H_2 \rightarrow C_6H_6 + HCl \tag{4}$$

radicals, which are more often problematic than are calculations on closed-shell molecules. For this reaction, the HF prediction of $\Delta_r H°_{298}$ = -74 kJ/mol agrees well with the experimental value of -64 ± 1 kJ/mol; the error is only -10 ± 1 kJ/mol. (B3LYP energies yield $\Delta_r H°_{298}$ = -50 kJ/mol.)

The important "higher-level correction" in G2-type theories originated from considering processes such as reaction 2 (*48*). Explicit calculations on H and H_2 are unnecessary; the (constant) energy for the hypothetical reaction H → ½ H_2 is applied once for each unpaired electron in the molecule.

Isodesmic Reactions. An *isodesmic reaction* is one in which the numbers of chemical bonds of each formal type (e.g., C-C, C=O, C-H) do not change (*30,49*). For example, there are six aromatic C-C bonds, nine C-H bonds, and one C-Cl bond on each side of reaction 5. The inexpensive HF calculations predict $\Delta_r H°_{298}$ = 10 kJ/mol for reaction

$$C_6H_5Cl + CH_4 \rightarrow C_6H_6 + CH_3Cl \tag{5}$$

5. The experimental value is 21 ± 1 kJ/mol, for a small error of -11 ± 1 kJ/mol. (B3LYP energies yield $\Delta_r H°_{298}$ = 21 kJ/mol.) One may balance the chemistry at an even finer level of detail. Reaction 6 is isodesmic even if one distinguishes between sp³- and sp²-

$$C_6H_5Cl + C_2H_4 \rightarrow C_6H_6 + C_2H_3Cl \tag{6}$$

hybridized carbon atoms. In this case the calculated enthalpy change of -3 kJ/mol is in perfect agreement with the experimental $\Delta_r H°_{298}$ = -3 ± 2 kJ/mol. (B3LYP energies also yield $\Delta_r H°_{298}$ = -3 kJ/mol.) All the *ab initio* calculations for reactions 1-6 (geometry optimizations, vibrational frequency calculations, and energy evaluations) were done on a desktop personal computer.

As mentioned earlier, chemically-balanced reaction schemes are useful not only for converting theoretical energies into enthalpies of formation but also for making purely empirical estimates. For example, if one simply *assumed* reaction 6 to be thermoneutral, the corresponding prediction for $\Delta_r H°_{298}(C_6H_5Cl)$ would be in error by only -3 ± 2 kJ/mol. Sometimes the experimental data needed to construct well-balanced reaction schemes are not available. One must then rely on chemically similar functional groups, such as in isovalent homologous or congeneric series (e.g., Cl instead of Br).

Current Challenges and Outlook

There are many challenges in improving the state of computational thermochemistry. Empirical estimation methods are still evolving and continue to be constrained by experimental databases that grow too slowly. Quantum chemistry is promising, but many potential users still find it too expensive, too intimidating, too difficult, or too approximate. These challenges are being accepted primarily by academic researchers. There are also technological issues, such as improving the quality of experimental databases and computer software, that are being addressed mostly in the industrial and government sectors. Finally, there is the cultural problem of acceptance; many potential users remain skeptical of the reliability of the newer methods.

Computational Technology. Computer speed is increasing rapidly (*50*); a doubling time of 18-24 months is often quoted. In addition, since algorithms are constantly being improved, raw computing power can be used to greater effect. As a result, the cost of a given computational prediction is decreasing very quickly. In contrast, experimental thermochemistry is not getting cheaper and the number of well-equipped laboratories is stagnant or declining. Thus, one can conclude that computational approaches will become increasingly important in thermochemistry.

One can also expect the computer programs to become more robust and easier to use. This is driven partly by the emergence of commercial vendors of software for quantum chemistry; customers expect software tools to be accessible to non-specialists with limited time for training. It is also driven by increasing demand for user-friendly software throughout the software industry. Computer users now expect every program to have a convenient, graphical interface.

Scientific Developments. Research in the methods of MO theory and DFT is progressing in many directions, including many of obvious benefit for predicting thermochemistry. Improved algorithms are being developed to implement existing theories much more efficiently, thus allowing larger molecules to be studied. New basis sets extend current methods to additional chemical elements. Improved functionals increase the reliability of DFT predictions. More sophisticated theories of electron correlation improve the reliability of MO theory. Widespread testing provides quantitative information about the reliability and range of applicability of new methods and theories. Better theories of molecular solvation are essential if quantum chemistry is to trusted for predicting thermochemistry in solution and not just in the gas phase (see the chapter by Giesen et al.). Many of the advances in *ab initio* theory are also applicable to semiempirical methods. In addition, the underlying semiempirical theory is being extended by the development of more rigorous Hamiltonians that lead to substantially more accurate results (see the chapter by Thiel).

There are also some less-developed areas of theory that will affect computational thermochemistry as they mature. For example, high-level MO theory still has trouble with some "difficult" molecules. As mentioned above, multireference theory does well with these problems but depends upon the user's judgment, thus requiring an expert user. There has been some progress in the search for automatic ways to make the necessary choices (*51,52*), but not enough to allow multireference methods to be used widely or by non-specialists.

Another emerging area involves efficient, sophisticated treatments of molecular vibration (*53,54*), which determines the vibrational zero-point energy and also the molecular partition function. The major errors in *ab initio* thermochemistry are currently presumed to be in a molecule's electronic energy. Calculations of electronic energies are becoming very accurate, however, and errors in vibrational energy may soon limit the accuracy of predictions, especially for floppy molecules.

Symposium Panel Discussion. A panel discussion was held as part of the symposium upon which this book is based. The topic was the needs for and applications of accurate thermochemical data, with an emphasis on prediction. Panel members represented the users of thermochemical data in areas such as chemical processing, microelectronics, academia, and the military. The discussion offered an open forum for both the users and developers of techniques to discuss common problems and goals. The proceedings of the symposium, including the panel discussion, are available in print (*55*) and also electronically from NIST (http://www.nist.gov/). The discussion reached four major conclusions, which follow.

Benchmark Data. A critically evaluated, experimental database is needed to provide a standard metric for evaluating, comparing, and parametrizing the performance of different predictive methods. This was the principal item that the discussion participants thought would help both the developers and users of new techniques. Most of the technique developers currently have their own databases, sometimes culled uncritically from data compilations or the open literature. Curtiss (Argonne NL) has offered for public use his new, larger database compiled for testing the G2 method and its variants. He plans to make it available on the Internet. Others who offered to share their databases are Bozzelli (NJIT), Melius (Sandia NL), and Dixon (Pacific Northwest NL). We discussed that perhaps the best institution for compiling these data is NIST.

Wider Range of Properties and Molecular Types. Accurate predictions are needed for engineering properties in addition to $\Delta_f H°_{298}$ and also for a wider variety of molecular types. Whereas thermochemical properties are industrially important, there are many more properties (e.g., kinetics, vapor-liquid equilibria, critical properties) which are essential for making a process economically successful. It was pointed out many times during the discussion that work is being done on other properties; some papers on the topic were presented at the symposium (e.g., those by Politzer (*56*) and by Carreira (*57*)). Extending predictive methods to organometallic compounds and to the properties of molecules on surfaces is also important.

Acceptance among Engineers. It appeared clear that although a few companies (e.g., Dow, DuPont, and Amoco) have strong and active molecular modeling departments, many do not. The broader thermochemistry community appears to be unaware that new predictive techniques exist and are beneficial. Some such education can be achieved by this symposium and book, but continuing efforts are essential. One point of consensus is that the techniques must be easy to use to gain widespread acceptance and use. To ensure a reliable "black box," robustness must accompany ease of use.

Hidden Experimental Data. Heavily-parametrized methods are currently limited by the available experimental data (*9*). Many experimental measurements are done in industrial labs, but most corporations have clearance processes that do not encourage the dissemination of private but not necessarily proprietary data. Keeping data private sometimes provides a competitive advantage, but often this is not the case. The first step in encouraging the publication of such data is probably to identify and advertise conspicuous "holes" in the data tables.

Acknowledgments

We are grateful for generous financial support for the symposium from the following sponsors: the National Institute of Standards and Technology (Physical and Chemical Properties Division); International Business Machines (Scientific and Technical Systems and Solutions Group); the Donors of the Petroleum Research Fund (administered by the ACS); Gaussian, Inc.; Cray Research, Inc. We also thank those of our colleagues who discussed and critiqued this overview.

Literature Cited

1.   Whitten, J. L.; Yang, H. *Surf. Sci. Rep.* **1996**, *24*, 59-124.
2.   van Santen, R. A. *J. Mol. Catal. A* **1997**, *115*, 405-419.
3.   Lyman, W. J.; Reehl, W. F.; Rosenblatt, D. H. In *Handbook of Chemical Property Estimation Methods: Environmental Behavior of Organic Compounds*; Lyman, W. J.; Reehl, W. F.; Rosenblatt, D. H., Eds.; ACS: Washington DC, 1990.
4.   Reid, R. C.; Prausnitz, J. M.; Poling, B. E. *The Properties of Gases and Liquids*, 4th ed.; McGraw-Hill: New York, 1987.
5.   Pedley, J. B. *Thermochemical Data and Structures of Organic Compounds*; Thermodynamics Research Center: College Station, Texas, 1994; Vol. 1.
6.   Cox, J. D.; Pilcher, G. *Thermochemistry of Organic and Organometallic Compounds*; Academic: London, 1970.
7.   *CRC Handbook of Chemistry and Physics*, 76th ed.; Lide, D. R.; Frederikse, H. P. R., Eds.; CRC Press: Boca Raton, 1995.
8.   Benson, S. W.; Buss, J. H. *J. Chem. Phys.* **1958**, *29*, 546-572.
9.   Cohen, N. *J. Phys. Chem. Ref. Data* **1996**, *25*, 1411-1481.
10. Cohen, N.; Benson, S. W. *Chem. Rev.* **1993**, *93*, 2419-2438.
11. Benson, S. W. *Thermochemical Kinetics*, 2nd ed.; Wiley: New York, 1976.
12. Yoneda, Y. *Bull. Chem. Soc. Japan* **1979**, *52*, 1297-1314.
13. Lias, S. G.; Bartmess, J. E.; Liebman, J. F.; Holmes, J. L.; Levin, R. D.; Mallard, W. G. *J. Phys. Chem. Ref. Data* **1988**, *17*, Suppl. 1.
14. Lossing, F. P.; Holmes, J. L. *J. Am. Chem. Soc.* **1984**, *106*, 6917-6920.

15. Martinho Simões, J. A. In *Estimates of thermochemical data for organometallic compounds*; Martinho Simões, J. A., Ed.; Kluwer: Dordrecht, 1992; Vol. 367.

16. Myers, K. H.; Danner, R. P. *J. Chem. Eng. Data* **1993**, *38*, 175-200.

17. Minas da Piedade, M. E.; Martinho Simões, J. A. *J. Organomet. Chem.* **1996**, *518*, 167-180.

18. Pilcher, G.; Skinner, H. A. In *Thermochemistry of organometallic compounds*; Hartley, F. R.; Patai, S., Eds.; John Wiley & Sons, 1982, pp 43-90.

19. Martinho Simões, J. A. *NIST Organometallic Thermochemistry Database*; National Institute of Standards and Technology, 1997.

20. Allinger, N. L.; Chen, K.; Lii, J.-H. *J. Comput. Chem.* **1996**, *17*, 642-668.

21. Nevins, N.; Lii, J.-H.; Allinger, N. L. *J. Comput. Chem.* **1996**, *17*, 695-729.

22. Foresman, J. B.; Frisch, Æ. *Exploring Chemistry with Electronic Structure Methods: A Guide to Using Gaussian*; Gaussian, Inc.: Pittsburgh, 1993.

23. Dixon, S. L.; Merz, K. M., Jr. *J. Chem. Phys.* **1996**, *104*, 6643-6649.

24. Lee, T.-S.; York, D. M.; Yang, W. *J. Chem. Phys.* **1996**, *105*, 2744-2750.

25. Parr, R. G.; Yang, W. T. *Annu. Rev. Phys. Chem.* **1995**, *46*, 701-728.

26. Schwegler, E.; Challacombe, M. *J. Chem. Phys.* **1996**, *105*, 2726-2734.

27. Strain, M. C.; Scuseria, G. E.; Frisch, M. J. *Science* **1996**, *271*, 51-53.

28. White, C. A.; Johnson, B. G.; Gill, P. M. W.; Head-Gordon, M. *Chem. Phys. Lett.* **1996**, *253*, 268-278.

29. Brown, R. D.; Boggs, J. E.; Hilderbrandt, R.; Lim, K.; Mills, I. M.; Nikitin, E.; Palmer, M. H. *Pure Appl. Chem.* **1996**, *68*, 387-456.

30. Hehre, W. J.; Radom, L.; Schleyer, P. v. R.; Pople, J. A. *Ab Initio Molecular Orbital Theory*; Wiley: New York, 1986.

31. Szabo, A.; Ostlund, N. S. *Modern Quantum Chemistry: Introduction to Advanced Electronic Structure Theory*, 1st (revised) ed.; McGraw-Hill: New York, 1982.

32. Parr, R. G.; Yang, W. *Density-Functional Theory of Atoms and Molecules*; Oxford University: New York, 1989.

33. Lüchow, A.; Anderson, J. B. *J. Chem. Phys.* **1996**, *105*, 4636-4640.

34. Wiberg, K. B. *J. Comput. Chem.* **1984**, *5*, 197-199.

35. Ibrahim, M. R.; Schleyer, P. v. R. *J. Comput. Chem.* **1985**, *6*, 157-167.

36. Dewar, M. J. S.; O'Connor, B. M. *Chem. Phys. Lett.* **1987**, *138*, 141-145.

37. Herndon, W. C. *Chem. Phys. Lett.* **1995**, *234*, 82-86.

38. Mole, S. J.; Zhou, X.; Liu, R. *J. Phys. Chem.* **1996**, *100*, 14665-14671.

39. Kafafi, S. "Computation of accurate atomization energies of molecules from density functional calculations"; 212th ACS National Meeting, August 27, 1996, Orlando, Florida, paper COMP 116.

40. Stephens, P. J.; Devlin, F. J.; Chabalowski, C. F.; Frisch, M. J. *J. Phys. Chem.* **1994**, *98*, 11623-11627.

41. Becke, A. D. *J. Chem. Phys.* **1993**, *98*, 5648-5652.

42. Gordon, M. S.; Truhlar, D. G. *J. Am. Chem. Soc.* **1986**, *108*, 5412-5419.

43. Rossi, I.; Truhlar, D. G. *Chem. Phys. Lett.* **1995**, *234*, 64-70.

44. Siegbahn, P. E. M.; Svensson, M.; Boussard, P. J. E. *J. Chem. Phys.* **1995**, *102*, 5377-5386.

45. Hassanzadeh, P.; Irikura, K. K. *J. Phys. Chem. A* **1997**, *101*, 1580-1587.

46. Lias, S. G.; Liebman, J. F.; Levin, R. D.; Kafafi, S. A. *Structures and Properties, vers. 2.02*; NIST Standard Reference Database 25; software by Stein, S. E.; National Institute of Standards and Technology, 1994.

47. Pople, J. A.; Luke, B. T.; Frisch, M. J.; Binkley, J. S. *J. Phys. Chem.* **1985**, *89*, 2198-2203.

48. Pople, J. A.; Head-Gordon, M.; Fox, D. J.; Raghavachari, K.; Curtiss, L. A. *J. Chem. Phys.* **1989**, *90*, 5622-5629.

49. Hehre, W. J.; Ditchfield, R.; Radom, L.; Pople, J. A. *J. Am. Chem. Soc.* **1970**, *92*, 4796-4801.

50. Brenner, A. E. *Physics Today* **1996**, *49*, 24-30.

51. Bone, R. D.; Pulay, P. *Int. J. Quantum Chem.* **1993**, *45*, 133-166.

52. Langlois, J.-M.; Yamasaki, T.; Muller, R. P.; Goddard, W. A., III *J. Phys. Chem.* **1994**, *98*, 13498-13505.

53. Balint-Kurti, G. G.; Pulay, P. *J. Mol. Struct.* **1995**, *341*, 1-11.

54. Norris, L. S.; Ratner, M. A.; Roitberg, A. E.; Gerber, R. B. *J. Chem. Phys.* **1996**, *105*, 11261-11267.

55. Irikura, K. K.; Frurip, D. J. *Summary Report: Symposium on Computational Thermochemistry*; NISTIR 5973; National Institute of Standards and Technology, 1997 (in press).

56. Politzer, P. "Density functional calculations of enthalpies of formation and reaction energetics"; 212th ACS National Meeting, August 25-29, 1996, Orlando, Florida, paper COMP 178.

57. Carreira, L. A.; Karickhoff, S. W. "Vapor pressure, boiling point, and activity coefficient calculations by SPARC"; 212th ACS National Meeting, August 25-29, 1996, Orlando, Florida, paper COMP 114.

EMPIRICAL METHODS: GROUP CONTRIBUTIONS

Chapter 2

Current Status of Group Additivity

S. W. Benson[1] and Norman Cohen[2]

[1]Department of Chemistry, Donald P. and Katherine B. Loker Hydrocarbon
Institute, University of Southern California, University Park,
Los Angeles, CA 90089–1661
[2]Thermochemical Kinetics Research, 6507 SE 31st Avenue,
Portland, OR 97202–8627

Group Additivity is currently the most widely used method for
estimating thermochemical data for molecules and radicals in either
ideal gas or liquid state. Like all empirical schemes, its reliability
depends on the reliability of the database from which it is derived.
In the past twenty years this database has not grown much and so it
has become very important to devise alternate methods to estimate
group contributions. One such method involves use of a new electronegativity scale to estimate accurately the effects of branching on
$\Delta_f H^\circ$ of functionalized hydrocarbons. Another uses data on heats of
hydrogenation to estimate unknown groups. A third method makes
selective use of key molecules for which data are more reliable than
for their homologs in order to reduce gross scatter in the database.
These procedures will be discussed, together with some recent
developments in estimating $\Delta_f H^\circ$ of crystalline solids.

The principal goal of physical chemistry is to provide a theoretical framework with
which to predict the physical properties as well as chemical reactivity of all matter.
Implicitly the Schrödinger-Dirac equations do this, but their utilization for accurate
estimation (± 1 kcal/mol) of properties of polyatomic systems is greatly restricted.
This barrier has led to the development of empirical schemes to accomplish the
same goals.

The quantitative prediction of chemical reactivity requires access to reliable
thermochemical data, namely heats of formation, entropies, and heat capacities. It
is accepted that the range of molecules of possible interest will, for the foreseeable
future, exceed by orders of magnitude the number of compounds that can be

measured in the laboratory. Hence, simple but reliable techniques are desirable for estimating thermochemical properties of species for which experimental data are not available. A number of methods have been developed in recent years; while there is not unanimous agreement regarding the best method, it seems generally agreed that one of the best (if not *the* best) is that of Group Additivities (GA), especially as developed by S. W. Benson and co-workers (*1*). Not only is the method fairly easy to apply, but it usually can estimate properties with an uncertainty no larger than typical experimental uncertainties. To provide the group values (GAVs) for GA, a reliable set of key experimental values must be available.

The basic premise of GA is that one can subdivide an arbitrary chemical compound into a set of smaller structural units in such a way that the thermochemical properties of that compound can be calculated from constants associated with the smaller units.

The earliest atempts at an enthalpy additivity scheme were prompted by the recognition that if one listed the enthalpies of straight chain saturated hydrocarbons--methane, ethane, propane, butane . . . etc., the difference between two successive enthalpies was very nearly a constant value of approximately 5.0 kcal/mol. Since the difference between successive straight chain hydrocarbons is the incremental $-CH_2-$ group, this 5.0 kcal/mol is to be associated somehow with the CH_2 structural unit. It is not possible to generalize this observation to other structural units without some ancillary assumptions, inasmuch as one cannot form a succession of compounds that differ only, say, in the number of CH_3 groups and nothing else. Nevertheless, Pitzer (*2*) developed a set of additivity parameters for calculating enthalpy functions $[(H^o(T) - H^o(0)]/T$ as a function of number of C atoms and various constants. Successive early developments in GA have been summarized elsewhere (*3*).

In 1958, Benson and Buss (*4*) proposed a hierarchy of additivity schemes for molecular properties and established a conceptual framework that provided a physical justification for the approach (*5*). In this hierarchy, atomic additivity is the first level of approximation; the second and third are bond additivity and group additivity, respectively. Atom additivity, as observed above, is valid for such simple properties as molecular weights, but certainly not for thermochemical properties. Bond additivities thus constitute the first non-trivial level in the Benson/Buss hierarchical scheme. Although not much effort has gone into bond additivity schemes, the method offers some advantages.

A simple example will illustrate: A molecule of CH_3CH_2Cl has the same number and kinds of bonds as 1 CH_3Cl plus 1 C_2H_6 minus 1 CH_4 molecule. CH_3CH_2Cl should then have, according to the principle of bond additivity, the same Δ_fH^o, S^o (after correcting for symmetry), and C_p^o as $CH_3Cl + C_2H_6 - CH_4$. The following table (data taken from Ref. *1*, Appendix) compares the results at 298 K (parenthetical entropy values are intrinsic entropies--*i.e.*, without corrections for symmetry contributions). The deviation in Δ_fH^o is 4.8 kcal/mol; rather large, but not too bad, considering the level of approximation. However, the values for S^o and C_p^o are within the experimental uncertainties.

Molecule	$\Delta_f H^o$ (kcal/mol)	S^o (cal/mol-K)		C_p^o (cal/mol-K)
CH_3Cl = **a**	-19.6	56.0	(58.2)	9.7
CH_3CH_3 = **b**	-20.2	54.9	(60.6)	12.7
CH_4 = **c**	-17.9	44.5	(49.4)	8.5
a + **b** - **c**	-21.9	66.4	(69.4)	13.9
CH_3CH_2Cl	-26.7	66.1	(68.3)	15.1

Group Additivity: Description

A group is defined by Benson (6) as a polyvalent atom of ligancy $\geq$ 2 in a molecule together with all of its ligands. A group is written as $X\text{-}(A)_i(B)_j(C)_k(D)_l$, where X is the central atom attached to i A atoms, j B atoms, etc. For example, isooctane (see below) consists of five $C\text{-}(C)(H)_3$ groups, one $C\text{-}(C)_2(H)_2$ group, one $C\text{-}(C)_3(H)$ group, and one $C\text{-}(C)_4$ group. In contrast, another octane isomer, 2,2,3,3-tetramethylbutane, consists of six $C\text{-}(C)(H)_3$ groups and two $C\text{-}(C)_4$ groups.

Isooctane	2,2,3,3-Tetramethylbutane

A group additivity scheme would thus provide the basis for differentiating between these two isomers. Note, however, that the groups of some octane isomers--e.g., 2-methylheptane, 3-methylheptane, and 4-methylheptane--are the same: three $C\text{-}(C)(H)_3$ groups, four $C\text{-}(C)_2(H)_2$ groups, and one $C\text{-}(C)_3(H)$ group. We could not, then, expect to differentiate between these isomers on the basis of groups alone as defined above.

The four groups just enumerated--$C\text{-}(C)(H)_3$, $C\text{-}(C)_2(H)_2$, $C\text{-}(C)_3(H)$, and $C\text{-}(C)_4$--suffice to characterize all alkanes. We abbreviate them P, S, T, and Q, respectively, for "primary," "secondary," "tertiary," and "quaternary." However, the number of groups proliferates rapidly as one expands the parameters to encompass the entire range of organic compounds. Benson tabulates thermochemical values for 37 hydrocarbon groups, 61 oxygen-containing groups, 59 nitrogen-containing groups, 46 halogen-containing groups, 53 sulfur-containing groups, 57 organometallic groups, and 65 organophosphorus and organoboron groups (7).

An examination of the tabulated groups in Ref. 1 or elsewhere reveals that there are separate groups for $C\text{-}(C)_2(H)_2$, $C\text{-}(C_d)(C)(H)_2$, $C\text{-}(C_B)(C)(H)_2$, and $C\text{-}(C_t)(C)(H)_2$, where C_d specifies a double-bonded C atom (i.e., an sp^2 hybrid C atom), C_t specifies a triple-bonded C atom (sp hybrid), and C_B specifies a C atom in a benzene ring. At first glance, these groups would seem to violate the premises of GA--unless one regards C, C_d, C_B, and C_t carbon atoms as different species. In developing the GA method, Benson and coworkers started with the assumption that all such groups could indeed be regarded as identical unless the empirical evidence demanded otherwise. The tabulated group additivity values (GAVs) for enthalpy contributions (298 K) of these four groups are, respectively, -4.93, -4.76, -4.86, and -4.73 kcal/mol; the differences are small--probably just barely outside the range of experimental uncertainties. This modification of the GA postulates has become well-accepted into the standard procedure. In addition, there are corrections for various interactions that cannot be treated strictly within the definition of groups.

One such correction is required because of spatial interactions that are not directly defined in terms of a series of chemical bonds. These include the "1,4," or *gauche*, interaction of two methyl groups, the 1,5 interaction of side chains, the *cis* interaction between two substituents on the same side of a double bond or ring structure; and the *ortho* interaction for adjacent substituents on an aromatic ring. The presence or absence of these interactions may depend on non-bonded but close interactions of parts of a molecule whose proximity is not implicit in bonding alone. These correction terms are lumped together under the heading of Non-Nearest Neighbor Interactions (NNIs), and will be discussed further in a later section. In most cases, such NNI corrections are small (a few kcal/mol or less in the case of enthalpy).

Additionally, there are Ring Strain Corrections (RSCs): additive contributions to thermochemical properties of compounds with ring structures that cannot be described strictly in group additivity terms. While such contributions are wholly consistent with intuitive ideas of ring strain, or stabilization effects, their *a priori* prediction is outside the scope of GA (or of any other empirical prediction scheme, for that matter). RSCs will be discussed further on.

Determination of group values. There are at least two different approaches to the determination of the numerical values of group contributions to different thermochemical properties. The first depends heavily on the experimental values for the smaller molecules, which are known to higher accuracy than those for larger analogs. Consider, for example, the enthalpy contributions for hydrocarbons. The four groups, P, S, T, and Q, have already been defined. Four relationships, involving experimental values for four small alkanes, suffice to define the numerical values of the enthalpy contributions of the four groups:

$$\Delta_f H(C_2H_6) \quad = \quad -20.03 \quad = \quad 2P$$
$$\Delta_f H(C_3H_8) \quad = \quad -25.02 \quad = \quad 2P + S$$
$$\Delta_f H(i\text{-}C_4H_{10}) \quad = \quad -32.07 \quad = \quad 3P + T$$
$$\Delta_f H(neo\text{-}C_5H_{12}) \quad = \quad -40.18 \quad = \quad 4P + Q$$

From these equations, one obtains

$$P \quad = \quad 0.5\, \Delta_f H(C_2H_6) \qquad\qquad\qquad = -10.01_5 \text{ kcal/mol,}$$
$$S \quad = \quad \Delta_f H(C_3H_8) - \Delta_f H(C_2H_6) \qquad = -4.99 \text{ kcal/mol,}$$
$$T \quad = \quad \Delta_f H(i\text{-}C_4H_{10}) - 1.5\, \Delta_f H(C_2H_6) \quad = -2.03 \text{ kcal/mol,}$$
$$Q \quad = \quad \Delta_f H(neo\text{-}C_5H_{12}) - 2\, \Delta_f H(C_2H_6) \quad = -0.12 \text{ kcal/mol.}$$

Is the advantage of greater precision and accuracy in the experimental data for these small alkanes offset by the possibility that these small molecules might be slightly irregular? In other words, do the P groups in, say, hexadecane, look more like the P groups in decane than in those in ethane? One could take this likelihood into account by redetermining the values of P, S, T, and Q on the basis of a larger database than just the four alkanes just considered. There is more than one way this can be done. For example, one could first determine the S group by looking at the list of all straight-chain alkanes and taking the average of the incremental enthalpies, since $\Delta_f H^\circ(C_n H_{2n+2}) = 2P + (n-2)S$. Experimental values are available only for straight-chain alkanes from ethane through dodecane and three higher alkanes. If we minimize the average error for all 19 alkanes, the best value for S is -4.94; on the other hand, if we include *all* alkanes for which experimental values of $\Delta_f H^\circ(298)$ are available, the best value is -5.00. A difference of 0.05 kcal/mol seems negligible; yet for a large hydrocarbon we can easily accumulate a discrepancy on the order of 1 kcal/mol. On the other hand, we have elsewhere presented arguments that experimental enthalpy measurements are probably no better than ± 0.1 kcal/mol per carbon atom (*8*). Thus, since errors tend to increase with molecular weight, it very well may be more reliable to base GAVs on a smaller number of (lower molecular weight) compounds for which experimental measurements are better established, rather than to rely on a least squares fit and be misled by the apparent security of a larger database. The possibility of the smaller molecules not being representative is the lesser of the two dangers.

Unevaluated or Unreliably Evaluated Groups

The problems of the preceding paragraph ensue when one has a plethora of experimental data at hand. At the other end of the spectrum are those cases where a particular group occurs in only a single compound for which experimental data are available. In such cases, GAV uncertainties can easily be several kcal/mol or larger. Still more problematic are those groups for which no direct experimental data are available.

In Ref. *1* (Table A.2), 98 GAVs were evaluated for the gas phase for C-H and C-H-O compounds. In a more recent reevaluation, 140 GAVs were defined for this same set (*9*). While this would appear to represent a remarkable advance in GA, it must be confessed that almost all the new group evaluations depend on a single compound; and many of those are based on older thermochemical data which are of limited reliability. This brings us to an important conclusion regarding the current and future status of GA. The growth rate of the experimental thermochemical database has slowed considerably in recent decades; there are few experimental programs devoted to the evaluation of new compounds. This means that the number of *reliably established* GAVs will not increase significantly in the future if we are forced to depend on experimental data, which were the *sine qua non* of GAVs when the method was first worked out. Hopefully, theoretical contributions can begin to fill this gap.

Using Group Additivity to Check the Database

Group Additivity has now reached the degree of maturity where one can use it as a check on the reliability of experimental data. For example, following are data for enthalpies (in this and all successive tabulations, data are in kcal/mol) for the liquid phase for three compounds (taken from Ref. 9):

Compound	Expt.	Calc.	(Expt. - Calc.)
Methylcyclopropane	0.4	-2.9	3.3
Cyclohexane	-37.4	-37.3	-0.1
Cyclohexylcyclopropane	-75.2	-20.8	-54.4

Without going into the details of the group values, it is apparent that parameters that fit the first two entries fail for cyclohexylcyclopropane. Since there is no apparent reason why GA should not work in these cases, the discrepancies between experimental and calculated values are a good indication that the former are in error rather than the latter.

As another example, consider the cycloalkanes from cyclodecane to cyclotetradecane. The third column below lists the RSC (ring strain correction) term necessary to bring the calculated enthalpy of formation into agreement with the experimental gas phase value (taken from Ref. 5):

Compound	Expt.	RSC
Cyclodecane	-36.9	13.1
Cycloundecane	-42.9	12.1
Cyclododecane	-55.0	5.0
Cyclotridecane	-58.9	6.1
Cyclotetradecane	-57.2	12.8
Cyclopentadecane	-72.0	3.0
Cyclohexadecane	-81.1	3.1

The trend down the list suggests that the experimental value for cyclotetradecane is too positive by about 6 - 7 kcal/mol. (In fact, this one datum was not determined by the same laboratory that reported all the other accepted values for the cycloalkanes.)

Non-Bonded Interactions (NNIs AND RSCs)

Another area in which data are insufficient to draw definitive conclusions is that of non-nearest neighbor interactions: *e. g., cis, ortho*, and *gauche* effects, which involve atoms separated by at least two other atoms (normal group additivity relations take into account interactions between atoms with at most one atom between them). Some workers have tried to resolve these difficulties by introducing a non-additive hindrance correction such as a buttress effect--an additional strain resulting from the presence of three or more adjacent substituents (10). But the experimental data at hand do not provide consistent justification for such an effect. For example, consider the following methyl-substituted benzenes in the liquid phase. If there are buttressing effects, one would expect the error, when the calculations take into account only standard groups and additive *ortho* corrections, to increase with increasing number of vicinal (neighboring) methyl groups. The data in the table below (taken from Ref. *9*) do not support this prediction.

Compound	Expt. $\triangle_f H$	Calc. $\triangle_f H$	Vicinal CH₃ Groups
1,2-dimethylbenzene	-5.8	-5.3	1
1,3-dimethylbenzene	-6.1	-5.8	0
1,4-dimethylbenzene	-5.8	-5.8	0
1,2,3-trimethylbenzene	-14.0	-13.7	2
1,2,4-trimethylbenzene	-14.8	-14.1	1
1,3,5-trimethylbenzene	-15.2	-14.5	0
1,2,3,4-tetramethylbenzene	-21.6	-22.1	3
1,2,3,5-tetramethylbenzene	-23.0	-22.4	2

Other steric effects present similar problems resulting from inadequate or inconsistent data. Consider next the ether *gauche* effect (the 1-4 interaction across an O atom bridging two alkyl groups). In Ref. *1*, a value of 0.5 kcal/mol was recommended. Reanalysis of the best currently available data suggests that 0.8 kcal/mol may be a better value, though, as the table below comparing the differences between experimental and calculated enthalpies (with a 0.8 kcal/mol ether *gauche* correction) shows, there is considerable scatter in the data. The middle column is the number of *gauche* interactions.

Compound	Gauche	$\triangle_f H$(Expt - Calc)
Methyl isopropyl ether	1	0.0
Ethyl s-butyl ether	1	-1.7
Di-s-butyl ether	2	0.3
Methyl t-butyl ether	2	1.0
Diisopropyl ether	2	0.3
Isopropyl t-butyl ether	3	-0.3

Another non-nearest neighbor interaction that has been suggested is hydrogen bonding between vicinal OH groups in diols (*11*). Several polyols for which gas or liquid phase data are available are shown below. The differences between experimental and calculated values for the gas and liquid phases are based on GAVs taken (with slight modifications) from Ref. *3*. The values for heats of vaporization, however, are taken from an independent evaluation by Ducros *et al.* (*12*). The last column gives the number of neighboring OH groups.

Is there some additional stabilization? Taken at face value, the data suggest possibly some stabilization effect in the gas, but probably less in the liquid. A hydrogen bond of 2-4 kcal/mol is not unreasonable; but the magnitude of the effect is clearly not constant. (Such a value for the *intramolecular* hydrogen bond strength is slightly smaller than the value of 6.1 kcal/mol deduced by Benson for *intermolecular* hydrogen bonding in alcohols; see below.)

Compound	Δ_fH(gas)	Expt - Calc Δ_fH(liquid)	Δ_vH	Vicinal OH groups
1,3-Propanediol	3.3	4.2	0.1	0
1,3-Butanediol	2.6	5.1	0.2	0
1,2-Ethanediol	-0.6	0.4	0.1	1
1,2-Propanediol	0.4	2.6	0.2	1
1,2-Butanediol		-0.3		1
2,3-Butanediol	-5.1	-1.2	-0.1	1
1,2,3-Propanetriol	-2.2	1.9	1.4	2
1,2,3,4-Butanetetrol	-3.1	2.2		3

Because of the overall inconsistency between calculated and experimental values--especially in the case of the two compounds with no vicinal OH groups--it is not possible to assign a correction term for an H-bonding effect.

The comparisons for enthalpies of vaporization shown in the preceding table and based on the parameters of Ref. *12* illustrate another important point: the accuracy with which one can predict heats of vaporization is presently much greater than that for enthalpies of formation. In other words, since tabulated enthalpies of formation for liquid and gas phases invariably depend on one experimental measurement of enthalpy of formation and one measurement of enthalpy of vaporization, the difference between the two enthalpies of formation is known independent of either value alone. This is simply a reflection of the greater accuracy in the enthalpy of vaporization measurement than in the enthalpy of formation measurement(s). It also reinforces the conclusion that the experimental data are just not precise enough to quantify the effects of intramolecular hydrogen bonding on enthalpies of formation.

All of these examples demonstrate the difficulty in extracting values for all but the simplest non-nearest neighbor effects: we are searching for effects of a few kcal/mol in relatively large molecules, for which the uncertainties in the experimental enthalpies are several kcal/mol.

There are other intramolecular interactions that pose problems. Consideration of the halogenated compounds--fluorinated ones in particular--raises a complication not amenable to group additivity treatment: we refer to the effects of highly polarizing substituents. The problem is illustrated by consideration of three compounds: C_2H_6, CH_3CF_3, and C_2F_6. Their enthalpies are reasonably well established as -20, -179, and -321 kcal/mol, respectively. Only two groups comprise these three compounds: C-(H)$_3$ and C-(F)$_3$. If group additivity were valid, we would expect Δ_fH°(CH$_3$CF$_3$) = ½[Δ_fH°(C$_2$H$_6$) + Δ_fH°(C$_2$F$_6$)] = -170.5 kcal/mol. In fact, CH$_3$CF$_3$ is 8.5 kcal/mol more stable than that, owing to the additional C--C bond strengthening resulting from the fluorine atoms inducing a positive charge on one carbon atom while the H atoms induce a slight negative charge on the other.

At the present maturity level of the GA method, the most likely obstacle to a worker trying to estimate the enthalpy of some compound of interest for which experimental data are not available will be the absence of the necessary RSC. Drawing again on the experience of one of us (NC), efforts to expand the enthalpy database for C-H and C-H-O compounds have more than doubled the number of RSCs identified in Ref. *1*. Again, the disconcerting side to this revelation is that almost all of these additions are based on but a single compound, and generally one whose measurements are subject to some question. Unlike the NNIs discussed in the preceding paragraphs, an unknown RSC can make an enormous contribution to the calculated enthalpy; not knowing an RSC therefore can render a GA calculation practically meaningless.

In the absence of experimental data, GA's accepted starting point, advances in the method will require some novel techniques. The remainder of this survey is devoted to several such techniques that have provided alternative approaches to expanding the range of utility of GA.

Group Families

If groups are collected into families that ignore the differences between C, C_d, C_t, *etc.*, some interesting relations reveal themselves. For example, consider the following groups where the central species is the carbonyl ($>C=O$) functional group:

Group	GAV (gas phase)	GAV (liquid phase)
$CO\text{-}(C)_2$	-31.7	-36.2
$CO\text{-}(C_b)_2$	-27.0	-28.0
$CO\text{-}(C_d)_2$	-30.0	-31.0
$CO\text{-}(C_t)_2$	(-29.5?)	(-32.2?)
$CO\text{-}(C)(O)$	-35.2	-38.1
$CO\text{-}(C_b)(O)$	-34.4	-36.9
$CO\text{-}(C_d)(O)$	(-36.9?)	-32.7
$CO\text{-}(C_t)(O)$	-39.5	-35.2

In each case, the spread among the three known values is about 5 kcal/mol. It would seem quite reasonable, therefore, that the fourth GAV is going to lie within the range of values of the other three. If we choose the middle of the range, the probable uncertainty is about $\pm$ 2.5 kcal/mol.

This approach can be used in assigning values to groups of the form $C\text{-}(C_i)(C_j)(C_k)(C_l)$, where the four ligands are chosen from among C, C_d C_b, and C_t. Of the 35 possible groups of this type, gas phase GAVs are known for only 7, and they range from -0.1 to 7.3 kcal/mol. While this spread is larger than one would like, it still suggests a range of values for the unknown 28 groups that would be considerably better than no estimate at all.

It should also not escape notice that the difference between the liquid and gas phase GAVs is not large: generally only a few kcal/mol. This observation can provide estimates of GAVs for one phase when a value is available only for another phase.

Bond Additivity

Although the method of group additivities has proved gratifyingly successful, it cannot be denied that the large number of groups required can be a deterrent to routine application of the technique. For this reason, we should not fail to acknowledge that the much simpler method of bond additivities, while not nearly so accurate, can still be useful in the estimation of properties of molecules possessing unusual groups whose values have not yet been determined. We illustrate the method by deriving the bond additivity values (BAVs) for enthalpy of formation for the C-H, C-C, C-O, and O-H bonds. It should be stressed that these values are not the same as the bond dissociation enthalpies of the respective bonds.

BAVs can be derived from gas phase GAVs; in the following discussion we use the GAVs evaluated in Ref. 3. We note first that the method requires that all C-H bonds be treated as equivalent, and similarly for other bonds. Consider first the bonds of normal straight chain alkanes, namely C-H and C-C. Any straight chain alkane, C_nH_{2n+2}, can be decomposed into a sum of groups or a sum of bonds:

$$2P + (n-2)S = (n-1)[C-C] + (2n+2)[C-H],$$

where [C-C] and [C-H] designate the BAVs for the C-C and C-H bonds, respectively, and P and S designate the GAVs for the primary and secondary groups, as discussed above. By taking any two values of $n > 1$, we can write two linear equations in the two unknowns, [C-C] and [C-H], which can then be solved to yield

$$[C-C] + 2[C-H] = S = -5.00$$
$$[C-C] + 6[C-H] = 2P = -20.00$$

whence

$$[C-C] = 2.5$$
$$[C-H] = -3.75$$

The value for [C-O] can be evaluated by considering an arbitrary ether, ROR. The difference in enthalpies between the ether and the alkane of the same number of carbons is, in terms of groups:

$$\Delta_fH(ROR) - \Delta_fH(RR) = 2\,C\text{-}(H)_2(O)(C) + O\text{-}(C)_2 - 2S$$
$$= 2\,(-8.1) + (-23.5) - 2(-5.0) = -29.7$$

and in terms of bonds,

$$\Delta_fH(ROR) - \Delta_fH(RR) = 2[C-O] - [C-C]$$
$$= 2[C-O] - 2.5$$

whence

$$[C-O] = \tfrac{1}{2}(-29.7 + 2.5) = -13.6$$

In Ref. *1* BAVs for 32 different bonds common in organic compounds are tabulated. There are, though, some special cases that do not exactly meet the requirements of strict bond additivity. For example, the benzene ring is treated as a single species with a valency of 6, and the aldehydic CO-H bond is treated as different from the simple [C-H] bond. Bond additivity can predict enthalpies to within ± 2 kcal/mol for some carbon compounds, but is less reliable in the case of heavy branching.

A striking example is provided by the species $HC \equiv N$ and $HC \equiv CH$, which can be looked upon as both having $HC \equiv$ bonds. The derivatives, cyanogen ($N \equiv C\text{-}C \equiv N$), diacetylene ($HC \equiv C\text{-}C \equiv CH$) and cyanoacetylene ($N \equiv C\text{-}C \equiv CH$) can be regarded as related by an exchange of $N \equiv$ for $HC \equiv$. The heats of formation fit this simple view to within 1 kcal/mol: The average of $\Delta_f H^o(C_2N_2) = 73.9$ and $\Delta_f H^o(C_4H_2) = 106$ is 90 kcal/mol, while the observed value of $\Delta_f H^o(HC \equiv C\text{-}C \equiv N)$ is 90 kcal/mol. (Alternatively, one can regard C_2N_2 as consisting of 1 C-C and 2 $C \equiv N$ bonds, and C_4H_2 as consisting of 2 C-H, 1 C-C, and 2 $C \equiv C$ bonds; the sum of all these is 2 C-C + 2 $C \equiv N$ + 2 C-H + 2 $C \equiv C$ bonds, while $HC \equiv C\text{-}C \equiv N$ consists of 1 C-C + 1 $C \equiv N$ + 1 C-H + 1 $C \equiv C$ bond.)

Homothermal Pairs

Some time ago, Benson (*13*) observed that certain pairs of monovalent species were very similar in bond lengths and electronegativity. They might be similar in their contributions to heats of formation, and he proposed the name "homothermal pairs." F and OH, on the basis of available data, form such a pair; Br and SH are another. In organic compounds, the replacement of F by OH would lead to a constant difference in $\Delta_f H^o$. Taking CH_3OH and CH_3F as examples, $\Delta_f H^o(RF) - \Delta_f H^o(ROH) = -7$ kcal/mol. Similarly, $\Delta_f H^o(RBr) - \Delta_f H^o(RSH) = -3.7$ kcal/mol. The uncertainties in extending these values to other alkyls is estimated at about ± 1.5 kcal/mol. The method has frequently been referred to as the "replacement" or "difference" method (DM). For example, if one wishes to estimate $\Delta_f H^o$(2-fluorohexane, g, 298 K), one can start with the corresponding value for 2-hexanol and make a correction for the difference between an OH and an F substituent. In the language of GA, one replaces the C-(O)(H)$_2$(C) and O-(H)(C) groups by the C-(F)(H)$_2$(C) group.

$$C\text{-}(F)(H)_2(C) = C\text{-}(O)(H)_2(C) \text{ and } O\text{-}(H)(C) + \text{Constant}$$
$$-54.0 = -8.1 + (-37.9) + (-7.2)$$

DM suggests itself as particularly useful if one is interested in a number of structurally similar but complex molecules for none of which has the enthalpy been measured. One then carries out measurements for one of them and uses DM to estimate the corresponding properties of the others. The method has also been used to advantage in estimating thermochemical properties of free radicals (*14, 15*). In general, the method will be applicable if the unknown molecule and the model differ only by a series of substitutions of groups whose numerical values are well established.

Use of Liquid Phase Data to Obtain Gas Phase Enthalpies

In the absence of direct gas phase enthalpy data, one can use data from the liquid phase and make appropriate corrections, as exemplified in a study by Benson and Garland (*16*). The two quantities are related by the equation,

$$\Delta_f H^o(g, 298 \text{ K}) = \Delta_f H(l, 298 \text{ K}) + \Delta_v H(298 \text{ K}) \tag{1}$$

The required correction is the last term--the heat of vaporization at 298 K. However, what is usually measured and tabulated is the heat of vaporization at the boiling point, $\Delta_v H(T_b)$. The latter can be corrected to the former by making an approximate temperature correction:

$$\Delta_v H(298 \text{ K}) = \Delta_v H(T_b) + (298 - T_b) <\Delta_v C_p(T)> \tag{2}$$

here the last term is the average difference between heat capacities of the liquid and of the saturated vapor. An approximate expression for the temperature dependence of this term can be derived from four relationships:

(1) By definition, at the critical point

$$\Delta_v C_p(T_c) = 0 \tag{3}$$

(2) According to the Guye-Guldberg rule (*17*), for most regular liquids the critical point is related to the boiling point by

$$T_b/T_c = 0.625 \tag{4}$$

(3) For many small hydrocarbons,

$$\Delta_v C_p(T_b) = -10.5 \text{ cal-mol}^{-1}\text{-K}^{-1} \tag{5}$$

(4) $\Delta_v C_p$ varies approximately linearly with temperature:

$$\Delta_v C_p = a + bT \tag{6}$$

From these four expressions, one can derive

$$-\Delta_v C_p(T) = 28.0 - 17.5(T/T_b) \text{ cal-mol}^{-1}\text{-K}^{-1} \tag{7}$$

and one can obtain $\Delta_v H^o(298)$ knowing only the values of $\Delta_v H^o(T_b)$ and T_b. Benson and Garland were thus able to assemble a database of 25 acetylenes and polyacetylenes from which to derive group values for alkynes. Table 1 illustrates the gas phase enthalpies of formation calculated from the liquid phase data and the enthalpies calculated from the GAs derived from those "experimental" gas phase enthalpies.

COMPUTATIONAL THERMOCHEMISTRY

Table I. An Example of the Derivation of Gas Phase Enthalpies from Liquid
Phase Experimental Data (all data in kcal/mol) (16)

Compound	$\Delta_f H°(298)$		$\Delta\Delta_f H°(298)$
	Expt.	GA	(Expt. - GA)
acetylene	54.51	54.4	0.11
propyne	44.17	44.6	-0.43
1-butyne	39.46	39.9	-0.44
2-butyne	34.81	34.8	0.01
1-pentyne	34.50	34.9	-0.40
2-pentyne	29.88	30.1	-0.22
3-methyl-1-butyne	32.87	32.9	0.03
3,3-dimethyl-1-butyne	25.42	25.2	0.22
trans-3-penten-1-yne	61.61	61.1	0.54
cis-3-penten-1-yne	60.98	60.5	0.51
1,5-hexadiyne	100.14	99.8	0.34
phenylacetylene	73.3	73.7	-0.4
1,7-octadiyne	90.14	89.8	0.34
1-octen-3-yne	43.72	44.2	-0.52
trans-3-decen-1-yne	36.16	36.3	-0.15
cis-3-decen-1-yne	35.44	35.7	-0.27
5,7-dodecadiyne[a]	56.98	57.0	(-0.02)
3,9-dodecadiyne	60.79	60.8	-0.01
3,3,6,6-tetramethyl-1,7-octadiyne	61.86	62.0	-0.14
3,3-dimethyl-1,4-pentadiyne[a]	91.23	91.2	(0.03)
1-phenyl-1-propyne	64.1	63.9	0.2
1-phenyl-1-butyne	59.7	59.2	0.5
1-phenyl-1-hexyne	49.5	49.2	0.3
diphenylacetylene	93.8	93.0	0.8
diphenylbutadiyne	142.5	144.6	-2.1

[a]Compound was used to derive unique group contribution.

In order to apply equations 1 - 7, one needs to know the boiling point, T_b, of the compound of interest. If the structure is known, there are several empirical methods for estimating boiling points. These are described in some detail by Reid, Prausnitz, and Poling (18), and are usually accurate to about 5 - 10 K. An error of this magnitude, in conjunction with Trouton's constant, 20.5 cal/mol-K, will introduce an uncertainty of ±(0.1 - 0.2) kcal/mol in $\Delta_v H(T_b)$. An error of 5% in Trouton's constant (±1 cal/mol-K) will introduce an additional corresponding uncertainty of ±1 x T_b cal/mol in the estimate of $\Delta_v H(T_b)$. For example, if T_b = 500 ± 25 K, the uncertainty in $\Delta_v H(T_b)$ will be ±0.5 kcal/mol.

Alternatively, there are a number of empirical methods which have been proposed for estimating $\Delta_v H(T_b)$ and $\Delta_v H(298 K)$. Some of these are described in Ref. 13 and have errors on the order of about ±5%.

In a series of papers, Ducros et al. (19-21) have shown that GA can be applied to a series of compounds to estimate $\Delta_v H(298 K)$, whence it is possible to estimate $\Delta_f H^o$(gas). For hydrocarbons, average deviations are the order of 0.1 kcal/mol. For alcohols, which would be expected to provide a severe test because of their structural hydrogen bonding, average deviations are about 0.2 kcal/mol. For the halides, nitriles, and other mono-functionalized organics, the authors chose an algebraic formula to represent the effect of the chain length on primary-, secondary-, and tertiary-substituted carbons. In their fourth paper, Ducros *et al.* extend the method to metalorganic compounds (*21*). Here, deviations are much larger. These have proven to be the most useful methods of going from liquid to gas for enthalpies of formation.

The large majority of reactions studied by chemists are studied in the liquid phase. Thus, in order to make use of the thermochemical data discussed here it is important to be able to relate them to solution values. This has been discussed in some detail (*22-24*).

Where $\Delta_f H^o$(liq.) has not been measured directly, it can be estimated from $\Delta_f H^o$(gas) using the GAVs of Ducros (*12, 18*) for $\Delta_v H(298 \text{ K})$.

If we are dealing with non-structured liquids, they mix to form what are known as "regular" solutions, for which the heats of mixing are usually negligible compared to 1 kcal/mol and for which entropies of solution are essentially the ideal entropies of mixing. For structured liquids such as alcohols, acids, amides, and nitriles, this is not the case and heats of mixing can be of the order of several kcal/mol. Such structured liquids are generally self-associated in the liquid state and much less in the gas state. Their entropies of vaporization at the boiling point reflect this in having values appreciably larger than the Trouton value of 20.5 cal/mol-K. Benson has shown that alcohols exist in the liquid state in the form of cyclic tetramers, quite independent of the extent of branching (*25*). As a consequence of this, the heat of vaporization at 298 K of an alcohol ROH, $\Delta_v H^o$(ROH, 298), is quantitatively related to that of its analogous hydrocarbon RCH_3, $\Delta_v H(RCH_3, 298)$, by the relation

$$\Delta_v H(ROH, 298) - \Delta_v H(RCH_3, 298) = 6.1 \pm 0.2 \text{ kcal/mol.}$$

It is surmised that the energy of the hydrogen bond between alcohol molecules is the same 6.1 kcal/mol. Dihydroxy alcohols, $R(OH)_2$, which are capable of forming two H-bonds per molecule, differ from their $R(CH_3)_2$ analogues by having $\Delta_v H$ about twice this value, namely 11 to 12 kcal/mol.

Thermochemical Analogies Used to Predict Enthalpies: Electronegativity

Bond additivity is less accurate than group additivity in estimating $\Delta_f H^o$. However, it is quite good at estimating S^o and C_p^o. The database requirement for a bond additivity scheme is orders of magnitude less than that required for the more accurate group additivity scheme, and so it is tempting to ask if there is not some way of estimating groups from bonds. What is involved in going from bond additivity to group additivity is the effect of next-nearest neighbors on the property in question. In the case of $\Delta_f H^o$, we are concerned with the total energy of the compound, and this involves basically the interactions between valence electrons and the nuclei. The replacement of an H atom in a hydrocarbon by a group X involves most significantly the loss of a C-H bond and the gain of a C-X bond. At the bond additivity level this would be a constant for each such replacement. However, this is not an accurate description, and the deviations from constancy have to do with the differences in interactions of H and X with other ligands (H or C) attached to the carbon atom being substituted. Such interactions are small--on the order of 1 - 6 kcal/mol--and a simple scheme that can estimate them would be very attractive.

Many schemes have been suggested to estimate such interactions, some empirical and some with some theoretical foundation. Among the latter, the concept of electronegativity has played an important role. A very promising approach has been provided based on a new definition of electronegativity. Luo and Benson (26) have shown that a linear relation exists between $\Delta_f H^o(MeX) - \Delta_f H^o(RX)$ and V_x, where V_x is the newly defined electronegativity. Figure 1, taken from Ref. 25, shows three such relations for R = ethyl, isopropyl, and t-butyl, while X ranges over the main group elements. V_x is defined as n_x/r_x, where n_x is the number of valence electrons of X (1 to 7), and r_x is the covalent radius of X. Deviations from the linear relation are generally within the precision of the value of $\Delta_f H^o$, and better than 0.5 kcal/mol. With the availability of $\Delta_f H^o$ for, let us say, CH_3X and for two different RX molecules, say, for example, RH and ROH, it is possible to evaluate the slope and intercept of the line and then to estimate values of $\Delta_f H^o(RX)$ for all other compounds RX where $\Delta_f H^o(CH_3X)$ is known.

Such a scheme has been shown to work very well for substituted organosilicon compounds (27, 28). In the case of alkyl fluorides, this scheme has revealed large inconsistencies in reported values of $\Delta_f H^o$ for ethyl, isopropyl, and t-butyl fluorides (26). It has been extended to many other series where it would be expected to be useful. For example, one would expect that although carbon atoms in saturated compounds differ from those in unsaturates, there should be similar linear relations in their $\Delta_f H^o$'s upon substitution. Thus, it seems plausible that $\Delta_f H^o(C_2H_5X) - \Delta_f H^o(C_2H_3-X)$ would show a linear relation when plotted against V_x. Luo and Holmes (29) have shown that there is in fact a nearly linear relation in $[\Delta_f H^o(CH_3X) - \Delta_f H^o(C_2H_3X)]$ plotted against V_x. However, only for X = halogens or hydrogen do the values fall on the line within the uncertainty of the data, while X = F, OH, and NH_2 compounds seem to lie on a separate line, suggesting that in the sequence C_2H_3X, with X ranging over the isoelectronic sequence CH_3, OH, and F, the relative stability of C_2H_3X to CH_3X decreases with increasing electronegativity (Figure 2).

A more accurate relation explored by Luo and Holmes (30) is that $\Delta_f H^o(phenyl-X) - \Delta_f H^o(vinyl-X) = 7.1 \pm 1.5$ kcal/mol for a broad spectrum of compounds. This uncertainty is probably within the accuracy of the data (Figure 3). They do note a small number of exceptions (four) to this rule for which the deviations are large. However, in every case one of the $\Delta_f H^o$ values is in error. One expects better agreement in this case than in the preceding example inasmuch as both benzene and ethylene have similar carbon atoms in sp^2 valence states. Luo

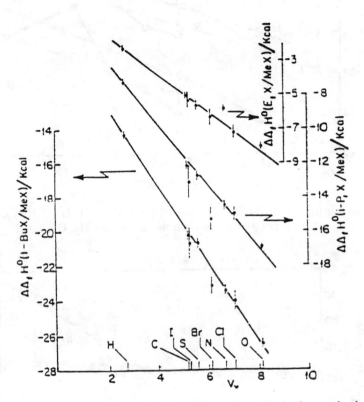

Figure 1. $\triangle_f H^o(RX) - \triangle_f H^o(MeX)$ *vs.* V_x, where V_x is the newly defined electronegativity and R is t-butyl, i-propyl, or ethyl.

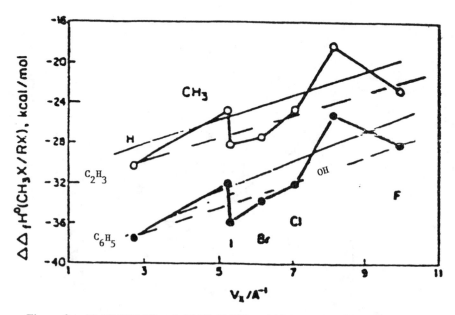

Figure 2. $[\Delta_f H^\circ(CH_3X) - \Delta_f H^\circ(C_2H_3X)]$ and $[\Delta_f H^\circ(CH_3X) - \Delta_f H^\circ(C_6H_5X)]$ *vs.* V_x, where X ranges over the isoelectronic sequence CH_3, OH, and F.

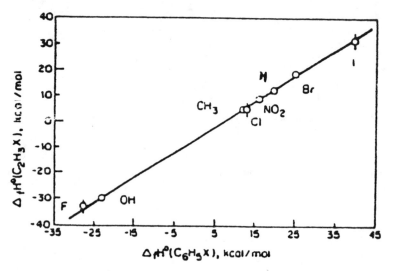

Figure 3.   Linear relation between $\Delta_f H^\circ$(phenyl-X) and $\Delta_f H^\circ$(vinyl-X).

and Holmes have demonstrated that the following relations are obeyed to within the accuracy of available experimental data:

$$\Delta_f H°(\text{allyl-X}) - \Delta_f H°(\text{ethyl-X}) = 25.2 \pm 0.5 \text{ kcal/mol}$$
$$\Delta_f H°(\text{benzyl-X}) - \Delta_f H°(\text{ethyl-X}) = 32.0 \pm 1.0 \text{ kcal/mol}$$

Since there are many more $\Delta_f H°$ known for ethyl-X than for allyl-X or benzyl-X, this permits the estimation of many unknown allyl or benzyl compounds from $\Delta_f H°(\text{ethyl-X})$. This has also led to changes in values for allyl-I and benzyl-I that had been based on questionable experimental reports. It is likely that a relation similar to the above holds for $\Delta_f H°(\text{propargyl-X})$ and $\Delta_f H°(\text{ethyl-X})$.

Many similar relations remain to be explored. For example: from the standpoint of *bond* additivity, the CH_2 groups in the following unsaturated compounds are the same:

$$CH_2 = CH_2 \qquad CH_2 = NH \qquad CH_2 = O$$

This isoelectronic sequence, however, shows large changes in dipole moment and increasing positive character of the CH_2 group going from ethylene to formaldehyde. One would expect this to be reflected in an increasing stability on substituting a methyl group in place of an H atom in the CH_2. In fact, one finds that $\Delta_f H°(CH_3CH=CH_2) - \Delta_f H°(C_2H_4) = -7.6$ kcal/mol and $\Delta_f H°(CH_3CHO) - \Delta_f H°(CH_2O) = -13.6$ kcal/mol. Hence, one might guess that $\Delta_f H°(CH_3CHNH) - \Delta_f H°(CH_2NH)$ would be close to the average, or -10.6 kcal/mol. $\Delta_f H°(CH_2NH)$ is +22.0 kcal/mol, but $\Delta_f H°(CH_3CHNH)$ has not been measured. It would be surprising if its value differs from 11.4 kcal/mol, estimated from the average, by more than 1 kcal/mol.

In similar fashion, substituting CH_3 for H in acetylene (C_2H_2) leads to 10.5 kcal/mol greater stability, compared to 7.6 kcal/mol for similar substitution in ethylene (C_2H_4). We anticipate that the CH group in acetylene is more electropositive than the CH_2 group in ethylene, in agreement with the above. In the case of HCN, one expects the CH group to be more electropositive than in C_2H_2, and thus it is not surprising that CH_3CN is more stable than HCN by 13 kcal/mol.

Some Applications.   The most common application of estimation methods is the prediction of equilibrium constants, K_{eq}, for specified stoichiometric reactions. This in turn requires an estimation of the values of $\Delta_f H°$, $S°$, and $C_p°(T)$ for each species--products and reagents--occurring in the reaction. One reason thermochemistry has received relatively limited attention from chemists has been that in facing problems such as predicting a given K_{eq}, one or more of the needed thermochemical values, generally a value of $\Delta_f H°$, has been missing. GA has provided only limited relief for such problems since GAV tables are incomplete. In the absence of pertinent groups, we can turn to the much more general database available for bond additivity, tempered by "judicious" considerations of electronegativity. Let us explore some examples.

$\Delta_f H°(CH_3NO)$, the heat of formation of nitrosomethane, is available from kinetic studies as 16.6 kcal/mol. Can we deduce from this a value for

$\Delta_f H^\circ(C_2H_5NO)$? To do this we need a value for the group $[C-(H)_2(C)(NO)]$. Alternatively, and less abstractly, this is equivalent to asking the question, "What is the difference between $\Delta_f H^\circ$ of MeX and EtX?" On the assumption that the best guide is the electronegativity of the NO group, we can look at the following differences in $\Delta_f H^\circ$, which are known:

$$\Delta\Delta_f H^\circ[(CH_3CH=CH_2) - (C_2H_5CH=CH_2)] = 5.0 \text{ kcal/mol}$$
$$\Delta\Delta_f H^\circ[(CH_3CH=O) - (C_2H_5CH=O)] = 5.3 \text{ kcal/mol}$$

Both the vinyl group ($CH=CH_2$) and the aldehyde group ($CH=O$) are isoelectronic with the (NO) group. We would expect NO to be intermediate between them in electronegativity, and thus a value

$$\Delta\Delta_f H^\circ[(CH_3NO) - (C_2H_5NO)] = 5.2 \pm 0.2 \text{ kcal/mol}$$

seems entirely appropriate. This immediately leads to a value for $\Delta_f H^\circ(C_2H_5NO)$ of 11.4 kcal/mol, and since

$$\Delta_f H^\circ(C_2H_5NO) = \Delta_f H^\circ[C-(H)_3(C)] + \Delta_f H^\circ[C-(H)_2(C)(NO)]$$

we find, for the unknown group,

$$\Delta_f H^\circ[C-(H)_2(C)(NO)] = 21.6 \text{ kcal/mol}$$

Another example is provided by the modeling of combustion chemistry where NO is a minor product and the compound $CH_2=CHNO$ is of potential importance. We can start with the "known" compound, nitrosoethane (C_2H_5NO), and ask the question, "What is the enthalpy of dehydrogenation, $\Delta_{dehyd}H$, to $CH_2=CHNO$?" As before, isoelectronic sequences are of interest:

$$\Delta_{hyd}H^\circ[(CH_2=CHCH=CH_2) \rightarrow (C_2H_5CH=CH_2)] = -26 \text{ kcal/mol}$$
$$\Delta_{hyd}H^\circ[(CH_2=CHCH=O) \rightarrow (C_2H_5CH=O)] = -27.2 \text{ kcal/mol}$$

Again, it seems reasonable to assign an intermediate value of -26.6 ± 0.6 kcal/mol to the enthalpy of hydrogenation of $CH_2=CHNO$, and so from our derived value of $\Delta_f H^\circ(C_2H_5NO) = 11.4$ kcal/mol, we estimate $\Delta_f H^\circ(CH_2CHNO) = 38.0$ kcal/mol. From this, in turn, we can further estimate the group value,

$$\Delta_f H^\circ[C_d-(H)(NO)] = 31.7 \text{ kcal/mol}$$

It is tempting to see how far such analogies can be extended. For example, it is known that $\Delta_f H^\circ(CH_2=CHC\equiv CH)$ (vinyl-acetylene) is 68.7 kcal/mol (31). Can we make an estimate of $\Delta_f H^\circ(HC\equiv CNO)$ using the above estimated value for $\Delta_f H^\circ$(vinyl-NO)?

As before, using hydrogenation as a starting point, we have

$$\Delta_{hyd}H^\circ[CH_2=CHC\equiv CH \rightarrow CH_2=CHCH=CH_2] = -42.7 \text{ kcal/mol}$$

Note that the second π-bond in acetylene is at right angles to the first and cannot conjugate with the vinyl group in vinyl acetylene. On this basis, we might conclude that the same will be true of the vinyl-NO and vinyl-CHO compounds, and so the -42.7 kcal/mol might be directly transferable. This would then give a value of 80.7 kcal/mol for $\Delta_f H^o(HC \equiv CNO)$ and 24.9 kcal/mol for $\Delta_f H^o(HC \equiv CCHO)$. The uncertainty here is greater than in the previous estimates since we have only one value from which to extrapolate.

A final example of interest involves hydrogenation. The enthalpy of hydrogenation of ethylene to ethane is -32.5 kcal/mol. That of formaldehyde (CH_2O) to methanol (CH_3OH) is -22.2 kcal/mol, reflecting the stronger π-bond in formaldehyde relative to that in $CH_2=CH_2$. $\Delta_f H^o(CH_2=NH)$ has been estimated at about 22.2 ± 2 kcal/mol by various authors. We might expect that its π-bond should be intermediate between those in C_2H_4 and CH_2O. Thus we could estimate from the arithmetic average of the enthalpies of hydrogenation that the value for $CH_2=NH \rightarrow CH_3NH_2$ should be 27.4 kcal/mol. Since $\Delta_f H^o(CH_3NH_2, 298\ K) = -5.5$ kcal/mol, this would yield 21.9 kcal/mol for $\Delta_f H^o(CH_3=NH)$, in excellent agreement with the estimated values.

Heat Capacities, Entropies, and Temperature Dependence of Enthalpies. The discussion so far has been confined to the single temperature of 298 K. The evaluation of entropy, $S^o(T)$, or the evaluation of enthalpy at some other temperature, both require measurements of heat capacity--in the case of entropy, from 0 K up to the temperature of interest; in the case of enthalpy, between 298 K and the temperature of interest. (Alternatively, entropy can be calculated from the temperature dependence of an equilibrium constant measurement--the "Second Law method.") Here too, however, the number of laboratories involved in the experimental measurement of heat capacity data has greatly diminished in recent decades. Fortunately, GA has proven very successful at estimating heat capacities; most of the references that have been cited in connection with GAVs for $\Delta H^o(298\ K)$ also evaluate and discuss GAVs for C_p--either for a discrete range of temperatures from 298 K up to 1500 K or at least 1000 K, or as coefficients for C_p as a polynomial function of T. Past experience indicates that GA calculations of S or C_p are generally reliable to within approximately ± 1 cal/mol-K.

Free Radicals

Free radicals can be partitioned into groups in the same manner as can stable molecules. Thus, the *s*-butyl radical, $CH_3CHCH_2CH_3$, consists of the groups C-$(C\cdot)(H)_3$, $C\cdot-(C)_2(H)$, S, and P. There are two classes of radical groups: radical-centered groups, such as $C\cdot-(C)_2(H)$, and radical-adjacent groups, such as C-$(C\cdot)(H)_3$. Every radical with more than one group contains one of the former and at least one of the latter groups. The method of group additivities is assumed to apply equally well to free radicals, but the persisting uncertainties in radical enthalpies limit the accuracy of GA applications (*32*).

Dilling (*33*) has proposed a useful relationship between GAVs involving radical-centered groups:

$$[C \cdot -(X)(Y)(Z)] - [C \cdot -(H)(Y)(Z)] =$$
$$(0.86 \pm 0.03)\{[C-(H)(X)(Y)(Z)] - [C-(H)_2(Y)(Z)]\} - (4.64 \pm 0.62)$$

That is, replacing an H atom ligand in a radical with some other ligand, X, is almost energetically equivalent to replacing one of 2 H ligands in a molecule with the same X ligand. The relationship was derived by examining 89 sets of data each involving four related groups. He then applies the relationship to derive new GAVs for some three dozen groups not previously evaluated.

Solids

We have refrained so far in discussing the thermochemistry of solids. This owes in large measure to the fact that until now we lack a quantitative theory of melting or sublimation. Solid thermochemistry also lacks general applicability for chemists, since the first law of chemistry is that in order for chemicals to react, their molecules must be in contact. This reduced the rates of chemical reaction in solids to the rates of interdiffusion in solids, which suggests times of the order of geological epochs. Despite these barriers, there is potential interest for geochemists in the chemical stability of solids and particularly of salts. Recently, some very surprising relations have been discovered which now make it possible to estimate the value of $\Delta_f H^{\circ}(298 \text{ K})$ for any solid salt as well as its hydrate, alcoholate, and ammoniate (*34*). These relations are all of the form:

$$\Delta_f H^{\circ}(MX_n) - \Delta_f H^{\circ}(MY_n) = a[\Delta_f H^{\circ}(MZ_n) - \Delta_f H^{\circ}(MY_n)] + b,$$

where a and b are constants for any metal M of formal valence n and X, Y, and Z can be any three anions. Comparable relations hold for anions with different metals or cations. Figure 4 illustrates such a relation between oxides, chlorides, and bromides of the alkali metal family. Figure 5 illustrates such a relationship between carbonates, sulfates, and oxides for a broad spectrum of cations for almost all the groups of the periodic table. These relations lead to equations of the type,

$$\Delta_f H^{\circ}(MX_n) = a \, \Delta_f H^{\circ}(MZ_n) - (a-1) \, \Delta_f H^{\circ}(MY_n) + b$$

When the coefficients a and b are known, it is possible to estimate $\Delta_f H^{\circ}$ of any salt from two known salts to within the experimental uncertainty of the class. This is about 1 to 2 kcal/mol for most salts. The generality of such relations is shown in Figure 6 for solid halides.

It can be shown that these relations require a database in between that of atom additivity and that of bond additivity and suggests a simplicity in the energetics of solid salts that could not have been anticipated. More recently, the same authors (*35*) were able to show that an even more accurate relation could be found between metal halides by employing the quantity $\Delta(MX_n)$, defined as

ENTHALPIES OF FORMATION OF SOLID SALTS

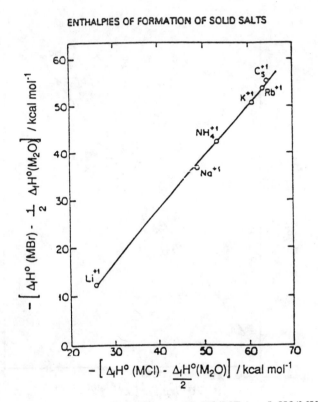

Figure 4. $\triangle_f H^o(MX_n) - \triangle_f H^o(MY_n) = a[\triangle_f H^o(MZ_n) - \triangle_f H^o(MY_n)] + b$, where a and b are constants for alkali metals M of formal valence n and X, Y, and Z are chlorides, bromides, and oxides.

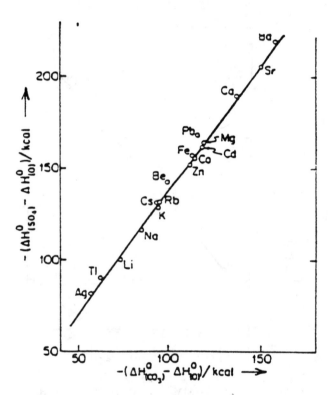

Figure 5. $\Delta_f H^o$ relationships (per mol of cation) among carbonates, sulfates, and oxides.

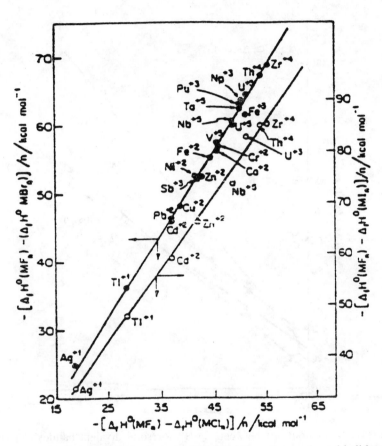

Figure 6.   Relationships among $\triangle_f H^o(MX_n)$ for subgroup metal halides.

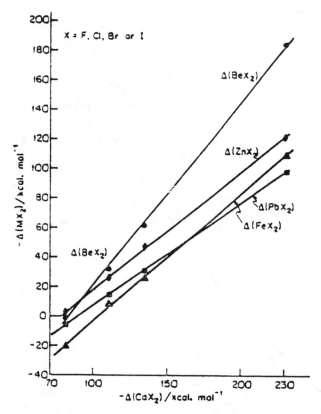

Figure 7.   Relationships between $\Delta_f H^o$ for solid divalent halides (X) and $\Delta_f H^o(X^-, g)$.

$$\Delta(MX_n) = \Delta_f H^o(MX_n, cr) - \Delta_f H^o(X^-, g)$$

Here $\Delta_f H^o(X^-, g)$ is the very precisely known value of the enthalpy of formation of the gas phase halide ion X^-. The resulting relations are simpler than those listed earlier, and take the form,

$$\Delta(MX_n) = a \Delta(M'X_n) + b$$

Figure 7 illustrates the relation for divalent halides MX_2 using $\Delta(CaX_2)$ as a common base. No theoretical explanation has been suggested to account for such a relation.

Closing Remarks

Non-bonded interactions (in particular RSCs) are probably the ultimate factor limiting the accuracy of straight-forward group additivity methods in predicting thermochemical properties of large molecules. However, the handful of cases in which the disagreement between calculation and experiment exceeds a few kcal/mol can be accounted for either by non-bonded interactions or, just as plausibly, by experimental error, since in most of those cases the assumed experimental values depend on a single laboratory measurement. The extent to which non-bonded interactions can be systematically treated will not be known unless we have a better database to rely upon. Parenthetically, we might note that of those groups for which values are already derived, it is unlikely that future measurements will change them significantly. There are very few cases for which the values of 1976 have changed by more than ½ - 1 kcal/mol. We have discussed above strategies for estimating gas phase enthalpies when direct experimental data are not available, and suggested other approximative approaches. The methods discussed deserve wider use, and can considerably expand the roster of available GAVs.

Thus, we come back to the problem of the limitations imposed by the accuracy and extent of the experimental database. GA was developed to afford a simple means of estimating unmeasured thermochemical properties. It will be superannuated if and when the database includes all molecules of interest or when *a priori* methods become so reliable and convenient as to become the procedure of choice. The unlikelihood of this happening soon insures that GA will remain a useful tool for making rapid property estimates or for checking the likely reliability of existing measurements. The rate of progress of theoretical methods has been encouraging, and is elaborated upon in other chapters in this volume.

Acknowledgements

Support, in part, by a contract to one of us (NC) from the National Institute of Standards and Technology (Contract No. 43NANB420595), is gratefully acknowledged. The senior author is indebted to the donors of the Petroleum Research Fund administered by the ACS for their support in this work.

Literature Cited

1. See Benson, S. W. *Thermochemical Kinetics*, 2nd ed.; Wiley: New York, NY, 1976; and the references cited therein.
2. Pitzer, K. S. *J. Chem. Phys.* **1940**, *8*, 711.
3. Cohen, N.; Benson, S. W. *Chem. Rev.* **1993**, *93*, 2419.
4. Benson, S. W.; Buss, J. H. *J. Chem. Phys.* **1958**, *29*, 546.
5. For a recent résumé of the theory, see Cohen, N.; Benson, S. W. In *The Chemistry of Alkanes and Cycloalkanes*; Patai, S; Rappoport, Z., Eds.; Wiley: New York, NY, 1992; pp 215-287.
6. Ref. 1, p 26.
7. Ref. 1, Appendix.
8. Ref. 5.
9. Cohen, N. *J. Phys. Chem. Ref. Data* **1996**, *25*, 1411.
10. See, for example, Wu, Y. G.; Patel, S. N.; Ritter, E. R.; Bozzelli, J. W. *Thermochim. Acta* **1993**, *222*, 153.
11. Eigenmann, H. K.; Golden, D. M.; Benson, S. W. *J. Phys. Chem.* **1973**, *77*, 1687.
12. Ducros, M.; Gruson, J. F.; H. Sannier, H. *Thermochim. Acta*, **1980**, *36*, 39.
13. Benson, S. W. *Chem. Rev.* **1978**, *78*, 231.
14. O'Neal, H. E.; Benson, S. W. In *Free Radicals,* Kochi, J. K., Ed.; Wiley: New York, NY, 1973; pp 275-359.
15. Cohen, N. *J. Phys. Chem.* **1992**, *96*, 9052.
16. Benson, S. W.; Garland, L. J. *J. Phys. Chem.* **1991**, *95*, 4915.
17. Guldberg, J. *Z. Physik. Chem.* **1910**, *116*, 32.
18. Reid, R. C.; Prausnitz, J. M.; Poling, B. E. *The Properties of Gases and Liquids*, 4th ed.; McGraw-Hill: New York, NY, 1987; Chap. 6.
19. Ducros, M.; Gruson, J. F.; Sannier, H.; Velasco, I. *Thermochim. Acta* **1981**, *44*, 131.
20. Ducros, M.; Sannier, H. *Thermochim. Acta* **1982**, *54*, 153.
21. Ducros, M.; Sannier, H. *Thermochim. Acta* **1984**, *75*, 329.
22. Benson, S. W.; Golden, D. M. In *Physical Chemistry*; Eyring, H.; Henderson, D.; Jost, W., Eds.; Academic Press: New York, NY, 1975; Chap. 2.
23. Benson, S. W.; Seibert, E. D. *J. Am. Chem. Soc.* **1992**, *114*, 4269.
24. Mendenhall, D.; Benson, S. W. *J. Am. Chem. Soc.* **1976**, *98*, 2046.
25. Benson, S. W. *J. Am. Chem. Soc.* 118, 10645 (1996).
26. Luo, Y.-R.; Benson, S. W. *J. Phys. Chem.* **1988**, *92*, 5255.
27. Luo, Y.-R.; Benson, S. W. *J. Phys. Chem.* **1989**, *93*, 1674.
28. Luo, Y.-R.; Benson, S. W. *J. Phys. Chem.* **1989**, *93*, 3791.
29. Luo, Y.-R.; Holmes, J. L. *J. Phys. Org. Chem.* **1994**, *7*, 403.
30. Luo, Y.-R.; Holmes, J. L. *J. Phys. Chem.* **1992**, *96*, 9568.
31. Ref. 5.
32. For more details, see Refs. 5 and 13.
33. Dilling, W. L. *J. Org. Chem.* **1990**, *55*, 3286.
34. Hisham, M. W.; Benson, S. W. In *From Atoms to Polymers,* Liebman, J. F.; Greenberg, A., Eds.; VCH, New York, NY: 1989; pp 394-465.
35. Hisham, M. W.; Benson, S. W. *J. Chem. Eng. Data* **1992**, *37*, 194.

Chapter 3

Estimation of Enthalpies of Formation of Organic Compounds at Infinite Dilution in Water at 298.15 K

A Pathway for Estimation of Enthalpies of Solution

Eugene S. Domalski

Physical and Chemical Properties Division, National Institute of Standards and Technology, Gaithersburg, MD 20899

A method has been developed for estimation of enthalpies of formation of organic compounds at infinite dilution in water at 298.15 K. This method provides a pathway for estimation of enthalpies of solution at infinite dilution when enthalpies of formation of the compound in the gas, liquid, or solid phase are also available. The approach taken follows the well-established predictive techniques devised by S.W. Benson and coworkers (*1-5*). These techniques have had wide usage in group additivity calculations of thermodynamic properties and reactions for organic compounds, primarily in the gas phase. In this chapter, the coverage of organic compounds is limited to those compounds which contain the elements carbon, hydrogen, and oxygen. A tabular comparison is provided which shows the difference between experimental and estimated values for enthalpies of solution at infinite dilution in water at 298.15 K.

Because of the recent extension to include the liquid and solid phases in group additivity schemes for the prediction of thermodynamic properties of organic compounds (*6*), it appeared reasonable to develop the capability of these schemes further to accommodate prediction for infinitely dilute aqueous solutions. Other researchers as well have developed additivity schemes for predicting the thermodynamic properties of organic compounds in dilute aqueous solution. Nichols et al. (*7*) derived a simple group additivity scheme for partial molar heat capacities of non-electrolytes in infinitely dilute aqueous solutions to facilitate the interpretation of results from thermodynamic studies on biochemical systems. Cabani et al. (*8*) developed group contribution parameters for thermodynamic properties of hydration associated with the transfer of non-ionic organic compounds from the gas phase to dilute aqueous solution.

The Estimation Approach

From a search of the literature, experimental data for enthalpies of solution at 298.15 K at infinite dilution and experimental data for enthalpies of formation at 298.15 K for organic compounds in the gas, liquid, or solid phase were summed to yield the corresponding enthalpies of formation at 298.15 K for the infinitely dilute solute. From this collection of data, group values were derived which permit the estimation of the enthalpy of formation for the solute at infinite dilution at 298.15 K. The difference between the enthalpy of formation for a given compound in aqueous solution at infinite dilution and its enthalpy of formation in the gas, liquid, or solid phase, both at 298.15 K, is the enthalpy of solution at infinite dilution; i.e., $\Delta_f H°(aq) - [\Delta_f H°(gas), \Delta_f H°(liquid),$ or $\Delta_f H°(solid)] = \Delta_{sol} H°(aq)$.

Numerically, the enthalpy of solution for an organic compound in water is a small quantity in comparison to its enthalpy of formation in aqueous solution or its enthalpy of formation for a specific physical phase. Hence, extreme accuracy is needed for both enthalpies of formation. Unfortunately, the number of calorimetric data reported for enthalpies of solution at infinite dilution in water are limited in comparison to the data available for enthalpies of formation of gases, liquids, or solids which are usually derived from combustion or reaction calorimetry. Hence, the coverage provided here shows comparisons between experimental and estimated values of the standard molar enthalpy of solution at infinite dilution for only 17 hydrocarbons and 46 organic oxygen compounds. Upon occasion, in addition to calorimetric data, good quality data for the enthalpy of solution of a substance can be derived from a knowledge of its solubility as a function of temperature. However, calorimetric data are preferred because high accuracies and small uncertainties are more readily achievable using this technique.

Development of Group Values

In the notation for molecular groups developed by Benson and coworkers, C denotes an aliphatic carbon atom, C_d is a doubly-bonded carbon atom, C_t is a triply-bonded carbon atom, and C_B is a benzene-type carbon atom. The development of group values begins with aliphatic hydrocarbons in order to establish the group values $C\text{-}(H)_3(C)$, $C\text{-}(H)_2(C)_2$, $C\text{-}(H)(C)_3$, and $C\text{-}(C)_4$ upon which further development can be made for other families of compounds such as alcohols, ethers, aldehydes, ketones, acids, and esters. Because of the limited amount and uneven quality of data for enthalpies of solution, no global least squares, least sums, or regression-type fit of the data was carried out. The generation of group values was in part manual and in part computer-assisted. Some computations of average values or average deviations were performed using a desk-top calculator. Others were made using computer spread-sheet analysis. Group values generated for hydrocarbons were held fixed for the generation of non-hydrocarbon values. The value derived for $C\text{-}(H)_3(C) = -51.65$ kJ/mol for hydrocarbons, was kept constant for a methyl group attached to any other element or functional group, i.e., $C\text{-}(H)_3(C) = C\text{-}(H)_3(O) = C\text{-}(H)_3(CO)$. Experimental data for some compounds are used to form the basis for a missing group and group value. Some examples are: ethane which forms the basis for the $C\text{-}(H)_3(C)$ group, 2-methylpropane which forms the basis for the $C\text{-}(H)(C)_3$, and methanol which forms the basis for the $O\text{-}(H)(C)$ group. Consequently, differences between experimental and estimated $\Delta_{sol}H°(aq)$ values for such compounds will be zero. Table I offers an example of how the estimation procedure is performed for 1-hexanol. Group values are summed to give $\Delta_fH°(aq) = -383.90$ kJ/mol. Data also provided are: the equation for the solution reaction, experimental values for $\Delta_{sol}H°(aq)$ and *references*, a selected value for $\Delta_{sol}H°(aq)$, and the estimated $\Delta_{sol}H°(aq)$. The reader interested in additional detail concerning the second-order group-additivity method or definitions of the group notation is referred to earlier publications (*1-6*).

Table II provides the values for organic groups and group values so that the enthalpy of formation of the infinitely dilute solute in water, $\Delta_fH°(aq)$, at 298.15 K can be estimated. Group values applicable to hydrocarbons and hydrocarbon moieties are listed under *CH group* while group values applicable to organic oxygen compounds and related moieties are listed under *CHO group*. Cyclic compounds require an additional group value for a ring strain correction (rsc) and are shown at the end of the table.

Estimation of Enthalpies of Solution for Hydrocarbons (CH) and Organic Oxygen Compounds (CHO)

Table III provides a comparison of a selected experimental $\Delta_{sol}H°(aq)$ with the estimated $\Delta_{sol}H°(aq)$. In this comparison, each organic compound is identified by: (a) its name and group notation obtained from Table II, (b) the equation for the solution reaction specifying the physical phase of the reactant, the experimental $\Delta_{sol}H°(aq)$ values for the reaction, the corresponding (*references*), and a selected value, SV, for $\Delta_{sol}H°(aq)$. The last entry for each compound, (c), gives the estimated $\Delta_{sol}H°(aq)$ which is derived from the difference between $\Delta_fH°(aq)$ and $\Delta_fH°$ (gas), (liquid), or (solid). The corresponding (*reference*) for the latter value is also provided.

Hydrocarbons

The agreement between experimental and estimated enthalpies of solution for the hydrocarbons listed in Table III appears to be quite satisfactory. Calorimetric data are available for both the gas and liquid phases for pentane, hexane, and cyclohexane. Differences between the selected $\Delta_{sol}H°(aq)$ based on experiment and the estimated $\Delta_{sol}H°(aq)$ are within the experimental uncertainities for these hydrocarbons except for liquid hexane which is -0.84 kJ/mol. Another significant difference appears to be that between experimental and estimated values for n-propylbenzene, amounting to 1.64 kJ/mol; otherwise, agreement is usually within ±0.20 kJ/mol. High quality data for both experimental enthalpies of solution and enthalpies of formation in the gas and liquid phases are essential if good agreement between selected and estimated $\Delta_{sol}H°$ values is expected.

Alcohols

Agreement between experimental and estimated enthalpies of solution for aliphatic alcohols is reasonably good; the absolute value of most differences is less than 1 kJ/mol; the average deviation is 0.60 kJ/mol.

Table I. Estimation of the Enthalpy of Aqueous Solution of 1-Hexanol at 298.15 K
1-Hexanol: CH_3-CH_2-CH_2-CH_2-CH_2-CH_2-OH

group kJ/mol	$\Delta_f H\,°(aq)$, kJ/mol	groups	group sum,
C-(H)$_3$(C)	-51.65	1	-51.65
C-(H)$_2$(C)$_2$	-24.05	4	-96.20
C-(H)$_2$(C)(O)	-41.95	1	-41.95
O-(C)(H)	-194.10	1	-194.10
total sum ($\Delta_f H°(aq)$)			-383.90

$C_6H_{14}O(liq) = C_6H_{14}O(aq)$: -4.64±0.33 (*9*), -6.56±0.13 (*10*),
-6.41±0.04 (*11*), -5.77±0.10 (*12*)   Selected Value = -6.41
est'd $\Delta_{sol}H° = (\Delta_f H°(aq),-383.90) - (\Delta_f H°(liq),-377.50$ (*13*))   = -6.40

Table II. Group Values for Estimating Enthalpies of Formation
in Water at Infinite Dilution at 298.15 K

CH group	$\Delta_f H\,°(aq)$ kJ/mol	CHO group	$\Delta_f H\,°(aq)$ kJ/mol
C-(H)$_3$(C)	-51.65	C-(H)$_3$(O)	-51.65
C-(H)$_2$(C)$_2$	-24.05	C-(H)$_2$(C)(O)	-41.95
C-(H)(C)$_3$	-3.42	C-(H)(C)$_2$(O)	-33.70
C-(C)$_4$	13.56	C-(C)$_3$(O)	-27.60
C$_d$-(H)$_2$	18.02	C-(H)$_2$(C$_d$)(O)	-33.86
C$_d$-(H)(C)	31.75	C-(H)$_2$(C$_B$)(O)	-31.46
C-(H$_3$)(C$_d$)	-51.65	O-(H)(C)	-194.10
C-(H)$_2$(C)(C$_d$)	-23.54	O-(H)(CO)	-284.90
C$_t$-(H)	106.79	O-(CO)(CH$_3$)	-201.00
C$_B$-(H)(C$_B$)$_2$	8.50	O-(CO)(C)	-194.00
C$_B$-(C)(C$_B$)$_2$	22.89	O-(C)$_2$	-115.30
C-(H$_3$)(C$_B$)	-51.65	CO-(C)$_2$	-155.00
C-(H)$_2$(C)(C$_B$)	-24.05	CO-(C)(O)	-149.10
cyclopropane (rsc)	102.15	CO-(H)(C)	-158.89
cyclohexane (rsc)	-11.95	CO-(H)(O)	-141.30
		C-(H)$_3$(CO)	-51.65
		C-(H)$_2$(C)(CO)	-24.55
		C-(H)(C)$_2$(CO)	-2.60
		C-(C)$_3$(CO)	26.96
		C$_B$-(O)(C$_B$)$_2$	-0.60
		cyclopentanol (rsc)	13.63
		cyclohexanol (rsc)	-8.35

α,ω-Alkanediols

The values for the molar enthalpies of solution and the enthalpies of formation of the liquid and solid phases are not always well-behaved for α,ω-alkanediols. The measurements of $\Delta_{sol}H°(aq)$ for C_2 through C_5 diols by Nichols et al. (*49*) are of very high quality in contrast to the $\Delta_{sol}H°(aq)$ measurements of Corkill et al. (*47*) which are not. Also, general agreement of $\Delta_f H°$ for liquid and solid α,ω-alkanediols is lacking; their measurement has been a challenge to the thermochemist. Lastly, it is likely that a correction term is needed for OH-OH interactions. The magnitude of this correction term seems to be in the vicinity of 5-15 kJ/mol, however, until general agreement among the $\Delta_f H°$ values for α,ω-alkanediols in the liquid and solid phases is achieved as well as a set of high quality $\Delta_{sol}H°(aq)$ values, one cannot proceed much further with estimation and prediction procedures for this class of compounds. In addition, one should also expect a diminution of the correction term as the number of carbon atoms increases for α,ω-alkanediol homologues. However, this diminution is not apparent from an examination of the current data.

Ethers, Aldehydes, and Ketones

Data on the enthalpies of aqueous solution for ethers, aldehydes, and ketones are limited in that precise and accurate measurements have not been made for a sufficient number of compounds in a homologous series so that good quality group values may be derived for key groups such as $O-(C)_2$, $CO-(H)(C)$, and $CO-(C)_2$. Values for these and related groups have been derived, however, their values may have uncertainties which are significantly large, i.e., ± 1-4 kJ/mol.

Monocarboxylic Acids

The experimental data for $\Delta_{sol}H°(aq)$ of the aliphatic monocarboxylic acids at 298.15 K represent the solute in an unionized standard state. Agreement between experimental and estimated values for $\Delta_{sol}H°$ of these acids is reasonably good and is usually better than ± 0.5 kJ/mol. For some monocarboxylic acids, we have listed two estimated $\Delta_{sol}H°(aq)$ values to show that reasonably close agreement between $\Delta_f H°(liq)$ data of 1-3 kJ/mol can result in significant differences with the selected value (SV) derived from experimental values for $\Delta_{sol}H°(aq)$. For example, two values for $\Delta H°(liq)$ are listed with heptanoic acid: -611.41 ± 0.71 (*68*) and -608.27 ± 0.88 (*70*) kJ/mol. The latter $\Delta_f H°(liq)$ offers better agreement when combined with the estimated $\Delta_f H°(aq)$ value of -606.40 kJ/mol to calculate an estimated $\Delta_{sol}H°(aq)$, (1.76 vs. 1.87 kJ/mol) than the former $\Delta_f H°(liq)$, (1.76 vs. 5.01 kJ/mol).

Esters

In order to establish as small a difference as possible between the experimental and estimated $\Delta_{sol}H°(aq)$ for esters, separate values were derived for the $\Delta_f H°(aq)$ group for methyl and other kinds of esters. For methyl esters the $O-(CO)(CH_3)$ group value is -201.00 kJ/mol while for other esters the $O-(CO)(C)$ group value is -194.00 kJ/mol.

**Table III. Estimated Enthalpies of Solution for Hydrocarbons and CHO
Compounds at Infinite Dilution at 298.15 K (all values are in kJ/mol)**

(a) *compound name and group notation*
(b) *equation for solution reaction, expt'l $\Delta_{sol}H°$ values(references), selected value SV*
(c) *est'd $\Delta_{sol}H°(aq) = (est'd \Delta_f H°(aq)$ from Table II) - (expt'l $\Delta_f H°(g/liq/s)$ (ref))*

1(a) ethane   (2×C-(H)$_3$(C))
1(b) $C_2H_6(g) = C_2H_6(aq)$: -19.52±0.12 (*14*), -19.30±0.12 (*15*), -19.50±0.10 (*16*):
                                                                          SV = -19.45
1(c) $\Delta_{sol}H°(aq) = (\Delta_f H°(aq),-103.30) - (\Delta_f H°(g), -83.85$ (*17*))   = -19.45

2(a) propane   (2×C-(H)$_3$(C))+(1×C-(H)$_2$(C)$_2$)
2(b) $C_3H_8(g) = C_3H_8(aq)$:$\Delta_{soln}H°(aq)$: -23.27±0.26 (*14*), -22.90±0.08 (*15*),
       -22.61±0.06 (*18*),   SV = -22.93
2(c) $\Delta_{sol}H°(aq) = (\Delta_f H°(aq),-127.35) - (\Delta_f H°(g),-104.68$(*17*))   = -22.67

3(a) n-butane (2×C-(H)$_3$(C))+(2×C-(H)$_2$(C)$_2$)
3(b) $C_4H_{10}(g) = C_4H_{10}(aq)$:-25.92±0.17 (*14*),-25.93±0.08 (*15*):   SV = -25.92
3(c) $\Delta_{sol}H°(aq) = (\Delta_f H°(aq),-151.40) - (\Delta_f H°(g),-125.65$ (*17*))   = -25.75

4(a) 2-methylpropane (3×C-(H)$_3$(C))+(1×C-(H)(C)$_3$)
4(b) $C_4H_{10}(g) = C_4H_{10}(aq)$: -24.19±0.25 (*14*),   SV = -24.19
4(c) $\Delta_{sol}H°(aq) = (\Delta_f H°-158.37(aq)) - (\Delta_f H°,-134.18(g)$ (*17*))   = -24.19

5(a) n-pentane   (2×C-(H)$_3$(C))+(3×C-(H)$_2$(C)$_2$)
5(b) $C_5H_{12}(g) = C_5H_{12}(aq)$: -28.3 ±1.1 (*14*),   SV = -28.30
5(c) $\Delta_{sol}H°(aq) = (\Delta_f H°$ (aq),-175.45) - ($\Delta_f H°(g)$,-146.77 (*19*))   = -28.68

5(b') $C_5H_{12}(liq) = C_5H_{12}(aq)$: -2.0 ±0.2 (*20*),   SV = -2.00
5(c') $\Delta_{sol}H°(aq) = (\Delta_f H°(aq),-175.45) - (\Delta_f H°(liq),-173.51$ (*19*))   = -1.94

6(a) 2,2-dimethylpropane   (4×C-(H)$_3$(C))+(1×C-(C)$_4$)
6(b) $C_5H_{12}(g) = C_5H_{12}(aq)$: -25.11±0.17 (*14*),   SV = -25.11
6(c) $\Delta_{sol}H°(aq) = (\Delta_f H°(aq),-193.06) - (\Delta_f H°(g),-167.95$ (*19*))   = -25.11

7(a) n-hexane (2×C-(H)$_3$(C))+(4×C-(H)$_2$(C)$_2$)
7(b) $C_6H_{14}(g) = C_6H_{14}(aq)$: -31.1 ±2.8 (*14*),   SV = -31.10
7(c) $\Delta_{sol}H°(aq) = (\Delta_f H°(aq),-199.50) - (\Delta_f H°(g),-167.28$ (*21,22*))   = -32.22

7(b') $C_6H_{14}(liq) = C_6H_{14}(aq)$: 0.0 ±0.2 (*20*), 0.47±0.12 (*23*),   SV = 0.00
7(c') $\Delta_{sol}H°(aq) = (\Delta_f H°(aq),-199.50 - (\Delta_f H°(liq),-198.66$ (*21*)   = -0.84

8(a) ethylene   (2×C$_d$-(H)$_2$)
8(b) $C_2H_4(g) = C_2H_4(aq)$: -16.46±0.07 (*14*),   SV = -16.46
8(c) $\Delta_{sol}H°(aq) = (\Delta_f H°(aq),36.04) - (\Delta_f H°(g),52.50$ (*24*))   = -16.46

Table III (continued). Estimated Enthalpies of Solution for Hydrocarbons and CHO Compounds at Infinite Dilution at 298.15 K (all values are in kJ/mol)

(a) *compound name and group notation*
(b) *equation for solution reaction, expt'l $\Delta_{sol}H°$ values(references), selected value SV*
(c) *est'd $\Delta_{sol}H°(aq) = (est'd \Delta_fH°(aq)$ from Table II) - (expt'l $\Delta_fH°(g/liq/s)$ (ref))*

9(a) propene $(1×C-(H)_3(C_d))+(1×C_d-(H)(C))+(1×C_d-(H)_2)$
9(b) $C_3H_6(g) = C_3H_6(aq)$: -21.64±0.12 (*14*),   SV = -21.64
9(c) $\Delta_{sol}H°(aq)$ ($\Delta_fH°(aq)$,-1.88) - ($\Delta_fH°(g)$,19.76 (*24*))   = -21.64

10(a) 1-butene $(1×C-(H)_3(C))+(1×C-(H)_2(C)(C_d))+(1×C_d-(H)(C))+(1×C_d-(H)_2)$
10(b) $C_4H_8(g) = C_4H_8(aq)$: -24.88±0.11 (*14*),   SV= -24.88
10(c) $\Delta_{sol}H°(aq)$ = ($\Delta_fH°(aq)$,-25.42) - ($\Delta_fH°(g)$,-0.54 (*25*))   = -24.88

11(a) acetylene $(2×C_t-(H))$
11(b) $C_2H_2(g) = C_2H_2(aq)$: -14.62±0.02 (*14*),   SV= -14.62
11(c) $\Delta_{sol}H°(aq)$ =($\Delta_fH°(aq)$,213.58) - ($\Delta_fH°(g)$,228.19 (*26*))   = -14.61

12(a) cyclopropane $(3×C-(H)_2(C)_2)+(1×$ cyclopropane (rsc))
12(b) $C_3H_6(g) = C_3H_6(aq)$: -23.26±0.06 (*14*),   SV= -23.26
12(c) $\Delta_{sol}H°(aq)$ = ($\Delta_fH°(aq)$,30.00) - ($\Delta_fH°(g)$,53.26 (*27*))   = -23.26

13(a) cyclohexane $(6×C-(H)_2(C)_2)+(1×$ cyclohexane (rsc))
13(b) $C_6H_{12}(g) = C_6H_{12}(aq)$: -32.8 ±2.3 (*14*),   SV= -32.80
13(c) $\Delta_{sol}H°(aq)$ = ($\Delta_fH°(aq)$,-156.25) - ($\Delta_fH°(g)$,-123.10 (*21,22*))   = -33.15

13(b') $C_6H_{12}(liq) = C_6H_{12}(aq)$: 0.26±0.08 (*23*), -0.1±0.1 (*20*),   SV = -0.10
13(c') $\Delta_{sol}H°(aq)$ = ($\Delta_fH°(aq)$,-156.25) - ($\Delta_fH°(liq)$,-156.15 (*21*))   = -0.10

14(a) benzene $(6×C_B-(H)(C_B)_2)$
14(b) $C_6H_6(liq) = C_6H_6(aq)$: 2.08±0.04 (*28*), 0.80±0.12 (*23*), 2.21±0.01 (*18*), SV = 2.10
14(c) $\Delta_{sol}H°(aq)$ = ($\Delta_fH°(aq)$,51.00) - ($\Delta_fH°(liq)$,48.95 (*21*))   = 2.05

15(a) toluene $(5×C_B-(H)(C_B)_2)+(1×C_B-(C)(C_B)_2)+(1×C-(H)_3(C_B))$
15(b) $C_7H_8(liq) = C_7H_8(aq)$: 1.73±0.04 (*20*),   SV = 1.73
15(c) $\Delta_{sol}H°(aq)$ = ($\Delta_fH°(aq)$,13.71) - ($\Delta_fH°(liq)$,12.01 (*21*))   SV = 1.70

16(a) ethylbenzene $(5×C_B-(H)(C_B)_2)+(1×C_B-(C)(C_B)_2)+(1×C-(H)_2(C)(C_B))+(1×C-(H)_3(C))$
16(b) $C_8H_{10}(liq) = C_8H_{10}(aq)$: 2.02±0.04 (*20*)   SV = 2.02
16(c) $\Delta_{sol}H°(aq)$ = ($\Delta_fH°(aq)$,-10.34) - ($\Delta_fH°(liq)$,-12.34 (*29*))   SV = 2.00

17(a) n-propylbenzene $(5×C_B-(H)(C_B)_2)+(1×C_B-(C)(C_B)_2)+(1×C-(H)_2(C)(C_B))+$
     $(1×C-(H)_2(C)_2)+(1×C-(H)_3(C))$
17(b) $C_9H_{12}(liq) = C_9H_{12}(aq)$: 2.3±0.1 (*20*),   SV = 2.30
17(c) $\Delta_{sol}H°(aq)$ = ($\Delta_fH°(aq)$,-34.39) - ($\Delta_fH°(liq)$,-38.33 (*29*))   SV = 3.94

Continued on next page.

Table III (continued). Estimated Enthalpies of Solution for Hydrocarbons and CHO Compounds at Infinite Dilution at 298.15 K (all values are in kJ/mol)

(a) *compound name and group notation*
(b) *equation for solution reaction, expt'l $\Delta_{sol}H°$ values(references), selected value SV*
(c) *est'd $\Delta_{sol}H°(aq) = (est'd \Delta_f H°(aq)$ from Table II) - (expt'l $\Delta_f H°(g/liq/s)$ (ref))*

18(a) methanol   (1×C-(H)$_3$(O))+(1×O-(H)(C))
18(b) CH$_3$OH(liq) = CH$_3$OH(aq): -7.32±0.13 (*9*), -7.25±0.06 (*30*),
  -7.104±0.085 (*23*), -7.29±0.04 (*11*), -7.34±0.03 (*31*),   SV = -7.25
18(c) $\Delta_{sol}H°$(aq) = ($\Delta_f H°$(aq),-245.75)- ($\Delta_f H°$(liq),-238.50 (*32,33*))   = -7.25

19(a) ethanol (1×C-(H)$_3$(C))+(1×C-(H)$_2$(C)(O))+(1×O-(H)(C))
19(b) C$_2$H$_5$OH(liq) = C$_2$H$_5$OH(aq): -10.00±0.08 (*9*), -11.20±0.08 (*34*), -10.15±0.05 (*11*),
  -10.10±0.08 (*30*),-10.13±0.10 (*12*), -10.125±0.085 (*23*), -10.42±0.38 (*35*),
  -10.18±0.05 (*31*),   SV = -10.10
19(c) $\Delta_{sol}H°$(aq) = ($\Delta_f H°$(aq),-287.70)- ($\Delta_f H°$(liq),-277.60 (*32,33*)) = -10.10

20(a) 1-propanol (1×C-(H)$_3$(C))+(1×C-(H)$_2$(C)$_2$)+(1×C-(H)$_2$(C)(O))+(1×O-(H)(C))
20(b) C$_3$H$_8$O(liq) = C$_3$H$_8$O(aq): -9.20±0.13 (*9*), -10.16±0.02 (*11*),-10.13±0.08 (*30*),
  -9.20±0.10 (*12*), -10.12±0.03 (*31*),   SV = -9.35
20(c) $\Delta_{sol}H°$(aq) = ($\Delta_f H°$(aq),-311.75) - ($\Delta_f H°$(liq),-302.60 (*36*)) = -9.15

21(a) 2-propanol (2×C-(H)$_3$(C))+(1×C-(H)(C)$_2$(O))+(1×O-(H)(C))
21(b) C$_3$H$_8$O(liq) = C$_3$H$_8$O(aq): -12.98±0.10 (*30*), -13.07±0.04 (*31*), SV = -13.00
21(c) $\Delta_{sol}H°$(aq) = ($\Delta_f H°$(aq),-331.10(aq)) - ($\Delta_f H°$(liq),-318.10(*36*)) = -13.00

22(a) 2-propen-1-ol (1×C$_d$-(H)$_2$)+(1×C$_d$-(H)(C))+(1×C-(H)$_2$(C$_d$)(O))+(1×O-(H)(C))
22(b) C$_3$H$_6$O(liq) = C$_3$H$_6$O(aq): -7.09±0.03 (*31*),   SV = -7.09
22(c) $\Delta_{sol}H°$(aq) = ($\Delta_f H°$(aq),-178.19) - ($\Delta_f H°$(liq),Δ-171.10 (*37*)) = -7.09

23(a) 1-butanol (1×C-(H)$_3$(C))+(2×C-(H)$_2$(C)$_2$)+(1×C-(H)$_2$(C)(O))+(1O-(H)(C))
23(b) C$_4$H$_{10}$O(liq) = C$_4$H$_{10}$(aq): -8.16±0.13 (*9*), -9.25 (*38*),-10.25±0.07 (*34*),
  -9.32±0.02 (*11*), -9.28±0.08 (*30*),-9.00±0.10 (*12*), -9.41±0.04 (*31*) SV = -9.25
23(c) $\Delta_{sol}H°$(aq) = ($\Delta_f H°$(aq),-335.80) - ($\Delta_f H°$(liq),-327.20 (*39*)) = -8.60

24(a) 2-butanol (2×C-(H)$_3$(C))+(1×C-(H)$_2$(C)$_2$)+(1×C-(H)(C)$_2$(O))+(1×O-(H)(C))
24(b) C$_4$H$_{10}$O(liq) = C$_4$H$_{10}$O(aq): -13.18±0.04 (*31*),   SV = -13.18
24(c) $\Delta_{sol}H°$(aq) = ($\Delta_f H°$(aq),-355.15) - ($\Delta_f H°$(liq),-342.50 (*40*)) = -12.65

25(a) 2-methyl-1-propanol (2×C-(H)$_3$(C))+(1×C-(H)(C)$_3$)+(1×C-(H)$_2$(C)(O))+
  (1×O-(H)(C))
25(b) C$_4$H$_{10}$O(liq) = C$_4$H$_{10}$O(aq): -9.31±0.04 (*31*),   SV = -9.31
25(c) $\Delta_{sol}H°$(aq) = ($\Delta_f H°$(aq),-342.77) - ($\Delta_f H°$(liq),-334.70 (*40*)) = -8.07

Table III (continued). Estimated Enthalpies of Solution for Hydrocarbons and CHO Compounds at Infinite Dilution at 298.15 K (all values are in kJ/mol)

(a) *compound name and group notation*
(b) *equation for solution reaction, expt'l $\Delta_{sol}H°$ values(references), selected value SV*
(c) *est'd $\Delta_{sol}H°(aq) = (est'd \Delta_f H°(aq)$ from Table II) - (expt'l $\Delta_f H°(g/liq/s)$ (ref))*

26(a) 2-methyl-2-propanol $(3\times C-(H)_3(C))+(1\times C-(C)_3(O))+(1\times O-(H)(C))$
26(b) $C_4H_{10}O(liq) = C_4H_{10}O(aq)$: -17.44±0.02 (*41*), -17.31±0.14 (*30*), -17.46±0.06 (*31*),
$\qquad\qquad\qquad\qquad\qquad\qquad\qquad\qquad\qquad\qquad\qquad\qquad\qquad\qquad$ SV = -17.45
26(c) $\Delta_{sol}H°(aq) = (\Delta_f H°(aq),-376.65) - (\Delta_f H°(liq),-359.20$ (*40*))$\qquad$ = -17.45

27(a) 1-pentanol $(1\times C-(H)_3(C))+(1\times C-(H)_2(C)(O))+(3\times C-(H)_2(C)_2)+(1\times O-(H)(C))$
27(b) $C_5H_{12}O(liq) = C_5H_{12}O(aq)$: -6.36±0.25 (*9*), -7.99±0.05 (*11*), -8.08±0.10 (*12*),
$\qquad$ -7.82±0.05 (*31*), $\qquad\qquad\qquad\qquad\qquad\qquad\qquad\qquad\qquad\qquad$ SV = -7.99
27(c) $\Delta_{sol}H°(aq) = (\Delta_f H°(aq),-359.85) - (\Delta_f H°(liq),-351.60$ (*13*))$\qquad$ = -8.25

28(a) 2-methyl-2-butanol $(3\times C-(H)_3(C))+(1\times C-(C)_3(O))+(1\times C-(H)_2(C)_2)+(1\times C-(H)(O))$
28(b) $C_5H_{12}O(liq) = C_5H_{12}O(aq)$: -18.51±0.42 (*31*), $\qquad\qquad\qquad$ SV = -18.51
28(c) $\Delta_{sol}H°(aq) = (\Delta_f H°(aq),-400.70) - (\Delta_f H°(liq),-379.50$ (*42*))$\qquad$ = -21.20

29(a) 1-hexanol $(1\times C-(H)_3(C))+(1\times C-(H)_2(C)(O))+(4\times C-(H)_2(C)_2)+(1\times O-(H)(C))$
29(b) $C_6H_{14}O(liq) = C_6H_{14}O(aq)$: -4.64±0.33 (*9*), -6.56±0.13 (*10*),
$\qquad$ -6.41±0.04 (*11*), -5.77±0.10 (*12*), $\qquad\qquad\qquad\qquad\qquad$ SV = -6.41
29(c) $\Delta_{sol}H°(aq) = (\Delta_f H°(aq),-383.90) - (\Delta_f H°(liq),-377.50$ (*13*))$\qquad$ = -6.40

30(a) 1-heptanol $(1\times C-(H)_3(C))+(1\times C-(H)_2(C)(O))+(5\times C-(H)_2(C)_2)+(1\times O-(H)(C))$
30(b) $C_7H_{16}O(liq) = C_7H_{16}O(aq)$: -5.37±0.15 (*10*),-4.88±0.07 (*11*), $\qquad$ SV = -4.88
30(c) $\Delta_{sol}H°(aq) = (\Delta_f H°(aq),-407.95) - (\Delta_f H°(liq),-403.30$ (*13*))$\qquad$ = -4.65

31(a) 1-octanol $(1\times C-(H)_3(C))+(1\times C-(H)_2(C)(O))+(6\times C-(H)_2(C)_2)+(1\times O-(H)(C))$
31(b) $C_8H_{18}O(liq) = C_8H_{18}O(aq)$: -3.37±0.09 (*11*), -3.40±0.09 (*18*), $\qquad$ SV = -3.39
31(c) $\Delta_{sol}H°(aq) = (\Delta_f H°(aq),-432.00) - (\Delta_f H°(liq),-428.04$ (*13*))$\qquad$ = -3.96

32(a) cyclopentanol $(4\times C-(H)_2(C)_2)+(1\times C-(H)(C)_2(O))+(1\times O-(H)(C))+$
$\qquad$ (1× cyclopentanol (rsc))
32(b) $C_5H_{10}O(liq) \rightarrow C_5H_{10}O(aq)$: -10.34±0.03 (*31*), $\qquad\qquad\qquad$ SV = -10.34
32(c) $\Delta_{sol}H°(aq) = (\Delta_f H°(aq),-310.37) - (\Delta_f H°(liq),-300.03$ (*43*))$\qquad$ = -10.34

33(a) cyclohexanol $(5\times C-(H)_2(C)_2)+(1\times C-(H)(C)_2(O))+(1\times O-(H)(C))+$
$\qquad$ (1×cyclohexanol (rsc))
33(b) $C_6H_{12}O(liq) = C_6H_{12}O(aq)$: -9.01 (*31*), $\qquad\qquad\qquad\qquad\qquad$ SV = -9.00
33(c) $\Delta_{sol}H°(aq) = (\Delta_f H°(aq),-356.40) - (\Delta_f H°(liq),-347.40$ (*43*))$\qquad$ = -9.00

34(a) phenol $(5\times C_B-(H)(C_B)_2)+(1\times C_B-(O)(C_B)_2)+(1\times O-(H)(C))$
34(b) $C_6H_6O(s) = C_6H_6O(aq)$: 12.93±0.03 (*44*), $\qquad\qquad\qquad\qquad$ SV = 12.90
34(c) $\Delta_{sol}H°(aq) = (\Delta_f H°(aq),-152.20) - (\Delta_f H°(s),-165.10$ (*45*))$\qquad$ = 12.90

Continued on next page.

Table III (continued). Estimated Enthalpies of Solution for Hydrocarbons and CHO Compounds at Infinite Dilution at 298.15 K (all values are in kJ/mol)

(a) *compound name and group notation*
(b) *equation for solution reaction, expt'l $\Delta_{sol}H°$ values(references), selected value SV*
(c) *est'd $\Delta_{sol}H°(aq) = (est'd \Delta_f H°(aq) from Table II) - (expt'l \Delta_f H°(g/liq/s) (ref))*

35(a) benzyl alcohol $(5\times C_B-(H)(C_B)_2)+(1\times C_B-(C)(C_B)_2)+(1\times C-(H)_2(C_B)(O))+$
$\quad (1\times O-(H)(C))$
35(b) $C_7H_8O(liq) = C_7H_8O(aq)$: 0.51 ± 0.03 *(44)*, 0.55 ± 0.06 *(31)*,　　　　SV = 0.53
35(c) $\Delta_{sol}H°(aq) = (\Delta_f H°(aq),-160.17) - (\Delta_f H°(liq),-160.70$ *(46)*)　　　= 0.53

36(a) 1,2-ethanediol　$(2\times C-(H)_2(C)(O))+(2\times O-(H)(C))$
36(b) $C_2H_6O_2(liq) = C_2H_6O_2(aq)$:$-6.69\pm0.21$ *(47)*, -6.28 ± 0.21 *(48)*,-6.87 ± 0.02 *(49)*,
　　　　　　　　　　　　　　　　　　　　　　　　　　　　　　SV = -6.87
36(c) $\Delta_{sol}H°(aq) = (\Delta_f H°(aq),-472.10) - (\Delta_f H°(liq),-453.10$ *(50)*)　　= -19.00
36(c) $\Delta_{sol}H°(aq) = (\Delta_f H°(aq),-472.10) - (\Delta_f H°(liq),-454.92$ *(51)*)　　= -17.18
36(c) $\Delta_{sol}H°(aq) = (\Delta_f H°(aq),-472.10) - (\Delta_f H°(liq),-455.64$ *(52)*)　　= -16.46
36(c) $\Delta_{sol}H°(aq) = (\Delta_f H°(aq),-472.10) - (\Delta_f H°(liq),-460.00$ *(53)*)　　= -12.10

37(a) 1,3-propanediol $(2\times C-(H)_2(C)(O))+(1\times C-(H)_2(C)_2)+(2\times O-(H)(C))$
37(b) $C_3H_8O_2(liq) = C_3H_8O_2(aq)$: -8.79 ± 0.21 *(47)*, -8.67 ± 0.02 *(49)*,　　SV = -8.67
37(c) $\Delta_{sol}H°(aq) = (\Delta_f H°(aq),-496.15) - (\Delta_f H°(liq),-464.84$ *(52)*)　　=-31.31
37(c) $\Delta_{sol}H°(aq) = (\Delta_f H°(aq),-496.15) - (\Delta_f H°(liq),-480.80$ *(53)*)　　= -15.35

38(a) 1,4-butanediol $(2\times C-(H)_2(C)(O))+(2\times C-(H)_2(C)_2)+(2\times O-(H)(C))$
38(b) $C_4H_{10}O_2(liq) = C_4H_{10}O_2(aq)$:$-10.04\pm0.42$ *(47)*,-10.46 ± 0.01 *(49)*,　SV =-10.46
38(c) $\Delta_{sol}H°(aq) = (\Delta_f H°(aq),-520.20) - (\Delta_f H°(liq),-503.34$ *(52)*)　　= -16.86
38(c) $\Delta_{sol}H°(aq) = (\Delta_f H°(aq),-520.20) - (\Delta_f H°(liq),-505.30$ *(53)*)　　= -14.90

39(a) 1,5-pentanediol $(2\times C-(H)_2(C)(O))+(3\times C-(H)_2(C)_2)+(2\times O-(H)(C))$
39(b) $C_5H_{12}O_2(liq) = C_5H_{12}O_2(aq)$: -10.88 ± 0.42 *(47)*, -10.59 ± 0.01 *(49)*,　SV = -10.59
39(c) $\Delta_{sol}H°(aq) = (\Delta_f H°(aq),-544.25) - (\Delta_f H°(liq),-531.37$ *(52)*)　　= -12.88
39(c) $\Delta_{sol}H°(aq) = (\Delta_f H°(aq),-544.25) - (\Delta_f H°(liq),-528.80$ *(53)*)　　= -15.45

40(a) 1,6-hexanediol $(2\times C-(H)_2(C)(O))+(4\times C-(H)_2(C)_2)+(2\times O-(H)(C))$
40(b) $C_6H_{14}O_2(s) = C_6H_{14}O_2(aq)$: 13.39 ± 0.29 *(47)*,　　　　　　　　SV = 13.39
40(c) $\Delta_{sol}H°(aq) = (\Delta_f H°(aq),-568.30) - (\Delta_f H°(s),-569.86$ *(52)*)　　　= 1.56
40(c) $\Delta_{sol}H°(aq) = (\Delta_f H°(aq),-568.30) - (\Delta_f H°(s),-574.10$ *(54)*)　　　= 5.80
40(c) $\Delta_{sol}H°(aq) = (\Delta_f H°(aq),-568.30) - (\Delta_f H°(s),-583.86$ *(55)*)　　　=15.56

41(a) 1,7-heptanediol $(2\times C-(H)_2(C)(O))+(5\times C-(H)_2(C)_2)+(2\times O-(H)(C))$
41(b) $C_7H_{16}O(liq) = C_7H_{16}O(aq)$: -9.62 ± 0.33 *(47)*,　　　　　　　　SV = -9.62
41(c) $\Delta_{sol}H°(aq) = (\Delta_f H°(aq),-592.35) - (\Delta_f H°(liq),-574.20$ *(54)*)　　= -18.15

Table III (continued). Estimated Enthalpies of Solution for Hydrocarbons and CHO Compounds at Infinite Dilution at 298.15 K (all values are in kJ/mol)

(a) *compound name and group notation*

(b) *equation for solution reaction, expt'l $\Delta_{sol}H°$ values(references), selected value SV*

(c) *est'd $\Delta_{sol}H°$(aq) = (est'd $\Delta_f H°$(aq) from Table II) - (expt'l $\Delta_f H°$(g/liq/s) (ref))*

42(a) 1,8-octanediol (2×C-(H)$_2$(C)(O))+(6×C-(H)$_2$(C)$_2$)+(2×O-(H)(C))

42(a) C$_8$H$_{18}$O$_2$(s) = C$_8$H$_{18}$O$_2$(aq): 21.76±0.42 (*47*),   SV = 21.76

42(c) $\Delta_{sol}H°$(aq) = ($\Delta_f H°$(aq),-616.40) - ($\Delta_f H°$(s),-626.60 (*54*))   = 10.20

43(a) 1,9-nonanediol (2×C-(H)$_2$(C)(O))+(7×C-(H)$_2$(C)$_2$)+(2×O-(H)(C))

43(b) C$_9$H$_{20}$O(s) = C$_9$H$_{20}$O(aq): 28.03±0.42 (*47*),   SV = 28.03

43(c) $\Delta_{sol}H°$(aq) = ($\Delta_f H°$(aq),-640.45) - ($\Delta_f H°$(s),-657.60 (*54*))   = 17.15

44(a) dimethyl ether   (2×C-(H)$_3$(O))+(1×O-(C)$_2$)

44(b) C$_2$H$_6$O(g) = C$_2$H$_6$O(aq): -34.5±2.0 (*56*),   SV = -34.50

44(c) $\Delta_{sol}H°$(aq) = ($\Delta_f H°$(aq),-218.60) - ($\Delta_f H°$(g),-184.10 (*57*))   = -34.50

45(a) diethyl ether   (2×C-(H)$_3$(C))+(2×C-(H)$_2$(C)(O))+(1×O-(C)$_2$)

45(b) C$_4$H$_{10}$O(liq) = C$_4$H$_{10}$(aq): -19.25 (*35*), -26.11 (*58*,)   SV = -22.68

45(c) $\Delta_{sol}H°$(aq) = ($\Delta_f H°$(aq),-302.50) - ($\Delta_f H°$(liq),-279.40 (*59*))   = -23.10

46(a) ethanal (1×C-(H)$_3$(CO))+(1×CO-(H)(C))

46(b) C$_2$H$_4$O(liq) = C$_2$H$_4$O(aq): -18.74 (*60*),   SV = -18.74

46(c) $\Delta_{sol}H°$(aq) = ($\Delta_f H°$(aq),-210.54) - ($\Delta_f H°$(liq),-191.80 (*61,62*))   = -18.74

47(a) propanone   (2×C-(H)$_3$(CO))+(1×CO-(C)$_2$)

47(b) C$_3$H$_6$O(liq) = C$_3$H$_6$O(aq): -10.21±0.21 (*35*),   SV = -10.21

47(c) $\Delta_{sol}H°$(aq) = ($\Delta_f H°$(aq),-258.30) - ($\Delta_f H°$(liq),-248.10 (*63*))   = -10.20

48(a) 2-pentanone   (1×C-(H)$_3$(C))+(1×C-(H)$_3$(CO))+(1×C-(H)$_2$(C)(CO))+
        (1×C-(H)$_2$(C)$_2$)+(1×CO-(C)$_2$)

48(b) C$_5$H$_{10}$O(liq) = C$_5$H$_{10}$O(aq): -11.51±0.42 (*64*),   SV = -11.51

48(c) $\Delta_{sol}H°$(aq) = ($\Delta_f H°$(aq),-306.90) - ($\Delta_f H°$(liq),-297.29 (*65*))   = -9.61

49(a) 3-pentanone   (2×C-(H)$_3$(C))+(2×C-(H)$_2$(C)(CO))+(1×CO-(C)$_2$)

49(b) C$_5$H$_{10}$O(liq) = C$_5$H$_{10}$O(aq): -10.88±0.42 (*64*),   SV = -10.88

49(c) $\Delta_{sol}H°$(aq) = ($\Delta_f H°$(aq),-307.40) - ($\Delta_f H°$(liq),-296.51 (*65*))   = -10.89

50(a) methanoic acid (1×CO-(H)(O))+(1×O-(H)(CO))

50(b) CH$_2$O$_2$(liq) = CH$_2$O$_2$(aq): -0.678±0.001 (*66*),   SV = -0.68

50(c) $\Delta_{sol}H°$(aq) = ($\Delta_f H°$(aq),-426.20) - ($\Delta_f H°$(liq),-425.50 (*68*))   = -0.70

Continued on next page.

Table III (continued). Estimated Enthalpies of Solution for Hydrocarbons and CHO Compounds at Infinite Dilution at 298.15 K (all values are in kJ/mol)

(a) *compound name and group notation*
(b) *equation for solution reaction, expt'l $\Delta_{sol}H°$ values(references), selected value SV*
(c) *est'd $\Delta_{sol}H°(aq)$ = (est'd $\Delta_f H°(aq)$ from Table II) - (expt'l $\Delta_f H°(g/liq/s)$ (ref))*

51(a)ethanoic acid $(1\times C-(H)_3(CO))+(1\times CO-(C)(O))+(1\times O-(H)(CO))$
51(b) $C_2H_4O_2(liq)=C_2H_4O_2(aq)$:-1.176±0.004 (66),-1.25±0.05 (12),-1.17±0.04 (67),
$\hspace{10cm}$ SV = -1.20
51(c)$\Delta_{sol}H°(aq) = (\Delta_f H°(aq),-485.65) - (\Delta_f H°(liq),-484.50$ (68))$\hspace{2cm}$ = -1.15

52(a) propanoic acid$(1\times C-(H)_3(C))+(1\times C-(H)_2(C)(CO))+(1\times CO-(C)(O))+$
$\hspace{1.5cm}(1\times O-(H)(CO))$
52(b) $C_3H_6O_2(liq) = C_3H_6O_2(aq)$: -1.544±0.004 (66), -1.67±0.05 (12),$\hspace{1cm}$ SV = -1.60
52(c) $\Delta_{sol}H°(aq) = (\Delta_f H°(aq),-510.20) - (\Delta_f H°(liq),-510.70$ (68))$\hspace{1.5cm}$ = 0.50
52(c) $\Delta_{sol}H°(aq) = (\Delta_f H°(aq),-510.20) - (\Delta_f H°(liq),-508.50$ (69))$\hspace{1.5cm}$ = -1.70

53(a) butanoic acid $(1\times C-(H)_3(C))+(1\times C-(H)_2(C)_2)+(1\times C-(H)_2(C)(CO))+(1\times CO-(C)(O))+$
$\hspace{1.5cm}(1\times O-(H)(CO))$
53(b) $C_4H_8O_2(liq) \rightarrow C_4H_8O_2(aq)$: -1.460±0.004 (66), -1.67±0.05 (12),$\hspace{1cm}$ SV = -1.56
53(c) $\Delta_{sol}H°(aq) = (\Delta_f H°(aq),-534.25) - (\Delta_f H°(liq),-533.80$ (68))$\hspace{1.5cm}$ = -0.45

54(a) pentanoic acid $(1\times C-(H)_3(C))+(2\times C-(H)_2(C)_2)+(1\times C-(H)_2(C)(CO))+$
$\hspace{1.5cm}(1\times CO-(C)(O))+(1\times O-(H)(CO))$
54(b) $C_5H_{10}O_2(liq) = C_5H_{10}O_2(aq)$ -0.50±0.05 (12),$\hspace{3cm}$ SV = -0.50
54(c) $\Delta_{sol}H°(aq) = (\Delta_f H°(aq),-558.30) - (\Delta_f H°(liq),-560.20$ (68))$\hspace{1.5cm}$ = 1.90
54(c) $\Delta_{sol}H°(aq) = (\Delta_f H°(aq),-558.30) - (\Delta_f H°(liq),-558.20$ (70))$\hspace{1.5cm}$ = -0.10

55(a) 2,2-dimethylpropanoic acid $(3\times C-(H)_3(C))+(1\times C-(C)_3(CO))+(1\times CO-(C)(O))+$
$\hspace{1.5cm}(1\times O-(H)(CO))$
55(b) $C_5H_{10}O_2(cr) = C_5H_{10}O_2(aq)$: 2.51±0.04 (66),$\hspace{3.5cm}$ SV = 2.51
55(c) $\Delta_{sol}H°(aq) = (\Delta_f H°(aq),-561.99) - (\Delta_f H°(liq),-564.50$ (71))$\hspace{1.5cm}$ = 2.51

56(a) hexanoic acid $(1\times C-(H)_3(C))+(3\times C-(H)_2(C)_2)+(1\times C-(H)_2(C)(CO))+$
$\hspace{1.5cm}(1\times CO-(C)(O))+(1\times O-(H)(CO))$
56(b) $C_6H_{12}O_2(liq) = C_6H_{12}O_2(aq)$: 0.92±0.05 (12),$\hspace{3.5cm}$ SV = 0.92
56(c) $\Delta_{sol}H°(aq) = (\Delta_f H°(aq),-582.35) - (\Delta_f H°(liq),-585.60$ (68))$\hspace{1.5cm}$ = 3.25
56(c) $\Delta_{sol}H°(aq) = (\Delta_f H°(aq),-582.35) - (\Delta_f H°(liq),-583.60$ (70))$\hspace{1.5cm}$ = 1.25

57(a) heptanoic acid $(1\times C-(H)_3(C))+(4\times C-(H)_2(C)_2)+(1\times C-(H)_2(C)(CO))+$
$\hspace{1.5cm}(1\times CO-(C)(O))+(1\times O-(H)(CO))$
57(b) $C_7H_{14}O_2(liq) = C_7H_{14}O_2(aq)$: 1.76±0.10 (12),$\hspace{3.5cm}$ SV = 1.76
57(c) $\Delta_{sol}H°(aq) = (\Delta_f H°(aq),-606.40) - (\Delta_f H°(liq),-611.41$ (68))$\hspace{1.5cm}$ = 5.01
57(c) $\Delta_{sol}H°(aq) = (\Delta_f H°(aq),-606.40) - (\Delta_f H°(liq),-608.27$ (70))$\hspace{1.5cm}$ = 1.87

Table III (continued). Estimated Enthalpies of Solution for Hydrocarbons and CHO Compounds at Infinite Dilution at 298.15 K (all values are in kJ/mol)

(a) *compound name and group notation*
(b) *equation for solution reaction, expt'l $\Delta_{sol}H°$ values(references), selected value SV*
(c) *est'd $\Delta_{sol}H°(aq)$ = (est'd $\Delta_f H°(aq)$ from Table II) - (expt'l $\Delta_f H°(g/liq/s)$ (ref))*

58(a) methyl ethanoate $(1×C-(H)_3(CO))+(1×CO-(C)(O))+(1×O-(CO)(CH_3))+$
  $(1×C-(H)_3(O))$
58(b) $C_3H_6O_2(liq) = C_3H_6O_2(aq)$: $-7.60±0.03$ (72),$-7.81±0.07$ (73),$-7.11±0.13$ (74),
  SV = -7.65
58(c) $\Delta_{sol}H°(aq) = (\Delta_f H°(aq),-453.40) - (\Delta_f H°(liq),-445.80$ (75))   = -7.60

59(a) methyl hexanoate $(1×C-(H)_3(C))+(3×C-(H)_2(C)_2)+(1×C-(H)_2(C)(CO))+$
  $(1×CO-(C)(O))+(1×O-(H)(CH_3))+(1×C-(H)_3(O))$
59(b) $C_7H_{14}O_2(liq) = C_7H_{14}O_2(aq)$; $-6.70±0.03$ (48),   SV = -6.70
59(c) $\Delta_{sol}H°(aq) = (\Delta_f H°(aq),-550.10) - (\Delta_f H°(liq),-540.20$ (76))   = -9.90

60(a) ethyl ethanoate $(1×C-(H)_3(CO))+(1×CO-(C)(O))+(1×O-((CO)(C))+$
  $(1×C-(H)_2(C)(O))+(1×C-(H)_3(C))$
60(b) $C_4H_8O_2(liq) = C_4H_8O_2(aq)$: $-10.46(35)$, $-9.33±0.17$ (77), $-8.34±0.13$ (78),
  SV = -9.33
60(c)$\Delta_{sol}H°(aq) = (\Delta_f H°(aq),-488.35) - (\Delta_f H°(liq),-478.80$ (79))   = -9.55

61(a) ethyl propanoate $(2×C-(H)_3(C))+(1×C-(H)_2(C)(CO))+(1×CO-(C)(O))+$
  $(1×O-(CO)(C))+(1×C-(H)_2(C)(O))$
61(b) $C_5H_{10}O_2(liq) → C_5H_{10}O_2(aq)$: $-10.25±0.04$ (72),   SV = -10.25
61(c) $\Delta_{sol}H°(aq) = (\Delta_f H°(aq),-512.90) - (\Delta_f H°(liq),-502.70$ (80))   = -10.20

62(a) ethyl pentanoate $(2C-(H)_3(C))+(2×C-(H)_2(C)_2)+(1×C-(H)_2(C)(CO))+$
  $(1×CO-(C)(O))+(1×O-(CO)(C))+(1×C-(H)_2(C)(O))$
62(b) $C_7H_{14}O_2(liq) = C_7H_{14}O_2(aq)$: $-9.51±0.09$ (72),   SV = -9.51
62(c) $\Delta_{sol}H°(aq) = (\Delta_f H°(aq),-561.00) - (\Delta_f H°(liq),-553.00$ (81))   = -8.00

63(a) n-butyl ethanoate $(1×C-(H)_3(CO))+(1×C-(H)_3(C))+(1×C-(H)_2(C)(O))+$
  $(2×C-(H)_2(C)_2)+(1×CO-(C)(O))+(1×O-(CO)(C))$
63(b) $C_6H_{12}O_2(liq) = C_6H_{12}O_2(aq)$: $-7.83±0.13$ (74),   SV = -7.83
63(c) $\Delta_{sol}H°(aq) = (\Delta_f H°(aq),-536.45) - (\Delta_f H°(liq),-529.20$ (82))   = -7.25

Summary and Conclusions

We have shown that the second-order group-additivity method which was used earlier for the estimation of the enthalpy of formation, entropy, and heat capacity of organic compounds in the gas, liquid, and solid phases at 298.15 K is applicable to aqueous solutions at infinite dilution. Although experimental data for $\Delta_{sol}H°$(aq) at 298.15 K for organic compounds at infinite dilution are not as numerous as data for $\Delta_f H°$ at 298.15K for the gas, liquid, and solid phases, good agreement between experimental and estimated $\Delta_{sol}H°$(aq) has been obtained for hydrocarbons, alcohols, and monocarboxylic acids; fair agreement was obtained for ethers, ketones, and esters. On the other hand, α,ω-alkanediols did not show satisfactory agreement between experimental and estimated $\Delta_{sol}H°$(aq) because of poor agreement among the experimental data in the liquid or solid phase for $\Delta_f H°$, or because of the availability of only poor quality experimental data for $\Delta_{sol}H°$(aq). The attempt to apply this estimation method to α,ω-alkanediols for the prediction of $\Delta_{sol}H°$(aq) strongly exemplifies the necessity for high quality experimental results not only for $\Delta_f H°$ in the gas, liquid, and solid phases but also for measurements of $\Delta_{sol}H°$(aq), and the consequences of such a deficiency.

The development of group values for $\Delta_f H°$(aq) of organic nitrogen compounds is likely to be the next step in the continuation of the extension of group additivity to aqueous solutions at infinite dilution. A significant body of good quality experimental data are available in the literature for $\Delta_f H°$ in the gas, liquid, and solid phases and $\Delta_{sol}H°$(aq) for this class of compounds.

Literature Cited

1. Benson, S.W.; Buss, J.H. *J. Chem. Phys.* **1958**, *29*, 546.
2. Benson, S.W.; Cruickshank, F.R.; Golden, D.M.; Haugen, G.R.; O'Neal, H.E.; Rodgers, A.S.; Shaw, R.; Walsh, R. *Chem. Rev.* **1969**, *69*, 269.
3. Benson, S.W. *Thermochemical Kinetics*, Second Edition; J. Wiley & Sons: New York, NY, 1976.
4. Cohen, N.; Benson, S.W. in *Chemistry of Alkanes and Cycloalkanes*, Patai. S.; Rappoport, Z., Eds.; J. Wiley and Sons: Chichester, UK, 1992; Chapter 4, *The Thermochemistry of Alkanes and Cycloalkanes*, pp. 215-287.
5. Cohen, N.; Benson, S.W. *Chem. Rev.* **1993**, *93*, 2419.
6. Domalski, E.S.; Hearing, E.D. *J. Phys. Chem. Ref. Data* **1993**, *22*, 805.
7. Nichols, N.; Skold, R.; Spink, C.; Suurkuusk, J.; Wadsö, I. *J. Chem. Thermodyn.* **1976**, *8*, 1081.
8. Cabani, S.; Gianni, P.; Mollica, V.; Lepori, L. *J. Solution Chem.* **1981**, *10*, 563.
9. Aveyard, R.; Lawrence, A.C.S. *Trans. Faraday Soc.* **1964**, *60*, 2265.
10. Hill, D.J.T.; White, L.R. *Australian J. Chem.* **1974**, *27*, 1905.
11. Hallen, D.; Nilsson, S.-O.; Rothchild, W.; Wadsö, I. *J. Chem. Thermodyn.* **1986**, *18*, 429.
12. Aveyard, R.; Mitchell, R.W. *Trans. Faraday Soc.* **1968**, *64*, 1757.
13. Mosselman, C.; Dekker, H. *J. Chem. Soc., Faraday Trans. I* **1975**, *71*, 417.
14. Dec, S.F.; Gill, S.J. *J. Solution Chem.* **1984**, *13*, 27.

15. Olofsson, G.; Oshodj, A.A.; Qvarnstrom, E.; Wadsö, I. *J. Chem. Thermodyn.* **1984**, *16*, 1041.
16. Rettlich, T.R.; Handa, Y.P.; Battino, R.; Wilhelm, E. *J. Phys. Chem.* **1981**, *85*, 3230.
17. Pittam, D.A.; Pilcher, G. *Trans. Faraday Soc. I* **1972**, *68*, 2224.
18. Hallen, D.; Nilsson, S.-O.; Wadsö, I. *J. Chem. Thermodyn.* **1989**, *21*, 529.
19. Good, W.D. *J. Chem. Thermodynam.* **1970**, *2*, 237.
20. Gill, S.J.; Nichols, N.F.; Wadsö, I. *J. Chem. Thermodyn.* **1976**, *8*, 445.
21. Good, W.D.; Smith, N.K. *J. Chem. Eng. Data* **1969**, *14*, 102.
22. Osborne, N.S.; Ginnings, D.C. *J. Res. Nat. Bur. Stds.* **1947**, *39*, 543.
23. Reid, D.S.; Quickenden, M.A.J.; Franks, F. *Nature* **1969**, *224*, 1293.
24. Rossini, F.D.; Knowlton, J.W. *J. Res. Nat. Bur. Stds.* **1937**, *19*, 249.
25. Prosen, E.J.; Maron, F.W.; Rossini, F.D. *J. Res. Nat. Bur. Stds.* **1951**, *46*, 106.
26. Conn, J.B.; Kistiakowsky, G.B.; Smith, E.A. *J. Am. Chem. Soc.* **1939**, *61*, 1868.
27. Knowlton, J.W.; Rossini, F.D. *J. Res. Nat. Bur. Stds.* **1949**, *43*, 155.
28. Gill, S.J.; Nichols, N.F.; Wadsö, I. *J. Chem. Thermodyn.* **1975**, *7*, 175.
29. Prosen, E.J.; Gilmont, R.; Rossini, F.D. *J. Res. Nat. Bur. Stds.* **1945**, *34*, 65.
30. Alexander, D.M.; Hill, D.J.T. *Australian J. Chem.* **1969**, *22*, 347.
31. Arnett, E.M.; Kover, W.B.; Carter, J.V. *J. Am. Chem. Soc.* **1969**, *91*, 4028.
32. Rossini, F.D. *J. Res. Nat. Bur. Stds.* **1932**, *8*, 119.
33. Majer, V.; Svoboda, V. *Enthalpies of Vaporization of Organic Compounds*; Blackwell Scientific Publishers: Oxford, UK, 1985, 300pp.
34. Hill, D.J.T.; Malar, C. *Australian J. Chem.* **1975**, *28*, 7.
35. Arnett, E.M.; Burke, J.J.; Carter, J.V.; Douty, C.F. *J. Am. Chem. Soc.* **1972**, *94*, 7837.
36. Snelson, A.; Skinner, H.A. *Trans. Faraday Soc.* **1961**, *57*, 2125.
37. Gellner, O.H.; Skinner, H.A. *J. Chem. Soc.* **1949**, 1145.
38. Nwankwo, S.; Wadsö, I. *J. Chem. Thermodyn.* **1980**, *12*, 1167.
39. Mosselman, C.; Dekker, H. *Rec. Trav. Chim.* **1969**, 88, 161.
40. Skinner, H.A.; Snelson, A. *Trans. Faraday Soc.* **1960**, *56*, 1776.
41. Skold, R.; Suurkuusk, J.; Wadsö, I. *J. Chem. Thermodyn.* **1976**, *8*, 1075.
42. Chao, J.; Rossini, F.D. *J. Chem. Eng. Data* **1965**, *10*, 374.
43. Sellers, P.; Sunner, S. *Acta Chem. Scand.* **1962**, *16*, 46.
44. Nichols, N.; Wadsö, I. *J. Chem. Thermodyn.* **1975**, *7*, 329.
45. Andon, R.J.L.; Biddiscombe, D.P.; Cox, J.D.; Handley, R.; Harrop, D.; Herington, E.F.G.; Martin, J.F. *J. Chem. Soc.* **1960**, 5246.
46. Parks, G.S.; Manchester, K.E.; Vaughan, L.M. *J. Chem. Phys.* **1954**, *22*, 2089.
47. Corkill, J.M.; Goodman, J.F.; Tate, J.R. *Trans. Faraday Soc.* **1969**, *65*, 1742.
48. Finch, A.; Gardner, P.J.; McNamara, P.M.; Wellum, G.R. *J. Chem. Soc. A* **1970**, 3339.
49. Nichols, N.; Skold, R.; Spink, C.; Wadsö, I. *J. Chem. Thermodyn.* **1976**, *8*, 993.
50. Moureu, H.; Dode, M. *Bull. Soc. Chim. France* **1937**, *4*, 637.
51. Parks, G.S.; West, T.J.; Naylor, B.F.; Fujii, P.S.; McClaine, L.A. *J. Am. Chem. Soc.* **1946**, *68*, 2524.
52. Gardner, P.J.; Hussain, K.S. *J. Chem. Thermodyn.* **1972**, *4*, 819.

53. Knauth, P.; Sabbah, R. *J. Chem. Thermodyn.* **1989**, *21*, 203.
54. Knauth, P.; Sabbah, R. *J. Chem. Thermodyn.* **1989**, 21, 779.
55. Steele, W.V.; Chirico, R.D.; Nguyen, A.; Hossenlopp, I.A.; Smith, N.K., *DIPPR Data Series* 1991, *No. 1*, 101.
56. Berthelot, M.P.E. *Ann. Chim. Phys.* **1881**, *[5] 23*, 176.
57. Pilcher, G.; Pell, A.S.; Coleman, D.J. *Trans. Faraday Soc.* **1964**, *60*, 499.
58. Cabani, S.; Conti, G.; Lepori, L. *Trans. Faraday Soc.* **1971**, *67*, 1943.
59. Counsell, J.F.; Lee, D.A.; Martin, J.F. *J. Chem. Soc. A* **1971**, 313.
60. Blaschko, H. *Biochem. Z.* **1925**, *158*, 428.
61. Dolliver, M.A.; Gresham, T.L.; Kistiakowsky, G.B.; Smith, E.A.; Vaughan, W.E. *J. Am. Chem. Soc.* **1938**, *60*, 440.
62. Coleman, C.F.; De Vries, T. *J. Am. Chem. Soc.* **1949**, *71*, 2839.
63. Pennington, R.E.; Kobe, K.A. *J. Am. Chem. Soc.* **1957**, *79*, 300.
64. Bury, R.; Luca, M.; Barberi, P. *J. Chim. Phys., Phys. Chim. Biol.* **1978**, *75*, 575.
65. Harrop, D.; Head, A.J.; Lewis, G.B. *J. Chem. Thermodyn.* **1970**, *2*, 203.
66. Konicek, J.; Wadsö, I. *Acta Chem. Scand.* **1971**, *25*, 1541.
67. Stern, J.H.; Sandstrom, J.P.; Hermann, A. *J. Phys. Chem.* **1967**, *71*, 3623.
68. Lebedeva, N.D. *Zh. Fiz. Khim.* **1964**, *38*, 2648.
69. Konicek, J.; Wadsö, I. *Acta Chem. Scand.* **1970**, *24*, 2612.
70. Adriaanse, N.; Dekker, H.; Coops, J. *Rec. Trav. Chim.* **1965**, *84*, 393.
71. Hancock, C.K.; Watson, G.M.; Gilby, R.F. *J. Phys. Chem.* **1954**, *58*, 127.
72. Nilsson, S.-O.; Wadsö, I. *J. Chem. Thermodyn.* **1986**, *18*, 673.
73. Della Gatta, G.; Stradella, L.; Venturello, P. *J. Solution Chem.* **1981**, *10*, 203.
74. Stern, J.H.; O'Connor, M.E. *J. Chem. Eng. Data* **1972**, *17*, 185.
75. Hall, H.K., Jr.; Baldt, J.H. *J. Am. Chem. Soc.* **1971**, *93*, 140.
76. Adriaanse, N.; Dekker, H.; Coops, J. *Rec. Trav. Chim.* **1965**, *84*, 393.
77. Stern, J.H.; Hermann, A. *J. Phys. Chem.* **1967**, *71*, 306.
78. Cross, R.F.; McTigue, P.T. *Australian J. Chem.* **1977**, *30*, 2957.
79. Fenwick, J.O.; Harrop, D.; Head, A.J. *J. Chem. Thermodyn.* **1978**, *10*, 687.
80. Månsson, M. *J. Chem. Thermodyn.* **1972**, *4*, 865.
81. Schjånberg, E. *Z. Physik. Chem.* **1937**, *A178*, 274.
82. Wadsö, I. *Acta Chem. Scand.* **1958**, *12*, 630.

Chapter 4

Estimating Phase-Change Enthalpies and Entropies

James S. Chickos[1] William E. Acree, Jr.[2], and Joel F. Liebman[3]

[1]Department of Chemistry, University of Missouri, St. Louis, MO 63121
[2]Department of Chemistry, University of North Texas, Denton, TX 76203
[3]Department of Chemistry and Biochemistry, University of Maryland
Baltimore County, 1000 Hilltop Circle, Baltimore, MD 21250

A group additivity method based on molecular structure is described that can be used to estimate total phase change entropies and enthalpies of organic molecules. Together with vaporization enthalpies which are estimated by a similar technique, this provides an indirect method to estimate sublimation enthalpies. The estimations of these phase changes are described and examples are provided to guide the user in evaluating these properties for a broad spectrum of organic structures.

Fusion, vaporization and sublimation enthalpies are important physical properties of the condensed phase. They are essential in studies referencing the gas phase as a standard state and are extremely useful in any investigation that requires information regarding the magnitude of molecular interactions in the condensed phase (*1-4*). The divergence in quantity between the many new organic compounds prepared and the few thermochemical measurements reported annually has encouraged the development of empirical relationships that can be used to estimate these properties.

We have found that techniques for estimating fusion, vaporization and sublimation enthalpies can play several useful roles (*5-7*). Perhaps most importantly, they provide a numerical value that can be used in cases when there are no experimental data. In addition we have used an estimated value to select the best experimental value in cases where two or more values are in significant disagreement and in cases where only one measurement is available, to assess whether the experimental value is reasonable. Given the choice between an estimated or experimental value, selection of the experimental value is clearly preferable. However, large discrepancies between estimated and calculated values can also identify experiments worth repeating. Finally, the parameters generated from such a treatment permit an investigation of inter and intramolecular interactions that are not well understood.

Fusion Enthalpies

There are very few general techniques reported for directly estimating fusion enthalpies. Fusion enthalpies are most frequently calculated from fusion entropies and the experimental melting temperature of the solid, T_{fus}. One of the earliest applications of this is the use of Walden's Rule (8). The application of Walden's Rule provides a remarkably good approximation of $\Delta_{fus}H_m$, if one considers that the estimation is independent of molecular structure and based on only two parameters. Recent modifications of this rule have also been reported (9-10).

Walden's Rule: $\Delta_{fus}H_m(T_{fus})/T_{fus} \approx 13 \text{ cal·K}^{-1}\text{·mol}^{-1} = 54.4 \text{ J·mol}^{-1}\text{·K}^{-1}$.   (1)

Estimations of fusion entropies. A general method was reported recently for estimating fusion entropies and enthalpies based on the principles of group additivity (11, 12). This method has been developed to estimate the total phase change entropy and enthalpy of a substance associated in going from a solid at 0 K to a liquid at the melting point, T_{fus}. Many solids undergo a variety of phase changes prior to melting, which affects the magnitude of the fusion entropy. The total phase change entropy and enthalpy, $\Delta_0^{T_{fus}} S_{tpce}$ and $\Delta_0^{T_{fus}} H_{tpce}$, in most instances provide a good estimate of the entropy and enthalpy of fusion, $\Delta_{fus}S_m(T_{fus})$ and $\Delta_{fus}H_m(T_{fus})$. If there are no additional solid phase transitions then $\Delta_0^{T_{fus}} S_{tpce}$ and $\Delta_0^{T_{fus}} H_{tpce}$ become numerically equal to $\Delta_{fus}S_m(T_{fus})$ and $\Delta_{fus}H_m(T_{fus})$.

An abbreviated listing of the group parameters that can be used to estimate these phase change properties is included in Tables I and III. The group values in these tables have been updated from previous versions (11, 12) by the inclusion of new experimental data in the parameterizations. Before describing the application of these parameters to the estimation of $\Delta_0^{T_{fus}} S_{tpce}$ and $\Delta_0^{T_{fus}} H_{tpce}$, the conventions used to describe these group values need to be defined. Primary, secondary, tertiary and quaternary centers, as found on atoms of carbon and silicon and their congeners, are defined solely on the basis of the number of hydrogens attached to the central atom, 3, 2, 1, 0, respectively. This convention is used throughout this chapter. In addition, compounds whose liquid phase is not isotropic at the melting point are not modeled properly by these estimations. Those forming liquid crystal or cholesteric phases as well amphiphilic compounds are currently overestimated by the parameters and should also be excluded from these estimations. A large discrepancy between the estimated total phase change enthalpy and experimental fusion enthalpy is a good indication of undetected solid-solid phase transitions or non-isotropic liquid behavior. Finally, it should be pointed out that the experimental melting point along with an estimated value of $\Delta_0^{T_{fus}} S_{tpce}$ is necessary to estimate the fusion enthalpy of a compound.

The parameters used for estimating $\Delta_0^{T_{fus}} S_{tpce}$ of hydrocarbons and the hydrocarbon portions of more complex molecules are listed in Table I. The group

Table I A. Contributions of the Hydrocarbon Portion of Acyclic and Aromatic Molecules

Aliphatic and Aromatic Carbon Groups		Group Value G_i, $J \cdot mol^{-1} \cdot K^{-1}$	Group Coefficients C_i
primary sp^3	CH$_3$-	17.6	
secondary sp^3	>CH$_2$	7.1	1.31[a]
tertiary sp^3	-CH<	-16.4	0.60
quaternary sp^3	>C<	-34.8	0.66
secondary sp^2	=CH$_2$	17.3	
tertiary sp^2	=CH-	5.3	0.75
quaternary sp^2	=C(R)-	-10.7	
tertiary sp	H-C≡	14.9	
quaternary sp	-C≡	-2.8	
aromatic tertiary sp^2	=C$_a$H-	7.4	
quaternary aromatic sp^2 carbon adjacent to an sp^3 atom	=C$_a$(R)-	-9.6	
peripheral quaternary aromatic sp^2 carbon adjacent to an sp^2 atom	=C$_a$(R)-	-7.5	
internal quaternary aromatic sp^2 carbon adjacent to an sp^2 atom	=C$_a$(R)-	-0.7	

[a]The group coefficient of 1.31 for C_{CH_2} is applied only when the number of consecutive methylene groups exceeds the sum of the remaining groups; see equation 2 in text.

Table I B. Contributions of the Cyclic Hydrocarbon Portions of the Molecule

Contributions of Cyclic Carbons		Group Value (G_i) $J \cdot mol^{-1} \cdot K^{-1}$	Group Coefficient C_i
cyclic tertiary sp^3	>C$_c$H(R)	-14.7	
cyclic quaternary sp^3	>C$_c$(R)$_2$	-34.6	
cyclic tertiary sp^2	=C$_c$H-	-1.6	1.92
cyclic quaternary sp^2	=C$_c$(R)-	-12.3	
cyclic quaternary sp	=C$_c$=; R-C$_c$≡	-4.7	

value, G_i, associated with a molecular fragment is identified in the third column of the table. The group coefficients, C_i, are listed in column 4 of the table. These group coefficients are used to modify G_i whenever a functional group is attached to the carbon in question. Functional groups are defined in Table III. All values of C_i and C_k that are not specifically defined in both Tables I and III are to be assumed equal to 1.0. The group coefficient for a methylene group in Table I, C_{CH_2}, is applied differently from the rest. The group coefficient for a methylene group is used whenever the total number of consecutive methylene groups in a molecule, n_{CH_2}, equals or exceeds the sum of the other remaining groups, Σn_i. This applies to both hydrocarbons and all derivatives. Introduction of this coefficient is new and differentiates this protocol from previous versions (*11, 12*). The application of this group coefficient is illustrated below.

Acyclic and Aromatic Hydrocarbons. Estimation of $\Delta_0^{T_{fus}} S_{tpce}$ for acyclic and aromatic hydrocarbons (*aah*) can be achieved by summing the group values consistent with the structure of the molecule as illustrated in the following equation:

$$\Delta_0^{T_{fus}} S_{tpce}(aah) = \sum_i n_i G_i + n_{CH_2} C_{CH_2} G_{CH_2}; \quad C_{CH_2} = 1.31 \text{ when } n_{CH_2} \geq \Sigma n_i; i \neq CH_2$$
$$\text{otherwise } C_{CH_2} = 1. \tag{2}$$

Some examples illustrating the use of both the groups in Table I A and equation 2 are given in Table II. Entries for each estimation include the melting point, T_{fus}, and all transition temperatures, T_t, for which there is a substantial enthalpy change. The estimated and experimental (in parentheses) phase change entropies follow. Similarly, the total phase change enthalpy calculated as the product of $\Delta_0^{T_{fus}} S_{tpce}$ and T_{fus} is followed by the experimental total phase change enthalpy (or fusion enthalpy). Finally, details in estimating $\Delta_0^{T_{fus}} S_{tpce}$ for each compound are included as the last entry.

n-Butylbenzene. The estimation of the fusion entropy of n-butylbenzene is an example of an estimation of a typical aromatic hydrocarbon. Identification of the appropriate groups in Table I A results in an entropy of fusion of 66.3 $J \cdot mol^{-1} \cdot K^{-1}$ and together with the experimental melting point, an enthalpy of fusion of 12.3 $kJ \cdot mol^{-1}$ is estimated. This can be compared to the experimental value of 11.3 $kJ \cdot mol^{-1}$. It should be pointed out that the group values for aromatic molecules are purely additive while the group values for other cyclic sp^2 atoms are treated as corrections to the ring equation. This will be discussed in more detail below.

n-Heptacosane. The fusion entropy of n-heptacosane is obtained in a similar fashion. In this case, the number of consecutive methylene groups in the molecule exceeds the sum of the remaining terms in the estimation and this necessitates the use of the group coefficient, C_{CH_2}, of 1.31. Heptacosane exhibits two additional phase transitions below its melting point. These are shown in parentheses for both $\Delta_0^{T_{fus}} S_{tpce}$ and $\Delta_0^{T_{fus}} H_{tpce}$ following the estimated value for each, respectively. For a molecule such as 4-methylhexacosane (estimation not shown), the group coefficient of 1.31 would be applied to the 21 consecutive methylene groups. The remaining two methylene groups would be treated normally ($C_{CH_2} = 1.0$) but would not be counted in Σn_i.

Ovalene. Estimation of the phase change entropy of ovalene provides an example of a molecule containing both peripheral and internal quaternary sp^2 carbon atoms adjacent to an sp^2 atom. The carbon atoms in graphite are another example of internal quaternary sp^2 carbon atoms. In the application of these group values to obtain the phase change properties of other aromatic molecules, it is important to remember that aromatic molecules are defined in these estimations as molecules containing only

Table II. Estimations of Total Phase Change Entropies and Enthalpies of Hydrocarbons[a]

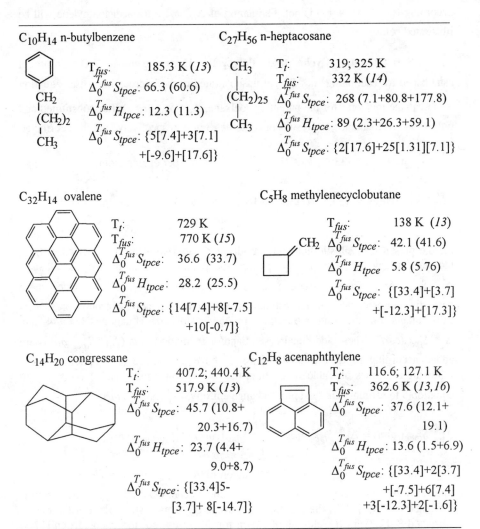

$C_{10}H_{14}$ n-butylbenzene

T_{fus}: 185.3 K (*13*)
$\Delta_0^{T_{fus}} S_{tpce}$: 66.3 (60.6)
$\Delta_0^{T_{fus}} H_{tpce}$: 12.3 (11.3)
$\Delta_0^{T_{fus}} S_{tpce}$: {5[7.4]+3[7.1]
      +[-9.6]+[17.6]}

$C_{27}H_{56}$ n-heptacosane

T_t: 319; 325 K
T_{fus}: 332 K (*14*)
$\Delta_0^{T_{fus}} S_{tpce}$: 268 (7.1+80.8+177.8)
$\Delta_0^{T_{fus}} H_{tpce}$: 89 (2.3+26.3+59.1)
$\Delta_0^{T_{fus}} S_{tpce}$: {2[17.6]+25[1.31][7.1]}

$C_{32}H_{14}$ ovalene

T_t: 729 K
T_{fus}: 770 K (*15*)
$\Delta_0^{T_{fus}} S_{tpce}$: 36.6 (33.7)
$\Delta_0^{T_{fus}} H_{tpce}$: 28.2 (25.5)
$\Delta_0^{T_{fus}} S_{tpce}$: {14[7.4]+8[-7.5]
      +10[-0.7]}

C_5H_8 methylenecyclobutane

T_{fus}: 138 K (*13*)
$\Delta_0^{T_{fus}} S_{tpce}$: 42.1 (41.6)
$\Delta_0^{T_{fus}} H_{tpce}$ 5.8 (5.76)
$\Delta_0^{T_{fus}} S_{tpce}$: {[33.4]+[3.7]
      +[-12.3]+[17.3]}

$C_{14}H_{20}$ congressane

T_t: 407.2; 440.4 K
T_{fus}: 517.9 K (*13*)
$\Delta_0^{T_{fus}} S_{tpce}$: 45.7 (10.8+
      20.3+16.7)
$\Delta_0^{T_{fus}} H_{tpce}$: 23.7 (4.4+
      9.0+8.7)
$\Delta_0^{T_{fus}} S_{tpce}$: {[33.4]5-
      [3.7]+ 8[-14.7]}

$C_{12}H_8$ acenaphthylene

T_t: 116.6; 127.1 K
T_{fus}: 362.6 K (*13,16*)
$\Delta_0^{T_{fus}} S_{tpce}$: 37.6 (12.1+
      19.1)
$\Delta_0^{T_{fus}} H_{tpce}$: 13.6 (1.5+6.9)
$\Delta_0^{T_{fus}} S_{tpce}$: {[33.4]+2[3.7]
      +[-7.5]+6[7.4]
      +3[-12.3]+2[-1.6]}

[a]Units for $\Delta_0^{T_{fus}} S_{tpce}$ and $\Delta_0^{T_{fus}} H_{tpce}$ are J·mol⁻¹·K⁻¹ and kJ·mol⁻¹, respectively; experimental values are included in parentheses following the calculated value (in cases where additional solid-solid transitions are involved, the first term given is the total property associated with the transition(s) and the second term represents the fusion property). A reference to the experimental data is included in parentheses following T_{fus}.

benzenoid carbons and the corresponding nitrogen heterocycles. While a molecule like 1,2-benzacenaphthene (fluoranthene) would be considered aromatic, acenaphthylene, according to this definition is not. Estimation of $\Delta_0^{T_{fus}} S_{tpce}$ for acenaphthylene will be illustrated below.

Non-aromatic Cyclic and Polycyclic Hydrocarbons. The protocol established for estimating $\Delta_0^{T_{fus}} S_{tpce}$ of unsubstituted cyclic hydrocarbons uses equation 3 to evaluate this term for the parent cycloalkane, $\Delta_0^{T_{fus}} S_{tpce}(ring)$. For substituted and polycyclic cycloalkanes, the results of equations 3 or 4, respectively, are then corrected

$$\Delta_0^{T_{fus}} S_{tpce}(ring) = [33.4 \ ] + [3.7][n\text{-}3] \ ; \qquad n = \text{number of ring atoms} \qquad (3)$$

$$\Delta_0^{T_{fus}} S_{tpce}(ring) = [33.4 \ ]N + [3.7][R\text{-}3N]; \qquad R = \text{total number of ring atoms};$$
$$N = \text{ number of rings} \qquad (4)$$

for the presence of substitution and hybridization patterns in the ring that differ from the standard cyclic secondary sp^3 pattern found in the parent monocyclic alkanes, $\Delta_0^{T_{fus}} S_{tpce}(corr)$. These correction terms can be found in Table I B. Once these corrections are included in the estimation, any additional acyclic groups present as substitutents on the ring are added to the results of equations 3 or 4 and $\Delta_0^{T_{fus}} S_{tpce}(corr)$. These additional acyclic and/or aromatic terms ($\Delta_0^{T_{fus}} S_{tpce}(aah)$) are added according to the protocol discussed above in the use of equation 2. The following examples of Table II illustrate the use of equations 3 and 4 according to equation 5 to estimate the total phase change entropy, $\Delta_0^{T_{fus}} S_{tpce}(total)$.

$$\Delta_0^{T_{fus}} S_{tpce}(total) = \Delta_0^{T_{fus}} S_{tpce}(ring) + \Delta_0^{T_{fus}} S_{tpce}(corr) + \Delta_0^{T_{fus}} S_{tpce}(aah). \qquad (5)$$

Methylenecyclobutane. The estimation of $\Delta_0^{T_{fus}} S_{tpce}$ for methylenecyclobutane illustrates the use of equation 5 for a monocyclic alkene. Once the cyclobutane ring is estimated ([33.4]+[3.7]), the presence of a cyclic quaternary sp^2 carbon in the ring is corrected ([-12.3]) next. Addition of a term for the acyclic sp^2 methylene group [17.3] completes this estimation.

Congressane. Congressane, a pentacyclic hydrocarbon, provides an example of how equation 4 is used in conjunction with equation 5. The usual criterion, the minimum number of bonds that need to be broken to form a completely acyclic molecule, is used to determine the number of rings. Application of equation 4 to congressane [[33.4]5+3.7[14-15]] provides $\Delta_0^{T_{fus}} S_{tpce}(ring)$. Addition of the contribution of the eight cyclic tertiary sp^3 carbons to the results of equation 4, $\Delta_0^{T_{fus}} S_{tpce}(corr)$, completes the estimation.

Acenaphthylene. Estimation of $\Delta_0^{T_{fus}} S_{tpce}$ and $\Delta_0^{T_{fus}} H_{tpce}$ for acenaphthylene completes this section on cyclic hydrocarbons. Molecules that contain rings fused to aromatic rings but are not completely aromatic, according the definition provided above, are estimated by first calculating $\Delta_0^{T_{fus}} S_{tpce}(ring)$ for the contributions of the non-aromatic ring according to equations 3 or 4. This is then followed by adding the corrections and contributions of the remaining aromatic groups and any other acyclic substitutents. The five membered ring in acenaphthylene $\{\Delta_0^{T_{fus}} S_{tpce}(ring)$ [33.4]+2[3.7]$\}$ is first corrected for each non-secondary sp^3 carbon atom $\{\Delta_0^{T_{fus}} S_{tpce}(corr): +2[-1.6]+3[-12.3]\}$, and then the remainder of the aromatic portion of the molecule $(\Delta_0^{T_{fus}} S_{tpce}(aah): [-7.5] +6[7.4]\}$ is estimated as illustrated above.

Hydrocarbon Derivatives. Estimations involving derivatives of hydrocarbons are performed in a fashion similar to hydrocarbons. The estimation consists of three parts: the contribution of the hydrocarbon component, that of the carbon(s) bearing the functional group(s), $\sum_i n_i C_i G_i$, and the contribution of the functional group(s), $\sum_k n_k C_j G_k$. The symbols n_i, n_k refer to the number of groups of type i and k. Acyclic and cyclic compounds are treated separately as before. For acyclic and aromatic molecules, the hydrocarbon portion is estimated using equation 2; cyclic or polycyclic molecules are estimated using equations 3 and 4, respectively. Similarly, the contribution of the carbon(s) bearing the functional group(s) is evaluated from Table I A or Table I B modified by the appropriate group coefficient, C_i, as will be illustrated below. The group values of the functional groups, G_k, are listed in Table III A-C. The corresponding group coefficient, C_j is equal to one for all functional groups except those listed in Table III B. Selection of the appropriate value of C_j from Table III B is based on the total number of functional groups and is discussed below. Functional groups that make up a portion of a ring are listed in Table III C. The use of these values in estimations will be illustrated separately. Equations 6 and 7 summarize the protocol developed to estimate $\Delta_0^{T_{fus}} S_{tpce}(total)$ for acyclic and aromatic derivatives and for cyclic and polycyclic hydrocarbon derivatives, respectively.

$$\Delta_0^{T_{fus}} S_{tpce}(total) = \Delta_0^{T_{fus}} S_{tpce}(aah) + \sum_i n_i C_i G_i + \sum_k n_k C_j G_k, \tag{6}$$

$$\Delta_0^{T_{fus}} S_{tpce}(total) = \Delta_0^{T_{fus}} S_{tpce}(ring) + \Delta_0^{T_{fus}} S_{tpce}(corr) + \sum_i n_i C_i G_i + \sum_k n_k C_j G_k, \tag{7}$$

where: $C_j = \sum_k n_k.$

In view of the large number of group values listed in Table III A-C, selection of the appropriate functional group(s) is particularly important. The four functional groups of Table III B are dependent on the total substitution pattern in the molecule. Coefficients

Table III A. Functional Group Values[a]

Functional Groups		Group Value (G_k) $J \cdot mol^{-1} \cdot K^{-1}$	Functional Groups		Group Value (G_k) $J \cdot mol^{-1} \cdot K^{-1}$
bromine	-Br	17.5	tetrasubst. urea	>NC(=O)N<	[-19.3]
fluorine on an			1,1-disubst. urea	>NC(=O)NH$_2$	[19.5]
sp^2 carbon,	=CF-	19.5	1,3-disubst. urea	-NHC(=O)NH-	[1.5]
aromatic			monosubst. urea	-NHC(=O)NH$_2$	[22.5]
fluorine	=C$_a$F-	16.6	carbamate	-OC(=O)NH$_2$	[27.9]
3-fluorines on			N-subst. carbamate	-OC(=O)NH-	10.6
an sp^3 carbon	CF$_3$-	13.3	imide	>(C=O)$_2$NH	[7.7]
2-fluorines on			phosphine	-P<	[-20.7]
an sp^3 carbon	>CF$_2$	16.4	phosphate ester	P(=O)(OR)$_3$	[-10.0]
1-fluorine on			phosphonyl halide	-P(=O)X$_2$	[4.8]
an sp^3 carbon	-CF<	12.7	phosphorothioate ester	(RO)$_3$P=S	1.1
fluorine on an			phosphorodithioate ester	-S-P(=S)(OR)$_2$	-9.6
sp^3 ring carbon	>CHF; CF$_2$	[17.5]	phosphonothioate ester	-P(=S)(OR)$_2$	[5.2]
iodine	-I	19.4	phosphoroamidothioate		
phenol	=C(OH)-	20.3	ester	-NHP(=S)(OR)$_2$	[16.0]
ether	>O	4.71	sulfide	>S	2.1
aldehyde	-CH(=O)	21.5	disulfide	-SS-	9.6
ketone	>C(=O)	4.6	thiol	-SH	23.0
ester	-(C=O)O-	7.7	sulfone	>S(O)$_2$	0.3
heterocyclic			sulfonate ester	-S(O)$_2$O-	[7.9]
aromatic amine	=N$_a$-	10.9	N,N-disubst.		
acyclic sp^2			sulfonamide	-S(O)$_2$N<,	[-11.3]
nitrogen	=N-	[-1.8]	N-subst. sulfonamide	-S(O)$_2$NH-	6.3
tert. amine	-N<	-22.2	sulfonamide	-S(O)$_2$NH$_2$	[28.4]
sec. amine	-NH-	-5.3	aluminum	-Al<	[-24.7]
primary amine	-NH$_2$	21.4	arsenic	-As<	[-6.5]
aliphatic tert.			boron	-B<	[-17.2]
nitramine	>N-NO$_2$	5.39	gallium	-Ga<	[-11.9]
nitro group	-NO$_2$	17.7	quat. germanium	>Ge<	[-35.2]
oxime	=N-OH	[13.6]	sec. germanium	>GeH$_2$	[-14.7]
azoxy nitrogen	N=N(O)-	[6.8]	quat. lead	>Pb<	[-30.2]
nitrile	-C≡N	17.7	selenium	>Se	[6.0]
tert. amide	-C(=O)N<	-11.2	quat. silicon	>Si<	-27.1
sec. amide	-C(=O)NH-	1.5	quat. tin	>Sn<	-24.2
primary amide	-CONH$_2$	27.9	zinc	>Zn	[11.1]

[a]Values in brackets are tentative assignments; R refers to alkyl and aryl groups.

Table III B. Functional Group Values Dependent on the Degree of Substitution[a]

Functional Group		Group Value (G_k) $J \cdot mol^{-1} \cdot K^{-1}$	Group Coefficient C_j				
			2	3	4	5	6
chlorine	-Cl	10.8	1.5	1.5	1.5	1.5	1.5
hydroxyl group	-OH	1.7	10.4	9.7	13.1	12.1	13.1
carboxylic acid	-C(=O)OH	13.4	1.21	2.25	2.25	2.25	2.25
1,1,3-trisubst urea	>NC(=O)NH-	[0.2]	-12.8	-24	6		

[a]Values in brackets are tentative assignments

Table III C. Heteroatoms and Functional Groups Within a Ring[a]

Cyclic Functional Group		Group Value, G_k $J \cdot mol^{-1} \cdot K^{-1}$	Cyclic Functional Group		Group Value, G_k $J \cdot mol^{-1} \cdot K^{-1}$
cyclic ether	>O$_c$	1.2	cyclic tert. amide	-C(=O)NR-	-21.7
cyclic ketone	>C$_c$(=O)	-1.4	cyclic carbamate	-OC(=O)N-	[-5.2]
cyclic ester	-C(=O)O-	3.1	cyclic anhydride	-C(=O)OC(=O)-	2.3
cyclic sp^2 nitrogen	=N$_c$-	0.5	N-substituted		
cyclic tert. amine	-N$_c$<	-19.3	cyclic imide	-C(=O)NRC(=O)-	[1.1]
cyclic tert. amine			cyclic imide	-C(=O)NHC(=O)-	[1.4]
-N-nitro	>N$_c$(NO$_2$)	-27.1	cyclic sulfide	>S$_c$	2.9
cyclic tert. amine			cyclic disulfide	-SS-	[-6.4]
-N-nitroso	>N$_c$(N=O)	-27.1	cyclic disulfide		
cyclic sec. amine	>N$_c$H	2.2	S-oxide	-SS(O)-	[1.9]
cyclic tert. amine			cyclic sulphone	>S$_c$(O)$_2$	[-10.4]
-N-oxide	>N$_c$(O)-	[-22.2]	cyclic		
cyclic azoxy group	N=N(O)-	[2.9]	thiocarbonate	-OC(=O)S-	[14.2]
cyclic sec. amide	-C(=O)NH-	2.7	cyclic quat. Si	>Si$_c$<	-34.7

[a]Values in brackets are tentative assignments; R refers to alkyl and aryl groups.

for these four groups, C_j, are available for molecules containing up to six functional groups. Selection of the appropriate value of C_j for one of these four functional groups is based on the total number of functional groups in the molecule. All available evidence suggests that the group coefficient for C_6 in Table III B, is adequate for molecules containing more than a total of six functional groups (*17*).

Acyclic and Aromatic Hydrocarbon Derivatives. The estimations of 2,2',3,3',5,5'-hexachlorobiphenyl, 3-heptylamino-1,2-propanediol, trifluoromethanethiol and 2,3-dimethylpyridine, shown in Table IV A, illustrate the estimations of substituted aromatic and acyclic hydrocarbon derivatives.

Table IV. Estimations of Total Phase Change Entropies and Enthalpies

A. Substituted Aromatic and Aliphatic Molecules[a]

$C_{12}H_4Cl_6$ 2,2',3,3',5,5'-hexachlorobiphenyl

T_{fus}: 424.9 K (*18*)

$\Delta_0^{T_{fus}} S_{tpce}$: 66.8 (68.7)

$\Delta_0^{T_{fus}} H_{tpce}$: 28.4 (28.2)

$\Delta_0^{T_{fus}} S_{tpce}$: {6[1.5][10.8] +8[-7.5]+4[7.4]}

$C_{10}H_{23}NO_2$ 3-heptylamino-1,2-propanediol

—OH
—OH
—NH
 |
$CH_3-(CH_2)_6$

T_{fus}: 324.9 K (*19*)

$\Delta_0^{T_{fus}} S_{tpce}$: 105.4 (88.6)

$\Delta_0^{T_{fus}} H_{tpce}$: 34.2 (28.8)

$\Delta_0^{T_{fus}} S_{tpce}$: {2[9.7][1.7]+ [-5.3]+2[7.1]+[17.6]+ 6[1.31][7.1]+[-16.4][.6]}

CHF_3S trifluoromethanethiol

CF_3SH

T_{fus}: 116.0 K (*13*)

$\Delta_0^{T_{fus}} S_{tpce}$: 39.9 (42.4)

$\Delta_0^{T_{fus}} H_{tpce}$: 4.6 (4.9)

$\Delta_0^{T_{fus}} S_{tpce}$: {[-34.8][.66] +3[13.3]+[23.0]}

C_7H_9N 2,3-dimethylpyridine

T_{fus}: 258.6 K (*20*)

$\Delta_0^{T_{fus}} S_{tpce}$: 49.1 (52.1)

$\Delta_0^{T_{fus}} H_{tpce}$: 12.7 (13.5)

$\Delta_0^{T_{fus}} S_{tpce}$: {2[17.6]+[10.9] +3[7.4]+2[-9.6]}

B. Substituted Cyclic Molecules[a]

$C_{12}H_7ClO_2$ 1-chlorodibenzodioxin

T_{fus}: 378.2 K (*21*)

$\Delta_0^{T_{fus}} S_{tpce}$: 58.2 (61.3)

$\Delta_0^{T_{fus}} H_{tpce}$: 22.0 (23.2)

$\Delta_0^{T_{fus}} S_{tpce}$: {[33.4]+3[3.7] +2[1.2]+4[-12.3]+7[7.4]+ [-7.5]+[1.5][10.8]}

C_3H_3NS thiazole

T_{fus}: 239.5 K (*13*)

$\Delta_0^{T_{fus}} S_{tpce}$: 35.0 (40.0)

$\Delta_0^{T_{fus}} H_{tpce}$: 8.4 (9.6)

$\Delta_0^{T_{fus}} S_{tpce}$: {[33.4]+2[3.7] +[2.9]+[0.5]+ 3[-1.6][1.92]}

$C_6H_8N_2O_2$ 1,3-dimethyluracil

T_{fus}: 398 K (*13*)

$\Delta_0^{T_{fus}} S_{tpce}$:30.2 (36.7)

$\Delta_0^{T_{fus}} H_{tpce}$:12.0(14.6)

$\Delta_0^{T_{fus}} S_{tpce}$: {[33.4]+ 3[3.7]+2[17.6]+ 2[-1.6][1.92] + 2[-21.7]}

$C_{21}H_{28}O_5$ prednisolone

T_{fus}: 513 K (*22*)

$\Delta_0^{T_{fus}} S_{tpce}$: 76.7 (75.8)

$\Delta_0^{T_{fus}} H_{tpce}$: 39.3 (38.9)

$\Delta_0^{T_{fus}} S_{tpce}$:{4[33.4]+[4.6] +5[3.7]+2[17.6]+[7.1] +2[-1.6][1.92]+[-1.6]+ [-12.3]+4[-14.7]+[-1.4]+ 3[-34.6]+3[1.7][12.1]}

---

[a]Units for $\Delta_0^{T_{fus}} S_{tpce}$ and $\Delta_0^{T_{fus}} H_{tpce}$ are J·mol⁻¹·K⁻¹ and kJ·mol⁻¹, respectively; experimental values are given in parentheses and references are in italics.

2,2',3,3',5,5'-Hexachlorobiphenyl. The estimation of 2,2',3,3',5,5'-hexachlorobiphenyl illustrates an estimation of a substituted aromatic molecule. Selection of the appropriate value for a quaternary aromatic sp^2 carbon from Table IA depends on the nature of the functional group. If the functional group at the point of attachment is sp^2 hybridized or contains non-bonding electrons, the value for a "peripheral aromatic sp^2 carbon adjacent to an sp^2 atom" is selected. The remainder of the estimation follows the guidelines outlined above with the exception that chlorine is one of the four functional groups whose group coefficient, C_j, depends on the degree of substitution (six in this example).

3-(n-Heptylamino)-1,2-propanediol. The estimation of 3-(n-heptylamino)-1,2-propanediol illustrates another example of a molecule where the number of consecutive methylene groups exceeds the number of other functional groups. As noted previously, the group coefficient for a methylene group, C_{CH_2}, is only applied to the consecutive methylene groups. The remaining two methylene groups are treated normally and are not counted in Σn_i (equation 2). One final comment about this estimation. The group coefficient for the hydroxyl group, C_3, was chosen despite the fact that the molecule contains two hydroxyl groups. In general, a C_j value is chosen based on the total number of functional groups present in the molecule and in this case $j_{OH}(3)$ is used.

Trifluoromethanethiol. The estimation of $\Delta_0^{T_{fus}} S_{tpce}$ for trifluoromethanethiol illustrates an example of a molecule containing fluorine. The group value for a fluorine on a trifluoromethyl group in Table III A is given per fluorine atom. The contribution of the quaternary carbon atom when attached to functional groups is attenuated by the group coefficient, C_i. Inclusion of the group value for a thiol completes this estimation.

2,3-Dimethylpyridine. The estimation of $\Delta_0^{T_{fus}} S_{tpce}$ for 2,3-dimethylpyridine in Table IV provides an example of a calculation for a heterocyclic aromatic compound. Other aromatic heterocyclic molecules related to pyridine are estimated similarly, regardless of the number of nitrogens in the aromatic ring and their location. Molecules that can exist in two tautomeric forms such as dihydroxypyrimidine (uracil), should be calculated on the basis of the form which dominates the equilibrium.

Cyclic and Polycyclic Hydrocarbon Derivatives. The protocol for estimating the total phase change properties of cyclic and polycyclic molecules also follows from the procedure described above for the corresponding hydrocarbons. In cyclic molecules, the substituent or functional group may be attached to the ring or it may be part of the ring. If the functional group is part of the ring, the group values listed in Table III C are to be used. The procedure first involves estimating $\Delta_0^{T_{fus}} S_{tpce}$ for the corresponding hydrocarbon ring, then correcting for the heterocyclic component(s), and if necessary, correcting the ring carbons attached to the cyclic functional group by the appropriate group coefficients. This is illustrated in Table IV B by the following examples.

1-Chlorodibenzodioxin. 1-Chlorodibenzodioxin is treated as being a derivative of cyclohexane. According to equation 7, the ring equation is first used to estimate the contributions of the dioxane ring. This ring contains two cyclic ether oxygens and four quaternary cyclic sp^2 carbon atoms and must be modified accordingly. The remaining 8 carbon atoms are treated as aromatic carbons and values appropriate to their substitution pattern are chosen. The addition of the contribution of the chlorine completes the estimation.

Thiazole. Thiazole is estimated in a similar fashion. The ring equation (equation 3) is used first to generate the contribution of the five membered ring. In this instance the ring has been modified by the addition of a cyclic sulfur atom and a cyclic sp^2 hybridized nitrogen atom. Both substitutions require appropriate corrections. The hybridization pattern of the remaining three ring carbon atoms have likewise been changed from the hybridization and substitution pattern found in cyclopentane and these changes must also be included in $\Delta_0^{T_{fus}} S_{tpce}(corr)$. Each cyclic sp^2 hybridized carbon atom is attached directly to one of the functional groups. The group coefficient, which in this case differs from 1.0, must also be included in evaluating the contributions of the ring carbons.

1,3-Dimethyluracil. Estimations of 1,3-dimethyluracil reqiure some thought in properly identifying the functional groups in the molecule. The functional group that makes up a portion of the ring in this molecule can not be found directly in Table III C. It must therefore be simplified and this simplification can be accommodated in various ways. The functional group can be considered to be a combination of either an adjacent cyclic imide (-CONRCO-) and cyclic amide nitrogen (-NR-), a cyclic urea (-NRCONR-) and amide carbonyl (-CO-), or two cyclic tertiary amides. An examination of the available groups in Table III C will reveal that although a group value for a N-substituted cyclic imide is available, there is no appropriate group available for an N-substituted cyclic nitrogen of an amide. Similarly, group values for a cyclic urea and amide carbonyl are not available. The most appropriate group value that is available is for a cyclic tertiary amide. Once the appropriate group is identified, the procedure follows the same protocol established for thiazole.

Prednisolone. The estimation of the fusion enthalpy of prednisolone illustrates an example of an estimation of a complex polycyclic compound. This tetracyclic 17 atom ring system (4[33.4]+5[3.7]) contains three cyclic quaternary centers (3[-34.6], four cyclic tertiary sp^3 centers, (4[-14.7]), three cyclic tertiary sp^2 centers, two of which are attached to a functional group (2(1.92)+1)[-1.6], a quaternary sp^2 center ([-12.2]) as well as a cyclic carbonyl group ([-1.4]). Addition of these modifications to the ring equation estimates the contributions of the ring. Addition of the contributions of the substituents which include three hydroxyls ((3)(12.1)[1.7]), two methyls (2[17.6]), a methylene ([7.1]) and a carbonyl group of an acyclic ketone ([4.6]) completes the estimation. The molecule contains five functional groups, hence $j_{OH}(5)$ is used.

Polymers. In addition to the estimation of $\Delta_0^{T_{fus}} S_{tpce}$ of small molecules, the parameters of Tables I and III can be used to predict $\Delta_0^{T_{fus}} S_{tpce}$ and $\Delta_0^{T_{fus}} H_{tpce}$ of crystalline polymers when the experimental melting point is known. Since the parameters in Tables I and III differ somewhat from those reported previously, the predictions of equations 2-6 will likewise produce slightly different results than reported previously (*17*). However a similar overall correlation between experimental and calculated results should be obtained by these modified parameters. The protocol used to evaluate $\Delta_0^{T_{fus}} S_{tpce}$ of polymers is exactly the same as outlined above with the exception that the enthalpic or entropic value is calculated on the basis of the structure of the repeat unit of the polymer. As examples, the calculated and experimental values (in brackets) of $\Delta_0^{T_{fus}} S_{tpce}$ are provided for the following: polyethylene (CH_2), 9.3 [9.9]; polytetramethylene terephthalate: 61.8 [58.6], nylon [6,12]: 152.2 [154]. Experimental values have been taken from the literature (23).

Statistics of the Correlation. The group values included in Tables I and III were generated from the fusion entropies of a total of 1862 compounds. The absolute average and fractional errors between experimental and calculated $\Delta_0^{T_{fus}} S_{tpce}$ and $\Delta_0^{T_{fus}} H_{tpce}$ values for these 1862 compounds were 9.8 $J \cdot mol^{-1} \cdot K^{-1}$ and 3.48 $kJ \cdot mol^{-1}$, and 0.152 and 0.168, respectively. The standard deviations between experimental and calculated values for $\Delta_0^{T_{fus}} S_{tpce}$ and $\Delta_0^{T_{fus}} H_{tpce}$ were ±13.0 $J \cdot mol^{-1} \cdot K^{-1}$ and ± 4.84 $kJ \cdot mol^{-1}$, respectively. An additional 62 compounds with errors exceeding 3 standard deviations were excluded from the correlations and from the histogram of Figure 1. A

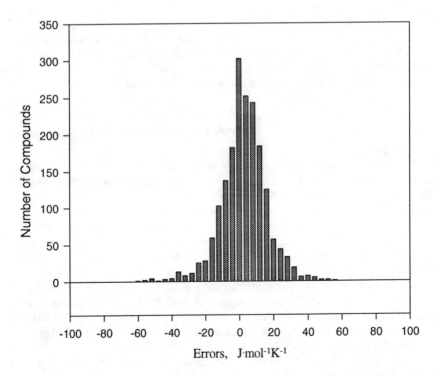

Figure 1. A histogram illustrating the distribution of errors in estimating $\Delta_0^{T_{fus}} S_{tpce}$.

similar histogram was obtained for $\Delta_0^{T_{fus}} H_{tpce}$ (not shown). Values reported in brackets in Table III should be considered as tentative assignments.

Vaporization Enthalpies

Vaporization enthalpy is a thermochemical property that can be estimated quite accurately. Many estimation methods in the chemical engineering literature, as reported by Rechsteiner, Jr. (24) and others (25), are reported accurate to a few %. Most require critical constants and other parameters which themselves may have to be estimated. In addition, many of these methods have been developed to provide vaporization enthalpies near or at the boiling point.

Numerous group additivity procedures have been reported for estimating the enthalpy of vaporization, $\Delta_{vap}H_m^{\circ}$ (298.15 K) (24-37). Most group methods have been developed to provide vaporization enthalpies at 298.15 K although more recent work has focused on the development of group methods applicable to a wider range of temperatures (25, 26). Similar accuracies of a few % have been reported by their developers. While some vaporization enthalpies are known quite accurately, the n-alkanes from C_5-C_{18} for example (38), most $\Delta_{vap}H_m^{\circ}$ (298.15 K) values in the literature are probably accurate to about 3-5% of the value reported. This 3-5% uncertainty reflects both experimental errors and errors introduced as a result of correcting the vaporization enthalpy from the mean temperature of measurement to 298.15 K. Consequently, any general estimation method which attempts to reproduce experimental data to better than a few % will obviously be affected by the limited amount of accurate experimental data available and the applicability of the method is likely to be highly focused. The 3-5% experimental uncertainty should serve as a useful lower limit of the typical error to be expected from an estimation technique developed to reproduce $\Delta_{vap}H_m^{\circ}$ (298.15 K) of a wide range of substances using a reasonable number of parameters.

Selection of an estimation technique will generally be guided by a number of factors. The method of choice will depend on the temperature or temperature range of interest, the level of user sophistication necessary to perform the estimation, the required accuracy of the estimation and the availability of appropriate group values or other parameters. A major limitation of most group methods is the lack of a sufficient number of group values that can be applied to cover the broad spectrum of molecular structures that are of interest. For this reason, we decided to develop an alternative method to a group additivity approach (34-37). This method described below uses fewer parameters than most group methods and is quite flexible with regards to the carbon architecture that it can successfully model. Recently we compared this method to those methods reported by Guthrie and Taylor (29) and Ducros et. al. (30-32) to a series of hydrocarbon derivatives containing a single functional group (37) and also for more complex molecules (36).

A series of 48 monosubstituted hydrocarbons was randomly selected from a database of 433 (39-40). Group values for these compounds were unavailable for 9 of the 48 compounds using Guthrie's method and 23 of the 48 compounds using Ducros' method. When group values were available, Ducros' method was generally the most accurate resulting in the best value 18 out of 25 times while Guthrie's method gave the best agreement for 8 of the 39 compounds. The method described below gave the best agreement 28 out of 48 times. Identical predictions were obtained in some instances for certain compounds after the values were rounded off to the nearest 0.1. The average absolute error of the 25 compounds estimated by Ducros' method was 0.9 kJ·mol⁻¹; 2.9 kJ·mol⁻¹ was the average absolute error for the 39 compounds estimated by Guthrie's method. This compares to an average absolute error of 1.76 kJ·mol⁻¹ for the 48 compounds estimated by the method to be described below.

A second set of 30 compounds containing two or more functional groups was also compared (*39-40*). We were not able to reproduce the precise values reported by Ducros for those compounds whose functional groups depended on the function γ. Vaporization enthalpy predictions for these compounds were obtained from the tables provided. Ducros' method resulted in the best value 13 out of 25 times with a standard deviation between experimental and calculated values of ± 2.9 kJ·mol⁻¹ while Guthrie's method gave the best agreement for 9 of the 28 compounds with a standard deviation of ± 5.0 kJ·mol⁻¹. The estimation method described below gave the best agreement 14 out of 30 times with a standard deviation of ± 3.9 kJ·mol⁻¹.

Hydrocarbons. A number of simple equations have been developed for the estimation of the vaporization of hydrocarbons. Equation 8, originally reported by Morawetz (A, B values) (*41*) was derived from the enthalpy of vaporization of the n-alkanes. The symbol n_{CH_2} refers to the number of methylene groups. This equation, recently modified to reflect both refinements and inclusion of additional vaporization data (A', B' values) (*42*), is capable of reproducing the known vaporization enthalpies of the n alkanes from pentane to triacontane with a standard error of ± 2.1 kJ·mol⁻¹ (1σ).

$$\Delta_{vap}H_m^\circ(298.15 \text{ K})/(\text{kJ·mol}^{-1}) = A \cdot n_{CH_2} + B; \quad A= 4.97; B=1.61; A'=5.43; B'=-3.3. \quad (8)$$

A similar equation, equation 9, containing only three parameters was found to reproduce the vaporization enthalpies of any hydrocarbon with 20 or fewer carbons, regardless of structure, with an error of approximately ± 4.2 kJ·mol⁻¹. The symbols n_C and n_Q refer to the total number of carbons and the total number of quaternary sp³ hybridized carbon atoms where the definition of quaternary is based, as above, on the number of hydrogens attached to carbon. This equation can also be used on molecules containing more than 20 carbons but the error appears to be larger (*7, 42*).

$$\Delta_{vap}H_m^\circ(298.15 \text{ K})/(\text{kJ·mol}^{-1}) = (4.69 \pm 0.08) \cdot (n_C - n_Q) + (1.3 \pm 0.2) \cdot n_Q + (3.0 \pm 0.2) \quad (9)$$

Simple Hydrocarbon Derivatives. Vaporization enthalpies of compounds that contain a single functional group can be estimated by using the functional group values in Tables V A-C, the correction terms of Table V D and equation 10:

$$\Delta_{vap}H_m^{\circ}(298.15 \text{ K})/(\text{kJ·mol}^{-1}) = 4.69·(n_C - n_Q) + 1.3·n_Q + (3.0) + n_M·M + b + C \quad (10)$$

Application of equation 10 to estimate $\Delta_{vap}H_m^{\circ}(298.15 \text{ K})$ of a particular hydrocarbon derivative is quite straightforward. Once the number and type of carbon atoms are properly identified, the contribution of the functional group is included next. Consider as an example the value of a carbonyl group in a ketone. The value b of 10.5 kJ·mol^{-1} represents the additional contribution of the carbonyl oxygen since the contribution of the carbon has already been accounted for by n_C. For compounds containing silicon, germanium and tin, the metal atom is treated like carbon but with a contribution, M, that depends on the number of such atoms, n_M, and the substitution pattern as indicated in the lower portion of Table V A. For compounds containing a single fluorine substitutent, the value of the fluorine is chosen on the basis of the hybridization of the atom to which it is attached. Values for fluorine attached to carbon and silicon are listed in Table V C.

Additional Correction Terms. The terms listed in Table V D were introduced primarily to correct for steric effects on the solvation of the functional group in the neat liquid. Carbon branching near the functional group generally increases steric interactions and reduces intermolecular solvation of the functional group as do *ortho* carbon branches on a ring. Carbon branching also decreases the solvent accessible surface area and this can result in a decrease in the magnitude of the vaporization enthalpy (7). Inclusion of a functional group as part of a ring generally decreases the steric environment around the group thereby allowing better intermolecular interaction. This structural feature usually results in an increase in vaporization enthalpy. These factors, while small, should be taken into consideration when applying these correction terms to molecules of interest.

Applications. Some applications of equation 10 are illustrated in the examples of Table VI. Group values for tertiary amines, fluorine, and the organometallic compounds are new. Details concerning these values will be published elsewhere. Group values given in brackets are considered tentative assignments.

Triisobutyl amine. Estimation of the vaporization enthalpy of triisobutyl amine in Table VI illustrates the use of the group values and correction terms listed in Table VA and V C. Application of equation 10 without the correction term affords a vaporization enthalpy of 65.9 kJ·mol^{-1} which would be the vaporization enthalpy calculated for tributyl amine (lit. 66.5 kJ·mol^{-1} (43) all vaporization enthalpies obtained from this source were calculated from vapor pressures given by the Antoine Equation over a 30 K temperature range from a ln P vs 1/T treatment according to the Clausius Clapeyron equation followed by correction of vaporization enthalpy to 298.15 K using equation 14 as is described below). Application of the branching correction for each branch completes the estimation for triisobutyl amine.

Bicyclo[3.3.0]octan-2-one. The estimation of *cis* and *trans* bicyclo[3.3.0]octan-2-one illustrates an example of an estimation where the functional group is part of a ring. Application of equation 10 without the correction terms results in a value of 51.0 kJ·mol^{-1}. Note that the carbonyl carbon according to our definition

Table V A. Functional Group Contributions to Vaporization Enthalpies

Class of Compounds	Functional Group Class		b	Class of Compounds	Functional Group Class		b
acid	-C(=O)OH	I	38.8	iodide	-I	I	18.0
alcohol	-OH	I	29.4	ketone	>C=O	II	10.5
aldehyde	-CHO	I	12.9	nitrile	-CN	I	16.7
amide [mono-				nitro	$-NO_2$	I	22.8
subst.]	-C(=O)NH-	II	42.5	heterocyclic aromatic			
amine [pri.]	$-NH_2$	I	14.8	nitrogen	=N-	II	[12.2]
amine [sec.]	-NH-	II	8.9	sulfide	>S	II	13.4
amine [tert.]	>N-	II	6.6	disulfide	-SS-	II	[22.3]
bromide	-Br	I	14.4	sulfoxide	>SO	II	[42.4]
chloride	-Cl	I	10.8	sulfone	$-SO_2-$	II	[53.0]
ester	-C(=O)O-	II	10.5	thiolester	-C(=O)S-	II	[16.9]
ether	>O	II	5.0	thiol	-SH	I	13.9

Organometallics			**M**				**M**
prim. silane	$-SiH_3$	I	7.8	prim. germane	$-GeH_3$	I	10.8
sec. silane	$>SiH_2$	II	3.9	sec. germane	$>GeH_2$	II	[9.8]
tert. silane	>SiH-	II	3.4	quat. germane	>Ge<	II	6.6
quat. silane	>Si<	II	1.8	quat. stannane	>Sn<	II	[10.9]

Table V B. Functional Group Contributions to Vaporization Enthalpies in Molecules with Multiple Functional Groups

Substitution Pattern[a]	Substitution Factor, F_i	Substitution Pattern	Substitution Factor, F_i
single substitution on a		1,1-disubstitution on a	
primary sp^3 atom	1.62[b]	secondary sp^3 atom	0.94
secondary sp^3 atom	1.08	tertiary sp^3 atom	0.78
tertiary sp^3 atom	0.60	quaternary sp^3 atom	0.55
quaternary sp^3 atom	0.79	quaternary sp^2 atom	0.56
tertiary sp^2 atom	0.69		
quaternary sp^2 atom	0.85	1,1,1-trisubstitution	
quaternary sp atom	0.3	tertiary sp^3 atom	0.81
		quaternary sp^3 atom	0.62
		1,1,1,1-tetrasubstitution	
		quaternary sp^3 atom	0.59

[a]Primary, secondary, tertiary, and quaternary positions are defined by the number of hydrogens attached to the atom bearing the substituent, 3,2,1,0, respectively. [b]A value of 0.79 replaces 1.62 for compounds containing silicon, germanium and tin.

Table V C. Fluorine Group Contributions to Vaporization Enthalpies

Fluorine as a single substituent	b	Fluorine as one of several substituents	b
a single fluorine		single or multiple fluorine atoms	
on any sp^2 C (or Si)	1.2	on a 1,1-disubstituted sp^3 C (or Si)	3.1
on any sp^3 C (or Si)	7.1	on a 1,1,1-trisubstituted C (or Si)	1.9
		on a 1,1,1,1-tetrasubstituted C	1.1
		on a 1,1,1,1-tetrasubstituted Si	3.2

Table V D.   Correction Terms for Monosubstituted and Multisubstituted Hydrocarbons

Nature of the Correction C	Correction $(kJ \cdot mol^{-1})$	Nature of the Correction C	Correction $(kJ \cdot mol^{-1})$
Ring correction for cyclic Class II functional groups including cyclic ethers, cyclic ketones, cyclic secondary amines, and cyclic sulfides	2.9 [a]	Alkyl branching on acyclic sp^3 carbons	-2.0[b]
		Ortho and vicinal alkyl branching on sp^2 and sp^3 carbons on 5 and 6 membered rings	-2.0[b]

Table V E.   Additional Corrections Term for Multisubstituted Compounds

Nature of the Correction C	Correction $(kJ \cdot mol^{-1})$	Nature of the Correction C	Correction $(kJ \cdot mol^{-1})$
Intramolecular hydrogen bonding for alcohols (5-9 membered rings)	-7.6	Intramolecular hydrogen bonding for β diketones	-18.0

[a]One correction per molecule; [b]branching and *ortho* alkyl branching corrections are applied for each carbon branch; branching due to an acyclic quaternary carbon center is counted as one branch; branching due to a cyclic quaternary carbon center is ignored; a branch resulting from attachment of a functional group is ignored.

is quaternary but also sp^2 hybridized and therefore is treated normally. The carbonyl group is also part of a ring. The ring correction increases the vaporization enthalpy to 53.9 kJ·mol⁻¹. In addition, the carbonyl group is *ortho* to a five membered ring. Since this second ring is part of a fused ring system, the *ortho* correction is not applied. A molecule like 2-methylcyclopentanone however should have both the ring and *ortho* correction applied.

Table VI. Estimation of Vaporization Enthalpies (kJ·mol^{-1})

$C_{12}H_{27}N$ triisobutylamine

$\Delta_{vap}H_m^\circ$(298.15 K)

lit: 56.4 (43)
calcd: 59.9
{[4.69]12+[3.0]+
[6.6] -3[2]}

$C_8H_{12}O$ bicyclo[3.3.0]octan-2-one

$\Delta_{vap}H_m^\circ$(298.15 K)

cis lit: 54.4 (39)
trans lit: 53.6 (39)
calcd: 53.9
{[4.69]8+[3.0]+[10.5]
+[2.9]}

$C_{11}H_{14}O$ 2,4,6-trimethylacetophenone

$\Delta_{vap}H_m^\circ$(298.15 K)

lit: 62.3 (39)
calcd: 61.1
{[4.69]11+[3.0]+
[10.5]-2[2.0]}

C_3H_9FSi fluorotrimethylsilane

$\Delta_{vap}H_m^\circ$(298.15 K)

lit: 25.7 (43)
calcd: 26.0
{[4.69]3+[3.0]+[1.8]+
[7.1]}

$C_5H_{12}O_2$ 2-isopropoxyethanol

$\Delta_{vap}H_m^\circ$(298.15 K)

lit: 50.2 (43)
calcd: 54.8
{[4.69]5+[3.0]+
[29.4](1.08)+[-7.6]
+[5.0](1.08+0.6)/2}

$C_8H_{15}ClO_2$ 2-methylpropyl 3-chlorobutanoate

$\Delta_{vap}H_m^\circ$(298.15 K)

lit: 52.3 (43)
calcd: 56.3
{[4.69]8+[3.0]+
[10.5]1.08+[-2.0]
[10.8](0.6)}

$C_6H_{12}O_2$ 5,5-dimethyl-1,3-dioxane

$\Delta_{vap}H_m^\circ$(298.15 K)

lit: 41.3 (43)
calcd: 40.8
{[4.69]5+[3.0]+[1.3]+
2[5.0](1.08+0.94)/2+[2.9]}

$C_6H_{16}O_2Si$ diethyldimethoxysilane

$\Delta_{vap}H_m^\circ$(298.15 K)

lit: 39.3 (43)
calcd: 39.6
{[4.69]6+[3.0]+[1.8]
+2[5.0](0.79+0.55)/2}

CCl_3F trichlorofluoromethane

$\Delta_{vap}H_m^\circ$(298.15 K)

lit.:   26.2 (43)
calcd: 24.1
{[3.0]+[1.3]+
0.59([1.1]+3[10.8])}

C_4BrF_6N bromo-N,N-*bis*(trifluoromethyl)ethynylamine

$\Delta_{vap}H_m^\circ$(298.15 K)

lit.:   31.9 (43)
calcd: 26.5
{[4.69]2+2[1.3]+[3.0]
+6[1.1](0.59)+
(0.3)[14.4]+
[6.6](0.3+2(0.59))/3}

2,4,6-Trimethylacetophenone. The estimation of 2,4,6-trimethylacetophenone illustrates an example with two *ortho* interactions. In this estimation, once the contributions of the carbons atoms and the functional group is evaluated, the *ortho* interaction can be evaluated in a straightforward fashion.

Fluorotrimethylsilane. The estimation of the organometallic compounds listed in the bottom of Table V A are estimated in the same manner as other organic molecules with one exception. The metals are not treated as functional groups but simply as replacements for carbon. Thus a molecule like fluorotrimethylsilane is treated like t-butyl fluoride, a molecule containing a single functional group. The branching correction has been incorporated into the group value for a quaternary silicon.

Polysubstituted Hydrocarbon Derivatives. The protocol established to estimate vaporization enthalpies of molecules containing two or more of the functional groups listed in Table V follows from the protocol established for hydrocarbons and singly substituted derivatives (*36*). Application of this protocol results in equation 11.

$$\Delta_{vap}H_m^{\circ}(298.15K)/(kJ \cdot mol^{-1}) = 4.69 \cdot (n_C - n_Q) + 1.3 \cdot n_Q + n_M \cdot M + \sum_i n_i \cdot F_i \cdot b_i + (3.0) + C \quad (11)$$

The contribution of carbon and any metal components is estimated as previously described for singly substituted compounds. The contribution of a functional group to the vaporization enthalpy of a multifunctional compound depends on both the nature (b) and location (F) of the functional group. The nature of a functional group, characterized by the constant b, has already been discussed. The substitution factors, F_i, reported in Table V B, take into account the location of the functional group in the molecule. Hybridization and substitution characteristics are used as identifiers of the steric environment of the functional group and its ability to interact intermolecularly. Most substitution factors attenuate the contribution of the functional group. Two classes of functional groups are identified in Table V A. Class I functional groups refer to monovalent groups while class II groups refer to multivalent groups. Substitution factors for class I functional groups depend solely on the hybridization and substitution of the carbon to which the functional group is attached. Substitution factors for class II functional groups are dependent on the hybridization and substitution pattern of two or more carbon atoms or their equivalent. The arithmetic mean of each of the substitution factors is used as the modifier to b in this case.

Most of the substitution factors reported in Table V B are identical to those reported previously (*36*) except in cases where a tentative value was reported. One substitution factor, the value for a primary sp³ carbon atom, has a value that depends on the chemical composition of the molecule in question. The value typically used for a primary sp^3 carbon atom, 1.62, is replaced by 0.79 in estimations of organosilanes. The same value, 0.79, should also be used for a primary sp^3 carbon in estimations of organogermanes and organostannanes, although this conclusion is based on far less experimental data.

The contribution of fluorine in organofluorine compounds containing multiple fluorines or other substituents can be obtained directly from the group values in Table V C. The contribution of fluorine to the vaporization enthalpy depends on the number of substituents attached to the same carbon or silicon atom. Once the appropriate b value is identified, the estimation of organofluorine compounds follows the same protocol as established for other functional groups. Values for fluorine substitution are new and some substitution factors that were tentatively assigned previously (36) have changed due to the inclusion of new data in the correlations.

Additional Correction Terms. In addition to the functional group values and substitution factors listed in Tables V A and B, additional correction terms applicable to polyfunctional compounds are also included in Table V D. An important correction applicable to polyfunctional molecules is for the formation of intramolecular hydrogen bonds. Two correction terms are available. One correction is applicable to any alcohol capable of forming an intramolecular hydrogen bond to oxygen by means of a 5-9 membered ring (including the hydrogen atom). A second correction is available specifically for intramolecular hydrogen bonds formed by the enolic form of β-diketones. Intramolecular hydrogen bonding corrections for other functional groups such as amines or thiols do not appear to be necessary. There is some evidence that suggests that this correction should be applied to hydroxyl groups intramolecularly hydrogen bonded to nitrogen in amines.

The inclusion of substitutent factors in equation 11 reduces the instances where branching corrections are necessary. Branching and the *ortho* correction are necessary in multifunctional compounds only when the branch occurs at a carbon atom that is not directly attached to any functional group but clearly affects intermolecular interactions of the functional group.

Applications. Some applications using equation 11 are illustrated in the last six examples of Table VI.

2-Isopropoxyethanol. 2-Isopropoxyethanol is an example of a molecule containing two functional groups and one correction term, a correction for intramolecular hydrogen bonding. The contributions of the carbons and the constant account for the first two terms in the estimation. The hydroxyl group is a class I functional group, and in this instance, is connected to an secondary sp^3 carbon. The contribution of the ether oxygen, a class II functional group, is obtained from the product of the group value for an ether and an averaged substitution factor based on the two carbon environments at the point the ether oxygen is attached. Correction for the intramolecular hydrogen bond completes the estimation. There is no branching correction applied because the branch is not remote but occurs at the point of substitution of a functional group and is corrected by the substitution factor for a tertiary sp^3 carbon (0.6).

2-Methylpropyl 3-chlorobutanoate. Estimation of the vaporization enthalpy of 2-methylpropyl 3-chlorobutanoate follows a similar protocol. The first two terms in the estimation account for the contributions of the carbon backbone and the constant. The third term accounts for the contribution of the ester group. In this case the structural environment at the atoms to which the -C(=O)O- group is attached is the same, both are secondary sp^3 carbons. The contribution of the chlorine is attenuated by its structural environment. Finally, this molecule contains a remote acyclic carbon branch which is not corrected by the substitution factor as it is in 2-isopropoxyethanol. This correction is included as the fourth term in the estimation. It should be pointed out that there are two functional groups that can be influenced by carbon branching. Based on the rationale presented earlier for justification of this correction term, it can be argued that this correction should be applied once for each functional group in the molecule. While the estimation would improve in this case if this correction were applied twice, there are not sufficient experimental data available at present to justify this argument. At present, we recommend applying this correction term once for each acyclic carbon branch in the molecule, regardless of the number of functional groups that are present.

5,5-Dimethyl-1,3-dioxane. The estimation of a cyclic molecule, 5,5-dimethyl-1,3-dioxane, follows the same protocol. In this instance the molecule contains a quaternary sp^3 carbon atom. The "ortho or vicinal" branching correction for cyclic molecules is not applicable here since the methyl groups are remote from the ether oxygens. Neither is the branching correction (see footnote b, Table V). Once the carbon atoms are accounted for, the contributions of the ether oxygens can be evaluated. In this instance, both oxygens are equivalent but are attached to two different environments, singly substituted and geminally substituted secondary sp^3 carbons. Finally, this molecule is a cyclic ether and requires a ring correction term. This correction term is applied once regardless of the number of oxygens in the ring.

Diethyldimethoxysilane. The first two terms in the estimation of diethyldimethoxy-silane account for the contribution of the carbon atoms and the constant and the third term accounts for the quaternary silicon. This molecule is considered to contain two functional groups. The contributions of the two ether oxygens are attenuated by their position of attachment. Note that the substitution factor used for a primary sp^3 carbon in an organosilane (0.79) is different than the value used for estimations in other compounds (1.62). Branching is not remote but occurs at the center of substitution by the functional groups and is corrected by the geminal substitution factor, 0.55.

Trichlorofluoromethane. The estimation of trichlorofluoromethane illustrates the use of the various group values for fluorine. The estimation consists of the contributions of the quaternary carbon and the constant, and the contributions of the three chlorines and single fluorine. All halogens are in the same structural environment. The group value selected for fluorine is the one for a tetrasubstituted carbon atom.

2-Bromo-N,N-bis(trifluoromethyl)ethynylamine. The estimation of the vaporization enthalpy of 2-bromo-N,N-*bis*(trifluoromethyl)ethynylamine is the last example of the diversity of molecular structure that can be handled by this approach. While all carbon atoms in this molecule are quaternary atoms based on our definition, only two are both quaternary and sp^3 hybridized. The contribution of the two pairs of carbon atoms to the vaporization enthalpy differ. The first two terms in the estimation are based on this distinction. The contributions of the functional groups, each attenuated by its location in the molecule as previously described, and the constant complete the estimation.

Statistics of the Correlation. The statistics of the correlation for hydrocarbons, monofunctional hydrocarbons and polyfunctional compounds have been reported previously (*34-36*). Typically, the vaporization enthalpies of hydrocarbons, monofunctional hydrocarbons, and polyfunctional compounds used in the data base (138, 433, 175 compounds, respectively) are reproduced to within 5% of the experimental data ($\Delta\Delta_{vap}H_m^{\circ}$(298.15 K)) (experimental - calculated) average deviation, hydrocarbons: ±2.5; monofunctional compounds: ±1.6; polyfunctional compounds: ±2.5 kJ·mol⁻¹). An additional group of compounds totaling 400 and containing the elements Si, Ge, Sn, F and N (in the form of tertiary amines) have been used to generate the parameters of the additional groups included in this discussion. These parameters were able to reproduce the experimental values of this data base within 8% (standard deviation ±4.0 kJ·mol⁻¹). These estimations have not yet been compared to those available from the group additivity approach described by Myers and Danner for organometallic compounds (*33*). The reader is encouraged to compare the predictions of alternative estimation methods whenever possible.

Sublimation Enthalpies

Several different empirical and theoretical approaches have been exploited in developing estimation techniques for sublimation enthalpies of solids. An optimized force field of general applicability for the calculation of crystal energies has been developed (*44*). Correlations have been found which allow an estimate of the sublimation enthalpy, $\Delta_{sub}H_m$ (298.15 K), from molecular parameters like the number of valence electrons and the van der Waals surface (*45*). Quantitative structure-sublimation enthalpy relationships have also been studied by neural networks (*46*), linear free energy relationships (*47*), and conformational force field analysis (CoMFA, (*48*)). Earlier work in this area also included a group additivity method reported by Bondi and various other related group incremental methods applicable for a related series of molecules (*49-51*).

The development of reliable means of estimating sublimation enthalpies is an extremely important goal in thermochemistry. Enthalpies of combustion can presently be measured with a precision of a few tenths of a percent. This in turn results in very precise heats of formation for many organic solids. Sublimation enthalpies are added to these enthalpies of formation to convert them to gas phase values. An examination of the sublimation literature reveals a situation where sublimation enthalpies are rarely accurate beyond 5% and for molecules with low volatility, discrepancies in sublimation enthalpies of 10 kJ·mol⁻¹ or more are not uncommon. In fact a survey of the reproducibility of experimental sublimation enthalpies of 44 compounds in the literature resulted in a standard deviation of the mean of 7.3 kJ·mol⁻¹ (*52*).

One of the most flexible approaches to estimating sublimation enthalpies is to take advantage of the thermodynamic cycle that relates sublimation enthalpy to the enthalpies of vaporization and fusion, equation 12. $\Delta_{fus}H_m(T_{fus})$ and $\Delta_{vap}H_m$ (298.15 K) refer to the molar enthalpy change in going from solid to liquid and from liquid to gas respectively. While this equation is an equality only for enthalpies measured at the

$$\Delta_{sub}H_m(298.15 \text{ K}) \approx \Delta_{fus}H_m(T_{fus}) + \Delta_{vap}H_m(298.15 \text{ K}) \tag{12}$$

same temperature, it generally serves as a good approximation when vaporization enthalpies at 298 K are used in conjunction with fusion enthalpies measured at the melting point. The ability to mix and match both experimental and estimated enthalpies, depending on availability, makes this approach particularly attractive. The application of equation 12 to estimate sublimation enthalpies of hydrocarbons has been documented previously (53). This section will attempt to illustrate various applications of equation 12 to estimate sublimation enthalpy.

Vaporization Enthalpies at 298.15 K. For many compounds that are solids at room temperature, experimental vaporization enthalpies or vapor pressures as a function of temperature are available above the melting point of the solid (43). Vaporization enthalpies of these compounds are often evaluated from the temperature dependence of vapor pressure (P) and are obtained from the slope of a ln P vs 1/T plot according to the Clausius-Clapeyron equation. These vaporization enthalpies are usually referenced to some mean temperature $\overline{T}$ evaluated either from the average value of the reciprocal, $2/[1/T_1+1/T_2]$ or from $(T_1+T_2)/2$ where T_1 and T_2 are the initial and final temperatures of the measurements, respectively. To use the vaporization enthalpy in equation 12, correction to 298.15 K may be necessary. Compendia of heat capacities for many organic liquids are available (54-57). However heat capacity data for the liquid state of the solid compound of interest at 298.15 K may be unavailable. Several estimation methods are available for estimating the heat capacities of the liquid and gas phases and these techniques can be used directly to arrive at a value for $\Delta_{vap}H_m$ (298.15 K)(58-60). An alternative and simpler method is through the use of equations 13 and 14. The term $\Delta_g^l C_p$(298.15 K) in equation 13 refers to the difference in heat capacities between the liquid and gas phases and the symbol C_{pl}(298.15 K) in

$$\Delta_g^l C_p(298.15 \text{ K}) = 10.58 \text{ J·mol}^{-1}\text{·K}^{-1} + 0.26 \, C_{pl}(\, 298.15 \text{ K}) \tag{13}$$

$$\Delta_{vap}H_m (298.15 \text{ K}) = \Delta_{vap}H_m (\overline{T}) + [10.58 + 0.26 C_{pl}(298.15 \text{ K})][\overline{T}-298.15] \text{ J·mol}^{-1} \tag{14}$$

equations 13 and 14 refers to the heat capacity of the liquid at 298.15 K which can be estimated by group additivity. An experimental value of C_{pl}(298.15 K) can be used if available. The relationship between $\Delta_g^l C_p$(298.15 K) and C_{pl}(298.15 K) was obtained by correlating differences in experimental heat capacities between the liquid and gas phases with the heat capacity of 289 organic liquids estimated by group additivity. This resulted in equation 13 (60, 61). While heat capacities of both the liquid and gas phases are temperature dependent, equation 14 is based on the assumption that $\Delta_g^l C_p$(298.15 K) will be independent of temperature. This assumption was tested by comparing the predictions of equation 14 to differences observed in experimental vaporization enthalpies of a series of compounds, each measured at temperature $\overline{T}$, and a reference temperature, usually 298.15 K. Vaporization enthalpies of a total of 126 organic compounds were examined. Vaporization enthalpies of these materials were reported over the temperature range 260-370 K and included a temperature near or at 298.15 K. Excluding compounds that form hydrogen bonds (15 of 126), the standard error associated with using equation 14 to correct the vaporization enthalpy measured at temperature $\overline{T}$ to the reference temperature, usually 298.15 K, was ± 490 J·mol⁻¹. This increased to ± 710 J·mol⁻¹ if molecules capable of hydrogen bonding were included in the correlation.

Sublimation enthalpies at 298.15 K. Although sublimation enthalpies at 298.15 K are necessary for correcting solid state enthalpy of formation data, the vapor pressure of many solids necessitate the measurement of sublimation enthalpies at other temperatures. This necessitates correcting these data back to 298.15 K. A number of simple equations have been used to adjust sublimation enthalpies to 298.15 K *(52)*. The term $\Delta_{sub}H_m(\overline{T})$ in equation 15 represents the sublimation enthalpy measured at some mean temperature $\overline{T}$ and R is the gas constant (8.31451 J·mol^{-1}·K^{-1}). Values for n of 2-6 have been used by various research groups *(52)*. An alternative approach to equation 15 and one which appears to give some improvement over equation 15 is equation 16. The symbol C_{pc}(298.15 K) refers to the heat capacity of the solid at 298.15 K in J·mol^{-1}·K^{-1}. Either experimental or estimated values of C_{pc}(298.15 K) can be used. The same assumption used in generating equation 13 was used here. Since

$$\Delta_{sub}H_m(298.15 \text{ K}) = \Delta_{sub}H_m(\overline{T}) + nR[\overline{T} - 298.15] \tag{15}$$

$$\Delta_{sub}H_m(298.15 \text{ K}) = \Delta_{sub}H_m(\overline{T}) + [0.75 + 0.15 \cdot C_{pc}(298.15 \text{ K})][\overline{T} - 298.15] \tag{16}$$

can be used. The same assumption used in generating equation 13 was used here. Since the heat capacity of a liquid is usually larger than for the corresponding solid, which in turn is larger than for the gas, heat capacity adjustments applied to experimental measurements conducted above 298.15 K increase the enthalpy of the corresponding phase change when corrected back down to 298.15 K. For a given temperature difference, the adjustment to the sublimation enthalpy is usually smaller than the adjustment to the vaporization enthalpy.

Applications. Application of equation 12 to estimate sublimation enthalpies is shown in the examples in Table VII. Examples were chosen to illustrate the use of most of the equations discussed in this presentation.

trans 1,2-Diphenylethene. The estimation of *trans* 1,2-diphenylethene illustrates the estimation of a hydrocarbon in a case where experimental fusion, vaporization and sublimation enthalpies are available. The vaporization enthalpy was obtained from the temperature dependence of vapor pressure at a mean temperature of 434 K and corrected back to 298.15 K using the experimental heat capacity of the liquid at 298.15 K and equation 14. The vaporization enthalpy was also estimated using equation 9. A direct measurement of the sublimation enthalpy of *trans* 1,2-diphenylethene has been reported by a number of workers and a partial list of available values is provided in Table VII. These values can be compared to the value obtained by addition of either the experimental or estimated latent enthalpies. Additional details describing the estimations of hydrocarbons have been reported previously *(53)* .

2,4,6-Trimethylbenzoic acid. This is an example of a molecule containing a single functional group. An experimental fusion enthalpy for this material is not currently available. However it can easily be estimated using the experimental melting point and the total phase change entropy as summarized in Table VII. The vaporization enthalpy can be estimated using equation 10. In this estimation there are two *ortho* alkyl branches to correct. Addition of these two estimated enthalpies produces a value, 103.8 kJ·mol^{-1}, which agrees favorably with the experimental value.

8-Hydroxy-5-nitroquinoline. The estimation of 8-hydroxy-5-nitroquinoline illustrates the use of equations 6 and 11 to estimate the fusion and vaporization

Table VII. Estimation of Sublimation Enthalpies[a]

$C_{14}H_{12}$ *trans* 1,2-diphenylethene

T_{fus}: 397.4 K	$\Delta_{vap}H_m$ (T/K)	$\Delta_{sub}H_m$ (T/K)
$\Delta_0^{T_{fus}} S_{tpce}$	exp: 65.5(434) (*43*)	exp:100.7(298.15)(*52*)
calcd.: 69.6	C_{pl}(298.15 K)/J·mol^{-1}	102.1 (*52*)
{10[7.4]+2[-7.5]	exp.: 235 (*55*)	99.2 (*52*)
+2[5.3]}	$\Delta_{vap}H_m$ (T/K)	99.6 (313) (*43*)
$\Delta_0^{T_{fus}} H_{tpce}$	calcd:	103.8 (315) (*64*)
	(eq 14): 75.2(298.15 K)	{27.4+75.2}= 102.6
exp.: 27.4 (*55*)	(eq 9): 68.7 (298.15 K)	calcd:
calcd.: 27.7	{14[4.69]+[3.0]}	27.7 + 75.2 = 102.9
		27.7 + 68.7 = 96.4

$C_{10}H_{12}O_2$ 2,4,6-trimethylbenzoic acid

T_{fus}: 428.15 K	$\Delta_{vap}H_m$ (T/K)	$\Delta_{sub}H_m$ (T/K)
$\Delta_0^{T_{fus}} S_{tpce}$	calcd: 84.7 (298.15)	exp: 103.6 (298.15)(*62*)
calcd: 44.7	{10[4.69]+[3.0]+	
{3[17.6]+[-7.5]+	[38.8]-2[2.0]}	calcd:
2[7.4]+3[-9.6]+[13.4]}		19.1 + 84.7 = 103.8
$\Delta_0^{T_{fus}} H_{tpce}$		
calcd: 19.1		

$C_9H_6N_2O_3$ 8-hydroxy-5-nitroquinoline

T_{fus}: 456.2 K	$\Delta_{vap}H_m$ (T/K)	$\Delta_{sub}H_m$ (T/K)
$\Delta_0^{T_{fus}} S_{tpce}$	calcd: 91.4 (298.15)	exp:114.1(298.15)(*63*)
calcd: 55.9	{9[4.69]+[3.0]+	
{5[7.4]+4[-7.5]+	([22.8]+[29.4])0.85	calcd:
[17.7]+[20.3]+[10.9]}	+[12.2](0.69+0.85)/2	25.5 + 91.4 = 116.9
$\Delta_0^{T_{fus}} H_{tpce}$	+[-7.6]}	
calcd: 25.5		

$C_6H_6Cl_6$ γ-hexachlorocyclohexane (Lindane)

T_{fus}: 386.8 K	$\Delta_{vap}H_m$ (T/K)	$\Delta_{sub}H_m$ (T/K)
$\Delta_0^{T_{fus}} S_{tpce}$	calcd: 70.0 (298.15)	exp: 90.8 (298.15) (*66*)
calcd.: 53.5	{6[4.69]+[3.0]+	106.6 (273) (*67*)
{[33.4]+3[3.7]+6[-14.7]	6[10.8]0.60}	99.2 (328) (*43*)
+6[1.5][10.8]}		88.9 (303) (*68*)
$\Delta_0^{T_{fus}} H_{tpce}$		115.5 (*69*)
calcd: 20.7		calcd:
exp: 22.1 (*65*); 25.9(*55*)		20.7 + 70.0 = 90.7
		24.0 + 70.0 = 94.0

---

[a]$\Delta_0^{T_{fus}} H_{tpce}$, $\Delta_{vap}H_m$ and $\Delta_{sub}H_m$ in kJ·mol^{-1}; $\Delta_0^{T_{fus}} S_{tpce}$ in J·mol^{-1}·K^{-1}

enthalpies, respectively. While different values for the hydroxyl group in phenols and alcohols are used for estimating solid-liquid phase change properties, the same group value is used for estimating vaporization enthalpies. The quinoline structure is treated as an aromatic system and the group value for a heterocyclic aromatic amine is used for nitrogen. In the estimation of the vaporization enthalpy, the pyridine group value is used for nitrogen and this group is treated like any other class II functional group in a multisubstituted compound. The vaporization enthalpy is corrected for an intramolecular hydrogen bond.

γ-Hexachlorocyclohexane. Lindane is another example of a molecule containing multiple substitution. In this instance, equation 7 is used in the estimation of the total phase change entropy. Chlorine is a functional group that is influenced by the presence of multiple substitutions in the evaluation of $\Delta_0^{T_{fus}} S_{tpce}$. Estimation of the vaporization enthalpy is accomplished using equation 11. An examination of the literature reveals a number of reports of the sublimation enthalpy of Lindane. The scatter in the experimental values observed here is not uncommon. The estimated value in this case illustrates how it can be used to select the most probable experimental value from a series of discordant measurements.

Statistics of the Correlation. Statistics to determine how well sublimation enthalpies can be estimated by this technique for functionalized molecules are not currently available. Statistics describing the correlation obtained when using equation 12 to estimate the sublimation enthalpies of hydrocarbons has been reported (53). A standard error of ± 10.6 kJ·mol^{-1} has been reported in the estimation of the sublimation enthalpies of 137 different hydrocarbons. We would expect this uncertainty to rise somewhat with increasing molecular complexity, but the magnitude of this number should serve as a useful guide in various applications.

Acknowledgments

The Research Board of the University of Missouri and the National Institute of Standards and Technology are gratefully acknowledged for support of portions of this work .

Literature Cited

1. Reid, R. C.; Prausnitz, J. M.; Poling, B. E. *The properties of Gases and Liquids*, 4th ed.; McGraw-Hill: New York, 1987.
2. Pedley, J. B.; Naylor, R. D.; Kirby, S. P. *Thermochemical Data of Organic Compounds*, 2nd ed.; Chapman and Hall: New York, 1986.
3. Cox, J. D.; Pilcher, G. *Thermochemistry of Organic and Organometallic Compounds*; Academic Press: New York, 1970.
4. Stull, D. R.; Westrum, E. F. Jr.; Sinke, G. C. *The Chemical Thermodynamics of Organic Compounds*; John Wiley: New York, 1969.
5. Chickos, J. S.; Annunziata, R.; Ladon, L. H.; Hyman, A. S.; Liebman, J. F. *J. Org. Chem.* **1986**, *51*, 4311.
6. Chickos, J. S.; Hesse, D. G.; Panshin, S. Y.; Rogers, D. W.; Saunders, M.; Uffer, P. M.; Liebman, J. F. *J. Org. Chem.* **1992**, *57*, 1897.

7. Chickos, J. S.; Hesse, D. G.; Hosseini, S.; Liebman, J. F.; Mendenhall, G. D.; Verevkin, S. P.; Rakus, K.; Beckhaus, H.-D.; Rüchardt, C. *J. Chem. Thermodyn.* **1995**, *27*, 693.
8. Walden, P. *Z. Elektrochem.* **1908**, *14*, 713.
9. Dannenfelser, R. M.; Surendren, N. Yalkowsky, S. H. *SAR QSAR Environ. Res.* **1994**, *1*, 273
10. Dannenfelser, R. M.; Yalkowsky, S. H. *Ind. Eng. Chem Res.* **1996**, *35*, 1483.
11. Chickos, J. S.; Hesse, D. G.; Liebman, J. F. *J. Org. Chem.* **1990**, *55*, 3833.
12. Chickos, J. S.; Braton, C. M.; Hesse, D. G.; Liebman, J. F. *J. Org. Chem.* **1991**, *56*, 927.
13. Domalski, E. S.; Hearing, E. D. *J. Phys. Chem. Ref. Data* **1996**, *25*, 1.
14. Lourdin, D.; Roux, A. H.; Grolier, J.-P. E.; Buisine, J. M. *Thermochim. Acta* **1992**, *204*, 99.
15. Smith, G. W. *Mol. Cryst. Liq. Cryst.* **1980**, *64*, 15.
16. Cheda, J. A. R.; Westrum, E. F. Jr. *J. Phys. Chem.* **1994**, *98*, 2482.
17. Chickos, J. S.; Sternberg, M. J. E. *Thermochim. Acta* **1995**, *264*, 13.
18. Acree, Jr., W. E. *Thermochim. Acta* **1991**, *189*, 37.
19. van Doren, H. A.; van der Geest, R.; Kellogg, R. M.; Wynberg, H. *Rec. Trav. Chim. Pays-Bas* **1990**, *109*, 197.
20. Chirico, R. D.; Hossenlopp, I. A.; Gammon, B. E.; Knipmeyer, S. E.; Steele. W. V. *J. Chem. Thermodyn.* **1994**, *26*, 1187.
21. Rordorf, B.F., *Proceedings of the 5th International Symposium on Chlorinated Dioxins and Related Compounds*, Bayreuth, FRG, Sept. 16-19, *Chemosphere* **1986**, *15*, 1325.
22. Regosz, A.; Chmielewska, A.; Pelplinska, T.; Kowalski, P. *Pharmazie* **1994**, *49*, 371.
23. Wunderlich, B. *Thermal Analysis*, Academic Press: New York, 1990, Chap.5; ATHAS Appendix, pp 417-431.
24. Rechsteiner, Jr., C. E. In *Handbook of Chemical Property Estimation Methods*, Lyman, W. J. L.; Reehl, W. F.; Rosenblatt, D. H., Eds.; ACS: Washington DC, 1990; Chapter 13.
25. Tu, C.-H.; Liu, C.-P. *Fluid Phase Equilibria* **1996**, *121*, 45.
26. (a) Basarova, P.; Svoboda, V. *Fluid Phase Equilibria* **1995**, *105*, 27; (b) Svoboda, V.; Smolova, H. *Fluid Phase Equilibria* **1994**, *97*, 1.
27. Vetere, A. *Fluid Phase Equilibria* **1995**, *106*, 1.
28. Laidler, K. *Can. J. Chem.* **1956**, *34*, 626.
29. Guthrie, J. P. and Taylor, K. F. *Can. J. Chem.* **1983**, *61*, 602.
30. Ducros, M.; Greison, J. F.; Sannier, H. *Thermochim. Acta* **1980**, *36*, 39.
31. Ducros, M.; Greison, J. F.; Sannier, H.; Velasco, I. *Thermochim. Acta* **1981**, *44*, 134.
32. Ducros, M.; Sannier, H. *Thermochim. Acta* **1981**, *54*, 153; *ibid.* **1984**, *75*, 329.
33. Myers, K. H. Danner, R. P. *J. Chem. Eng. Data* **1993**, *38*, 175.
34. Chickos, J. S., Hyman, A. S., Ladon, L. H., and Liebman, J. F. *J. Org. Chem.* **1981**, *46*, 4295.
35. Chickos, J. S., Hesse, D. G., Liebman, J. F., and Panshin, S. Y. *J. Org. Chem.* **1988**, *53*, 3424.
36. Chickos, J. S., Hesse, D. G., and Liebman, J. F. *J. Org. Chem.* **1989**, *54*, 5250.
37. Hesse, D. G. Ph. D. Thesis, The University of Missouri-St. Louis, St. Louis MO 63121.
38. Ruzicka, K.; Majer, V. *J. Phys. Chem. Ref. Data* **1994**, *23*, 1.
39. Pedley, J. B.; Rylance, J. *Sussex - N. P. L. Computer Analysed Thermochemical Data: Organic and Organometallic Compounds;* University of Sussex, Sussex, UK, 1977.

40. *Enthalpies of Vaporization of Organic Compounds;* Majer, V.; Svoboda, V. Eds.; IUPAC No. 32; Blackwell Scientific Publications: Oxford, UK, 1985.
41. Morawetz, E. *J. Chem. Thermodyn.* **1972**, *4*, 139.
42. Chickos, J. S.; Wilson, J. A. *J. Chem. Eng. Data* **1997**, *42*, 190.
43. Stephenson, R. M.; Malonowski, S. *Handbook of the Thermodynamics of Organic Compounds*, Elsevier: New York, N. Y., 1987.
44. Gavezzotti, A.; Fillippini, G. In *Computational Approaches in Supramolecular Chemistry*, Wipff, G., Ed.; Kluwer Academic Publishers: Dordrecht Netherlands, 1994; pp 51-62.
45. Gavezzotti, A. *Acc. Chem. Res.* **1994**, *27*, 309.
46. Charlton, M. H.; Docherty, R.; Hutchings, M. C. *J. Chem. Soc. Perkin Trans.* **1995**, 2023.
47. Nass, K.; Lenoir, D.; Kettrup, A. *Angew. Chem. Int. Ed. Engl.* **1995**, *34*, 1735.
48. Welsh, W. J.; Tong, W.; Collantes, E. R.; Chickos, J. S.; Gagarin, S. G. *Thermochim. Acta*, **1996**, *290*, 55.
49. Bondi, A. *J. Chem. Eng. Data*, **1963**, *8*, 371-380.
50. Morawetz, E. *J. Chem. Thermodyn.* **1972**, *4*, 461.
51. Davies, M. *J. Chem. Educ.* **1971**, *48*, 591.
52. Chickos, J. S. *In Molecular Structure and Energetics;* Liebman, J. F.; Greenberg, A., Eds.; VCH: New York, NY, 1987, Vol. 2; Chapter 3, pp 67-171.
53. Chickos, J. S. In *Energetics of Organometallic Species*; NATO ASI Series C: Mathematical and Physical Sciences, Simões, J. A. M. Ed.; Kluwer Academic Publishers: Boston, MA,1992, Vol. 367; Chapter 10, pp 159-169.
54. Zabransky, M.; Ruzicka, V. Jr.; Majer, V.; Domalski, E. S.; *Heat Capacity of Liquids*; J. Phys. Chem. Ref. Data, Monograph No. 6, ACS: Washington DC, Vol I and II, 1996.
55. Domalski, E. S.; Hearing, E. D. *J. Phys. Chem. Ref. Data* **1996**, *25*, 1.
56. Domalski, E. S.; Hearing, E. D. *J. Phys. Chem. Ref. Data* **1990**, *19*, 881.
57. Domalski, E. S.; Evans, W. H.; Hearing, E. D. *J. Phys. Chem. Ref. Data* **1984**, *13*, suppl. 1.
58. Benson, S. W. *Thermochemical Kinetics*, 2nd ed. ; Wiley: New York, 1978.
59. Domalski, E. S.; Hearing, E. D. *J. Phys. Chem. Ref. Data* **1988**, *17*, 1637.
60. Chickos, J. S.; Hesse, D. G.; Liebman, J. F. *Struct. Chem.* **1993**, *4*, 261.
61. Chickos, J. S.; Hosseini, S.; Hesse, D. G.; Liebman, J. F. *Struct. Chem.* **1993**, *4*, 271.
62. Colomina, M.; Jimenez, P.; Roux, M. V.; Turrion, C. *J. Chem. Thermodyn.* **1987**, *19*, 1139.
63. Ribeiro da Silva, M. A. V.; Monte, M. J. S.; Matos, M. A. R. *J. Chem. Thermodyn.* **1989**, *21*, 159.
64. Kratt, G.; Bechhaus, H.-D.; Bernlohr, W.; Rüchardt, C. *Thermochim. Acta* **1983**, *62*, 279.
65. Donnelly, J. R.; Drewes, L. A.; Johnson, R. L.; Munslow, W. D.; Knapp, K. K.; Sovocol, G. W. *Thermochim. Acta* **1990**, *167*, 155.
66. Sabbah, R.; An, X. W. *Thermochim. Acta* **1991**, *178*, 339.
67. Wania, F.; Shui, W.-Y.; Mackay, D. *J. Chem. Eng. Data* **1994**, *39*, 572.
68. Spencer, W. F.; Cliath, M. M. *Residue Reviews* **1983**, *85*, 57.
69. Jones, A. H. *J. Chem. Eng. Data* **1960**, *5*, 196.

EMPIRICAL METHODS: OTHER APPROACHES

Chapter 5

Estimation of the Enthalpies of Formation of Organic Componds in the Solid Phase

The Study of 2-Acetoxybenzoic Acid (Aspirin) and Its Isomers

Hussein Y. Afeefy and Joel F. Liebman

Department of Chemistry and Biochemistry, University of Maryland
Baltimore County, 1000 Hilltop Circle, Baltimore, MD 21250

Recent experiment shows that the archival value for the enthalpy of formation of solid 2-acetoxybenzoic acid is seriously in error. Although the source of error is relatively easy to find and to correct, this observation documents the importance of being able to estimate enthalpies of formation of solid phase species. Estimations of the enthalpies of formation of solid phase 2-, 3- and 4-acetoxybenzoic acids are offered and the first and last values favorably compared with those from experiment; the middle value is offered as a prediction to encourage its measurement.

We start with the observation that there are over ten million known organic compounds, of which enthalpies of formation are known for under ten thousand (*1-4*). Accordingly, it is usually necessary to estimate an enthalpy of formation when a value is desired, whether for industrial, academic, or theoretical reasons. This is *a fortiori* true for a new compound because it is regrettably unlikely that the compound will be of adequate quantity, purity and interest to excite the experimental calorimetrist to perform the necessary measurement. For relatively simple, i.e. unsubstituted or monofunctional, unstrained compounds found in the gas phase, calculation by Benson's group increments (*5*) is usually adequate to provide the missing number. However, when a group increment contribution is absent, other methods of estimation become imperative. Very often, the compound of interest is solely found in the condensed (liquid and/or solid) phase. Sometimes enthalpies of vaporization and/or sublimation can not be measured because of the low thermal stability of the compounds. Sometimes, the gas phase —so important and interesting for the theoretically inclined student of molecules — lacks relevance for the bench chemist. Parametrization of either the enthalpy of formation group increments (*6*) and/or of the phase change enthalpies (*7,8*) is thus essential.

Group Additivity, Isodesmic and Macroincrementation Reactions

Regardless of the desired phase of the compound, when there is more than one substituent and/or the molecule is strained, it is necessary to include correction terms in group additivity approaches (5) to account for otherwise ignored interactions. Many correction terms have been developed but they customarily lack generality, e.g. the ring correction (cf. strain energy) term for cyclopropane is not directly applicable to either its unsaturated or monooxygen derivative, cyclopropene or oxirane, respectively (5). Most often, the correction terms are derived from the enthalpy of formation of a single species. What is to be done when the desired correction term is absent?

It is important to note that our analysis is not limited to making predictions of the values of enthalpies of formation. It is not uncommon to find conflicting measurements of this quantity. Even more commonly, there may be reason for suspect the single value available for analysis. For examples of this last type of problem, it is desirable to emphasize that agreement between group additivity and experiment is *not* adequate because the value for the group may arise uniquely from the compound of interest.

An approach we have found most useful is that of isodesmic reactions (9) in which the various groups and interactions appear in equal number and type on both sides of the reaction. Numerous cancellations arise because large structural features such as benzene rings are common to both reactant and product. More properly, what the authors prefer are "macroincrementation reactions" (10) which are isodesmic reactions with some ancillary "verbal correction" included such as equating strain energies of comparably substituted cyclopropanes and cyclobutanes. As of now, no expert system or computer code exists to make these corrections. The power and weakness of "macroincrementation reactions" is the inherent subjectivity in its applications and assumptions.

Experimental Studies of the Energetics of 2-Acetoxybenzoic Acid

To demonstrate our reasoning, we will discuss the energetics of acetylsalicylic acid (*11*).

This compound is more properly named 2-acetoxybenzoic acid and is generally and colloquially called "aspirin". This compound is of considerable medicinal, and additionally of commercial, importance. It is a relatively low molecular weight (under 200), multifunctional species with pronounced nonpolar and polar parts (the benzene ring and ester linkage, and the carboxylic acid group respectively). 2-acetoxybenzoic acid is normally found as a solid although since its enthalpy of fusion is known, knowledge of its enthalpy of formation as a solid is immediately accompanied by the knowledge of enthalpy of formation as a liquid. However, no measured enthalpy of sublimation has been reported and so its gas phase energetics remain unavailable from experiment.

The standard thermochemical archive by Pedley (*1*) of the enthalpy of formation of organic compounds reports a solid phase value of -815.6 kJ mol^{-1} as determined from analysis of cited enthalpy of hydrolysis measurements. By contrast, Kirklin (*11*) recently reported the very dissonant value of -758.6 kJ mol^{-1} from direct enthalpy of combustion measurements; we note that Kirklin also gives us the value of -739.3 kJ mol^{-1} for the liquid by the use of his enthalpy of fusion results. We might ask which of these values is more plausible. Kirklin documented errors in Pedley's analysis that will be chronicled below. Nonetheless, we ask what theory would have predicted to help us decide. If our analysis is admittedly *a posteriori*, nonetheless, it gives us a chance to expand and expound on our thought processes and associated decisions about the interplay of structure and energetics.

Why are Pedley's and Kirklin's numbers are so different? Pedley, in his earlier edition (*12*), chronicles the literature hydrolysis reaction of 2-acetoxybenzoic acid to form salicylic (2-hydroxybenzoic) and acetic acids to be endothermic by 27 kJ mol^{-1},

$$COOH / OCOCH_3 \text{ (s)} + H_2O \text{ (lq)} \longrightarrow COOH / OH \text{ (s)} + CH_3COOH \text{ (lq)} \quad (1)$$

where we emphasize that this endothermicity does not refer to the reaction

$$COOH / OCOCH_3 \text{ (lq)} + H_2O \text{ (lq)} \longrightarrow COOH / OH \text{ (lq)} + CH_3COOH \text{ (aq)} \quad (2)$$

Upon careful reading of the original paper, Kirklin noted that the hydrolysis is *exo*thermic by 27 kJ mol^{-1} and thus there is a discrepancy or 27 -(-27) = 54 kJ mol^{-1} between the reported experimental value and that given by Pedley. Following our customary prejudice and preference we use the latest study for the desired data and the enthalpy of formation of salicylic acid (*13*). Using the archival values for acetic acid (*14*) and water (*15*), we find a revised value of the enthalpy of formation of solid 2-acetoxybenzoic acid of -766 kJ mol^{-1}.

Within a range of 7 kJ mol⁻¹ the values from enthalpy of combustion and enthalpy of reaction are equal; the discrepancy has thus effectively vanished. While this 7 kJ mol⁻¹ difference is not insignificant and its consequences should not be ignored, the difference tells us that we should not be disappointed if theoretically deduced and experimentally measured values of an enthalpy of formation of some solid phase organic compound fail to agree better than, say, 10 kJ mol⁻¹. More honestly and properly said, discrepancies of this magnitude are reasonable to expect because they may well be unavoidable in the absence of some new measurements. Indeed, had there not been a new value for the enthalpy of formation of 2-acetoxybenzoic acid, would we have had any reason to challenge the hydrolysis data and its subsequent analysis? That is, we find that we generally do not go back to the original source when we use compilations. Corresponding to customary practice as to the use of archival references, as research scientists, we generally assume that published numbers are valid unless there is reason for suspicion.

We could now turn to theoretical considerations, estimation techniques, and excruciating details as applied to the enthalpy of formation of solid and liquid 2-acetoxybenzoic acid. Rather, we commence by explaining how we could have known to challenge Pedley's archived value for the endothermicity of the hydrolysis of aspirin. In the current case, it is really quite easy. Bottles of aspirin that have sat in the cabinet too long often smell of vinegar, a household "compound" that we recognize as the common name for a dilute aqueous solution of acetic acid. This observation suggests that the hydrolysis of 2-acetoxybenzoic acid is thermodynamically favorable and indeed, this process must be facile if atmospheric moisture is enough to allow it to happen. Reactions solely in the condensed phase generally have rather small entropic contributions: we note the hydrolysis reaction takes us from two "particles" to two "particles" and so the entropy change would be expected to be small in the gas phase as well. The above analysis suggests we should expect the hydrolysis reaction of 2-acetoxybenzoic acid to be exothermic ($\Delta H < 0$) as well as exergic.($\Delta G < 0$) Equivalently, an ester hydrolysis enthalpy of +27 kJ mol⁻¹ appears not to be particularly plausible while a result of -27 kJ mol⁻¹ is consistent with other things we know.

Evaluation of the Energetics of 2-Acetoxybenzoic Acid

Suppose we did not have olfactory or other sensory information, after all, we generally dissuade our students and junior colleagues from smelling a sample of an arbitrary new organic compound. What could we then deduce? As said above, 2-acetoxybenzoic acid is a polyfunctional molecule. We recognize an ester functionality and, in turn, refine this realization to a carboxylic acid ester, an aryl ester and a hindered one at that (cf. the two rather large groups ortho or adjacent to each other). What do we expect about the sign of its enthalpy of hydrolysis? For a simple example, consider methyl acetate and its hydrolysis reaction;

$$CH_3COOCH_3(lq) + H_2O(lq) \longrightarrow CH_3OH(lq) + CH_3COOH(lq) \qquad (3)$$

Is this reaction exothermic? From literature (14,15) enthalpies of formation we find that it is endothermic by 8 kJ mol[-1]. Is that a surprise? We know ester hydrolysis reactions proceed in acid and basic solutions. If so, reaction rate aside, why does it not proceed spontaneously (in a thermodynamic sense) in essentially neutral media as well? We are not surprised that alkyl esters of strong oxy (i.e., hydroxylic) acids readily hydrolyze: dimethyl sulfate readily methylates numerous nucleophiles including water, cf. equation 4.

$$(CH_3O)_2SO_2(lq) + 2H_2O(lq) \longrightarrow 2CH_3OH(aq) + H_2SO_4(ai) \qquad (4)$$

where the "aq" is in aqueous media and the "ai" tells us that the acid product is in aqueous media and ionized. This reaction is exothermic by 93 kJ mol[-1]. Then we remember that concentrated sulfuric acid reacts violently with water because of its high exothermicity of solution and so the relevant reaction (equation 5) has not been considered so far.

$$(CH_3O)_2SO_2(lq) + 2H_2O(lq) \longrightarrow 2CH_3OH(lq) + H_2SO_4(lq) \qquad (5)$$

In fact, it is also endothermic, in this case by ca. 15 kJ mol[-1]. By contrast, alkyl esters of weak hydroxylic acids often hydrolyze exothermically. For example, the following, conventionally unobserved, reaction is exothermic (16) by 6.5 kJ mol[-1]

$$C_2H_5OC_2H_5(lq) + H_2O(lq) \longrightarrow 2C_2H_5OH(lq) \qquad (6)$$

Interpolating, the exothermicity of the hydrolysis of intermediate strength acid esters is hard to appraise.

Let's now change the alcohol. Acid anhydrides hydrolyze more readily, as well as exothermically, than ethers. For example, consider acetic anhydride which is formally the ester of acetic acid and "acetyl alcohol". We find the following reaction

$$CH_3COOCOCH_3(lq) + H_2O(lq) \longrightarrow 2CH_3COOH(lq) \qquad (7)$$

is exothermic by almost 60 kJ mol[-1]. What about phenols? Consider the reaction

$$C_6H_5OOCCH_3(lq) + H_2O(lq) \longrightarrow C_6H_5OH + CH_3COOH(lq) \qquad (8)$$

which is the same as the hydrolysis reaction (equation 1) except for the ortho or 2-carboxylic acid substituent. This reaction is also exothermic, here by nearly 29 kJ mol[-1], when phenol is taken to be in its standard solid state. However, we should be considering phenol as a liquid — or more properly said, our experience has shown predictions to be the most reliable when all of the species are in the same phase. We have also found that predictions are most reliable for gases, then liquids, then solids.

Lacking enthalpy of fusion data, we would still expect the enthalpy of fusion of phenol to be relatively low — or at least low enough to not reverse the energetics and make this last reaction endothermic. Why would we assume this? A simple response notes that 3-methylphenol is a liquid under standard conditions and thus suggests that the crystal energy in the parent species phenol is not that large. We also note the melting point of phenol is "only a few degrees" above 298 K (313 K)and so it is "almost a liquid". Alternatively, we can estimate the enthalpy of formation of liquid phenol by assuming that the following reaction is thermoneutral:

$$C_6H_5CH_3(lq) + 3\text{-}CH_3C_6H_4OH(lq) \longrightarrow 1,3\text{-}C_6H_4(CH_3)_2(lq) + C_6H_5OH(lq) \quad (9)$$

We make herein the simplest assumption that conservation of groups always results in thermoneutrality. The enthalpy of formation of liquid phenol, -156 kJ mol^{-1}, is found while assuming thermoneutrality for the related reaction

results in -152 kJ mol^{-1}. From archival enthalpies of formation and of fusion of solid phenol (*6*) and from the primary literature (*17*), the "correct" value for the enthalpy of formation of liquid phenol is ca. -154 kJ mol^{-1}, in good agreement with our suggested values. The following reaction

$$C_6H_5OOCCH_3(lq) + H_2O(lq) \longrightarrow C_6H_5OH(lq) + CH_3COOH(lq) \quad (11)$$

is exothermic by 18 kJ mol^{-1}. Since it is logical to assume that the carboxyl (COOH) group is larger than hydrogen (H), barring unexpected phase differences, "the earlier and most important" equation 1 — the hydrolysis of 2-acetoxybenzoic acid — should be even more exothermic.

Comparison of the Energetics of 2- and 4-Acetoxybenzoic Acids

Having convinced ourselves that the archival value is in error because the hydrolysis reaction of 2-acetoxybenzoic acid is, in fact, exothermic, we now endeavor to "predict" the desired enthalpy of formation. Let us start with the strain energy associated with the acetoxy and carboxyl groups being ortho or adjacent to each other. Consider the following reaction involving 4-acetoxybenzoic acid.

The enthalpy of reaction 12 is derived from the experimentally measured (*18*) enthalpies of hydrolysis of the 2- and 4-acetoxybenzoic acids upon correction of all of the enthalpies of formation to standard states. It is found that this reaction is exothermic by 7 kJ mol^{-1}. This does not mean that 2-acetoxybenzoic acid is strained by 7 kJ mol^{-1} but rather that the strain energy difference between the two acetoxybenzoic acids is 7 kJ mol^{-1} greater than between the two hydroxybenzoic acids. After all, what is the strain energy of 2-hydroxybenzoic acid? Besides, this last species has a stabilizing intramolecular hydrogen bond — a structural feature not found in any of the other compounds in this reaction. Nonetheless, from the enthalpies of formation of 2- and 4-hydroxybenzoic acids (*13*), and the hydrolysis enthalpies of 2- and 4-acetoxybenzoic acids (*18*), we deduce the value of the enthalpy of formation of solid 4-acetoxybenzoic acid to be -780.1 kJ mol^{-1}. This is presumably the enthalpy of formation of a strainless species and it is thus the difference between this last number and the recommended enthalpy of formation for the 2-isomer that defines the strain energy for 2-acetoxybenzoic acid. Numerically, this number is found to be ca. 21 kJ mol^{-1}.

We now ask the following questions about the isomeric acetoxybenzoic acids. First, is a strain energy of 21 kJ mol^{-1} "plausible" for 2-acetoxybenzoic acid? Secondly, how do theory and experiment correspond for the remaining isomer, the 3-acetoxybenzoic acid which is also presumably strainless? Finally, is this ca. -780 kJ mol^{-1} a reasonable value for the enthalpy of formation of 4-acetoxybenzoic acid?

The 2- vs. 4-isomer question is simply addressed by looking at the enthalpy of formation differences for other 2- and 4-substituted benzoic acids. The difference for solid methoxybenzoic acids is ca. 23 kJ mol^{-1} while that for solid ethylbenzoic acids is ca. 13 kJ mol^{-1}. A 21 kJ mol^{-1} difference for solid acetoxybenzoic acids is thus plausible. The last question is still unanswered — we know of no hydrolysis or combustion enthalpy measurement for 3-acetoxybenzoic acid.

Estimation of the Energetics of the Acetoxybenzoic Acids

Let us now try to estimate the enthalpy of formation of a strainless 2-acetoxybenzoic acid, confident that the above analysis will allow us to estimate the desired strain energy. Equivalently, we will estimate the enthalpy of 4-acetoxybenzoic acid and then correct this value to that of the 2-acetoxybenzoic acid of greatest interest. Ignoring all correction terms from strict additivity, the first method assumes thermoneutrality for

the following reaction

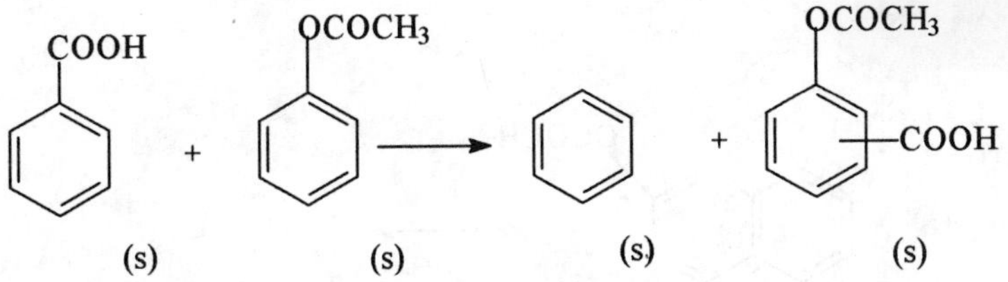

(15)

This reaction appears entirely plausible, but what about quinonoid resonance stabilization of the acetoxybenzoic acid as shown below?

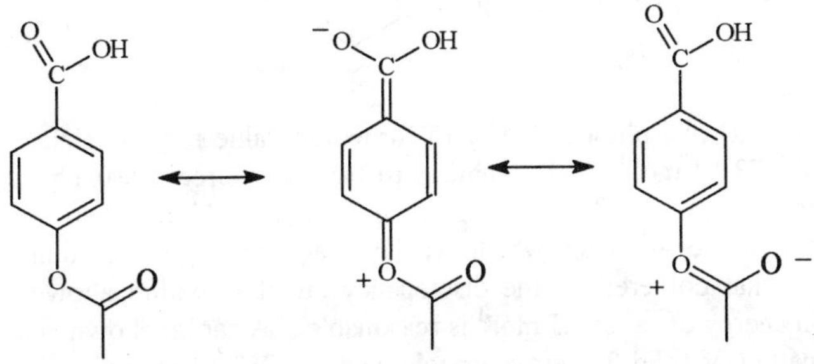

(16)

Additionally, we lack the enthalpy of formation of solid phenyl acetate. With regards to the first issue, we find the difference between the enthalpies of formation of 3- and 4-methoxybenzoic acid to be ca. 8 kJ mol^{-1}. Given the considerable resonance energy of esters (*19*) which is in "conflict" with this quinonoid stabilization mechanism, this 8 kJ mol^{-1} is an upper bound to the first consideration. With regards to the second, the entire reaction may be rewritten for liquid species from which the enthalpy of formation of liquid acetoxybenzoic acid may be deduced as -751 kJ mol^{-1}. Group additivity derived estimation (*4*) may be used for phenyl acetate because this species is monofunctional and so Benson-type predictions are expected to be reliable. This results in a value of -335 kJ mol^{-1} for the solid ester and so a predicted value of -759 kJ mol^{-1} for the solid acetoxybenzoic acid. This value seems unlikely if for no other reason than we view with suspicion the enthalpy of fusion of phenyl acetate: taking the difference of the enthalpies of formation of this species as a solid and as a liquid gives us a fusion enthalpy of but 9 kJ mol^{-1}. This value seems too little since it is less than the enthalpies of fusion (*4*) of the "smaller" ethyl acetate and methoxybenzene (ca. 10.5 and 17.0 kJ mol^{-1} respectively).

Alternatively, we note that there is a rather constant enthalpy of formation difference between gaseous naphthyl and phenyl derivatives (*20*), and more importantly, there is also near constancy for the difference between solid naphthyl and phenyl derivatives (21). So doing, we thus make the plausible assumption that the following reaction is thermoneutral as well

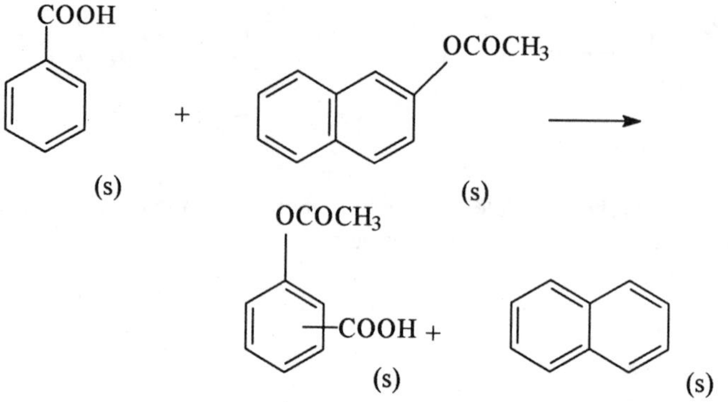

Accordingly, we derive a solid phase enthalpy of formation value for a strainless acetoxybenzoic acid of -773 kJ mol^{-1}. If this value is to have any direct meaning, it should be applied to the 4-isomer. This value is some 7 kJ mol^{-1} higher than the suggested value above for the 4-isomer which we have already argued is quite strainless. We are somewhat bothered by the discrepancy but it is within allowed tolerances. That a strain energy of ca. 20 kJ mol^{-1} is reasonable was earlier shown and so an enthalpy of formation of solid 2-acetoxybenzoic acid of -753 kJ mol^{-1} is thus also reasonable.

The final estimation approach assumes thermoneutrality for the reaction

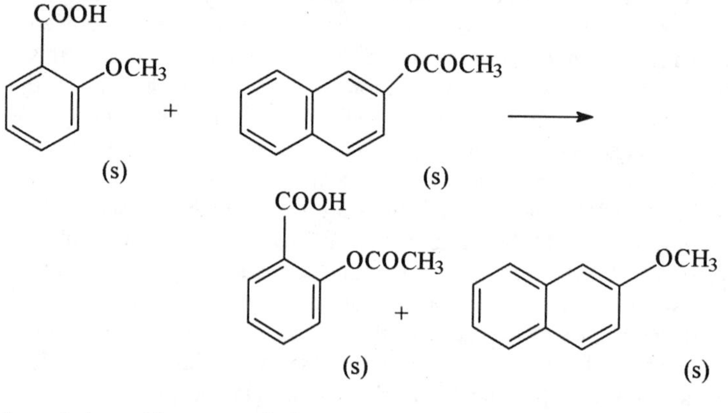

where it is tacitly assumed that the above phenyl/naphthyl enthalpy of formation constant difference is valid as well as that the strain and electronic effects of ortho methoxy and acetoxy substitution are the same. This results in a predicted value of

-750 kJ mol^{-1} for solid 2-acetoxybenzoic acid. This value is a bit higher than we should like compared to that found from Kirklin's direct experimental measurement of the enthalpy of combustion. Nonetheless, given the above assumptions, our results are "not bad" and completely corroborate Kirklin's value over that in Pedley's archive.

The remaining issue is that of the enthalpy of formation of solid phase 3-acetoxybenzoic acid. There is no experimental measurement of this quantity from either combustion or hydrolysis calroimetry. In order to encourage these experiments, we hereby give our prediction. In the absence of any quinonoid interaction and lattice effects that distinguish the isomers, this value would be the same as its 4-isomer, -780 kJ mol^{-1}. We remind the reader that solid 3-methoxybenzoic acid has an enthalpy of formation 8 kJ mol^{-1} higher than its 4-isomer. This suggests that solid 3-acetoxybenzoic acid has an enthalpy of formation no larger than 8 kJ mol^{-1} higher than its 4-isomer. The anticipated value is thus between -772 and -780 kJ mol^{-1}, or more "properly" written -776 ± 4 kJ mol^{-1}. We eagerly await confirmation of our prediction.

Acknowledgments. The authors wish to thank the U.S. National Institute of Standards and Technology for grant support, and Drs. Stephen E. Stein, W. Gary Mallard, and Duane R. Kirklin for their involvement in and encouragement of studies of the thermochemistry of organic compounds.

Literature Cited

1. Pedley, J. B. *Thermochemical Data of Organic Compounds*; Thermodynamics Research Center: College Station, TX, 1994; Vol. I.
2. Cox, J. D.; Pilcher, G. *Thermochemistry of Organic and Organometallic Compounds*; Academic Press: New York, NY, 1970.
3. Stull, D.R.; Westrum, E.F.; Sinke, G. C. *The Chemical Thermodynamics of Organic Compounds*; John Wiley and Sons, Inc.: New York, NY, 1969.
4. Domalski, E. S.; Hearing, E. D. *J. Phys. Chem. Ref. Data* **1993**, *22*, 805.
5. Benson, S. W. *Thermochemical Kinetics*; Second Edition, John Wiley and Sons, Inc.: New York, NY, 1976.
6. Domalski, E.S.; Hearing, E.D., *J. Phys. Chem. Ref. Data* **1996**, *25*, 1
7. Chickos, J. S.; Braton, C. M.; Hesse, D. G.; Liebman, J. F. *J. Org. Chem.* **1991**, *56*, 927.
8. Chickos, J. S.; Hesse, D. G.; Liebman, J. F.; Panshin, S. Y. *J. Org. Chem.* **1988**, *53*, 3424.
9. Hehre, W. J.; Radom, L.; Schleyer, P. v. R.; Pople, J. A. *Ab Initio Molecular Orbital Theory*, John Wiley, New York, NY, 1986.
10. Liebman, J. F. In *Molecular Structure and Energetics: Studies of Organic Molecules*; Liebman, J. F.; Greenberg, A., Eds. VCH Publishers, Inc.: New York, NY, 1986, Vol. 3; 267-328.
11. Kirklin, D.R. *J. Chem. Thermodyn.*, in press.

12. Pedley, J. B.; Naylor, R. D.; Kirby, S. P. *Thermochemical Data of Organic Compounds*; Second Edition, Chapman and Hall: New York, NY, 1986.
13. Sabbah, R.; Le, T. H. *Can. J. Chem.* **1993**, *71*, 1378.
14. Lebedeva, N. D. *Russ. J. Phys. Chem.* **1964**, *38*, 1435
15. Wagman, D. D.; Evans, W. H.; Parker, V. B.; Schumm, R. H.; Halow, I.; Bailey, S. M.; Churney, K. L.; Nuttall, R. L. The NBS tables of chemical thermodynamic properties; selected values for inorganic and C_1 and C_2 organic substances in SI units *J. Phys. Chem. Ref. Data* **1982**, *11* Suppl. 2.
16. Green, J. H. S. *Chem. Ind.(London)* **1960**, 1215.
17. Andon, R. J. L.; Counsell, J. F.; Herington, E. F. G.; Martin, J. F. *Trans. Faraday Soc.* **1963**, *59*, 830.
18. Nelander, L. *Acta Chem. Scand.* **1964**, *18*, 973.
19. Liebman, J. F.; Greenberg, A. *Biophys. Chem.* **1974**, *1*, 222.
20. Ribeiro da Silva, M. A. V.; Ferrao, M. L. C. C. H.; Lopes, A. J. M. *J. Chem. Thermodyn.* **1993**, *25*, 229.
21. Liebman, J. F. In *The Chemistry of the Sulphonic Acids, Esters and Their Derivatives*, Patai, S.; Rappoport, Z., Eds. John Wiley: Chichester, UK, 1991, 283-321.

Chapter 6

Electrostatic-Covalent Model Parameters for Molecular Modeling

R. S. Drago[1] and T. R. Cundari[2]

[1]Department of Chemistry, University of Florida, Gainesville, FL 32611–7200
[2]Department of Chemistry, University of Memphis, Memphis, TN 38152

In molecular modelling of physicochemical properties, the estimation of the magnitude of specific donor-acceptor interactions, e.g., hydrogen bonding, is most difficult. This article presents a compilation of the most recent set of parameters that allow prediction of neutral molecule, intermolecular, donor-acceptor interactions and a set of parameters that provides bond energies for organic, inorganic and organometallic compounds. Solvent polarity parameters are also discussed that provide a measure of non-specific solvation. The meaning of the parameters in the context of the electrostatic-covalent model are presented and their use illustrated. Examples are discussed to illustrate the value of these parameters when used in conjunction with quantum chemical calculations.

Understanding the trends and predicting values of bond energies are essential for many problems in chemistry ranging from chemical reactivity, to structure, to physicochemical measurements. The electrostatic-covalent model (*1*), without question, provides the most general set of parameters for they encompass enthalpies for neutral donor-neutral acceptor adducts, energies of gas phase ion-molecule reactions, homolytic bond dissociation energies and solvation energies. The elegance of the model is its simplicity. As the chemical reaction becomes more complex, terms are logically added, *a priori*, to account for the new energetic contributions. The purpose of this article is to illustrate the application of this model to the area of molecular modelling and to provide the most recent set of parameters for incorporation into modelling programs.

Neutral Donor-Acceptor Adducts

Correlations of physicochemical properties, like solubility parameter theory, often break down when contributions from specific donor-acceptor interactions exist because these energies are not adequately predicted. The correlation of over 500 enthalpies of donor-acceptor interactions in poorly solvating solvents leads to acceptor and donor parameters that can be used to determine the hydrogen bonding and other donor-acceptor contributions, resolving complex physicochemical properties into their energetic components. The electrostatic-covalent model of Pauling and Mulliken is used to write equation 1. Measured enthalpies, $-\Delta H$, for donor-acceptor reactions fit equation 1,

$$-\Delta H = E_A E_B + C_A C_B + W \qquad (1)$$

which consists of covalent, $C_A C_B$, and electrostatic, $E_A E_B$, components. Each acid is proposed to have a tendency to undergo electrostatic, E_A, and covalent, C_A, bonding as is each base (E_B and C_B). When the empirical E and C parameters (*1*) in Table I are combined according to equation 1 they produce the reaction enthalpy, that is, the adduct bond strength. In the molecular orbital description, E relates to the tendency of the acceptor or donor to undergo a charge-controlled reaction while C relates to the tendency to undergo a frontier-orbital controlled reaction. The W term, which is zero for most enthalpies, incorporates any constant contribution to the reaction of a particular acid (or base) that is independent of the base (or acid) it reacts with. For example, the enthalpy to cleave the Al_2Cl_6 dimer, when a $B-AlCl_3$ adduct is formed, would be included in the W term of the enthalpy fit.

The base parameters in Table I and the acceptor parameters in Table II can be used in equation 1 to predict over 8000 enthalpies of interaction. In those cases where donor-acceptor pairs have been subsequently measured, the predictions agree with experiment to 2%.

In molecular modelling, the parameters can be used as a scale of acid or base strength to correlate physicochemical measurements in energy units, $\Delta\chi$, and determine if the measurements are dominated by the donor- acceptor component. For physicochemical measurements, equation 1 takes the form shown in equation 2.

$$\Delta\chi = E_A^* E_B + C_A^* C_B + W \qquad (2)$$

Several physicochemical properties besides enthalpies have been correlated (*2*) to equation 2, the ECW model. In this application, weighted regression analyses with the weights in the tables should be used because the parameters are not all known to the same degree of certainty. When the measurement is that of an acceptor molecule studied with a series of donors whose E_B and C_B are known, the measured property, $\Delta\chi$, and the base parameters are substituted into equation 2,. The series of equations, one for each base, is solved for E_A^*, C_A^*, and W. Asterisks are placed on the acceptor parameters to indicate that they do not refer to enthalpies in poorly solvating solvents and that conversion units leading to $\Delta\chi$ are included. W is the value of $\Delta\chi$ when a donor is attached whose E_B and C_B value is zero. When a base property is measured with a series of acids, E_A and C_A are substituted and E_B^* and

C_B^* determined. The results for correlations to ECW of physicochemical properties other than enthalpies are summarized in Table III.

To illustrate the correlation of a physicochemical property to ECW, the CO stretching frequencies of a series of base, B, adducts of B-Rh(pfb)$_4$Rh-CO will be employed (pfb is a perfluorobutyrate anion that bridges two Rh centers). The question asked is: does the frequency shift parallel changes in the sigma donor strength of B? Instead of plotting the shift versus pK_B, a dual parameter correlation is made to E_B and C_B using equation 2. The independent variable ν_{CO} is correlated to the dependent variables E_B and C_B using a linear regression routine (e.g. NCSS, Kaysville, Utah). The E_B and C_B parameters are weighted according to Table I and solved for the parameters E_A^*, C_A^* and W. The resulting correlation equation ($r^2 = 0.935$) and data fit are given in Table IV. Since the ECW model only incorporates sigma bonding effects, one can conclude that the change in the CO stretching frequency is dominated by the sigma bond strength of the base with stronger bases leading to lower frequencies. Note the excellent agreement of the intercept, W, with the value of ν_{CO} with no base coordinated.

The good correlations summarized in Table III provide probes that can be combined with measured enthalpies to add new donors or acceptors to the model. To add a new donor, enthalpies of reaction with a series of acceptors whose E_A, C_A and W are known and other physicochemical measurements with acceptor probes whose E_A^*, C_A^* and W are known are substituted into equation 1. A series of equations results that can be solved for the E_B and C_B parameters for the new donor. Most of the bases whose statistical weights are low in Table I were determined from limited data in this manner. In most instances, the set of simultaneous equations solved to give these tentative parameters involves types of acids or bases whose electrostatic and covalent parameters are similar. The C_A / E_A ratios indicate the relative importance of these effects in acids and the C_B/E_B ratios indicate the base types. When the range of acids used to determine C_B and E_B for a donor is limited, the base parameters will provide a good fit of new data for similar acids but may not provide good estimates for acids with different ratios (*1*). When data become available for new acids that extend the ratio, the base parameters should be redetermined.

The ECW parameters provide an estimate of the sigma bond strength of the donor-acceptor interaction. The widespread applicability of these parameters to a variety of acids and bases and the subsequent use of these parameters to correlate many new systems (*1*) leads to confidence in the predictions made with this model. Correlations of data sets for a series of bases (or acids) with these parameters will indicate whether or not the physicochemical measurements are dominated by the same base (or acid) properties that determine donor- acceptor bond strengths. When poor correlations result, a systematic pattern in the deviations may reveal an unusual bonding effect. The deviant systems are omitted and the fit redetermined. To confidently omit systems, the deviations of the omitted system should be 2.5 times larger than the average deviation of the fit of well behaved donors. Deviations in data sets have been shown to provide estimates of the magnitudes of steric effects and pi-back bond stabilization (*2*). Unusual entropic effects also have

Table I. Base Parameter

Bases	E_B	C_B	Weight	C_B/E_B
NH$_3$	2.31	2.04	1	0.88
CH$_3$NH$_2$	2.16	3.12	1	1.4
(CH$_3$)$_2$NH	1.80	4.21	1	2.3
(CH$_3$)$_3$N	1.21	5.61	1	4.6
C$_2$H$_5$NH$_2$	2.34	3.30	0.6	1.4
(C$_2$H$_5$)$_2$NH	1.22	4.54	0.2	3.7
(C$_2$H$_5$)$_3$N	1.32	5.73	1	4.3
(CH$_2$)$_5$NH	1.44	4.93	1	3.4
HC(C$_2$H$_4$)$_3$N	0.80	6.72	1	8.4
N-CH$_3$Im	1.16	4.92	1	4.2
C$_5$H$_5$N[a]	1.78	3.54	1	2.0
3CH$_3$C$_5$H$_4$N	1.81	3.67	1	2.0
3ClC$_5$H$_4$N	1.66	3.08	1	1.9
3BrC$_5$H$_4$N	1.66	3.08	1	1.9
3IC$_5$H$_4$N	1.67	3.13	1	1.9
4CH$_3$C$_5$H$_4$N	1.83	3.73	1	2.0
4C$_2$H$_5$C$_5$H$_4$N	1.81	3.74	1	2.1
4CH$_3$OC$_5$H$_4$N	1.83	3.83	1	2.1
4-N(CH$_3$)$_2$C$_5$H$_4$N	1.92	4.43	1	2.3
4-CNC$_5$H$_4$N	1.53	2.94	1	1.9
quinoline	2.28	2.89	1	1.3
CH$_3$CN	1.64	0.71	1	0.43
n-C$_3$H$_7$CN	1.81	0.54	0.2	0.3
ClCH$_2$CN	1.67	0.33	1	0.2
(CH$_3$)$_2$NCN	1.92	0.92	0.6	0.48
C$_6$H$_5$CN	1.65	0.75	0.6	0.45
CH$_3$C(O)CH$_3$	1.74	1.26	1	0.72
CH$_3$CH$_2$C(O)CH$_3$	1.67	1.24	0.4	0.74
(CH$_2$)$_4$CO	2.02	0.88	0.4	0.43
CH$_3$C(O)OCH$_3$	1.63	0.95	1	0.58
CH$_3$C(O)OC$_2$H$_5$	1.62	0.98	1	0.61
(C$_6$H$_5$)C(O)CH$_3$	1.72	1.15	0.2	0.7
(C$_6$H$_5$)$_2$CO	2.01	0.55	0.2	0.3
Propylene Carbonate	1.51	1.32	0.4	0.9
CH$_3$C(O)N(CH$_3$)$_2$	2.35	1.31	1	0.56
HC(O)N(CH$_3$)$_2$	2.19	1.31	1	0.60
NMP	2.12	1.65	0.4	0.8
CO[N(CH$_3$)$_2$]$_2$	2.06	1.87	0.5	0.91
(C$_2$H$_5$)$_2$O	1.80	1.63	1	0.91
i-Pr$_2$O	1.95	1.60	0.2	0.85
(C$_4$H$_9$)$_2$O	1.89	1.67	1	0.88
O(C$_2$H$_4$)$_2$O	1.86	1.29	1	0.69
(CH$_2$)$_4$O	1.64	2.18	1	1.3
(CH$_2$)$_5$O	2.05	1.38	0.4	0.67
(CH$_3$)$_2$O	1.68	1.50	0.6	0.89

Table I (continued)

$(C_8H_{17})_2O$	1.77	1.95	0.2	1.1
$(CH_2)_4(CH)_2O$	1.45	2.14	1	1.5
$(CH_3)_2S$	0.25	3.75	1	15
$(C_2H_5)_2S$	0.24	3.92	1	16
$(CH_2)_4S$	0.26	4.07	1	16
$C_6H_5SCH_3$	0.21	3.13	0.6	15
$(CH_2)_5S$	0.34	3.81	0.6	11
$CH_3S_2CH_3$	0.55	2.34	0.2	4
$(CH_3)_2SO$	2.40	1.47	1	0.61
$(CH_2)_4SO$	2.44	1.64	1	0.67
$(CH_2)_4SO_2$	1.61	1.09	0.2	0.7
C_5H_5NO	2.29	2.33	0.2	0.33
$4\text{-}CH_3C_5H_4NO$	2.32	2.57	0.8	1.1
$4\text{-}CH_3OC_5H_4NO$	2.34	3.02	0.2	1.3
$(CH_3)_4C_5H_6NO(TEMP)$	1.46	3.20	1	2.2
$C_6H_5NO_2$	1.27	0.57	0.6	0.45
CH_3NO_2	1.09	0.70	0.2	0.6
$(C_6H_5)_3PO$	2.59	1.67	0.8	0.64
$(CH_3O)_3PO$	2.42	0.98	0.2	0.4
$(C_2H_5O)_3PO$	2.51	1.10	1	0.44
$[(CH_3)_2N]_3PO$	2.87	1.52	0.5	0.53
$(CH_3)_3P^b$	0.31	5.15	1	17
$(C_2H_5)_3P$	0.28	5.53	1	20
$(t\text{-butyl})_3P$	0.25	6.08	0.5	24
$(C_6H_5)P(CH_3)_2$	0.44	4.49	1	10
$(C_6H_5)_2PCH_3$	0.57	3.74	1	6.6
$(C_6H_5)_3P$	0.70	3.05	1	4.4
$(4\text{-}ClC_6H_5)_3P$	0.82	2.35	1	2.9
$(4\text{-}CH_3C_6H_5)_3P$	0.65	3.41	1	5.2
$(4\text{-}CF_3C_6H_5)_2P$	0.91	1.52	0.8	1.7
$(CH_3O)_3P$	0.50	3.32	1	6.6
$(C_2H_5O)_3P$	0.56	3.17	1	5.7
cage phosphite	0.09	4.85	0.7	54
$t\text{-}C_4H_9OH$	2.05	1.00	0.2	0.64
$(CH_3)_2Se$	0.05	4.24	0.4	83
C_6H_6	0.70	0.45	0.5	0.64
$(C_6H_5)_3PS$	0.35	3.65	0.2	10
$(C_6H_5)_3As$	0.90	2.16	0.2	2.4

[a]The E_B and C_B parameters in units of $(kcal\ mol^{-1})^{1/2}$ for 56 x-substituted pyridines can be obtained by using reported substituent constants ΔE^x and ΔC^x (Drago, R.S. *Organometallics* **1945**, *34*, 3453) in $E_b^x = 1.78 + \Delta E^x$ and $C_B^x = 3.54 + \Delta C^x$.

[b]For parameters for thirty-seven substituted phosphines see Joerg S.; Drago, R. S.; Seles, J. submitted. Most of the phosphines have been studied on systems that utilize only phosphine ligands. See the above reference for limitations on their use.

Table II. Enthalpy Acid Parameters

Acid	E_A	C_A	W	Weight	C_A/E_A
I_2	0.50	2.00		1.0	4.0
ICl	2.92	1.66		1.0	0.57
IBr	1.20	3.29		0.4	0.47
$C_6H_5OH^a$	2.27	1.07		1.0	0.47
4-FC$_6$H$_4$OH	2.30	1.11		1.0	0.48
3-FC$_6$H$_4$OH	2.37	1.17		1.0	0.49
3-CFC$_6$H$_4$OH	2.38	1.22		1.0	0.51
4-ClC$_6$H$_4$OH	2.34	1.14		1.0	0.49
4-CH$_3$C$_6$H$_4$OH	2.23	1.03		1.0	0.46
4-t-C$_4$H$_9$C$_6$H$_4$OH	2.22	1.03		1.0	0.46
CF$_3$CH$_2$OH	2.07	1.06		1.0	0.51
(CF$_3$)$_2$CHOH	2.89	1.33	-0.16	1.0	0.46
CH$_3$OH	1.27	0.74		1.0	0.58
C$_2$H$_5$OH^b	1.15	0.67		1.0	0.58
t-C$_4$H$_9$OH	1.14	0.66		0.8	0.58
(CF$_3$)$_3$COH	3.06	1.88	-0.87	0.8	0.61
C$_8$H$_{17}$OH^b	0.89	0.87		1.0	1.0
C$_6$H$_5$SH	0.58	0.37		0.2	0.6
H$_2$O	1.35	0.78		0.4	0.59
HCCl$_3$	1.56	0.44		1.0	0.28
C$_4$H$_4$NH	1.38	0.68		1.0	0.49
CF$_3$(CF$_2$)$_6$H	0.80	0.63		0.2	0.8
CH$_2$Cl$_2$	0.86	0.11		1.0	0.13
HNCS	2.85	0.70		0.6	0.2
HNCO	1.60	0.69		0.4	0.4
B(CH$_3$)$_3$	3.57	2.97		0.8	1.2
$^1/_2$[Al(CH$_3$)$_3$]$_2$	8.28	3.23	-8.46	1.0	0.39
Ga(C$_2$H$_5$)$_3$	6.95	1.48			0.21
ln(CH$_3$)$_3$	6.60	2.15		0.2	0.33
(CH$_3$)$_3$SnCl	2.87	0.71		1.0	0.25
SO$_2$	0.51	1.56		1.0	3.1
[Ni(TFAcCAM)$_2$]$_2^c$	1.55	1.32		1.0	0.85
Cu(HFAcAc)$_2^d$	1.82	2.86		1.0	1.6
Zn[N(Si(CH$_3$)$_3$)]$_2$	2.75	2.32		0.8	0.84
Cd[N(Si(CH$_3$)$_3$]$_2$	2.50	1.83		0.8	0.73
Mo$_2$PFB$_4^e$	3.15	1.05		1.0	0.33
ZnTPP^f	2.72	1.45		1.0	0.53
CoPPIXDME^g	2.32	1.34		1.0	0.58
$^1/_2$[MeCo(Hdmg)$_2$]$_2$	4.70	3.24	-5.84	1.0	0.69
$^1/_2$[Rh(CO)$_2$Cl]$_2$	4.32	4.13	-10.39	1.0	0.96

[a] Values for 56 x-substituted phenols can be obtained by substituting reported ΔE^X and ΔC^X substituent constants (Drago, R. S. *Organometallics* 1995, *34*, 3453) into E_A^X=2.27–0.817ΔE^X and C_A^X=1.07–0.225ΔC^X.

[b] Values for several other aliphatic alcohols have been reported: Joerg, S.; Drago, R. S. submitted.

[c] bis(3-trifluoroacetyl-d-camphorate)nickel(II)dimer

[d] bis(hexafluoroacetylacetonato)copper(II)

[e] molybdenum(II)perfluorobutyrate

[f] zinc tetraphenylporphyrin

[g] cobalt(II) protoporphyrin IX dimethyl ester

Table III. Probe Parameters

Acceptor Probe	E_A^*	C_A^*	W^*	C_A^*/E_A^*
$\Delta v_{OH}(C_6H_5OH)$[a]	167	109	-205	0.65
$\Delta v_{OH}(3\text{-}F_3CC_6H_5OH)$[a]	189	122	-220	0.65
$\Delta v_{OH}(t\text{-}C_4H_9OH)$[a]	89.9	55.0	-124	0.61
$\Delta v_{OH}(F_3CCH_2OH)$[a]	150	96.9	-193	0.65
$\Delta v_{IC}(ICN)$[a]	4.40	14.6	-2.0	3.3
$\Delta v_{IC}(I_2C_2)$[a]	4.59	6.05	-8.2	1.3
$\Delta v(I_2)$[b]	1080	1098	-1252	1.0
$\Delta v(4\text{-}NO_2C_6H_4NH_2)$[c,d]	-1.06	0.211	4.23	0.20
$\Delta v(4\text{-}NO_2C_6H_4OH)$[c,d]	-0.683	-0.140	1.14	0.20
$-\Delta H(BF_3/CH_2Cl_2)$[e]	7.23	4.93	0	0.68
$-\Delta H(HSO_3F/1,2Cl_2C_2H_4)$[e]	4.51	5.70	0.84	1.3

[a]Spectral shifts in the IR (cm^{-1}), parameter units cm^{-1}/(kcal mol^{-1})$^{1/2}$
[b]Spectral shifts in the UV-vis (cm^{-1}), parameter units cm^{-1}/(kcal mol^{-1})$^{1/2}$
[c]Spectral shfts in the UV-vis (kK; 1 kK = 1000 cm^{-1}), parameter units kK/(kcal mol^{-1})$^{1/2}$
[d]Δv is the 4-nitrophenol minus the 4-nitroanisole transition or the 4-nitroanaline minus the N,N-diethyl-4-nitroanaline transition.
[e]Parameter units (kcal mol^{-1})$^{1/2}$

Table IV. ECW Fit of the Carbonyl Frequencies for the Base Adducts of the Acceptor $Rh_2(pfb)_4CO$[a]

Base	$v_{CO}(EXP)$	$v_{CO}(ECW)$[b]
None	2135.8	2135.3
$CH_3C(O)OC_2H_5$	2130.8	2129.0
$(CH_3)_2CO$	2128.6	2128.2
$(C_2H_5O)_3PO$	2126.2	2126.3
$CH_3C(O)N(CH_3)_2$	2125.4	2126.4
$(CH_3)_2SO$	2126.2	2125.9
Bridged ether	2127.0	2127.3
MeIm	2124.7	2122.8
$(C_2H_5)_3N$	2121.6	2120.9
$HC[C_2H_4]_3N$	2120.8	2120.4
CH_3CN	2129.0	2129.5
Cage phosphite	2125.0	2125.9
$(C_2H_5)_2S$	2126.0	2127.2
C_5H_5N	2123.0	2123.7
$4\text{-}CH_3C_5H_4N$	2122.0	2123.2

[a]Data from Bilgrien, C; Drago, R.S.; Vogel, G. C. *Inorg Chem*, **1986**, *25*, 2864. Frequencies in cm^{-1}
[b]Calculated with v_{CO} = -2.751 (±0.829) - 1.892 (±0.147) + 2135.3(±0.8) R_2 = 0.935, x = 1.1, F-ratio 86.56

been detected (2) in the correlation of free energies of formation and activation to ECW.

In molecular modelling of properties that are not related to donor-acceptor interactions but are suspected of having minor contributions from this effect, the $E_A E_B + C_A C_B$ terms are added to the parameters being used to model the property. An improvement in the data fit would signify a donor-acceptor component.

Ion-molecule Interactions

Gas phase energies of reactions between donor molecules and ions do not correlate to equation 1 and one would not expect that they should in view of the different expressions for the energies of dipole-dipole and ion-dipole interactions. Furthermore, there are additional energy contributions to the ion-dipole interaction from the more extensive electron transfer that occurs between the donor and acceptor upon adduct formation to partially neutralize the positive charge on the cation. There also is a significant component to the energy of the gas phase reaction from the dispersion interaction between the ion and the donor. In solution, solvent molecules are displaced from both the donor and acceptor when an adduct is formed and this displacement cancels out most of the dispersion contribution in the adduct bond. It is important to remember that solution reactions are invariably displacement reactions and gas phase reactions are combination reactions. The consequences of these additional energy contributions to the correlation of gas phase ion-molecule reactions is to require the addition of a transfer term. In view of the increased extent of electron transfer and the increased dispersion interaction energy with an increase in the size of a donor molecule or ion, a single transfer term can incorporate both effects for the systems studied to date.

Equations 3a,b are found to provide accurate estimates of the energies of gas phase ion-molecule interactions. For reactions between cations and electron donors, the equation is:

$$-\Delta H = E_A^M E_B + C_A^M C_B + R_A^M T_B \qquad (3a)$$

The R parameter is called receptance and the T parameter transfer to indicate the direction of electron density transfer. For reactions between anions and electron the equation is:

$$-\Delta H = E_B^{an} E_A + C_B^{an} C_A + R_A T_B^{an} \qquad (3b)$$

The E_B and C_B or E_A and C_A parameters from the neutral donor- acceptor fit (Table I and II) are used for the neutral molecule. The R_A and T_B parameters for the neutral molecules and the cation parameters (E_A^M, C_A^M, R_A^M) for cation-donor adducts as well as the anion parameters $(E_B^{an}, C_B^{an}, T_B^{an})$ for acceptor adducts are given in the literature (1,3). It is particularly significant that the donor and acceptor parameters in Tables I and II can be used to interpret this set of data in view of the similar neutral molecule contribution to dipole-dipole and ion-dipole interaction equations.

Bond Energy Parameters

The fit of bond energies can be treated as reactions involving a cation and an anion or as reactions involving atoms. The former requires a knowledge of the dissociation energies, the ionization energies and electron affinities. Consequently, the decision was made (*4*) to fit the bond energies, which are dissociation energies with opposite sign. The parameters for an atom will depend on whether the atom becomes the positive or negative end of the bond dipole in the product molecule. The former are called catimers and described with catimer parameters. The latter are called animers and described with animer parameters. An excellent fit of bond energies for a wide variety of systems including organometallic compounds results (*4*) by substituting the parameters in Table V into equation 4. The E and C parameters have the same electrostatic and covalent

$$-\Delta H = E_{cat}E_{an} + C_{cat}C_{an} + R_{an}T_{cat} \qquad (4)$$

significance as in the ECW equation with subscripts referring to catimers and animers. T_{cat} is the catimer transference parameter and R_{an} is the animer receptance parameter. The product $R_{an}T_{cat}$ indicates the stabilization of the bond that occurs by electron transfer from the neutral atom catimer to the animer and by the dispersion interaction. For homonuclear diatomic molecules, the bond energy is predicted by combining the animer and catimer parameters for the same atom. For a polar molecule,e.g., ICl, an incorrect assignment of the atoms to animer and catimer results in a smaller calculated bond energy than that for the correct assignment. This difference provides a way of assessing bond polarity in systems where the choice is not obvious. As with the neutral adduct fit, deviations from the correlation are found in systems in which steric effects exist or in which pi-bonding exists.

The electrostatic-covalent parameters provide a much better fit of bond energies than the estimates of ionic and covalent contributions to bond energies from electronegativities (*4c*). The parameters can be used instead of electronegativities as a reactivity scale to fit physicochemical measurements. Reported correlations include chlorine quadrupole coupling constants in X-Cl compounds (*1*) and ^{13}C-H coupling constants (*4c*) in CH_3X compounds.

Solvation

The trends in properties of the systems treated above are not influenced by solvent polarity contributions. Recent advances have led to a Unified Solvation Model (*5*) for predicting the non-specific solvation contribution to chemical reactions and spectral shifts. The total specific and non-specific contributions to solvation are predicted by adding a PS' term to equation 1 to produce equation 5. The P parameter describes

$$\Delta\chi = E_A E_B + C_A C_B + PS' + W \qquad (5)$$

Table V. Catimer and Animer Parameters for use in Equation (4)

Catimer Parameters[a]

Catimer (wgt)	E_{ca}	C_{ca}	T_{ca}	Catimer	E_{ca}	C_{ca}	T_{ca}
H- (1)	7.84	13.00	0.52	Br- (0.4)	0.78	12.25	1.31
H_3C- (1)	4.00	11.83	3.37	I- (1)	0.95	8.94	3.37
CH_3CH_2- (1)	4.27	11.45	3.41	Mn- (0.5)	7.68	1.07	7.24
$(CH_3)_2$CH (1)	5.06	11.04	2.37	Co- (0.3)	8.37	2.88	5.31
$(CH_3)_3$C- (1)	4.78	10.77	3.16	Ni- (0.3)	7.44	4.51	6.57
H_5C_6- (0.3)	5.77	12.92	2.22	Sc- (0.3)	12.34	1.28	5.27
$C_6H_5CH_2$- (1)	2.36	10.22	4.59	Cr- (0.3)	8.55	2.05	5.31
CH_3C(O)- (0.3)	4.41	9.80	2.63	Zn- (0.3)	5.53	0.01	4.03
H_3Si- (0.3)	9.13	10.12	1.22	Cu- (1)	7.40	5.12	5.18
$(CH_3)_3$Si- (0.3)	10.29	9.89	1.00	Ag- (0.5)	6.30	4.11	5.63
Li- (1)	10.98	2.77	6.43	η^5-Cp*IrP$(CH_3)_3$H- (1)	3.77	6.42	9.96
Na- (1)	8.03	1.48	7.47	η^5CpRu[P$(CH_3)_3]_2$- (1)	2.92	3.29	4.62
K- (0.3)	9.99	0.26	8.00	DPPEPt- (1)	2.90	1.02	5.45
Rb- (0.4)	9.89	0.31	7.74	$(CO)_5$Mn- (0.5)	2.83	4.71	8.82
Cs- (0.4)	10.65	0.17	7.74	η^5-CpMo$(CO)_3$- (0.5)	0.55	5.31	12.83
Al- (0.5)	11.78	4.50	5.59	η^5-Cp$_2$Zr- (0.3)	7.63	7.06	8.74
In- (0.4)	9.42	3.14	6.81	η^5-Cp'$_3$U- (0.5)	7.02	3.67	5.54
Tl- (0.4)	8.07	1.81	6.69	η^5-Cp*ThOC$(CH_3)_3$ (0.5)	0.31	11.48	6.71
Bi- (0.3)	4.12	7.33	4.60	166-JH$_3$-^{13}C-[b](0.6)	-1.17	6.50	0.89
Cl- (0.3)	0.99	14.30	0.73				

Animer Parameters[a]

Animer (wgt)	E_{an}	C_{an}	R_{an}	Animer (wgt)	E_{an}	C_{an}	R_{an}
-H (1)	2.10	6.62	2.36	-OCH$_3$ (1)	5.07	4.97	1.14
-CH$_3$ (1)	2.36	6.66	0.60	-NH$_2$ (0.5)	3.98	5.78	0.53
-CH$_2$CH$_3$ (1)	1.77	6.70	0.20	-NO$_2$ (0.5)	4.13	3.47	1.28
-CH$_2$CH$_2$CH$_2$CH$_3$(0.3)	1.56	6.51	0.36	-SH (1)	3.44	4.78	1.11
-H$_2$CC$_6$H$_5$ (0.5)	0.89	6.12	0.18	-SCH$_3$ (0.5)	3.64	4.84	1.08
-H$_2$CSi$(CH_3)_3$ (0.3)	1.59	6.64	0.83	-F (1)	9.81	4.45	2.52
-H$_2$CC(O)CH$_3$ (0.3)	2.12	6.23	1.24	-Cl (1)	7.24	3.41	3.86
-(O)CCH$_3$ (0.5)	0.64	6.17	1.58	-Br (1)	6.12	2.91	3.53
-C$_6$H$_5$ (1)	3.43	6.54	2.63	-I (1)	5.14	2.23	3.26
-CH=CH$_2$ (1)	2.71	6.42	1.96	-Au (1)	4.90	2.50	1.13
-C°CC$_6$H$_5$ (0.5)	6.43	5.90	5.44	-Ag (1)	1.39	3.12	2.53
-CN (1)	5.52	6.19	5.87	-Cu (1)	2.29	2.63	2.33
-CF$_3$ (0.3)	4.18	5.44	5.88	-Bi (0.5)	1.45	5.05	1.07
-CCl$_3$ (0.3)	3.53	5.06	4.45	De^{2}Qq(Cl)[c] (0.5)[c]	5.51	-0.42	7.18
-OH (1)	7.12	4.76	2.55	14eV-IE[d] (0.2)[d]	0	0	1.25
-OC$_6$H$_5$ (0.3)	3.86	4.04	0.77				

[a]Enthalpy parameters in units of $(kcal\ mol^{-1})^{1/2}$

[b]The ^{13}C-H coupling constant for H$_3$C-X derivatives. The quantity calculated is 166-J^{13}CH$_3$. The value of 166 is for an sp^2 carbon so the larger the number the closer to sp^3 the carbon hybridization. Parameter units H$_3$/$(kcal\ mol^{-1})^{1/2}$

[c]The chlorine quadrupole coupling constant for M-Cl compounds where chlorine is the animer. The e^{2}Qq(Cl) value + 109.7 is fit where 109.7 is the value for a free chlorine atom in megahertz. Parameter units MHz/$(kcal\ mol^{-1})^{1/2}$

[d]14eV minus the ionization energy in electron volts. Parameter units MHz/$(kcal\ mol^{-1})^1$

the solute response to solvent polarity and S' is a measure of solvent polarity. Analysis with equation 5, produces the specific ($E_A E_B + C_A C_B$) and non-specific (PS') contributions to the solvation process. Again, it is important to emphasize that the E and C parameters used in these data fits (for the specific interaction) are those in Tables I and II. The reader is referred to the literature (*1,6*) for compilations of the solvation parameters, S' and P.

The ECT Model as a Tool to Accompany Quantum Chemical Calculations

The use of electrostatic-covalent parameters to accompany quantum mechanics in molecular modelling will be illustrated with the ECT model. Since our electrostatic-covalent parameter sets encompass all known sigma bond energies, they can be used in conjunction with molecular orbital calculations, e.g. ab initio quantum calculations, to compare calculated trends from theory against those for all known experimental data, *vide infra*. Furthermore, a major disadvantage of high-level calculations is that often a simpler, more chemically intuitive picture of bonding is sacrificed. Not only do ECT and ECW provide an intuitive picture, but the use of these models will be shown to suggest new avenues for inquiry in computational applications.

In a recent publication (*4c*), the Cl_3Ti-X and $(CO)_4Co$-X bond energies calculated (*7*) with density functional theory were analyzed with equation 4 using the parameters for X in Table 5. Average deviations of 3.6 and 3.1 kcal mol^{-1}, respectively, result between calculated and ECT predictions for bond energies that spanned a range of ~40 kcal mol^{-1}. The poor fit indicates that there is no known experimental data set that parallels these calculated trends to a reasonable (~1 kcal mol^{-1}) in the predicted bond energies. Either the calculations are at best good to ~10% or there are some unusual bonding effects in some of these compounds.

Below an application to organometallic compounds is presented to illustrate data fits to equation 4 and to highlight the ECT model as a valuable tool for the analysis of quantum chemical calculations.

Analysis of Metal-Ligand Bonding in Technetium Organometallics

Technetium is of importance as a radiometal in medical imaging (*8*) and also as a by-product of nuclear materials processing (*9*). A major challenge in the study of technetium compounds results from their wide diversity of formal oxidation states (from -1 to +7) and coordination numbers (from 4 to 7). Experimental results and theoretical computations point to M(=NR)$_3$ complexes of d^0 transition metals as being able to activate the C-H bonds of hydrocarbons including methane. On the other hand, d^2-analogues of XTc(=NR)$_3$ of C$_{3v}$ symmetry are stable enough to be isolated. Hence the strength and polarity of the interaction between X and M in these complexes is of interest in the design of a system to activate hydrocarbons. As part of a joint theory-experiment study, Tc-tris(imido) complexes of the form Tc(=NR)$_3$X, **1**, were studied.(*10-15*)

Three-coordinate tris(imido) species are crucial to understanding the bonding in **1**.(*14*) Calculations indicate that d^0-M(NH)$_3$ complexes (Mo and W) are pyramidal while Tc d^1 and Os d^2 analogues are trigonal planar. VSEPR considerations would predict a planar d^0 configuration instead of the calculated pyramidal structure. However, the strong π-bonding of the imido ligand is enhanced by going to a pyramidal structure. For d^1 and d^2 complexes, the singly- and doubly-occupied $d\sigma$ orbital, respectively, is greatly destabilized by pyramidalization, leading to a trigonal planar ground state.

The geometry of $XTc(NH)_3$ is sensitive to changes in the X-group. One can fit the X-Tc-N_{imido} angle determined by quantum calculation to equation 6. For X groups of lower symmetry than C_{3v}, the average X-Tc-N_{imido} angle is used. Substituting the E_{an}, C_{an} and R_{an} parameters from Table V and the corresponding angles into equation 5 leads to a series of simultaneous equations of the form of equation 6.

$$\theta_{X\text{-}Tc\text{--}N(0)} = E_{an}(X)*k_1 + C_{an}(X)*k_2 + R_{an}(X)*k_3 + W \qquad (6)$$

Using parameters for twelve X ligands and their calculated X-Tc-N_{imido} angles, a least squares best-fit is obtained for k_1, k_2, k_3 and W. The W value corresponds to the θ value in $XTc(NH)_3$ when $E_{an} = C_{an} = R_{an} = 0$ for X. Six additional X groups in the ECT database ($X = Et$, $C(O)CH_3$, vinyl, $C=CH$, OCH_3 and SCH_3) are of a size to be amenable to calculations on $Tc(=NH)_3X$. The X-Tc-N_{imido} angles predicted by substituting the parameters from the ECT fit with E_{an}, C_{an}, and R_{an} for these six groups into equation 6 are consistent with those subsequently determined by quantum methods with an average difference of 0.7Å.

The ECT parameters and weights in Table V are used in a fit of the quantum mechanical angles, θ, leading to the regression equation 7 with $r^2 = 0.90$, an average deviation of 0.7° and an F-ratio of 43.6. The ECT angles, θ_{ECT}, and θ are given in Table VI.

$$\theta_{ECT} = 0.77(\pm 0.11) E_{an} -0.59(\pm 0.17) C_{an} -0.56(\pm 0.42) R_{an} + 103.6 \qquad (7)$$

The parameters suggest that as the X-$Tc(=NH)_3$ bond becomes more $X^{\delta-}$-$Tc^{\delta+}$ polarized, the electrostatic interaction will increase by increasing the X-Tc-N_{imido} angle to above 103.6°. A larger electrostatic interaction results as nitrogen lone pair -X repulsions decrease by increasing the angle. As the X-Tc-bond becomes more covalent, the X becomes less anionic, repulsions decrease and d_z^2 is populated. Populating d_z^2, as in the d^1 complex, leads to a trigonal planar complex decreasing θ. The sign of the transfer term suggest that as Tc becomes more positive, the d-orbitals contract and π-bonding to the imido group increases causing θ to decrease. The X-Tc-N_{imido} angle is thus an important property for probing the bonding in $Tc(=NR)_3X$ (10).

High-level calculations are very time consuming for large complexes with heavy metals and it would be very difficult to calculate all animers whose ECT values are known. Thus, the ECT model can not only be used for analysis of data, but its predictive power provides an efficient tool in the search for new target compounds to study.

Table VI. ECT Fit of Calculated X-Tc-N Angles for Tc(=NR)$_3$X

X group	θ_{QM}	$\theta_{ECT}{}^a$	X group	θ_{QM}	$\theta_{ECT}{}^a$
H	98.9	100.1	Cl	105.0	105.0
CF$_3$	99.7	100.3	OH	105.4	104.8
CH$_3$	100.5	101.1	F	106.2	107.1
CN	101.2	100.9	C$_2$H$_5$	100.3	100.9
CCl$_3$	101.5	100.8	CH$_3$CO$_2$	100.7	99.5
SH	103.0	102.8	-C=CH$_2$	101.0	100.8
I	103.8	104.4	-C≡CH	102.4	102.0
NH$_2$	104.0	102.9	SCH$_3$	102.7	102.9
Br	104.6	104.6	OCH$_3$	105.2	103.9

$^a\theta$ in degrees calculated with: $\theta = 0.77 E_{an} - 0.59 C_{an} - 0.56 R_{an} + 103.6$

Apart from the observations of good statistical correlations there are several points of interest which reveal insights from ECT into the chemistry of the tris(imido) complexes, and their potential as metastable precursors to hydrocarbon C-H activating M(=NR)$_3$ species. First, observation of a reasonable correlation with animer parameters suggests that, for chemically diverse systems, the dominant description of Tc(-NR)$_3$X is X[Tc(=NR)$_3$]$^+$, with a d^0 configuration at the metal. Quantum calculations for methane activation by [2σ + 2π] addition across the metal-imido active sites show large kinetic barriers to C-H bond scission when the metal configuration is not d^0 due to a repulsive interaction between substrate and activating complex in the early stages of the activation event (*11a*). Second, the positive coefficient of E_{an} indicates that as the capacity for X to participate in electrostatic/ionic bonding increases so does the X-Tc-N$_{imido}$ angle. Third, since the ECT model does not include π-bonding effects, the existence of good correlations suggest that the primary influence of the X group is transmitted to the tris(imido) moiety through the σ framework. This is rationalized by the fact that the metal d-orbitals in these pseudotetrahedral complexes are monopolized by the strongly π-bonding imido ligand. The imido MN σ bond is insensitive to changes in the ligand environment (*14*). Hence, X should have a limited effect on tris(imido) reactivity in a metastable M(=NR)$_3$X. The X ligand thus quenches the reactivity of the M(=NR)$_3$X by engaging the σ acceptor orbital of d^0-M(=NR)$_3$. Experiments and calculations show the metal-based σ acceptor orbital plays a pivotal role in capturing CH bonds prior to scission (*11a,15*).

The foregoing analysis supports the proposal that it may be possible to design a metastable M(=NR)$_3$X for hydrocarbon activation by using X as a "place holder" for a methane substrate. Experimental data in support of this proposal can be found in the work of Wigley (*16*) who has shown that CH bonds of terminal alkynes can be activated by W(=NAr)$_3$(PMe$_3$) to yield W(=NAr)$_2$(NHR)(C≡CR). It is desirable to find X groups for Tc which are weakly bound and can be displaced by methane in either dissociative or associative mechanisms. The ECT analysis allows us to interpret the results of quantum chemical calculations with these factors in mind. It also provides a valuable tool for predicting systems which have the required electronic and structural characteristics, and hence ligands which engender the desired properties in the resulting organometallic complex.

Acknowledgments

The research by Thomas R. Cundari was supported by the National Science Foundation - CHE-9614346.

Literature Cited

1. Drago, R. S. *Applications of Electrostatic-Covalent Models in Chemistry;* Surfside Scientific: Gainesville, FL, 1994.
2. a) Drago, R. S. *Inorg. Chem.* **1990**, *29*, 1379.
    b) Drago, R. S.; Vogel, G. C. *J. Am. Chem. Soc.* **1992**, *114*, 9527.
    c) Drago, R. S.; Dadmun, A. P.; Vogel, G. C. *Inorg. Chem.* **1993**, *32*, 2473.
    d) Drago, R. S.; Vogel, G. C. *J. Chem. Edu.* **1996**, *73*, 701.
3. a) Drago, R. S.; Ferris, D. C.; Wong, N. *J. Am. Chem. Soc.* **1990**, *112*, 8953.
    b) Drago, R. S.; Wong, N. M.; Ferris, D. C *J. Am. Chem. Soc.* **1991**, *113*, 1970.
4. a) Drago, R. S. *J. Phys. Chem.* **1991**, *95*, 9800.
    b) Drago, R. S.; Wong, N. M.; Ferris, D. C. *J. Am. Chem. Soc.* **1992**, *114*, 91.
    c) Drago, R. S. Wong, N. M. *Inorg. Chem.* **1995**, *34*, 4004.
    d) Drago, R. S.; Wong, N. M. *J. Chem. Educ.* **1996**, *73*(2), 123.
5. a) Drago, R. S. *J. Chem. Soc. Perkin Trans. 2* **1992**, 1827.
    b) Drago, R. S. *J. Org. Chem.* **1992**, *57*, 6547.
6. a) Drago, R. S.; Hirsch, M. S.; Ferris, D. C.; Chronister, C. W. *J. Chem. Soc. Perkin Trans. 2* **1994**, 219.
    b) Ferris, D. C.; Drago, R. S. *J. Am. Chem. Soc.* **1994**, *116*, 7509.
    c) Drago, R. S.; Ferris, D. C. *J. Phys. Chem.* **1995**, *99*, 6563.
    d) George, J. E.; Drago, R. S. *Inorg. Chem.* **1996**, *35*, 239.
7. Ziegler, T.; Tschinke, V.; Versulius, L.; Baerends, E. J.; Ravene, W. *Polyhedron* **1988**, *7*, 1625.
8. Jurisson, S.; Berning, D.; Jia, W.; Ma, D. *Chem. Rev.* **1993**, *93*, 1137.
9. Bunker, B.; Virden, J.; Kuhn, B.; Quinn, R. In *Encyclopedia of Energy Technology and the Environment*; Wiley: New York, 1995.
10. Benson, M. T.; Bryan, J. C.; Burrell, A. K.; Cundari, T. R. *Inorg. Chem.* **1995**, *34*, 2348 and references therein.
11. a) Benson, M. T.; Cundari, T. R.; Moody, E. W. in *Aspects of C-H Activation*; special issue of *J. Organomet. Chem.* **1995**, *504*, 1
    b) Wolczanski, P. T. (Cornell Univ.) - personal communication.
12. Anhaus, J. T.; Kee, T. P.; Schofield, M. H.; Schrock, R. R. *J. Am. Chem. Soc.* **1990**, *112*, 1642.
13. Williams, D. S.; Anhaus, J. T.; Schofield, M. H.; Schrock, R. R.; Davis, W. M. *J. Am. Chem. Soc.* **1991**, *113*, 5480.
14. Cundari, T. R. *J. Am. Chem. Soc.* **1992**, *114*, 7879.
15. Schaller, C. P.; Cummins, C. C.; Wolczanski, P. T. *J. Am. Chem. Soc.* **1996**, *118*, 591.
16. Chao, Y. W.; Rodgers, P. M.; Wigley, D. E.; Alexander, S. J.; Rheingold, A. L. *J. Am. Chem. Soc.* **1991**, *113*, 6326.

Chapter 7

Molecular Mechanics in Computational Thermochemistry

Donald W. Rogers

Department of Chemistry, Long Island University, Brooklyn, NY 11201

Molecular mechanics (MM) is a fast, accurate computational method that uses a classical model of the molecule as a collection of point masses (atoms) connected by springs (chemical bonds) that obey or nearly obey Hooke's law. MM has long been successful in computing molecular structures and enthalpies of formation of common chemical compounds. Recent developments have extended use of MM to less common chemical species and to computation of entropies, Gibbs free energies and equilibrium constants. MM does not predict electronic properties of molecules.

One approach to molecular energy and structure determination is to regard the geometry of a molecule as the result of classical mechanics operating, through elastic chemical bonds, on the connected masses of the atoms that constitute the molecule. This approach is called molecular mechanics, (MM) (*1*). In its earliest form, molecular mechanics was a purely classical mechanical method but very soon, quantum mechanics entered into the mix, first as a subroutine for performing SCF (*2*) and VESCF (*3*) molecular orbital calculations and later in determining force constants by *ab initio* quantum-mechanical procedures (*4*).

The greatest limitation to molecular structure and energy calculations in general is the amount of computer time required to do them. Computation time is much less for MM calculations than it is for advanced molecular orbital (MO) methods. Aside from speed, MM calculations can be very accurate, for example, $\Delta_f H$ results for smaller alkanes compare favorably with the best experimental studies (*5, 6*). Root mean square errors of 0.5 kcal mol^{-1} have been found for alkanes and alkenes of complexity up to adamantane, 2,2-metaparacyclophane and bicyclo[2.2.2]octane (*5*). MM can be used to find and correct erroneous data in the literature because parameters drawn from many data are less subject to random errors than individual measurements (*7*).

The principal drawbacks of MM calculations, as contrasted to MO calculations, are that they tell us little about excited states or the course of chemical reactions (see, however, ref. 8) and that MM is not applicable to molecules containing moieties for which parameterization is incomplete. Like any fully-parameterized method (*9*) MM enables one to make predictions on the basis of a large mass of accumulated data, but does not predict phenomena that have not yet been observed.

The range of MM methods is extremely wide. Allinger's MM family of programs has been applied to many categories of the more common, medium sized, organic molecules including alkanes, alkenes, allenes and acetylenes (*5*), aromatics, polycyclics (*10*), deuterium compounds (*11*), nitrogen-containing heterocycles (*12*), amides, polypeptides and proteins (*13*), carboxylic acids, esters, and lactones (*14*). The programs and force fields CHARMm and AMBER are specifically defined for bio-organic molecules and pharmaceutical applications (*15*). Halgren (*4a*) takes the intended range of future force fields as "most organic structural types represented in the *Merck Index* or the *Fine Chemicals Directory*".

When MO calculations are desired, a strategy often seen in contemporary computational chemistry is to calculate the energy and geometry of the target molecule by MM and use the MM geometry to construct an input file for a semi-empirical MO calculation, an *ab initio* computation, or both. MM calculations with visualization are probably the easiest and fastest way to make sure one is working with the right conformer, which is not always a trivial problem. The formats of MM and MO input files are different, but conversion programs are available which reduce the burden of conversion or remove it altogether in simple cases (*16*).

Allinger's series of MM programs, MMx, (where x = 2, 3, or 4) produces thermochemical information in its most immediately useful form, as standard enthalpies of formation, absolute entropies, and Gibbs free energies of formation at 25° C. In contrast, *ab initio* programs, for example, yield total energies of the target molecules at 0 K, requiring ancillary heat capacity (C_p) calculations to reach a practical enthalpy of reaction, $\Delta_r H$. This is not merely an inconvenience; C_p may not be known and may be difficult to calculate accurately.

Many molecules are too large to encourage *ab initio* or even semi-empirical MO calculations using existing computers. For this reason MM calculations as complete computations, not as a preliminary calculation, are important, especially in biochemistry, molecular biology, pharmacology and pharmaceutical chemistry. Research-level MM calculations can be carried out using programs that are readily available and run on a minicomputer, at a workstation, or on an off-the-shelf microcomputer. This chapter gives a brief description of fundamentals and input data files and gives results for several simple illustrative MM problems involving enthalpy, entropy, Gibbs free energy, and equilibrium constants of hydrocarbons.

The focus of this chapter will be on force fields capable of producing accurate thermodynamic functions, primarily enthalpies of formation ($\Delta_f H$) at 298 K of organic molecules in the gas phase. We shall refer to MM3 and MM4 as up-to-date representatives of a large class of force fields, but one that excludes the MM force fields devoted to study of the geometric aspects of chemical reactions in biological systems.

Harmonic Motion

In the approximation of simple harmonic motion, Hooke's law gives

$$V = (k/2)(r - r_0)^2 \tag{1}$$

for the potential energy V of a mass m connected to the wall by a spring with a force constant k (17). The frequency of oscillation v is

$$v = (1/2\pi)(k/m)^{1/2} \tag{2}$$

The displacement r from an equilibrium position r_0 at any time is

$$r - r_0 = (2V/k)^{1/2} \tag{3}$$

If there are many masses connected by many springs, there will be an over-all equilibrium configuration that leads to an over-all potential energy minimum but this need not and, in general, will not be at the rest lengths of all or any of the springs. The equilibrium configuration will be a compromise of all forces acting on all masses.

Starting with an initial guess $\mathbf{R_0}$ as the matrix of all interparticle distances, the potential energy will be above its minimum because the guess will not place the masses at positions that satisfy the best compromise of forces among them. A systematic search like the Newton-Raphson method ($1a$) is used to find the composite minimum where all $dV/dx = 0$. The cumulative displacement of each mass that minimizes the potential energy constitutes a displacement matrix ($\mathbf{R - R_0}$). Knowing the x, y, and z coordinates of the initial guess, one has the x, y, and z coordinates of all masses after the displacements that brought about energy minimization. Knowing all the forces acting on each mass, one can add up the potential energy contributions to obtain the total potential energy of the system.

Molecular Mechanics

In molecular mechanics, physical springs of the macroscopic analog are replaced by mathematical functions describing the various interactions within the molecule. It turns out that Hooke's law is a rather good approximation to the stretching mode of a chemical bond. Bending forces, torsional (twisting) forces, and van der Waals attractions and repulsions must also be included. Force constants of the springs are represented by a collection of mathematical parameters. The parameters are called, collectively, a *force field*. The total potential energy of the system is called the *steric* energy.

Force constants can be obtained from vibrational spectra through equation 2. For example, from the C-H stretching frequency at 2900 cm^{-1} in the IR spectrum of hydrocarbons, we find a force constant of 4.57×10^2 Newtons per meter (N·m^{-1}). In the MM3 force field (18), this is given as 4.74 millidynes·angstrom^{-1} (md·Å^{-1}) = 474 N·m^{-1}. There is imperfect agreement between the two values for the C-H force constant because 2900 cm^{-1} is a single frequency arbitrarily selected from a range of C-H stretching frequencies in real molecules, and also because the MM force constants are fitted (at least partly by trial and error) to reproduce molecular structures and energies, not just frequencies.

We can look at the force concept for a system of many connected masses in another way. Expand the potential energy function about its minimum at r_0 in a Taylor series

$$dV(r) = V_0 + \Sigma\ (\partial V/\partial r_i)\ dr_i + (1/2)\Sigma\ (\partial^2 V/\partial r_i\ \partial r_j)\ dr_i\ dr_j + \text{higher terms} \qquad (4)$$

where dr is a small displacement from the equilibrium position. We define $V_0 = 0$ at r_0, and the slope of the potential function is zero at its minimum, so the first two terms of the expansion drop out. Within the harmonic oscillator approximation, the "higher terms" in equation 4 are zero and the potential function is a parabola open upward. The force constant is the second derivative of the harmonic potential with respect to displacement. The sums in equation 4 run over pairs of atom coordinates from i = 1 to 3n and from j = 1 to 3n, where n is the number of atoms in the molecule, each of which is specified by three coordinates. Hence the forces in the system can be arranged in a 3n x 3n matrix called a *hessian matrix* **G,** where the elements of **G** are $G_{ij} = \partial^2 V(x)/\partial x_i \partial x_j$ and g_i is the gradient vector $\partial V(x)/\partial x_i$. If the higher terms are important in equation 4, they contribute to the *anharmonicity* of the oscillator.

The origin of the term *coupling* can be seen by considering a system of a mass connected to a wall facing a mirror image mass connected to another wall by another spring,

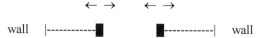

There are two equations of motion, one for each oscillator, leading to a 2 x 2 matrix of force constants (*19*). The motions of the two masses are uncoupled and the force constant matrix is diagonal. If a third spring is added to the system so that it connects the left and right hand masses, their motion is influenced by each other and by the force constant of the third spring. This is a system of coupled masses and springs. The force constant matrix has off-diagonal elements brought about by the presence of the coupling spring.

Along with force constants, it is also necessary to have an equilibrium or normal bond length to calculate the potential energy contribution to the steric energy of a molecule made by deformation of, for example, the C-H bond, using equation 1. The length of the C-H bond is commonly known to be about 1.1 Å (*20*). In the MM3 force field, r_0 is taken to be 1.1120 Å = 111.2 picometers. The reader has already noticed that there is considerable mixing of SI and non-SI notation and unit systems in the molecular mechanics literature. Much of this is caused by the embedded custom of expressing IR peaks in wavenumbers. We have tried to make the relationship between the systems decipherable, if not clear, by giving key results in both systems.

If we can calculate the potential energy contribution of a C-H stretch to the steric energy of a molecule, we can calculate the similar contribution made by other stretching modes of motion, those for C-C, C=C, C=O and so on. Moreover, if we can do this for stretching motions, we can do it for other kinds of motions such as bond angle bending, and torsional deformations. Even van der Waals interactions operating over the space between non-bonded atoms are included, though not as a harmonic function. If the sum of all interactions is taken into account, the energy of the molecule in question above that of the molecule without special steric interactions can be calculated. If the enthalpy of formation of the molecule is estimated as a bond energy sum taken from a basis set of molecules with "normal" steric effects, the enthalpy of formation of the target molecule can be obtained by adding the excess or *steric* energy. If the bond energy scheme used in the $\Delta_f H$ calculation is drawn from a different basis set, one comprised of molecules with steric effects adjudged to be negligible, then the difference between $\Delta_f H$ calculated from a normal set and $\Delta_f H$ calculated from the strainless set is the strain energy. Recall that once the distances and angles of all atoms in a molecule are known, the entire geometry is known, as is the energy relative to a known standard energy. Therefore we know the structure and energy of the target molecule to an accuracy determined by the accuracy of the force field and the bond-energy scheme.

Using an MM program, then, requires an initial, approximate geometry of the target molecule contained in an input file. Operation of the program involves many changes in the coordinates of each atom in the molecule, solution of the potential energy equations and a methodical (iterative) approach to the over-all energy minimum. Iterations continue until further small changes in the positions of the atoms no longer change the energy. Equations for the potential energy and its first and second derivatives (slope and curvature) are all rather simple, partly accounting for the speed of molecular mechanical calculations relative to advanced molecular orbital methods.

The Energy Equation. The energy equation may include a very simple set of interactions (*1c, 21*)

$$V = \Sigma\, V_{stretch} + \Sigma\, V_{bend} + \Sigma\, V_{tors} + \Sigma\, V_{vdW} + \cdots \qquad (5)$$

where the subscripted sums are taken over all bonds, simple angles, torsional angles and van der Waals repulsions, and each contribution to the sum requires an appropriate force constant. More recent force fields contain more terms in the energy equation, for example the MM3 equation (*18,22*) has 11 sums, the 4 given above plus 7 more for out-of-plane bending, stretch-bend interactions, dipole interactions and so on. The MM4 equation contains 6 new terms while retaining most of the terms in the MM3 equation with two slight modifications (*5b*). The new terms in MM3 and MM4 were added so that more kinds of molecules can be treated and so that more detailed information, especially spectral information and the vibrational contribution to molecular enthalpy and entropy can be calculated. The thrust of this article is computational *thermochemistry* so little more about structure and spectra will be said except where there are thermochemical implications.

Term Equations: The Stretching Term. Each term in the energy equation is a sum of terms for a contributor to the potential energy. A simple example is the energy contribution from a bond stretch, say the C-H bond stretch in ethane. As a part of the input file for MM3, for example, each atom type is designated by a number, in this case, 1 for an sp^3 carbon and 5 for a hydrogen. The stretching equation (essentially, equation 1 with an anharmonic term) is

$$ENERGY = 71.94(KS)(DR)(DR)(1-(CS)(DR)+(7/12)((CS)(DR))**2)$$
$$CS = 2.550$$

where $71.94 = 143.88/2$ is a conversion factor from $(md \cdot Å^{-1} \cdot molecule^{-1})$ to $(kcal \cdot mol^{-1} \cdot Å^{-1})$, the denominator 2 comes from $(1/2)KS$, and CS is the cubic stretch constant which is 2.55 in MM3. DR is the difference between the equilibrium bond length and the actual bond length at some displacement. KS for the CH bond (1-5) is 4.7400 $md \cdot Å^{-1}$ and r_0 is 1.112 Å.

Bending may be handled as a harmonic (or higher) potential

$$V = (k/2)(\theta - \theta_0)^2 \tag{6}$$

where θ is a simple angle. Internal rotation can be expressed

$$V_\phi = \Sigma K_\phi(1 + s \cos \phi) \tag{7}$$

where ϕ is the angle of rotation and s is the rotational barrier multiplicity, for example 3 for the threefold barrier in a methyl rotation (*23*).

Eqs. 1, 6, and 7 are only three of the possible expressions for the potential energy of stretching, bending and rotation. Indefinitely many other forms may be used, for example the Morse curve is sometimes substituted for equation 1. Two potential energy expressions used for van der Waals interactions are the Lennard-Jones 6/12 potential function

$$V_{vdW} = \varepsilon\{(r_0/r)^{12} - 2(r_0/r)^6 \} \tag{8}$$

or some modification thereof, and

$$V_{vdW} = A \exp(-Br) - C/r^6 \tag{9}$$

where A, B, and C are adjustable parameters. These equations are used in their modified forms so as to achieve more flexibility in fitting experimental data, for example by replacing r_0 in equation 8 by fitting parameters. The same is true for modification and fitting of some form of the coulombic equation

$$V_c = q_i q_j/\varepsilon r_{ij} \tag{10}$$

where ε is the dielectric constant, to bring the coulombic energy, V_c, of atom-dipole or dipole-dipole interactions into the sum representing steric energy (1c). Hydrogen bonding brings in another term, and so on.

A simple example of a cross term is the stretch-bend interaction. If the angle ABC, having elastic AB and BC bonds, is closed, the bonds stretch due to repulsion between atoms A and C. The opposite is true if the angle is opened. Thus, the stretch and bend of the system ABC are not independent, rather they are coupled. The stretch-bend coupling term might take the bilinear form

$$V_{SB} = K_{SB} (r - r_0) (\theta - \theta_0) \tag{11}$$

where r is the bond length and θ is the simple angle.

Aside, perhaps, from the simple stretch and simple bend, no term is obligatory in all force fields hence the marked differences in force fields. The energy deficit from a missing term can be made up in the parameterization of other terms. This rather arbitrary apportioning of energies among terms in the over-all energy equation (equation 5) means that undue weight should not be placed on small differences in the contributing energy terms; it is the total energy that counts.

To see how these potentials contribute to the steric energy and ultimately to the enthalpy of formation of the target molecule, fill in the numbers for a 0.0011 angstrom stretch in ethane, once again using MM3 as a specific example

$$E_s{}^{C\text{-}H} = 0.000413\{1 - 0.002805 + \sim 0.0\} = 00041 \text{ kcal mol}^{-1}$$

There are 6 C-H bonds in ethane, hence the C-H contribution to the potential energy in ethane is 6(0.0004) = 0.0024 kcal/mol, not much, but then ethane is not sterically hindered, and, except for a small methyl-methyl repulsion, the C-H bond length in ethane is almost r_0 (1.1131 vs. 1.1120 angstroms, to be exact). When the slight C-C stretching energy is calculated by the same method, and added to the total C-H energy, one has accounted for all the bond stretching (or compression) energy. The sum of the C-H and C-C energies appears in the output file as COMPRESSION 0.0267 which is the first line of a 9 line table

COMPRESSION	0.0267
BENDING	0.0452
BEND-BEND	-0.0058
STRETCH-BEND	0.0083
VANDERWAALS	
1,4 ENERGY	0.9003
OTHER	0.0000
TORSIONAL	0.0003
TORSION-STRETCH	0.0000

The final steric energy is 0.9750 kcal. This is the steric energy that will be carried forward to the enthalpy of formation calculation below.

Force Fields. In the history of molecular mechanics (see ref. 1) many force fields have been created. Some have appealing acronyms like CHARMm, (24) YETI, (25) and AMBER (26) and some are named in a more sedate way like MM3, (18, 22) MMX, (27) and CFF93 (28). (Citations to 16 force fields are given in ref. 4a) Force fields parameterized for biological molecules are usually rather simple as befits a program that must minimize energies for many bonds connecting many atoms to form large molecules. MM procedures exist that can handle thousands of atoms.

An ideal force field would enable one to calculate all the mechanical properties of any molecule with complete accuracy. The three major categories of properties we wish to calculate from real force fields, however, are structure, energy and vibrational spectroscopic resonances. These categories, respectively, place increasingly stringent demands on the force field. A relatively simple force field can be used to calculate molecular structures with good accuracy and several have been in use for some time (29). Such a field might operate on the assumption that harmonic motions are uncoupled, or it might include a few essential coupling terms.

Calculating both molecular structures and molecular energies requires a more refined force field. One might depart from the simple harmonic assumption to include anharmonic terms in the energy equations and to recognize that motions of atoms within molecules are coupled, thus more off-diagonal elements in the hessian matrix are required.

Calculating structure, energy, *and* spectroscopic resonances requires an even more complete force field. Spectroscopic resonances depend upon the fine detail of the shape of the energy surface in the immediate vicinity of the minimum. More coupling terms will be needed. Even if the harmonic approximation is made, one needs to know how "open" or "closed" the parabolic potential well is, for this determines the spacing of the energy levels within it.

These three increasingly sophisticated force fields have been called force fields of class 1, class 2, and class 3 by Hagler (28). Historically, force fields have not been developed along these logical lines. Instead, force fields were developed to calculate some specific property of molecules, for example, Lifson's early force field to calculate spectral peaks (30). Each force field so developed was successful for the class of properties it was designed to predict, but not very successful in predicting other molecular properties. The question of whether a perfect force field even exists arises naturally at this point. Perhaps real molecules are designed in such a way that the transferability postulate breaks down, that is, perhaps each bond is unique each time it appears. A decade ago it seemed as though transferability might ultimately fail, but force field development since then, in particular with the most recent in the MM family, MM4, makes it appear that a force field can be achieved that produces structure, enthalpy and vibrational energy-level separations (hence entropy) within present experimental thermochemical accuracy.

Programs. A growing list of about 100 molecular modeling programs for micro-mini- and supercomputers and sources (along with relevant graphics, data bases, etc.) has been compiled and is regularly updated by Boyd (31). Many packages contain more than one program. For example, PCMODEL (27) contains a graphical user interface, MMX, and file conversion (e.g., MM to MOPAC) capabilities (16a). MM3 contains a Saunders-type searching program (*32*) as part of the package (see **Local Minima**, below). PCMODEL has been upgraded from version V to version VI (*27*) and upgrades of MM3 to MM4 (commercial delivery anticipated at this writing) and of MMFF93 to MMFF94 have already been noted (*4a,5a*). Programs that can be obtained from QCPE deserve special note. See ref. 5a for sources.

Creating an Input File. Most MM files are created using a graphical user interface which permits the user to "draw" an approximate molecular geometry using a mouse. Coordinates of the approximate geometry are automatically calculated and entered into an MM input file along with the appropriate control information. In drawing hydrocarbons, for example, only the carbon framework need be drawn and double and triple bonds inserted. The file creation program inserts hydrogen atoms where they are needed as specified by the unused valences of carbon, taking into account the geometry appropriate to the hybridization of carbon. Input files can be created from scratch using an ordinary editor and it is worth while to do so to get a full understanding of input file structure so as to troubleshoot faulty input.

A minimal MM3 input file for ethene, for example, contains a block of digits specifying the number of atoms, number of connected atom lists, and attached atoms, followed by the Cartesian coordinates of a starting geometry along with digits that identify the atom types, 2 for sp^2 carbon and 5 for hydrogen (see users manuals). The starting geometry need be only a reasonable first guess at the true geometry of the molecule.

There are many options and even the most casual user will need some of them. For our purposes, a 1 in column 65 is obligatory because it brings about a $\Delta_f H$ calculation. Entering a 1 to activate an option is called "switching" the option on. Other common switches include *conjugated pi-system calculations*, in column 61, which necessitates further input specifying the nature of the pi system. Switches also control the *amount of output*, a number of *symmetry matrices* for simplifying long calculations, a number of *replacement cards* to be read in when constructing a new input file from an old one, and activation of the *dihedral driver* (see below). Many of these options require new lines of ancillary data as part of the input file. Groups of atoms can be *restricted to motion* along a line, motion in a plane, or they may be *frozen* entirely. Finally, new or *replacement parameters* can be entered into a file that is deleted at the end of a run. Adding or replacing parameters in effect creates a custom force field for the problem at hand. Permanent changes to the parameter files should not be made.

The number of atom types varies with the force field. MM3 currently lists 109 atom types. Many atom types are different for the same element in different molecular environments. For example, C(sp), C(sp^2), C(sp^3), C(sp^2, cyclobutene) C(sp^3, cyclobutane), etc. are different carbon atom types with different parameters and different atom type designators in the input file. A complete listing of the atom types comes with the program documentation.

Given all the options available, writing an input file can become a complicated process, but the beginner can leave most options idle and bring them in one at a time. Along with the graphical user interfaces available, there are also interactive question and answer programs to facilitate input file creation (e.g., MINP(92), part of the

MM3 package). Files can be converted from other formats like MOPAC, using conversion options in PCMODEL (*16a*) or the conversion program BABEL (*16*).

The Output File. Part of the output created by MM is a final optimized geometry, (which can be modified as the input for future runs). In general the bond lengths are not at their "normal" values, but have been stretched or compressed to suit the compromise necessary for the structure to have a minimum potential energy.

Knowing the positions of all atoms in a molecule, in either Cartesian or internal coordinates, completely describes its geometry. Graphical user interfaces permit visualization and presentation in stick diagram, PLUTO, ORTEP, etc. forms. Conversion between cartesian and z-matrix form is routine. Graphical user interfaces may have a QUERY option that allows the user to ask questions about specific internuclear distances for both bonded and non bonded pairs, simple 3-atom angles, and 4-atom dihedral angles.

An entire output file will not be reproduced here (see Clark, (*33*) and most user's manuals for extended printouts), rather some of its features will be discussed. The first MM3 output page for ethene consists of an echo of the input file that helps the user find errors in case of an unsuccessful run. The echo is followed by an initial steric energy of 251 kcal mol^{-1}. This is a measure of how good the initial geometry was. Usually, neither the energy nor the geometry is very good. The initial steric energy is the steric energy the molecule would have if it had the structure specified by the input file.

Following the block of information on the input file, there is a block of information on the iterative progress of the program. In ethene, the positions of the atoms and their maximum movement converge in a few iterations to within the criteria for a successful minimization. Energy gradients dV/dx etc. are also given along with the time necessary to reach convergence. The average movement of atoms converges on zero and the total steric energy converges on 2.60 kcal mol^{-1}.

The main output block starts with the final optimized atomic coordinates. No further geometric information is necessary (indeed, one *cannot* say anything more about the geometry), but a number of geometric calculations, angles, distances, etc., are carried out for the reader's convenience.

Following the final geometry, there is a block of information on the optimized bond lengths and stretching or compression energies associated with each. The final coordinates are printed along with bond lengths of 1.3375 angstroms for C=C and 1.1010 angstroms for C-H. Bond angles are 120.9 deg. Experimental values are 1.337, 1.103, and 121.4 respectively (*34*). Dihedral angles, expressed in the right-handed screw convention, are for the molecule viewed end-on as in a Newman projection. Thus, hydrogen atoms 4 and 5 exactly eclipse one another as do 3 and 6. Hydrogens 3 and 5 and 4 and 6 are exactly *anti* (180 deg.).

This is followed by a block of information on non-bonded interactions including hydrogen bonding where appropriate, and a block on simple bond angle bending and stretch-bend interactions (cross terms) with their contribution to the steric energy. Similar information is given for distortion of the dihedral angles (their energy is zero in ethene) and a block of information is given on dipole-dipole interaction. Following this, a final steric energy summary is given, along with the coordinates of the molecule with its origin shifted to its center of mass. The moments of inertia about the principle axes and the dipole moment are printed.

Thermochemistry

The Enthalpy of Formation. After the complete geometric output, an additional block of information headed HEAT OF FORMATION AND STRAIN ENERGY CALCULATIONS appears near the end of the output file. At the top of the heat of formation block, two sums are listed; one is a sum of normal bond enthalpies for ethene and the other is a sum   selected from a separate parameter set of strainless bonds. Both sets of bond enthalpies have been empirically chosen. Each bond is multiplied by the number of times it appears in the molecule to find its contribution to the total bond enthalpy. In ethene, $26.43 + 4(-4.59) = 8.07$ kcal mol^{-1}, which is given the symbol BE. The steric energy E is added to BE, as is the partition function energy contribution PFC = 2.40 kcal mol^{-1}. Total steric energy E = 2.60 kcal mol^{-1} comes from the previous block of output to yield $8.07 + 2.60 + 2.40 = 13.07$ kcal mol^{-1} for the enthalpy of formation $\Delta_f H$ of ethene in MM3 (experimental value: 12.55 kcal mol^{-1}) (*6*). The same thing is done with the strainless bond energies except that there is no steric energy E to add. In ethene, the strain energy is essentially zero; normal $\Delta_f H$ is the same as strainless $\Delta_f H$.

Each of these bond energy sums resembles an empirical BE method of the Benson type (*9*) but the underlying purpose is quite different. In MM, one wishes to find a normal or average enthalpy to which classical mechanical energies can be added to find the true enthalpy of the molecule. By contrast, pure bond energy schemes (*9*) seek to include steric   interactions in the bond enthalpy and to express different atomic environments as different enthalpy parameters, for example, -CH$_3$, -CH$_2$-, >CH, and >C< (*6*). (Especially in the earlier literature, the terms "energy" and "enthalpy" are sometimes used as synonyms.) Either MM or BE can be made to produce results that are at the same level as experimental uncertainty (but not less) by selecting a small enough set of target compounds and a large enough set of parameters. The difference is that MM, though its parameters are empirical, is based on a sound theoretical base of classical mechanics, while BE schemes are more or less *ad hoc*. This translates into a large difference in confidence (for this writer, at least) in excursions made outside the immediate circle of compounds for which paramaterization has·been carried out.

If we ignore vibrational motion for the moment, all translational and rotational modes of motion are fully excited, leading to an energy contribution of 3RT. Taking the ideal gas approximation for granted in calculating molar enthalpy on the basis of single-molecule energies, enthalpy H is RT higher than the energy. The total is 4RT. This accounts for the partition function contribution PFC = 2.40 kcal mol^{-1}.

The vibrational energy

$$U = \sum hv_i \{(1/2)+(1/(\exp(hv_i/kT) - 1)\} \tag{12}$$

would normally be added to this sum, but this term is difficult to evaluate, and, depending on the way in which the bond enthalpy parameters are obtained, it may be an over calculation. If bond energies are taken from $\Delta_f H_{298}$, as they are for MM2 and the default method in MM3, they already contain the thermal energy of vibration.

POP. There are two additive terms to the energy, POP and TORS, that have not been mentioned yet because they do not appear in ethene. The POP term comes from higher energy conformers. If the energy at the global minimum is not far removed from one or more higher conformational minima, molecules will be distributed over the conformers according to the Boltzmann distribution

$$N_i/N_0 = \exp(-(E_i - E_0)/kT) \tag{13}$$

where N_i/N_0 is the ratio of molecules in the ith high energy conformational state to molecules in the ground state. $E_i - E_0$ is the energy difference between the energetic conformer and the ground state. If there is degeneracy at any energy level, it is counted into the conformational mix. If there are several conformers not too far from the ground state, they will be simultaneously populated and all must be taken into account. In the end, the energy of the compound under investigation, as measured by experimental means, is that of the weighted average determined by the conformational mix at any selected temperature.

For example in but-1-ene, electron diffraction and microwave spectroscopic experiments show the eclipsed or *syn* rotamer to be present in a ratio of 17% of the total, the remaining 83% being the enantiomeric pair of skew rotamers (*35*). Higher enthalpy forms make a negligible contribution to the rotameric mix. This leads to an enthalpy difference of 0.53±0.42 kcal mol^{-1} between the two dominant conformers. PCMODEL V 6.0 calculations show the enantiomeric pair to be more stable than the *syn* rotamer by 0.49 kcal mol^{-1}. An *ab initio* calculation (*35*) yields 0.60 kcal mol^{-1} and MM3 gives 0.69 kcal mol^{-1}. In this case, the POP addition of energy to the enthalpy of the ground state should be about $(.17/.83) \times 0.53$ kcal mol$^{-1} \cong 0.10$ kcal mol^{-1} to be summed with the TORS correction (below) and the computed ground state $\Delta_f H$ of but-1-ene.

TORS. Many molecules have a low frequency torsional mode of motion, consequently the resulting energy levels are closely spaced and appreciably populated at room temperature. There is a torsional contribution to the energy of flexible molecules owing to the cumulative contribution of the several upper torsional states, but the contribution from each state is small because the states are close together. In rigid molecules, this low frequency contribution is zero. Rather than calculate the enthalpy contribution of the torsional states individually, an empirical sum that is an integral multiple of 0.42 kcal mol^{-1} per torsional degree of freedom is assigned to flexible molecules. Torsional motion of a methyl group is not added to a calculated $\Delta_f H$ because it is included in the methyl parameterization.

In propane, there are two low-frequency torsional motions at the C-C bonds but both have been accounted for in the empirical bond energy scheme by the methyl group increments. In butane, the central C-C bond contributes a torsional energy that is not carried into $\Delta_f H$ by an adjacent methyl group. Hence in butane, 1 unit of torsional energy is added to $\Delta_f H$. In pentane, 2 units are added, one for each internal C-C bond. In isobutane, there is no internal C-C bond contribution, isopentane has 1, and so on.

The value of the torsional energy increment has been variously estimated but TORS = 0.42 kcal mol^{-1} was settled on for the bond contribution method in MM3. In the full statistical method (see below), low frequency torsional motion should be calculated along with all the others so the empirical TORS increment should be zero. In fact TORS is not zero, indeed, it is larger than it is for the bond energy method: TORS = 0.72 kcal mol^{-1} in MM4 (5). It appears that the TORS increment is a repository for an energy error or errors in the method that are as yet unknown.

Pi Electron Calculations.   Pi electron conjugation and aromaticity are special problems. One approach is to recognize the difference between conjugated and non-conjugated carbon atoms and give them separate atom numbers (36). In parameterization, conjugated carbon atoms are assigned different parameter sets. This is an oversimplification in some cases because not all conjugated carbon atoms are the same, for example, conjugated bonds in benzene are less reactive than conjugated bonds in buta-1,3-diene. Nor are simple Huckel β values comparable between these two compounds.

A different approach is to carry out an MM minimization followed by a valence electron self consistent field (VESCF) calculation for the pi electrons (3). Self-consistent field calculations bring about a change in carbon-carbon bond orders which are no longer simply single or double. Force constants depend upon bond order, hence they are changed by the VESCF calculation. This requires a new MM minimization, which necessitates a new VESCF optimization and so on. This approach to self-consistency is cut off at some point when repeated calculations do not produce a significant change in the energy. The procedure is not as difficult as it sounds and with a fast workstation or microcomputer, conjugated and aromatic systems can be treated rather easily.

Finding a definition for *resonance energy* has never been difficult; everybody has ne. Unfortunately, they all differ and, resonance energy being a hypothetical construct, there is no "right" one. The basic idea is that benzene is not as reactive as it "ought" to be on the basis of one of its Kekule structures. Benzene is not unique in this respect, so there should be some general quantitative way of expressing extra stability or *aromaticity* found in some molecules. Some alternant hydrocarbons (e.g., cyclobuta-1,3-diene) are especially reactive (unstable), which leads to the concept of *antiaromaticity*. Cyclobuta-1,3-diene, for example,   is very difficult to prepare. Chemical stability is a thermodynamic property relative to a reference state; the question is, "Which reference state?"

If we take the resonance energy of *trans*- alternant polyenes (e.g., buta-1,3-diene) as zero, they become our reference state (37). We can calculate bond energies, E_A and E_B, for the double and single bonds respectively. The resonance energy then becomes the difference between the summed E_A and E_B for all bonds in a molecule and the electronic energy calculated by the VESCF method. The definition of resonance energy just given draws a distinction between resonance stabilization and conjugative stabilization (36). The older definition of resonance energy of benzene as the difference between the enthalpy change of hydrogenation relative to 3 times that of cyclohexene does not make this distinction.

A useful hypothetical concept that is not so difficult to define is the *strain energy* (*38*), which has already been discussed.

Local Minima. In trying alternate ways of writing input files, one soon discovers files that lead to different enthalpies of formation for the same molecule. For example, an input geometry for propene might lead to a normal minimization but result in a calculated enthalpy of formation that is about 2 kcal mol^{-1} above the experimental value. Does this mean that the arbitrary selection of starting geometry determines the outcome of an MM calculation? Yes and no.

In general, MM calculations seek the extremum closest to the starting geometry and may come to rest at a local minimum or (rarely) a saddle point on the potential energy surface. Different conformers satisfy the mathematical requirements for a minimum but are not necessarily at the absolute minimum. Butane, with its *anti* and *gauche* forms is a classic example which is treated in many elementary texts (*39*). Butane has three conformers, one of which, the *anti*, is at the global energy minimum.

Local minima are a pitfall for the unwary; a supposed ground-state structure may be calculated that does not represent the most stable conformer. If so, the calculated enthalpy of formation will always be higher than the ground state enthalpy of formation. Conversely, "false" minima can be just the thing one wishes to find in a conformational study. Drawing the optimized geometries of low and high-enthalpy forms of propene, for example, soon persuades us that the stable form has a methyl hydrogen that eclipses the double bond and that the high-enthalpy form has two hydrogens in a staggered conformation to it. This is contrary to the case for alkanes in which the staggered conformation is more stable than the eclipsed.

In the case of small molecules, one can determine the enthalpy of formation for all of the limited number of possible conformations simply by minimizing many different starting geometries and convincing oneself that all reasonable alternatives have been tried. More complicated molecules present a challenge to one's structural intuition for they often relax to local minima which may be numerous and similar (*40*). A good structural analysis presumes that enough starting geometries have been tried (however many that may be), that the lowest energy conformer is indeed at the global minimum of energy. While no mathematically rigorous proof exists that a conformer is at the global minimum, reasonable doubt can be made very small by judicious choice of starting geometries. Depending on molecular rigidity and symmetry, the number of conformational choices may increase rapidly for large molecules.

The opposite approach to judicious choice has been referred to as a "Saunders-type" search (*32*). This method employs a program that perturbs the starting geometry in a random way, then allows it to relax to the nearest minimum. After many random "kicks" of this kind, the output files are examined for the lowest steric energy. Although time consuming, the method has been shown to be effective in reproducing previously known global minima.

Dihedral Driver. A dihedral or "twist" angle $^A\backslash_{B-C}/^D$ that A makes with D can be driven around the B-C axis in arbitrarily chosen angular increments. At each step during the drive, the structure is minimized and an energy is recorded so that the locus of energies as a function of drive angle gives a profile of the potential energy hill a molecule goes over upon being driven away from a minimum. Especially in congested molecules, these potential energy loci should not be taken too literally as (within the model) rotated atoms or groups can "stick" during rotation, then suddenly "snap into place" giving a potential energy discontinuity that has no counterpart in the real world (*41*).

Results for $\Delta_f H_{298}$

Enthalpies of formation taken from recent evaluative articles of the molecular mechanics program MM4 (*5*), the semi-empirical program PM3 by Stewart, and the *ab initio* programs gathered under the title G2, evaluated by Curtiss *et al.*, are shown in Table I. The selection of compounds is arbitrary and absence of any compound from the list is does not imply that it cannot be treated by a given method, only that it is not present in the cited reference. The table does reflect the bias of MM4 (and earlier MM versions) toward organic molecules, the accuracy of MM and *ab initio* methods where they can be applied, relative to PM3, and one's inability to treat large molecules by the G2 *ab initio* protocol. PM3 can be used to calculate the enthalpies of formation of diatomic elements but these values were not included in the table because, in constrst with G2, PM3 is not really intended for that purpose.

Full Statistical Method

One reason early force fields and force fields designed for large-molecule biochemical applications have been kept as simple as possible is that inclusion of non-zero off diagonal terms in MM2 has been tried and found not to improve calculated structures very much. From a practical point of view, it was also advantageous to avoid off-diagonal elements in early force fields because of machine limitations. Machine limitations have become less severe in recent years but they continue to bedevil other molecular modeling techniques, in particular *ab initio* methods.

Questions of machine limitations aside, the great advantage of a force field that can be used to calculate geometry and $\Delta_f H$ and that also gives good results for vibrational spectra is twofold. First, the $\Delta_f H$ calculations can be put on a much more satisfactory theoretical base by calculating an enthalpy of formation at 0 K as in *ab initio* procedures, then adding various thermal energies by more rigorous means than simply lumping them together as an empirical bond enthalpy contribution to $\Delta_f H$. Aside from intellectual satisfaction, there is the practical advantage that a firm theoretical foundation lends confidence to predictions we might make. If only the structure and enthalpy can be calculated, even good results may conceal errors in the force field or cancellation of errors in the calculation, but if the spectra are also correct, the program is probably reliable.

Table I. Calculated and Experimental Enthalpies of Formation at 298 K.

Compound	MM4[a]	PM3[b]	G2[c]	Exp.[d]
Nitrogen	--	--	1.3	0.0
Oxygen	--	--	2.4	0.0
Fluorine	--	--	0.3	0.0
Ammonia	--	-3.1	10.8	11.0±0.1
Water	--	-53.4	-58.1	-57.8±0.0
Hydrogen Sulfide	--	-0.9	-4.8	-4.9±0.2
Sulfur Dioxide	--	--	-65.9	-71.0±0.1
Methane	-17.9	-13.0	-18.6	-17.9±0.1
Ethane	-19.7	-18.1	-20.6	-20.1±0.1
Propane	-25.0	-23.6	-25.4	-25.0±0.1
Bicyclobutane	52.1[e]	69.2	54.9	51.9±0.2
Spiropentane	44.2[e]	43.1	45.7	44.3±0.2
Cyclopentane	-18.6	-23.9	--	-18.3±0.2
Ethene	12.5	16.6	12.8	12.6±0.1
Propene	5.0	6.4	5.3	4.8±0.2
Cyclobutene	--	37.7	40.3	37.5±0.4
Benzene	19.9	23.5	23.7	19.7±0.2
Toluene	12.3	14.1	--	12.0±0.2
Styrene	35.8	39.2	--	35.3±0.4
Bicyclo[2,2,2]octane	-21.9	-27.8	--	-23.7±0.3
Anthracene	54.1	61.7	--	55.2±0.5
2,2-Metacyclophane	43.4	--	--	40.8±1.5
2,2-Metapara-				
cyclophane	52.2	--	--	52.2±0.4
Methanol	-48.3[f]	-51.9	-49.4	-48.2±0.1
Ethanol	-56.1[f]	-56.8	-57.2	-56.2±0.1
Methyl Amine	-5.0[g]	-5.2	-5.5	-5.5±0.1
Ethyl Amine	-11.9[g]	-12.5	-12.1	-11.3±0.2
Glycine	-94.7[h]	-96.0	--	-93.7±0.2
Absolute Mean difference from experiment	0.5	3.8	1.3	

[a] Reference 5. [b] Stewart, J. J. P. *J. Comp.-Aided Molecular Design* **1990**, *4*, 1.
[c] Curtiss, L. A.; Raghavachari, K.; Redfern, P. C.; Pople, J. A. *J. Chem. Phys.* **1997**, *106*, 1063. [d] Mostly from reference 6, see also footnote c. [e] MM3, Aped, P.; Allinger, N. L. *J. Amer. Chem. Soc.* **1992** *114* 1. [f] MM3, Allinger, N. L.; Rahman, M.; Lii, J.-H. *J. Amer. Chem. Soc.* **1990** *112* 8293. [g] MM3, Schmitz, L. R.; Allinger, N. L. *J. Amer. Chem. Soc.* **1990** *112* 8307. [h] MMX, Reference 16.

Calculation of $\Delta_f H$ for ethene by the more rigorous procedure yields for MM3 a sum of (a) energy, (b) steric energy, (c) vibrational heat content, including the zero point and thermal energies, and (d) structural features POP and TORS. With a force field that provides good vibrational spectra, we can calculate a molecular energy at 0 kelvins (equal to the enthalpy at 0 kelvins) by summing bond energies for, say the C-H bond from its constituent atoms. In complete calculations of enthalpy, bond energy parameters appear to be quite different from the conventional scheme above, for example, the C-H bond energy method from the MM3 default calculation described above. That is because zero point and thermal energies are not included in the parameters but are added later. The sum is the energy of the molecule with each constituent bond at the bottom of its potential well. To this is added the zero point energy, and a thermal energy calculated for fully excited translational and rotational motion at room temperature, *plus a vibrational contribution* to the thermal energy that is calculated from a rigorous statistical mechanical principles and is known as accurately as the vibrational spacings are known, which, as said, are known rather well if enough off-diagonal elements are involved in the hessian matrix.

Entropy. The second important advantage is that entropies, for example, that of ethene, can be calculated (*18b*).

Translational molecular motion has quanta that are too small to measure spectroscopically. Instead, one uses the Sakur-Tetrode equation (*42*) to determine the absolute entropy of a gas-phase substance at any temperature.

One can obtain the *rotational* entropy for ethene from the moments of inertia, I, through the rotational partition function

$$q_{rot} = 8\pi^2 (8\pi^3 I_x I_y I_z)^{1/2} (kT)^{3/2}/h^3 \sigma = 674.9 \tag{14}$$

[note the use of the cgs system of units in MMx programs]. MM3 produces an output block:

```
MOMENT OF INERTIA WITH THE PRINCIPAL AXES
(1) UNIT = 10**(-39) GM*CM**2
    IX= 0.5994   IY= 2.8021   IZ= 3.4015
```

Likewise, the *vibrational* contribution to the entropy can be calculated knowing the frequencies for each vibrational mode of motion, of which there are 12 in ethene. One must use a full-matrix optimization (option 5 in MM3) to obtain the necessary frequencies. When the entropy contributions are summed (*43*) according to the equation for each normal frequency

$$S_{vib} = \Sigma_i R\{(x_i/(\exp(x_i) - 1)) - \ln(1 - \exp(-x_i))\} \tag{15}$$

where $x_i = \theta_{v(i)}/T$, the vibrational temperature θ of the ith vibrational mode divided by the temperature, the result is 0.506 cal K^{-1} mol^{-1}. If there are conformers other than the ground state conformer, as there will be for, e.g., higher alkanes, one must determine their relative contributions to the conformational mix and add an entropy of mixing term

$$\Delta S_{mix} = -R \ \Sigma_j \ X_j \ \ln X_j \tag{16}$$

where X_j is the mol fraction of the jth conformer. The standard entropy (invoking the third law of thermodynamics at 0 K) is the weighted average of the conformational entropies

$$S = \Sigma_j \ X_j S_j + \Delta S_{mix} \tag{17}$$

There is no *electronic* contribution to the entropy of ethylene at this temperature.

The sum of these contributions, $S° = 52.34$ cal K^{-1} mol^{-1} with the MM3 force field and $S° = 52.44$ cal K^{-1} mol^{-1} using MM4 (*37*) compares with the experimental value of $S° = 52.45$ cal K^{-1} mol^{-1} for ethylene (*44*).

The *heat capacity* at constant pressure within the ideal gas approximation can be calculated in the same way as the entropy, a PV/T term, a translational term, a rotational term and a vibrational sum

$$C_p = R + 1.5R + 1.5 \ R + 2.127 = 10.075 \ cal \ K^{-1} \ mol^{-1}$$

where the experimental value is 10.41 cal K^{-1} mol^{-1}. Heat capacity calculations are more sensitive to failure of the harmonic approximation than entropies (*37*).

Free Energy and Equilibrium

Once having $\Delta_f H°$ and $S°$ for the participants in a chemical reaction, the obvious next step is to calculate the standard Gibbs free energy change of reaction $\Delta_r G°$ and the equilibrium constant, K_{eq}. In an early study, Allinger *et al.* (*45*) computed $\Delta_f H°$ and $S°$ and $\Delta_f G°$ at 298 K for the isomers of methylcyclohexene and methylenecyclohexane. Of these, 1-methylcyclohexene has the lowest $\Delta_f H°$ and $\Delta_f G°$ by more than 2 kcal mol^{-1}. Among the remaining three isomers, methylenecyclohexane has the lowest $\Delta_f H°$ at 298 K suggesting that it is most stable at that temperature. That is not the case, however. The 3- and 4-methylcyclohexene isomers exist as axial-equatorial pairs, each of which consists of a *dl* pair. Taking the appropriate entropies of mixing into account, the resulting calculated free energy differences favor the 3- isomer as the most stable of the three (*46*).

Kar, Lenz, and Vaughan (KLV) have calculated the $\Delta_f H°$ at 298 K of *cis*-but-2-ene and *trans*-but-2-ene as -1.621 kcal mol^{-1} and -2.85 kcal mol^{-1} respectively, (*47*) leading to $\Delta_r H° = -1.23$ kcal mol^{-1} for the reaction

cis-but-2-ene → *trans*-but-2-ene

The corresponding entropies are 71.83 cal K^{-1} mol^{-1} and 70.47 cal K^{-1} mol^{-1} leading to $\Delta_r S° = -1.36$ cal K^{-1} mol^{-1} and $\Delta_r G° = -0.825$ kcal mol^{-1}. The calculated equilibrium constant is $K_{eq} = 4.03$.

Kapeijn, van der Steen, and Mol (KSM) found $\Delta_r H° = -1.037$ kcal mol^{-1} and $\Delta_r S°$ = -1.219 cal K^{-1} mol^{-1} at 298 K by precise measurements of the equilibrium constant as a function of temperature (*48*). From these experimental values, one obtains $K_{eq} = 3.12$ and from an experimental study (*49*) by Akimoto, Sprung, and Pitts, $K_{eq} = 2.89$ at 298 K.

Equilibrium and temperature

But-2-ene isomerization.   Over the temperature range 298 K to 500 K, KLV (above) calculated $\Delta_r H$ differences of 0.01 kcal mol^{-1} or less for *cis-trans* isomerization of but-2-ene. The influence of temperature change upon entropy, not enthalpy, is dominant for this reaction up to 500 K. Taking $\Delta_r H° = -1.23$ kcal mol^{-1} at 500 K and $\Delta_r S = 1.30$ kcal mol^{-1} over the 298 to 500 K temperature range, $\Delta_r G° = -0.58$ kcal mol^{-1} leading to $K_{eq} = 1.79$ at that 500 K. The value for K_{eq} agrees well with the empirical equation developed by KSM at 298 K, but much less well at higher temperatures. Deviations of $(H - H_0)/T$ at higher temperatures and their correction procedures are discussed by KLV.

Butane Isomerization.   Low temperature acid catalyzed isomerization of lower alkanes has long been used in the petroleum industry to increase chain branching and, therewith, octane number. Olah and Molnar (*50*) cite experimental observation of monotonic decrease of the equilibrium constant (K_{eq}) for the reaction

$$n\text{-butane} \rightarrow \text{isobutane}$$

over the temperature interval 294 - 477 K from $K_{eq} = 5.66$ at the lower temperature to $K_{eq} \cong 1$ at the higher temperature.

Values of the free energy function $-(G - H_0)/T$ are generated by MM3, along with the enthalpy function $(H - H_0)/T$ at any temperature T. As described by Smith (*51*) the difference in free energy functions $-(\Delta G - \Delta H_0)/T$ for the isomerization at any temperature can be calculated and converted into the free energy change ΔG for the reaction at that temperature by subtracting ΔH_0, the enthalpy change for the reaction at 0 K. This leads to the equilibrium constant by $\Delta G = -RT \ln K_{eq}$.

The free energy function increases for isomerization of n-butane by about 1.5 cal mol^{-1} K^{-1} but the change in the enthalpy function for the reaction is rather small. $\Delta_r H$ for the reaction is about -1.5 kcal mol^{-1} and, as might be expected for an isomerization, it also changes by very little on being corrected to 0 K using the enthalpy function. Upon calculating K_{eq}, at the two temperatures, one obtains $K_{eq} = 6.1$ at 294 K and $K_{eq} = 2.3$ at 477 K, a rather good replica of the experimental observations over this temperature range.

Acknowledgment

We wish to thank the trustees of Long Island University and the L. I. U. Committee on Released Time for a grant of released time for research.

Literature Cited.

1. (a) Burkert, U.; Allinger, N. L. *Molecular Mechanics* ACS Publ. No. 177; ACS: Washington D C, 1982. (b) Allinger, N. L. *Adv. Phys. Org. Chem.* **1976**, *13*, 1. (c) Grant, G. H.; Richards, W. G. *Computational Chemistry*; Oxford Science: New York, 1995.
2. Parr, R. G. *The Quantum Theory of Molecular Electronic Structure*; W. A. Benjamin: New York, 1963.
3. (a) Brown, R. D.; Heffernan, M. L. *Trans. Far. Soc.* **1958**, *54*, 757. (b) Brown, R. D.; Heffernan, M. L. *Austral. J. Chem.* **1959**, *12*, 319.
4. (a) Halgren, T. A. *J. Comput. Chem.* **1996**, *17*, 490. (b) Maple, J.; Dinur, U.; Hagler, A. T. *Proc. Natl Acad. Sci. USA* **1989**, *85*, 5350. (b) Mc Gaughey, G. B.; Stewart, E. L.; Bowen, J. P. *J. Comput. Chem.* **1995**, *16*, 1250.
5. (a) Allinger,N. L.; Chen, K.; Lii, J. -H. *J. Comput Chem.,* **1996**, *17*, 642. (b) Nevins, N.; Chen, K.; Allinger, N. L. *J. Comput. Chem.* **1996**, *17*, 669.
6. (a) Cox, J. D.; Pilcher, G. *Thermochemistry of Organic and Organometallic Compounds*; Academic Press: London, 1970. (b) Pedley, J. B., Naylor, R. D.; Kirby, S. P. *Thermochemical Data of Organic Compounds*; Chapman and Hall: London, 1986.
7. Rogers, D. W.; von Voithenberg, H.; Allinger, N. L. *J. Org. Chem.* **1978**, *43*, 360.
8. Eksterowicz J, E.; Houk, K. N. *Chem. Rev.* **1993**, *93*, 2439.
9. Benson, S. W. *Thermochemical Kinetics 2nd ed.*; John Wiley & Sons: New York, 1976.
10. Allinger, N. L.; Pathiaseril, A. *J. Comput. Chem.* **1987**, *8*, 1225.
11. Allinger, N. L.; Flanagan, H. L. *J. Comput. Chem.* **1983**, *4*, 399.
12. Tai, J.; Allinger, N. L. *J. Am. Chem. Soc.* **1988**, *110*, 2050.
13. Lii, J. -H.; Allinger, N. L. *J. Comput. Chem.* **1991**, *17*, 186.
14. Allinger, N. L. *Pure Appl. Chem.* **1982**, *54*, 2515.
15. (a) Weiner, S. J.; Kollman, P. A.; Case, D. A.; Chandra Singh, U.; Ghio, C.; Algona, G.; Profeta, S.; Weiner, P. *J. Am. Chem. Soc.* **1984**, *106*, 765. (b) Weiner, S. J.; Kollman, P. A.; Nguyen, D. T.; Case, D. A. *J. Comput. Chem.* **1986**, *7*, 230. (c) Brooks, B. R.; Bruccoleri, B. D.; Olafson, B. D.; States, D. J.; Swaminathian, S.; Karplus, M. *J. Comput. Chem.* **1983**, *4*, 187.
16. (a) PCMODEL (Serena Software) has several WRITE FILE options; MOPAC has an **aigout** keyword for converting a MOPAC archive file (.ARC) to GAUSSIAN input format; see also program BABEL (babel@mercury.aichem.arizona.edu) for many input file conversion options.
17. Fowles, G. R. *Analytical Mechanics*; Holt, Reinhart and Winston: New York, 1962.
18. (a) Allinger, N. L.; Yuh, Y. H.; Lii, J. -H. *J. Am. Chem. Soc.* **1989**, *111*, 8551. (b) Allinger, N. L.; Lii, J. -H. *J. Am. Chem. Soc.* **1989**, *111*, 8566. (c) Lii, J. -H.;

Allinger, N. L. *J. Am. Chem. Soc.* **1989**, *111*, 8576. (d) MM3 for Microcomputers. Advance copy generously supplied to the author by Prof. Allinger and Dr. Yuh.

19. Dykstra, C. E. *Quantum Chemistry & Molecular Spectroscopy*; Prentice-Hall: Englewood Cliffs, NJ, 1992.

20. Ege, S. *Organic Chemistry, 2nd ed.*; D. C. Heath: Lexington, MA, 1989.

21. Smith, W. B. *Introduction to Theoretical Organic Chemistry and Molecular Modeling*; VCH: New York, 1996.

22. (a) Allinger, N. L.; Li, F.; Yan, L. *J. Comput. Chem.* **1990**, *11*, 848. (b) Allinger, N. L.; Li, F.; Yan, L.; Tai, J. C. *J. Comput. Chem.* **1990**, *11*, 868.

23. Lipkowitz, K. B. In *The Conformational Analysis of Cyclohexenes, Cyclohexadienes and related Hydroaromatic Compounds*; Rabideau, P. W., Ed.; VCH: New York, 1989; pp. 301 - 319.

24. Brooks, B. R.; Bruccoleri, R. E.; Olafson, B. D.; States, D. J.; Swaminathan, S.; Karplus, M. *J. Comput. Chem.* **1983**, *4*, 187. (b) Nilsson, L.; Karplus, M. *J. Comput. Chem.* **1986**, *7*, 591.

25. Verdani, A. *J. Comput. Chem.* **1988**, *9*, 269.

26. Cornell, W. D.; Cieplak, P.; Bayly, C. I.; Gould, I. R.; Mertz Jr., K. M.; Ferguson, D. M.; Spellmeyer, D. C.; Fox, T.; Caldwell, J. W.; Kollaman, P. A. *J. Am. Chem. Soc.* **1995**, *117*, 5179.

27. Serena Software. Box 3076, Bloomington, IN 47402-3076.

28. Hwang, M. J.; Stockfish, T. P.; Hagler, A. T. *J. Am. Chem. Soc.* **1994**, *116*, 2515.

29 (a) Warshel, A.; Karplus, M. *J. Am. Chem. Soc.* **1972**, *94*, 5612. (b) Wertz, D. H.; Allinger, N. L. *Tetrahedron* **1974**, *30*, 1579. (c) Lifson, S. Hagler, A. T.; Dauber, P. *J. Am. Chem. Soc.* **1979**, *101*, 5111.

30. Lifson, S. Warshel, A. *J. Chem. Phys.* **1968**, *49*, 5116. (b) Warshel, A.; Lifson, S. *J. Chem. Phys.* **1970**, *53*, 582. (c) Lifson, S. J.; Stern, P. S. *J. Chem. Phys.* **1982**, *77*, 4542.

31. Boyd, D. B. In *Reviews in Computational Chemistry I, III, IV* Lipkowitz, K. B.; Boyd, D. B., Eds.; VCH: 1991-4.

32. (a) Saunders, M. *J. Am. Chem. Soc.* **1987**, *109*, 3150. (b) Saunders, M. *J. Comput. Chem.* **1989**, *10*, 203.

33. Clark, T. *A Handbook of Computational Chemistry*; John Wiley & Sons: New York, 1985.

34. Bartell, L. S.; Roth, E. A.; Holowell, C. D.; Kuchitsu, K.; Young, J. E. *J. Chem. Phys.* **1965**, *42*, 2683.

35. van Hemelrijk, D.; van den Enden, L.; Geise, H. J.; Sellers, H. L.; Schaefer, L. *J. Am. Chem. Soc.* **1980**, *102*, 2189.

36. Roth, W. R.; Adamczak, O.; Breuckmann, R.; Lennartz, H.-W.; Boese, R. *Chem. Ber.* **1991**, *124*, 2449.

37. Nevins, N.; Lee, J.-H.; Allinger, N. L. *J. Comput. Chem.* **1996**, *17*, 695.

38. March, J. *Advanced Organic Chemistry*; John Wiley & Sons: New York, 1992.

39. Solomons, T. W. G. *Fundamentals of Organic Chemistry*; John Wiley & Sons: New York, 1997.

40. Ferguson, D. N.; Gould, I. R.; Glauser, W. A.; Schroeder, S.; Kollman, P. A. *J. Comput. Chem.* **1992**, *13*, 525.

41. Rogers, D. W.; Podosennin, A. V. unpublished.
42. (a) Barrow, G. M. *Physical Chemistry*; Mc Graw-Hill: New York, 1961. This is a clear, elementary discussion. (b) Lewis, G. N.; Randall, M. *Thermodynamics* 2nd ed. revised by Pitzer, K. S.; Brewer, L.; Mc Graw-Hill: New York, 1961, p. 419 ff. (c) Atkins, P. W., *Physical Chemistry*, 3rd ed.; Freeman, New York, 1986.
43. A short BASIC program makes the summing easy.
44. Stull, D. R.; Westrum, E. F.; Sinke, G. C. *The Chemical Thermodynamics of Organic Compounds*; John Wiley & Sons: New York, 1969.
45. Allinger, N. L.; Hirsch, J. A.; Miller, M. A.; Tyminski, I. J. *J. Am. Chem. Soc.* **1968**, *90*, 5773.
46. Lipkowitz, K. B. In *The Conformational Analysis of Cyclohexenes, Cyclohexadienes, and Related Hydroaromatic Compounds* Rabideau, P. W. Ed.; VCH: New York, 1989.
47. Kar, M.; Lenz, T. G.; Vaughan, J. D. *J. Comput. Chem.* **1994**, *15*, 1254.
48. Kapeijn,F.; van der Steen, A. J.; Mol, J. C. *J. Chem. Thermodyn.* **1983**, *15*, 137.
49. Akimoto, H.; Sprung, S. L.; Pitts, J. N. *J. Am. Chem. Soc.* **1972**, *94*, 4850.
50. Olah, G. A.; Molnar, A. *Hydrocarbon Chemistry*; John Wiley & Sons: New York, 1995, Chapter 4.
51. Smith, N. O. *Elementary Statistical Thermodynamics*; Plenum,: New York, 1982.

METHODS BASED ON MOLECULAR-ORBITAL OR DENSITY-FUNCTIONAL THEORY

Chapter 8

Thermochemistry from Semiempirical Molecular Orbital Theory

Walter Thiel

Organisch-Chemisches Institut, Universität Zürich, Winterthurestrasse 190,
CH–8057 Zürich, Switzerland

Semiempirical quantum-chemical methods are reviewed with regard to
their role as computational tools in thermochemistry. The theoretical
background and the conventions of the established methods are outlined,
particularly for MNDO, AM1, and PM3. The deviations between
calculated and observed heats of formation are evaluated statistically for
several large sets of test molecules to assess the performance of the
established methods. The results are compared with those from ab initio
and density functional approaches and from more recent semiempirical
developments.

The semiempirical molecular orbital (MO) methods of quantum chemistry (*1-12*) are
widely used in computational studies of large molecules, particularly in organic
chemistry and biochemistry. In their implementation, they neglect many of the less
important integrals that occur in the ab initio MO formalism. These severe
simplifications call for the need to represent the remaining integrals by suitable
parametric expressions and to calibrate them against reliable experimental or accurate
theoretical reference data. This strategy can only be successful if the semiempirical
model retains the essential physics to describe the properties of interest. Provided that
this is the case, the parametrization can account for all other effects in an average sense,
and it is then a matter of testing and validation to establish the numerical accuracy of a
given approach.

Different semiempirical methods are available to study different molecular
properties, both in the ground state and electronically excited states. The present article
will focus on the semiempirical calculation of thermochemical properties for ground

state molecules in the gas phase. Statistical evaluations are provided for the reference molecules in the G2 neutral test set (*13,14*) and in other larger test sets. Based on these results, the capabilities and limitations of current semiempirical methods are discussed with regard to thermochemical properties.

Theoretical Background

Quantum-chemical semiempirical treatments are defined by the following specifications:

(a) The underlying theoretical approach: Most current general-purpose semiempirical methods are based on MO theory and employ a minimal basis set for the valence electrons. Electron correlation is treated explicitly only if this is necessary for an appropriate zero-order description.

(b) The integral approximation and the types of interactions included: Traditionally there are three levels of integral approximation (*2,15*) - CNDO (complete neglect of differential overlap), INDO (intermediate neglect of differential overlap), and NDDO (neglect of diatomic differential overlap). NDDO is the best of these approximations since it retains the higher multipoles of charge distributions in the two-center interactions (unlike CNDO and INDO which truncate after the monopole).

(c) The integral evaluation: At a given level of integral approximation, the integrals are either determined directly from experimental data or calculated from the corresponding analytical formulas or computed from suitable parametric expressions. The first option is generally only feasible for the one-center integrals which may be derived from atomic spectroscopic data. The choice between the second and third option is influenced by the ease of implementation of the analytical formulas, but mainly depends on an assessment of how to model the essential interactions.

(d) The parametrization: Semiempirical MO methods are parametrized to reproduce experimental reference data (or, possibly, accurate high-level theoretical predictions as substitutes for experimental data). The reference properties are best selected such that they are representative for the intended applications. The quality of semiempirical results is strongly influenced by the effort put into the parametrization.

Several semiempirical methods are currently available for calculating ground state potential surfaces and thermochemical properties, e.g. MNDO (*16*), MNDOC (*17*), MNDO/d (*18-21*), AM1 (*22*), PM3 (*23*), SAM1 (*24,25*), MINDO/3 (*26*), and SINDO1 (*27,28*). The first six of the methods employ the NDDO integral approximation and are therefore conceptually preferable to the last two INDO-based approaches. MNDO, AM1, and PM3 are widely distributed in a number of software

packages, and they are probably the most popular semiempirical methods for thermochemical calculations. We shall therefore concentrate on these methods and discuss
other approaches only briefly, including some recent developments from our group
(*29,30*).

MNDO, AM1, and PM3 are valence-electron MO treatments which employ a
minimal basis set of atomic orbitals and the NDDO integral approximation. The
underlying theoretical formalism has been described in the literature (*9-12,16,22,23*).
The following energy terms and interactions are included:
- One-center one-electron energies,
- one-center two-electron repulsion integrals,
- two-center one-electron resonance integrals,
- two-center one-electron integrals representing electrostatic core-electron attractions,
- two-center two-electron repulsion integrals,
- two-center core-core repulsions composed of an electrostatic term and an additional
  effective atom-pair potential.

MNDO, AM1, and PM3 are based on the same semiempirical model (*12, 16*)
since they are identical with regard to the specifications (a)-(b) (see above). Their
implementation (c) is quite similar: Conceptually the one-center terms are taken from
atomic spectroscopic data, with the refinement that slight adjustments of the parameters
are allowed in the optimization to account for possible differences between free atoms
and atoms in a molecule. The one-center two-electron integrals derived from atomic
spectroscopic data are considerably smaller than their analytically calculated values
which is (at least partly) attributed to an average incorporation of electron correlation
effects. These integrals provide the one-center limit ($R_{AB} = 0$) of the two-center two-
electron integrals, whereas the asymptotic limit for $R_{AB} \to \infty$ is determined by classical
electrostatics. To conform with these limits, the two-center two-electron integrals are
evaluated from semiempirical multipole-multipole interactions (*31*) damped according
to the Klopman-Ohno formula (*32,33*). Therefore, at intermediate distances, the semiempirical two-electron integrals are smaller than their analytical counterparts which
again reflects some inclusion of electron correlation effects. Aiming for a reasonable
balance between electrostatic attractions and repulsions within a molecule, the core-
electron attractions and the core-core repulsions are treated in terms of the corresponding two-center two-electron integrals, neglecting e.g. penetration effects. The
additional effective atom-pair potential that is included in the core-core repulsions (with
an essentially exponential repulsion in MNDO and a more flexible parametric function
in AM1 and PM3) attempts to compensate for errors introduced by the above
assumptions, but mainly represents the Pauli exchange repulsions. Finally, following

semiempirical tradition, the resonance integrals are taken to be proportional to the corresponding overlap integrals.

The implementation (c) of MNDO, AM1, and PM3 is thus characterized by
- a realistic description of the atoms in a molecule,
- a balance between electrostatic attractions and repulsions,
- an average incorporation of dynamic electron correlation effects,
- a description of covalent bonding mainly through overlap-dependent resonance integrals,
- an inclusion of Pauli exchange repulsions and other corrections through an effective atom-pair function.

The parametrization (d) of MNDO, AM1, and PM3 has focused on ground state properties, mainly heats of formation and geometries, with the use of ionization potentials and dipole moments as additional reference data. Compared with MNDO, more effort has been spent on the parametrization of AM1 and PM3, and additional adjustable parameters have been introduced so that the number of optimized parameters per element has typically increased from 5-7 in MNDO to 18 in PM3. Hence, AM1 and PM3 may be regarded as methods which attempt to explore the limits of the MNDO model through careful and extensive parametrization.

MNDO, AM1, and PM3 employ an sp basis without d orbitals (*16,22,23*). Therefore, they cannot be applied to most transition metal compounds, and difficulties are expected for hypervalent compounds of main-group elements where the importance of d orbitals for quantitative accuracy is well documented at the ab initio level (*34*). To overcome these limitations, the MNDO formalism has been extended to d orbitals (*18-21*). In the resulting MNDO/d approach, the established MNDO formalism and parameters remain unchanged for hydrogen, helium, and the first-row elements. The inclusion of d orbitals for the heavier elements requires a generalized semiempirical treatment of the two-electron interactions. The two-center two-electron integrals are calculated by an extension (*18*) of the original MNDO point-charge model (*31*). All nonzero one-center two-electron integrals are retained and computed analytically from separately optimized orbital exponents. In all other aspects, MNDO/d is analogous to MNDO.

Having outlined the theoretical background of currently available semiempirical methods, it should be clear that these approaches use drastic simplifications (compared with ab initio methods), e.g. the formal neglect of electron correlation and of all three- and four-center integrals. It might therefore seem surprising that such approaches can yield results of useful accuracy, e.g. for thermochemical properties. In the following, we discuss this point using dynamic electron correlation effects as an example.

In ab initio theory, inclusion of electron correlation is essential to predict thermochemical properties such as atomization energies reliably (*13,14,34*). On the other hand, density functional theory (DFT) (*35*) is also rather successful in this regard (*14,36*) even though the density is constructed from a single determinant in the usual Kohn-Sham formalism. This success of DFT is mainly due to the incorporation of dynamic electron correlation effects during orbital optimization through the use of a suitable exchange-correlation functional. Conceptually, dynamic electron correlation is also built into the semiempirical SCF-MO methods in an average manner, by using effective two-electron interactions that are damped at small and intermediate distances (see above). The determination of the one-center two-electron integrals from atomic spectroscopic data ensures that the correlation effects in the atoms are represented as well as possible in the chosen semiempirical framework, while the Klopman-Ohno interpolation scheme yields a plausible representation at intermediate distances and guarantees convergence to the classical Coulomb limit for large distances (see above).

Correlation energies have been calculated for MNDO-type wave functions (*8,17*), both through configuration interaction and second-order perturbation theory. As expected from the preceding considerations, these semiempirical correlation energies are much smaller than their ab initio counterparts. Moreover, the correlation effects on energies and geometries (*8,17*) are bond-specific, transferable, and rather uniform, at least in the ground states of organic molecules. Such regularities are also found at the ab initio level where it has been shown (*37,38*) that the computed total correlation energies can often be reproduced quite well from additive bond correlation increments. Indirect support for the hypothesis of fairly uniform correlation energies in the ground states of organic molecules comes from attempts to derive heats of formation from ab initio SCF total energies by subtracting appropriate multiples of empirical atomic constants (*39-41*): The success of this approach indicates that the ab initio correlation corrections to the total energy can indeed be absorbed approximately into additive atomic constants.

In MNDO-type methods, the dynamic correlation effects are thus relatively small (due to the representation of the two-electron integrals) and rather uniform (in related molecules). It is therefore not surprising that these effects can be taken into account by a parametrization at the SCF level in an average manner. Hence, an explicit treatment of electron correlation as in MNDOC (*17*) is practically useful only in systems with specific correlation effects (*8,42-45*). In other systems, the MNDOC method (i.e. MNDO with explicit computation of correlation effects, parametrized at the correlated level) yields results of similar accuracy as the standard MNDO method (*16,17*).

These considerations suggest a general answer to the question of what to expect from semiempirical calculations, e.g. for thermochemical properties: The results can be reliable only to the extent that the relevant physical interactions are included in the semiempirical model (see above) and that the neglected features can be absorbed by the semiempirical parametrization (on the average). Whether a given semiempirical method is useful for quantitative predictions can only be established by careful validation against experimental data or high-level ab initio results.

Specific Conventions

In quantum-chemical calculations, the equilibrium atomization energy (D_e) of a given molecule is available from its total energy and from the electronic energies of the constituent atoms. Inclusion of zero-point vibrational corrections provides the ground state atomization energy (D_0). Subtracting D_0 from the sum of the experimental enthalpies of formation of the constituent atoms at 0 K then yields the heat of formation at 0 K (ΔH_f^0) which can be converted to the corresponding heat of formation at 298 K (ΔH_f^{298}) by incorporating suitable thermal corrections. The corrections for the atoms are taken from experiment, while those for the molecules are calculated theoretically (normally using the classical approximation for the translation and rotation, and the harmonic approximation for the vibration).

This procedure is followed in ab initio studies (see e.g. *(13,14)*) and could equally well be applied in semiempirical work. However, in MNDO-type methods, heats of formation at 298 K are traditionally derived in a simpler manner *(1,16)*: By formally neglecting the zero-point vibrational energies and the thermal corrections between 0 K and 298 K, the heats of formation at 298 K are obtained from the calculated total energies by subtracting the calculated electronic energies of the atoms in the molecule and by adding their experimental heats of formation at 298 K. This procedure implicitly assumes that the zero-point vibrational energies and the thermal corrections are composed of additive increments which can be absorbed by the semiempirical parametrization (see also *(39-41)*). There is some independent evidence for the validity of this assumption *(46,47)*. A more direct confirmation comes from a test *(48)* where the MNDO method has been reparametrized using atomization energies (D_e) rather than heats of formation (ΔH_f^{298}) as reference data and keeping everything else as in MNDO. The resulting parameters were very close to the MNDO ones, and the overall performance was quite similar *(48)*.

The semiempirical procedure for computing heats of formation at 298 K is computationally efficient, but not satisfactory theoretically. Problems will arise when

there are specific changes in the zero-point vibrational energies or the thermal corrections, e.g. during the course of a chemical reaction: The zero-point vibrational energy is
usually smaller in the transition state than in the reactants (because there is one less
vibrational mode) which reduces the relevant barrier. Since the parametrization can
certainly not account for such specific effects, it seems legitimate to compute these
changes in the barrier explicitly from the vibrational frequencies, both in ab initio and
semiempirical studies; the results are actually quite close to each other (*44*). When
calculating thermochemical properties of stable molecules, such specific problems are
fortunately less prominent.

According to the conventions in MNDO-type methods, the computed total
energies for a given molecule thus differ from the corresponding heats of formation at
298 K by a constant amount which is independent of the molecular geometry (see
above). In actual practice, the potential surface from an MNDO-type calculation is
normally discussed like a surface from any other quantum-chemical calculation (e.g. ab
initio), except that the energies at the minima are translated into heats of formation
(ΔH_f^{298}) without explicitly considering zero-point vibrational energies or thermal
corrections.

Applications of semiempirical methods must be performed in analogy to the
parametrization. This can be illustrated by a trivial example: The conversion of total
energies to heats of formation employs the experimental heats of formation of the atoms
in the molecule (see above). Since the original parametrization of MNDO, some of
these experimental values have changed. It would clearly be inconsistent to employ the
more recent and more accurate experimental values in MNDO without a reparame-
trization, because misleading systematic errors would be introduced in this manner.

Related remarks apply to the choice of the wave function in semiempirical
applications. The parametrization of MNDO-type methods has mostly been based on
closed-shell molecules described by a closed-shell RHF (restricted Hartree-Fock)
treatment. The same approach is clearly preferred in any application involving closed-
shell molecules. For open-shell species, it is likewise recommended to use an RHF
treatment (*49,50*) since the unrestricted UHF approach will generally lead to spin
contamination and tend to yield artificially low energies of open-shell species relative to
closed-shell species (*51,52*). The efficient implementation of analytical gradients for
semiempirical open-shell RHF treatments (*53*) has eliminated the practical arguments in
favor of UHF. In more complicated cases, it may be possible that the zero-order wave
function can no longer be represented at the RHF level, but only by a multiconfigurational function. To be consistent with the parametrization, it is then essential to
choose the multiconfigurational space as small as possible to include only the near-

degeneracy effects (and not the dynamic correlation effects that are taken into account by the parametrization, see above).

Performance for the G2 Neutral Test Set

The ultimate test of theoretical calculations is the comparison with experiment. This is true at any theoretical level (ab initio, DFT, semiempirical, etc.). It is essential that the evaluation and validation of theoretical methods is done with regard to reliable experimental reference data. In the development of the Gaussian-2 (G2) theory (*13*), a high-level composite ab initio approach, a standard set of small molecules has been assembled in order to evaluate the G2 performance in thermochemical calculations. This set has recently been extended (*14*) to 148 molecules for which reliable experimental heats of formation at 298 K are available (with a target accuracy of at least 1 kcal/mol). This "G2 neutral test set" provides a means for assessing theoretical methods, and detailed evaluations are already available (*14*) for the G2 method and its variants, for several DFT approaches (e.g. LDA(SVWN), BLYP, BP86 in the usual notation), and for various HF/DFT hybrid methods (e.g. B3LYP, B3P86).

The molecules in the G2 neutral test set are clearly much smaller than the molecules that are normally studied in semiempirical calculations. Nevertheless, they provide a welcome opportunity for an assessment of the accuracy of current semiempirical methods. Tables I and II compare the results for MNDO, AM1, PM3, and MNDO/d with the published results (*14*) on heats of formation at 298 K using exactly the same conventions as in the published work.

It is obvious from Table I that the G2 approach is the most accurate one (as expected) followed by G2 (MP2) and B3LYP. The semiempirical methods (especially PM3 and MNDO/d) show similar errors as BLYP, whereas BP86 and particularly LDA(SVWN) overbind strongly. To put these results into perspective, it should be noted that the complete geometry optimization for all 148 test molecules combined took less than 10 seconds of cpu time for MNDO, AM1, and PM3, and about 18 seconds for MNDO/d (using our current program (*54*) on an SGI R10000 workstation). The computational effort at the semiempirical level is clearly several orders of magnitude lower than at the G2 or DFT levels. Given this situation the performance of the semiempirical methods appears most acceptable.

Table II lists the statistical results for the proposed subgroups of the G2 neutral test set (*14*). The errors for G2 and B3LYP are fairly uniform, while those for the other

Table I. Summary of Mean Absolute Deviations Δ_{abs} and Maximum Deviations Δ_{max} (kcal/mol) for the Molecules from the G2 Neutral Test Set (a)

Method	Δ_{abs}	Δ_{max} (b)	Ref.
G2	1.58	8.2, -7.1	(14)
G2(MP2)	2.04	10.1, -5.3	(14)
LDA(SVWN)	91.16	228.7, (c)	(14)
BLYP	7.09	28.4, -24.8	(14)
BP86	20.19	49.7, -6.3	(14)
B3LYP	3.11	8.2, -20.1	(14)
MNDO	9.32	27.6, -116.7	
AM1	7.81	42.5, -58.2	
PM3	7.01	23.1, -32.2	
MNDO/d	7.26	27.6, -33.9	

(a) The G2 neutral test set contains 148 molecules (14). Due to the lack of parameters for certain elements, the data for MNDO, AM1, and PM3 refer to 146, 142, and 144 molecules, respectively. (b) Largest positive and negative deviations (experiment-theory). (c) No negative deviations.

methods scatter appreciably. This is well-known for the semiempirical methods: Hydrocarbons, substituted hydrocarbons, and inorganic hydrides are well described, whereas the "non-hydrogen" compounds and the radicals have larger errors. The former contain the few hypervalent second-row compounds from the test set, for which the semiempirical methods with an sp basis tend to fail (especially MNDO and AM1, with maximum errors of 116.7 and 58.2 kcal/mol for ClF_3). It should also be noted that the G2 neutral test set is dominated by first-row compounds where MNDO and MNDO/d are defined to be identical. The advantages of MNDO/d over the other semiempirical methods for second-row compounds are therefore not apparent from Tables I and II.

For further statistical analysis, we define other subgroups of the G2 neutral test set. We consider those 140 molecules (out of a total of 148) which only contain elements that have been parametrized in MNDO, AM1, PM3, and MNDO/d. Tables III and IV present statistical evaluations for the 93 first-row and 47 second-row compounds in this set, respectively, which are further subdivided according to the elements involved. This subdivision employs the convention that the elements are ordered

Table II. Mean Absolute Deviations (kcal/mol) for Different Types of Molecules from the G2 Neutral Test Set (a)

Method	Non-hydrogen	Hydrocarbons	Substituted Hydrocarbons	Radicals	Inorganic Hydrides
G2	2.53 (35)	1.29 (22)	1.48 (47)	1.16 (29)	0.95 (15)
BLYP	10.30	8.09	6.10	6.09	3.13
BP86	16.61	25.82	26.80	15.76	8.16
B3LYP	5.35	2.76	2.10	2.98	1.84
MNDO	15.57 (33)	4.78	5.35	12.33	4.86
AM1	11.01 (31)	4.83	4.75	13.23 (28)	4.89 (14)
PM3	10.50 (31)	4.06	3.83	11.31	5.80
MNDO/d	10.79	4.78	4.29	10.98	4.77

(a) The number of molecules in a subgroup is given in parentheses for G2 (*14*). For the other methods, these numbers are listed only if they deviate from the G2 value (due to missing parameters in MNDO, AM1, and PM3). Results for G2, BLYP, BP86, and B3LYP have been taken from (*14*).

(H,C,N,O,F,Cl,P,S,Si) and that a given molecule is assigned to the column of its "last" constituent atom, e.g. $SiCl_4$ is assigned to the column Si (not Cl). Apart from the mean absolute deviations, Tables III and IV also list mean signed deviations which are indicative of systematic errors.

In Table III the results in column "CH" include hydrocarbon radicals and are therefore different from those in column "Hydrocarbons" of Table II. According to Table III all methods except B3LYP tend to slightly underestimate the heats of formation of the first-row compounds (especially for CHNO). MNDO and MNDO/d are defined to be identical for first-row compounds and therefore show the same statistical results. Among the established semiempirical methods, PM3 is best for first-row compounds. However, the new semiempirical approaches with orthogonalization corrections in the one-center part of the core Hamiltonian (*29*) and in the complete core Hamiltonian (*30*) perform much better than the established methods, with mean absolute deviations below 5 kcal/mol (see Table III). These new approaches go beyond the MNDO model by including additional one-electron interactions and provide substantial improvements not only for thermochemical properties, but also in other areas (*29,30*).

Table III. Deviations (kcal/mol) for First-Row Molecules from the G2 Neutral Test Set (a)

Method	First-row	CH	CHN	CHNO	CHNOF
Mean absolute deviations					
G2	1.53 (93)	1.31 (30)	1.04 (17)	1.37 (33)	3.29 (12)
BLYP	7.38	7.16	5.41	8.34	8.70
B3LYP	2.42	2.66	2.23	2.18	2.86
MNDO	7.71	7.42	5.93	9.13	7.62
AM1	7.44	6.57	5.05	9.35	7.89
PM3	6.86	6.43	6.75	6.88	7.46
MNDO/d	7.71	7.42	5.93	9.13	7.62
Ref.29	4.64	3.20	3.67	5.89	6.37
Ref.30	3.36 (81)	2.37	3.25	4.30	(b)
Mean signed deviations (theory-experiment)					
G2	-0.23	1.00	0.71	-0.97	-2.55
BLYP	-1.25	6.71	-3.58	-6.13	-4.50
B3LYP	0.37	1.35	-1.41	0.05	1.41
MNDO	-1.35	-2.58	0.24	-2.78	3.24
AM1	-1.58	1.45	0.63	-4.17	-4.89
PM3	-0.75	-0.61	2.48	-2.40	-0.09
MNDO/d	-1.35	-2.58	0.24	-2.78	3.24
Ref.29	-0.84	-1.06	0.07	-2.74	3.37
Ref.30	-0.29	-0.66	1.37	-0.70	(b)

(a) The number of molecules in a subgroup is given in parentheses for G2. The H_2 molecule has not been included in any element-specific subgroup. Results for G2, BLYP, and B3LYP have been derived from the published data (14). (b) Not yet available.

Table IV indicates that all methods except G2 tend to overestimate the heats of formation for the second-row compounds. Among the semiempirical methods, MNDO/d is most accurate. This has been attributed (19-21) to the use of an spd basis which allows a balanced description of normalvalent and hypervalent molecules. This advantage is not documented well by Table IV since there are only 3 hypervalent molecules in the test set.

Table IV. Deviations (kcal/mol) for Second-Row Molecules from the G2 Neutral Test Set (a)

Method	Second-row	Cl	S	P	Si
Mean absolute deviations					
G2	1.72 (47)	1.41 (16)	1.55 (15)	1.93 (4)	2.32 (10)
BLYP	7.13	8.31	6.33	3.13	7.20
B3LYP	4.46	3.94	3.15	4.53	5.90
MNDO	12.44	13.82	14.16	2.82	13.70
AM1	8.86	8.59	7.64	4.96	14.09
PM3	7.15	6.99	7.62	7.56	5.93
MNDO/d	6.10	4.85	8.61	5.94	4.38
Mean signed deviations (theory-experiment)					
G2	-0.16	-1.07	0.54	1.83	-0.46
BLYP	1.61	-0.36	0.51	-3.13	6.36
B3LYP	3.19	3.06	2.51	-0.28	4.26
MNDO	7.06	6.78	11.33	1.88	4.90
AM1	0.88	1.62	3.14	0.10	-3.46
PM3	0.79	-1.19	4.61	-7.56	0.23
MNDO/d	1.62	-2.85	8.29	-3.05	2.21

(a) The number of molecules in a subgroup is given in parentheses for G2. AlF_3 and $AlCl_3$ have not been included in any element-specific subgroup. Results for G2, BLYP, and B3LYP have been derived from the published data (*14*).

In an overall assessment, the established semiempirical methods perform reasonably for the molecules in the G2 neutral test set. With an almost negligible computational effort (ca 10 seconds for the complete geometry optimization of all 148 molecules), they provide heats of formation with typical errors around 7 kcal/mol. The new semiempirical approaches that go beyond the MNDO model promise an improved accuracy (currently about 3-5 kcal/mol for first-row compounds).

Statistical Evaluations for Larger Test Sets

Even though the G2 neutral test set is very valuable, it is biased towards small molecules and does not cover all bonding situations that may arise for a given element. The validation of semiempirical methods has traditionally been done using larger test sets which, however, have the drawback that the experimental reference data are often less accurate than those in the G2 set. It is clearly cumbersome to check all the experimental reference data in large validation sets (see e.g. (9)), and casual inspection indeed shows that there are sometimes unreliable data included in such sets. Therefore, this section will focus on test sets that have been checked and are actually used in our group.

In our validation sets, experimental heats of formation are preferentially taken from recognized standard compilations (55-57). If there are enough experimental data for a given element, we normally only use reference values that are accurate to 2 kcal/mol. If there is a lack of reliable data, we may accept experimental heats of formation with a quoted experimental error of up to 5 kcal/mol. This choice is motivated by the target accuracy of the established semiempirical methods. If experimental data are missing for a small molecule of interest, we consider it legitimate (21) to employ theoretical heats of formation from the G2 method as substitutes (normally accurate to 2 kcal/mol, see above). Occasionally we also accept other ab initio heats of formation with an estimated accuracy of at least 5 kcal/mol.

Our primary validation set for first-row compounds is derived from the original MNDO development (58,59). It has been updated to include new experimental data for the reference molecules, but would clearly benefit from some extension. Nevertheless, this set still contains about twice as many molecules as the G2 neutral test set and there-fore remains useful (in spite of some overlap).

Table V shows the statistical evaluations for the primary validation set. MNDO/d values are not included (identical to MNDO for first-row compounds, see above). The results in Table V are consistent with those in Table III. In both cases, the mean absolute deviations decrease in the sequence MNDO > AM1 > PM3 > Ref.29 > Ref.30. Generally, the errors are smaller in Table V than in Table III which is probably due to the fact that our larger set contains a larger portion of "normal" organic molecules without "difficult" bonding characteristics.

The data in Table V refer almost exclusively to closed-shell molecules (except for triplet O_2). A second validation set for first-row compounds (60) contains 38 radicals and radical cations. The mean absolute errors for these species are higher than those in Table V. They amount to 11.08, 9.73, 9.41, 6.70, and 4.79 kcal/mol for MNDO, AM1, PM3, ref.29, and ref.30, respectively. Again, the errors for the new methods are significantly lower than for the established semiempirical methods.

Table V. Deviations (kcal/mol) for First-Row Reference Molecules (a)

Method	First-row	CH	CHN	CHNO	CHNOF
Mean absolute deviations					
MNDO	7.35 (181)	5.81 (58)	6.24 (32)	7.12 (48)	10.50 (43)
AM1	5.80	4.89	4.65	6.79	6.76
PM3	4.71	3.79	5.02	4.04	6.45
Ref.29	3.87	2.49	4.27	4.12	5.17
Ref.30	3.07 (138)	1.75	3.92	4.11	(b)
Mean signed deviations (theory-experiment)					
MNDO	-0.24	-3.08	-1.79	1.17	3.16
AM1	0.45	0.84	0.34	0.62	-0.17
PM3	0.47	-0.95	1.12	0.54	1.82
Ref.29	-0.01	-0.07	-0.16	0.07	0.11
Ref.30	0.19	0.07	0.04	0.45	(b)

(a) The number of molecules in each group is given in parentheses for MNDO and also applies to the other methods unless noted otherwise. The H_2 molecule has been included in column CH. (b) Not yet available.

In the course of the MNDO/d development (*18-21*) we have generated new validation sets for second-row and heavier elements. Those for Na, Mg, Al, Si, P, S, Cl, Br, I, Zn, Cd, and Hg have been published (*19-21*). The corresponding statistical evaluations for heats of formation (*21*) are summarized in Table VI.

It is obvious that MNDO/d shows by far the smallest errors followed by PM3 and AM1. All four semiempirical methods perform reasonably well for normalvalent compounds, especially when considering that more effort has traditionally been spent in the parametrization of the first-row elements (compared with the heavier elements). However, for hypervalent compounds, the errors are huge in MNDO and AM1, and still substantial in PM3, in spite of the determined attempt to reduce these errors in the PM3 parametrization (*23*). The improvements in MNDO/d have therefore been attributed to the use of an spd basis set (*19-21*).

The results in this section support the conclusions from the evaluations with the G2 neutral test set: Among the established semiempirical methods, PM3 and AM1 are best for first-row molecules, but they are outperformed by the new approaches with orthogonalization corrections (*29,30*). Currently MNDO/d appears to be the method of

Table VI. Mean Absolute Deviations (kcal/mol) for Compounds Containing Second-Row and Heavier Elements (a)

	All	Common (b)	Normal (c)	Hyper (d)
MNDO	29.2 (488)	29.2 (488)	11.0 (421)	143.2 (67)
AM1	15.3	15.3	8.0	61.3
PM3	10.9 (552)	10.0	9.6 (485)	19.9
MNDO/d	5.4 (575)	4.9	5.4 (508)	5.4

	Cl	Br	I	S
MNDO	39.4 (85)	16.2 (51)	25.4 (42)	48.4 (99)
AM1	29.1	15.2	21.7	10.3
PM3	10.4	8.1	13.4	7.5
MNDO/d	3.9	3.4	4.0	5.6

	P	Si	Al	Hg
MNDO	38.7 (43)	12.0 (84)	22.1 (29)	13.7 (37)
AM1	14.5	8.5	10.4	9.0
PM3	17.1	6.0	16.4	7.7
MNDO/d	7.6	6.3	4.9	2.2

(a) See (19-21) for details and more data (e.g. for Na, Mg, Zn, Cd). The number of molecules in each group is given in parentheses for MNDO and also applies to the other methods unless noted otherwise. (b) Common subset without Na, Mg, and Cd compounds. (c) Normalvalent compounds. (d) Hypervalent compounds.

choice for compounds with heavier atoms. These assessments refer to the semiempirical calculation of heats of formations at 298 K. They are based on statistical results which certainly depend on the choice of the test molecules. We believe, however, that the number and diversity of the test molecules are large enough to provide a fair qualitative assessment.

Concerning other semiempirical methods, the SAM1 approach (which has not yet been completely specified in the published literature) seems quite promising (24,25). For a large validation set of 285 neutral closed-shell first-row molecules (C,H,N,O,F), the mean absolute error for heats of formation is found to be 4.44 kcal/mol in SAM1, compared with 7.24 and 4.85 kcal/mol in AM1 and PM3, respectively (25). Using our own validation set for second-row and heavier molecules (Si, P, S, Cl, Br, I: 404 compounds), the mean absolute errors for heats of formation are 9.3, 16.2, 9.5, and 5.1 kcal/mol in SAM1, AM1, PM3, and MNDO/d, respectively (*21*). Inclusion of d orbitals for Cl and Br in the SAM1d approach improves the results significantly, in analogy to the MNDO/d case (*21*). Finally, for a set of 91 ions and radicals, the reported errors (*25*) are again slightly smaller for SAM1 than for AM1 and PM3 (9.01 vs. 11.32 and 9.61 kcal/mol).

Concerning other properties, there is a vast literature documenting the accuracy and the errors of existing semiempirical methods (see e.g. (*1-12,16-30, 42-45,58-65*)) which may be consulted for detailed information. Since semiempirical methods also provide reasonable molecular geometries and vibrational frequencies, they may be used to compute partition functions and thermodynamic properties other than heats of formation, e.g. entropies and heat capacities (via statistical thermodynamics). These applications are beyond the scope of the current review. This also applies to thermochemical properties that are related to differences in heats of formation, e.g. proton affinities, adiabatic ionization potentials, and electron affinities. The reader is referred to the literature for information on these topics (*1-12,16-30,42-45,58-65*).

Discussion

The statistical evaluations in the preceding sections indicate that semiempirical MO methods can predict heats of formation with useful accuracy and at very low computational costs. When comparing with ab initio or DFT methods, the following points should be kept in mind, however:

a) In general, errors tend to be more systematic at a given ab initio or DFT level and may therefore often be taken into account by suitable corrections. Errors in semiempirical calculations are normally less uniform and thus harder to correct.

b) The accuracy of the semiempirical results may be different for different classes of compounds, and there are elements that are more "difficult" than others. Such variations in the accuracy are again less pronounced in high-level ab initio and DFT calculations.

c) Semiempirical methods can only be applied to molecules containing elements that have been parametrized, while ab initio and DFT methods are generally applicable (apart from technical considerations such as basis set availability).

d) Semiempirical parametrizations require reliable experimental or theoretical reference data and are impeded by the lack of such data. For example, only very few accurate heats of formation are available for transition metal compounds which is one of the obstacles in the extension of MNDO/d to transition metals. Such problems do not occur in ab initio or DFT approaches.

In spite of these limitations (a)-(d), there are many areas where the established MNDO-type semiempirical methods (16-23) can be applied successfully for calculating thermochemical properties. This suggests that the underlying MNDO model includes the physically relevant interactions such that the parametrization can absorb the errors introduced by the MNDO approximations (in an average sense). In recent improvements of the MNDO model (29,30), the Pauli repulsions are treated explicitly through parametric contributions to the core Hamiltonian (rather than implicitly through an effective atom-pair potential in the core-core repulsion) which leads to significant improvements in the accuracy of the calculated thermochemical properties (see above). This supports our belief (12) that a theoretically guided search for better models offers the most promising perspective for general-purpose semiempirical methods with better overall performance.

For specific applications, there is an alternative to this approach: General-purpose semiempirical methods attempt to describe many classes of compounds and properties (e.g. heats of formation, geometries, dipole moments, and ionization potentials) simultaneously and equally well. It is obvious that compromises cannot be avoided in such an endeavour (e.g. during the parametrization). It would therefore seem legitimate to develop specialized semiempirical methods for certain classes of compounds or for specific properties because such methods ought to be more accurate in their area of applicability than the general-purpose methods. This advantage must be weighed against the additional parametrization work and the danger of parameter proliferation. Such specialized semiempirical treatments exist (12,66-70) but will not be reviewed here. It should be feasible to develop specific parametrizations also for thermochemical properties, if needed.

Conclusions

Semiempirical calculations are so much faster than ab initio or DFT calculations (by several orders of magnitude) that they will remain useful for survey studies, for establishing trends, and for scanning computational problems before proceeding with more expensive and more accurate theoretical studies. We advocate the combined use of ab initio, DFT, and semiempirical methods for calculating thermochemical properties. High-level methods should be applied whenever this is feasible, whereas semiempirical methods are recommended in cases where a complete ab initio or DFT study is impractical (e.g. for large molecules).

According to the available statistical evaluations, heats of formation are computed with an average accuracy of typically 5-7 kcal/mol by the established semiempirical methods based on the MNDO model (*16-23*). It is clearly desirable to improve this accuracy, which has already been accomplished for first-row compounds by approaches that go beyond the MNDO model (*29,30*).

Acknowledgments

The author wishes to thank his coworkers for their contributions, particularly M. Kolb, A. Voityuk, and W. Weber. Thanks are due to Dr. L.A. Curtiss for a preprint of ref. 14 and for providing the G2 neutral test set. Financial support from the Deutsche Forschungsgemeinschaft, the Alfried-Krupp-Stiftung, and the Schweizerischer Nationalfonds is gratefully acknowledged.

Literature Cited

1. Dewar, M.J.S. *The Molecular Orbital Theory of Organic Chemistry*; McGraw-Hill: New York, NY, 1969.
2. Pople, J.A.; Beveridge, D.L. *Approximate Molecular Orbital Theory*; Academic Press: New York, NY, 1970.
3. Murrell, J.N.; Harget, A.J. *Semiempirical Self-Consistent-Field Molecular Orbital Theory of Molecules*; Wiley: New York, NY, 1972.
4. *Modern Theoretical Chemistry*; Segal, G.A., Ed.; Plenum: New York, NY, 1977; Vols. 7-8.
5. Dewar, M.J.S. *Science* **1975**, *187*, 1037.
6. Jug, K. *Theor. Chim. Acta* **1980**, *54*, 263.
7. Dewar, M.J.S. *J. Phys. Chem.* **1985**, *89*, 2145.
8. Thiel, W. *Tetrahedron* **1988**, *44*, 7393.
9. Stewart, J.J.P. *J. Comp.-Aided Mol. Design* **1990**, *4*, 1.
10. Stewart, J.J.P. In *Reviews in Computational Chemistry*; Lipkowitz, K.B.; Boyd, D.B., Eds.; VCH Publishers: New York, NY, 1990, Vol. 1; pp. 45-81.
11. Zerner, M.C. In *Reviews in Computational Chemistry*; Lipkowitz, K.B.; Boyd, D.B., Eds.; VCH Publishers: New York, NY, 1991, Vol. 2; pp. 313-365.
12. Thiel, W. *Adv. Chem. Phys.* **1996**, *93*, 703.
13. Curtiss, L.A.; Raghavachari, K.; Trucks, G.W.; Pople, J.A. *J. Chem. Phys.* **1991**, *94*, 7221.
14. Curtiss, L.A.; Raghavachari, K.; Redfern, P.C.; Pople, J.A. *J. Chem. Phys.* **1997**, *106*, 1063.
15. Pople, J.A.; Santry, D.P.; Segal, G.A. *J. Chem. Phys.* **1965**, *43*, S 129.

16. Dewar, M.J.S.; Thiel, W. *J. Am. Chem. Soc.* **1977**, *99*, 4899.

17. Thiel, W. *J. Am. Chem. Soc.* **1981**, *103*, 1413.

18. Thiel, W.; Voityuk, A.A. *Theor. Chim. Acta* **1992**, *81*, 391; **1996**, *93*, 315.

19. Thiel, W.; Voityuk, A.A. *Int. J. Quantum Chem.* **1992**, *44*, 807.

20. Thiel, W.; Voityuk, A.A. *J. Mol. Struct.* **1994**, *313*, 141.

21. Thiel, W.; Voityuk, A.A. *J. Phys. Chem.* **1996**, *100*, 616.

22. Dewar, M.J.S.; Zoebisch, E.; Healy, E.F.; Stewart, J.J.P. *J. Am. Chem. Soc.* **1985**, *107*, 3902.

23. Stewart, J.J.P. *J. Comp. Chem.* **1989**, *10*, 209, 221.

24. Dewar, M.J.S.; Jie, C.; Yu, J. *Tetrahedron* **1993**, *49*, 5003.

25. Holder, A.J.; Dennington, R.D.; Jie, C. *Tetrahedron* **1994**, *50*, 627.

26. Bingham, R.C.; Dewar, M.J.S.; Lo, D.H. *J. Am. Chem. Soc.* **1975**, *97*, 1285.

27. Nanda, D.N.; Jug, K. *Theor. Chim. Acta* **1980**, *57*, 95.

28. Jug, K.; Iffert, R.; Schulz, J. *Int. J. Quantum Chem.* **1987**, *32*, 265.

29. Kolb, M.; Thiel, W. *J. Comp. Chem.* **1993**, *14*, 37.

30. Weber, W. Ph.D. Dissertation; Universität Zürich, 1996.

31. Dewar, M.J.S.; Thiel, W. *Theor. Chim. Acta* **1977**, *46*, 89.

32. Klopman, G. *J. Am. Chem. Soc.* **1964**, *86*, 4550.

33. Ohno, K. *Theor. Chem. Acta* **1964**, *2*, 219.

34. Hehre, W.J.; Radom, L.; Schleyer, P.v.R.; Pople, J.A. *Ab Initio Molecular Orbital Theory*; Wiley: New York, NY, 1986.

35. Parr, R.G.; Yang, W. *Density-Functional Theory of Atoms and Molecules*; Oxford University Press: Oxford, 1989.

36. Becke, A.D. *J. Chem. Phys.* **1993**, *98*, 5648.

37. Cremer, D. *J. Comp. Chem.* **1982**, *3*, 165.

38. Kellö, V.; Urban, M.; Noga, J., Diercksen, G.H.F. *J. Am. Chem. Soc.* **1984**, *106*, 5864.

39. Ibrahim, M.R.; Schleyer, P.v.R. *J. Comp. Chem.* **1985**, *6*, 157.

40. Baird, N.C.; Hadley, G.C. *Chem. Phys. Lett.* **1986**, *128*, 31.

41. Dewar, M.J.S.; O'Connor, B.M. *Chem. Phys. Lett.* **1987**, *138*, 141.

42. Thiel, W. *J. Am. Chem. Soc.* **1981**, *103*, 1420.

43. Schweig, A.; Thiel, W. *J. Am. Chem. Soc.* **1981**, *103*, 1425.

44. Schröder, S.; Thiel, W. *J. Am. Chem. Soc.* **1985**, *107*, 4422.

45. Schröder, S.; Thiel, W. *J. Am. Chem. Soc.* **1986**, *108*, 7985.

46. Flanigan, M.C.; Komornicki, A.; McIver, J.W. In Ref. 4, Vol. 8; pp. 1-47.

47. Schulman, J.M.; Disch, R.L. *Chem. Phys. Lett.* **1985**, *113*, 291.

48. Hicks, M.G.; Thiel, W. *J. Comp. Chem.* **1986**, *7*, 213.

49. Dewar, M.J.S.; Hashmall, J.A.; Venier, C.G. *J. Am. Chem. Soc.* **1968**, *90*, 1953.

50. Dewar, M.J.S.; Olivella, S. *J. Chem. Soc. Faraday II* **1979**, *75*, 829.

51. Dewar, M.J.S.; Olivella, S.; Stewart, J.J.P. *J. Am. Chem. Soc.* **1986**, *108*, 5771.

52. Pachkovski, S.; Thiel, W. *J. Am. Chem. Soc.* **1996**, *118*, 7164.

53. Pachkovski, S.; Thiel, W. *Theor. Chim. Acta* **1996**, *93*, 87.

54. Thiel, W. *Program MNDO96*; Zürich, 1996. Distributed as part of the UniChem software by Oxford Molecular Ltd.

55. Pedley, J.B.; Naylor, R.D.; Kirby, S.P. *Thermochemical Data of Organic Compounds*, 2nd ed.; Chapman Hall: London, 1986.

56. Chase, M.W.; Davies, C.A.; Downey, J.R.; Frurip, D.R.; McDonald, R.A.; Syverud, A.N. *JANAF Thermochemical Tables*, 3rd ed.; *J. Phys. Chem. Ref. Data* **1985**, 14, Suppl. 1.

57. Lias, S.G.; Bartmess, J.E.; Liebman, J.F.; Holmes, J.L.; Levin, R.D.; Mallard, W.G. *Gas Phase Ion and Neutral Thermochemistry*; *J. Phys. Chem. Ref. Data* **1988**, 17, Suppl.1.

58. Dewar, M.J.S.; Thiel, W. *J. Am. Chem. Soc.* **1977**, *99*, 4907.

59. Dewar, M.J.S.; Rzepa, H.S. *J. Am. Chem. Soc.* **1978**, *100*, 58.

60. Higgins, D.; Thomson, C.; Thiel, W. *J. Comp. Chem.* **1988**, *9*, 702.

61. Dewar, M.J.S.; Rzepa, H.S. *J. Am. Chem. Soc.* **1978**, *100*, 784.

62. Dewar, M.J.S.; Dieter, K.M. *J. Am. Chem. Soc.* **1986**, *108*, 8075.

63. Olivella, S.; Urpi, F.; Vilarassa, J. *J. Comp. Chem.* **1984**, *5*, 230.

64. Halim, H.; Heinrich, N.; Koch, W.; Schmidt, J.; Frenking, G. *J. Comp. Chem.* **1986**, *7*, 93.

65. Clark, T. *A Handbook of Computational Chemistry*; Wiley: New York, NY, 1985.

66. Slanina, Z.; Lee, S.-L.; Yu, C. In *Reviews in Computational Chemistry*; Lipkowitz, K.B.; Boyd, D.B., Eds.; VCH Publishers: New York, NY, 1996, Vol. 8; pp. 1-62.

67. Burstein, K.Y.; Isaev, A.N. *Theor. Chim. Acta* **1984**, *64*, 397.

68. Goldblum, A. *J. Comp. Chem.* **1987**, *8*, 835.

69. Rossi, I.; Truhlar, D.G. *Chem. Phys. Lett.* **1995**, *233*, 231.

70. Bash, P.A.; Ho, L.L.; MacKerell, A.D.; Levine, D.; Hallstrom, P. *Proc. Natl. Acad. Sci. USA* **1996**, *93*, 3698.

Chapter 9

Bond-Additivity Correction of Ab Initio Computations for Accurate Prediction of Thermochemistry

Michael R. Zachariah[1,3], and Carl F. Melius[2]

[1]National Institute of Standards and Technology, Gaithersburg, MD 20899
[2]Sandia National Laboratories, Livermore, CA 94551

This paper reviews a method for the correction of *ab-initio* derived molecular energy computations, for determination of thermochemical data of sufficient accuracy for application to chemical modeling. The basic concept is the use of a computationally inexpensive approach MP4/6-31G**//HF-6-31G*, which enables a large number of molecules to be computed, and combines this with a bond-additivity correction scheme. The basic concept behind the bond additivity approach is that errors in the energy computation are systematic with the type of bond, and that calibration for each bond class from comparison with molecules of known energy allow for a correction to the computed energy.

The rapid increase in computer power has enabled the development of ever more sophisticated "Computational Reacting Fluid Dynamics" (CRFD) models. These models use detailed chemical kinetic mechanisms to solve chemistry-flow interactions in a wide variety of complex problems; e.g. chemical vapor deposition, combustion, plasma processing, atmospheric processes. As computer models become more sophisticated, the need for the fundamental data that feed such models has become a major limiting factor. Thermochemistry (enthalpy, entropy, heat capacity) and kinetics (diffusion coefficients, thermal transport coefficients, reaction rate constants) are the life blood of such models, and while the models, which in a sense are nothing more than sophisticated bookkeeping schemes for mass and energy conservation can be adapted from one problem to another, the chemistry is unique.

In seeking a solution to the problem of lack of data, it is unreasonable to assume that we can measure our way out of the problem. Fortunately modern computational chemistry methods offer a means to circumvent the need to measure everything. The task at hand is to provide to users methods that afford both accuracy sufficient for most CRFD computations and computational cost suitable for the task of generating the large data bases necessary to undertake useful CRFD computations.

In this paper we will describe a procedure that aims to satisfy the basic requirements set forth in the paragraph above.

[3]Email address: mr2@tiber.nist.gov

Computational Approach

While it is true that given enough computer time one can compute molecular properties to great precision, a simpler and cheaper approach that sacrifices some of the scientific rigor may ultimately yield the same accuracy in a much more tractable format. The BAC procedure involves the systematic correction of ab-initio molecular orbital computations, which enables the use of a computationally cheaper level of theory with the same level of potential accuracy.

The BAC procedure is based on the observation that errors in electronic structure computations are systematic and therefore amenable to correction. Furthermore, BAC procedures assume that these errors are localized and are therefore additive. The best analogy can be drawn from the well known bond and group-additivity methods developed by Benson, which state that the energy of a molecule can be constructed as a sum of the energies of its constituent parts, bonds or groups *(1)* (see also the chapter by Benson in this volume). The BAC procedure assumes that the error in computing a molecular energy is equivalent to the sum of the errors in computing the constituent parts of the molecules, namely the bonds. As such the BAC approach is basically an extension of the isodesmic reaction approach. Since the error in computing a given type of bond can usually be determined by comparison with known bond energies, such a procedure enables the correction of the molecular energy in a relatively straightforward manner, as will be illustrated.

BAC-MP4 Methodology

The computational methodology is based on an MP4/6-31G**//HF/6-31G* computation *(2)*. Ab-initio computations in general tend to underestimate bond strengths as a result of finite basis set effects and the finite number of configurations used to treat the electron correlation. As previously stated, the errors can be considered in large part to be bond-wise additive and depend on bond type, bond distance and a small correction due to nearest neighbor bonds.

For the bond between A_i and A_j in the molecule A_k-A_i-A_j-A_l, the error in calculating the electronic energy can be estimated through a pair-wise additive empirical bond correction E_{BAC} of the form.

$$E_{BAC} (A_i-A_j) = f_{ij} \, g_{kij} \, g_{ijl} \tag{1}$$

where

$$f_{ij} = A_{ij} \, \exp(-\alpha_{ij} \, r_{ij}), \tag{2}$$

A_{ij} and α_{ij} are calibration constants (shown in Table I) that depend on bond type, and r_{ij} is the bond length at the Hartree-Fock level.

The multiplicative factor g_{kij},

$$g_{kij} = (1 - h_{ik} \, h_{ij}) \tag{3}$$

is the nearest-neighbor bond correction (to the bond being considered),

where

Table I. BAC Parameters

Bond Type	a_{ij}	A_{ij} (kcal/mol)	Atom Type	B_k
H-H	2	18.98	H	0
B-H	2	31.1	B	0.2
C-H	2	38.6	C	0.31
C-C	3.8	1444.1	N	0.2
N-H	2	70.1	O	0.225
B-N	2.84	370	Cl	0.42
C-N	2.8	462.3	F	0.33
N-N	2.6	472.6	Si	0.2
O-H	2	72.45	S	0.56
B-O	2.65	206.9		
C-O	2.14	175.6		
N-O	2.1	226		
O-O	2	169.8		
F-H	2	84.2		
C-F	2.1	143.3		
N-F	2	170		
O-F	2	189.7		
F-F	2	129.2		
Si-H	2	38.6		
Si-C	2.5	893.7		
Si-N	2.5	848		
Si-O	2.42	628.3		
Si-F	2	260.6		
Si-Si	2.7	3330.2		
P-H	2	137.5		
P-C	2	260		
P-N	2	290		
P-O	2	305		
P-F	2	320		
S-H	2	119.5		
C-S	2	281.9		
N-S	2	350		
S-O	2	455		
S-F	2	400		
S-S	2	100		
H-Cl	2	116.4		
B-Cl	2	172.5		
C-Cl	2	304.3		
N-Cl	2	340		
O-Cl	2	355.1		
F-Cl	2	380		
Si-Cl	2	721.9		
Cl-Cl	2	980.2		
P-Si	2.91	7085		

$$h_{ik} = B_k \exp (-\alpha_{ik} (r_{ik} - 1.4 \text{ Å})) \qquad (4)$$

The functional form implies that as the bond gets shorter (e.g. multiple bonds), the energy correction, E_{BAC}, increases exponentially and primarily depends on the factor f_{ij}. The parameter g_{kij} is an empirically determined correction due to the neighboring bond A_k-A_i, which reduces the size of the bond correction. This was found to be a necessary correction to prevent the effects of double counting of the BAC due to superposition of the basis sets. The reduction in bond error results from the fact that the basis functions for atom A_k help describe the A_k-A_i bond.

The parameters A_{ij}, α_{ij}, and B_k are empirically determined by fitting the heats of formation calculated at the BAC-MP4 level of theory for selected molecules to experimental or in some cases, high-quality ab-initio computations of the heat of formation. They are summarized for a wide variety of bond types in Table I.

For open-shell molecules, an additional correction is needed due to contamination of the wavefunction from higher spin states in the unrestricted Hartree-Fock (UHF) framework. This error is estimated using an approach developed by Schlegel *(3)* in which the spin energy correction E_{spin} (UHF) is obtained from:

$$E_{spin} (UHF) = E(UMP3) - E(PUMP3) \qquad (5)$$

where UMP3 refers to unrestricted wavefunction at the MP3 level (third-order Moller-Plesset perturbation theory) and the P refers to the projected UMP3 energy. For many radicals, E_{spin} is relatively small (e.g., 0.21 kcal/mol for OH), but it can become large for highly unsaturated molecules (e.g., 12.2 kcal/mol for C_2H).

Because the MP4 method is a single-reference calculation, molecules with multi-reference character, such as O_3, are not well treated. Each closed-shell molecule computed is checked for any UHF instability in the HF wavefunction. If found to be unstable, the UHF wavefunction is calculated along with its spin contamination, defined by $S(S+1)$, corrected by:

$$E_{spin} (S^2) = K \, S(S+1) \qquad (6)$$

where K = 10 kcal/mol, based on obtaining reasonable heats of formation for O_3 and $^1CH_2$.

The resulting total BAC energy correction for a given molecule.

$$E_{BAC} (Total) = \Sigma_{ij} E_{BAC} (A_i\text{-}A_j) + E_{spin} (UHF) + E_{spin} (S^2) \qquad (7)$$

where ij is summed over all chemical bonds in the molecule. This energy is then subtracted from the ab-initio computed heat of formation.

A basic summary of the methodology is shown schematically in Figure. 1 and a sample script using the Gaussian series of programs is shown in the Appendix. Note that the methodology is not limited to minima on potential energy surfaces but can be used for defining thermochemistry at transition states for chemical reactions *(2j)*.

BAC-MP4 Error Estimation

Like any computational approach, BAC-MP4 can give more accurate results for some species than for others. Having a method that allows for a rough estimate of the

Bond Additivity Corrected Quantum Chemical Calculation Method

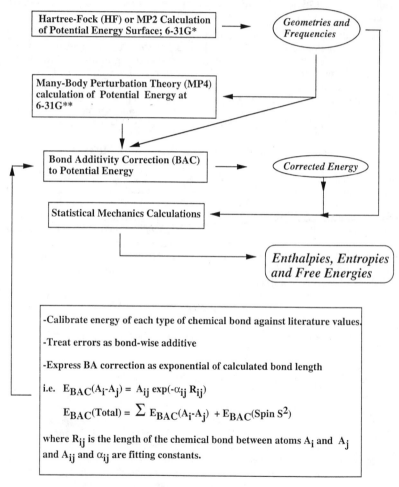

Figure 1. BAC-MP4 computational procedure.

accuracy of the computed enthalpy provides for a more intelligent approach to using the results of computations from a large number of species. The error estimation is based on calculation of the heat of formation at lower levels of perturbation theory, i.e. BAC-MP2, BAC-MP3, BAC-MP4SDQ, each with its own set of BAC correction parameters. The general trend in the correction terms f_{ij} are:

$$f_{ij}(MP2) < f_{ij}(MP4) < f_{ij}(MP3) \approx f_{ij}(MP4SDQ)$$

However, all the correction terms are similar implying that the primary error in the Møller-Plesset perturbation theory calculation is due to the finite basis set. By comparing the resulting heat of formation at the various levels one can estimate the extent of the error. If the results between the various levels are consistent then the error should be small, while differences, or lack of convergence, would indicate a larger error. The BAC-MP4 error estimate is defined as:

$$\text{Error(BAC-MP4)} = \{ 1.0 \text{ kcal-mol}^{-1} + \Delta H_{BAC\text{-}MP4} - \Delta H_{BAC\text{-}MP3} \}^2$$
$$+ (\Delta H_{BAC\text{-}MP4} - \Delta H_{BAC\text{-}MP4SDQ})^2$$
$$+ 0.25 \ (E_{BAC}(\text{Spin } _{S2}) \text{ or } E_{BAC}(\text{Spin }_{UHF\text{-}I}))^2 \}^{1/2} \quad (8)$$

Note that the error estimate includes terms arising from spin contamination and UHF instability and a 1.0 kcal-mol^{-1} inherent error estimate it also requires knowledge of BAC parameters at lower levels of theory in order to estimate the error.

Results

A large body of data have appeared in the literature using the BAC procedure *(2)*. Most of these computations have been directed to carbon-based compounds, particularly as they are related to combustion chemistry; however, a non-trivial number of other first- and second-row elements have also been computed *(2)*. In general, the bond energy correction for H bonded to atoms of the first row of the periodic table increases with atomic number with a near exponential behavior. A similar behavior is also seen for first row elements with each other, although there is more scatter. Such trends help in the assignment of bond correction parameters when data are either unavailable or are of questionable accuracy. This latter point is particularly relevant as one proceeds to the second row where experimental data are in many cases suspect.

Heats of formation accurate enough for chemical computation and modeling applications generally require a target of uncertainty within 2.5 kcal/mol to be of practical value. Comparisons between the BAC method and literature values show that one can achieve such accuracies for first-row elements.

In Table II we present results from the computations for various hydrocarbons. The results show that the BAC approach seems to work well for both stable and radical species and can treat the resonance energy of the aromatic ring. The BAC approach also has been shown to be able to treat molecules with nonconventional hypervalent or dative bonding (e.g. N_2O and nitro compounds).

We have recently completed an extensive study on 110 C_1 and C_2 hydrofluorocarbons (C/H/F/O) in which 70 literature values were compared. The results showed an average deviation of about 2.1 kcal/mol (9 kJ/mol), shown in Figure. 2 *(4)*. Such large comparisons on a single chemical system enable one to have confidence in the methodology.

COMPUTATIONAL THERMOCHEMISTRY

Table II. BAC-MP4 heats of formation at 298 K and 1 atm standard state for various hydrocarbons along with the BAC error estimate. Energies are in kcal-mol[-1].

CxHy	Molecule	$\Delta_f H^0{}_{298}$	Error Estimate	Theor. -Exp.
[2]CH$_3$	Methyl	34.9	1.2	0.1
CH$_4$	Methane	-17.9	1.0	0.0
[2]C$_2$H	C$_2$H $^2\Sigma$+	132.2	6.4	4.5
C$_2$H$_2$	Acetvlene	54.2	1.0	0.0
[2]C$_2$H$_3$	Vinyl	71.0	3.5	2.6
C$_2$H$_4$	Ethylene	12.3	1.0	-0.2
[2]C$_2$H$_5$	Ethyl	28.8	1.3	0.5
C$_2$H$_6$	Ethane	-20.8	1.0	-0.7
C$_3$H$_2$	H$_2$CCC ^{1}A$_1$	133.4	2.5	
	HCCCH ^{1}A'	141.4	4.7	
[3]C$_3$H$_2$	HCCCH ^{3}B	129.4	8.4	
	H$_2$CCC ^{3}B$_1$	160.7	7.8	
[2]C$_3$H$_3$	H$_2$CCCH	83.0	5.8	1.2
	Cycloprop-2-enyl	117.0	3.6	
	CH$_3$CC	123.8	6.3	
	Cycloprop-1-enyl	126.0	3.0	
C$_3$H$_4$	CH$_3$CCH	45.8	2.6	1.4
	CH$_2$CCH$_2$	47.7	3.4	2.0
	Cyclopropene	68.0	2.7	1.8
[2]C$_3$H$_5$	CH$_2$CHCH$_2$	38.7	4.2	-0.7
	CH$_3$CCH$_2$	61.0	3.5	
	CH$_3$CHCH, HCCH cis	64.7	3.5	
	CH$_3$CHCH HCCH trans	65.0	3.5	
	Cyclopropyl	69.3	2.1	2.4
C$_3$H$_6$	CH$_3$CHCH$_2$	5.3	1.2	0.5
	Cyclopropane	11.8	1.5	- 1.0
[2]C$_3$H$_7$	CH$_3$CHCH$_3$	21.4	1.5	1.2
	CH$_3$CH$_2$CH$_2$	24.8	1.3	
C$_3$H$_8$	CH$_3$CH$_2$CH$_3$	-25.7	1.0	-0.8
C$_4$H$_2$	HCCCCH	111.5	5.6	-1.5
[2]C$_4$H$_3$	H2CCCCH	111.3	15.9	
	HCCHCCH,HCCH cis	129.9	8.6	
	HCCHCCH,HCCH trans	130.2	8.3	
	CH$_2$CHCC	144.5	14.1	
C$_4$H$_4$	CH$_2$CHCCH	69.1	3.3	-3.7
	CH$_2$CCCH$_2$	75.5	6.1	
	Methylene-cyclopropene	95.9	5.6	
	1,3-Cyclobutadiene	98.7	4.8	
	Methyl-cyclo-propenylidene	109.7	4.4	
	1,2-Cyclobutadiene (Bicyclo)	123.4	1.3	
	Tetrahedrane	132.3	5.4	

Table II. (continued)

CxHy	Molecule	$\Delta_f H^0_{298}$	Error Estimate	Theor. -Exp.
$^2C_4H_5$	CH_2CHCCH_2	74.1	7.3	
	CH_3CCCH_2	74.3	6.6	5.1
	CH_3CHCCH	75.9	5.8	
	trans-$CH_2CHCHCH$,HCCH cis	86.1	9.1	
	trans-$CH_2CHCHCH$,HCCH trans	86.5	9.1	
	cis-$CH_2CHCHCH$,HCCH cis	86.6	9.6	
	cis-$CH_2CHCHCH$,HCCH trans	89.1	9.6	
C_4H_6	$CH_2CHCHCH_2$	24.7	2.3	- 1.4
	CH_3CCCH_3	37.6	4.3	2.9
	CH_3CH_2CCH	41.2	2.2	1.7
	Cyclobutene	40.6	1.1	3.2
	CH_2CCHCH_3	41.5	3.6	2.8
	Methylene-cyclopropane	46.8	3.0	
	1-Methyl-cyclopropene	58.1	3.5	
	3-Methyl-cyclopropene	60.9	2.5	
C_4H_8	trans-$CH_3CHCHCH_3$	-1.7	1.5	1.3
	$CH_3CH_2CHCH_2$	1.0	1.1	1.2
	CH_3-Cyclopropane	5.0	1.4	
	Cyclobutane	7.0	1.6	0.2
$^2C_4H_9$	$t-C_4H_9$	13.2	1.5	1.3
C_4H_{10}	$i-C_4H_{10}$	-32.0	1.3	0.4
	$n-C_4H_{10}$	-30.6	1.3	-0.3
$^2C_5H_3$	$H_2CCCCCH$	128.2	18.5	
	HCCHCCH	135.0	10.4	
	Cyclopenta- 1 -yn-3-enyl	168.9	16.1	
C_5H_4	$CH_2CCHCCH$	106.1	5.9	
	$HCCCH_2CCH$	111.1	4.3	
	1,2,4-Cyclopentatriene	131.8	5.6	
$^2C_5H_5$	Cyclopentadienyl	63.8	4.8	
C_5H_6	Cyclopentadiene	31.6	2.2	-0.4
$^2C_5H_7$	Cyclopentenyl	40.6	3.9	
	trans,trans-$CH_2CHCHCHCH_2$	45. 3	8.3	
	cis,trans-$CH_2CHCHCHCH_2$	47.6	8.4	
C_5H_8	Cyclopentene	9.9	1.1	1.6
	$CH_2CHCHCHCH_3$	17.4	2.4	
C_5H_{10}	$C(CH_3)_4$	-39.3	1.7	1.0
C_6H_4	Ortho-benzyne	101.6	8.1	
$^2C_6H_5$	Phenyl	79.4	11.0	0.8
	HCCCHCHCHCH	140.6	15.3	
C_6H_6	Benzene	17.0	2.5	-2.8
	Methylene-cyclopentadiene	52.2	2.3	
	$CH_3CHCCHCCH$	99.7	6.0	
$^2C_6H_7$	Benzyl	47.9	8.3	
C_6H_{12}	Cyclohexane	-29.4	2.0	0.1

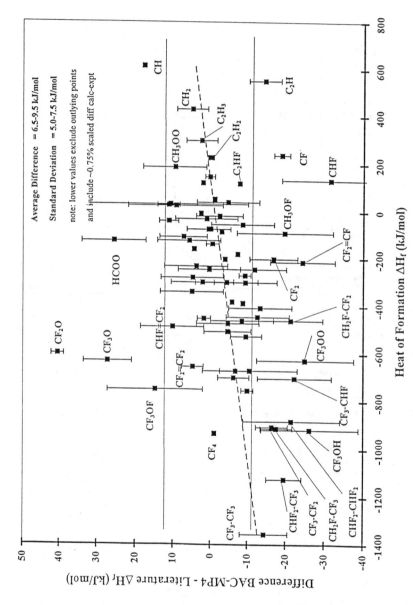

Figure 2. Comparison of BAC-MP4 and literature values for the heats of formation of fluorocarbons.

Example BAC-MP4 correction calculations

In this section, we use two examples to illustrate the BAC corrections. In the first example, we consider the isodesmic reaction,

$$CH_3NO + NH_3 = HNO + CH_3NH_2 \qquad (9)$$

Table III shows the BAC corrections for each of the molecules. The table is an extraction of the output presented on the internet at http://herzberg.ca.sandia.gov/~melius in file ~melius/bac/bac.tar.gz. The energies are given in kcal-mol[-1] For each molecule, if there is a spin correction (equation. 6, since these examples are closed-shell species), it is presented as Espin. Next, for each of the bonds in the molecule, the atom numbers, the bond type, the multiplicative factor (the product of the g_{ij}'s defined by equation (3)), and the bond's BAC (equation. 1) are given. If there is a "hr" followed by a number i at the end of the line, it indicates that the bond is a hindered rotor represented by vibrational frequency number i. Next are listed the heats of formation of the molecule at 0 K corresponding to the raw MP4 electronic energies and the resulting BAC-MP4. The BAC-MP4 number corresponds to the MP4 raw energy which includes the zero-point energy plus the BAC corrections (equation 7). The next line for each molecule provides the heats of formation at 0K and 298K as well as the free energies at 300K, 600K, 1000K,1500K, and 2000K. Finally, the last line gives the BAC-MP4 heat of formation at 300K followed by the error estimate, the difference with experiment (if available) and then the differences between the BAC-MP4 and BAC-MP4(SDQ), BAC-MP3, and BAC-MP2 respectively. One can see that for each of these molecules, all the methods (BAC-MP4, BAC-MP4(SDQ), BAC-MP3, and BAC-MP2) are giving similar results, indicating that one should have confidence in the calculated heats of formation.

For the isodesmic reaction 9, we see that there is a corresponding BAC correction for each bond type. For instance, the BAC correction for the CN bond is 6.95 kcal-mol[-1] for CH_3NO and 7.90 kcal-mol[-1] for CH_3NH_2. Likewise, the NO correction is 17.32 kcal-mol[-1] for CH_3NO and 19.16 kcal-mol[-1] for HNO. The NH and CH corrections also balance. Even the UHF instability for CH_3NO and HNO (2.29 kcal-mol[-1] and 3.66 kcal-mol[-1]) tend to balance. The net BAC corrections for each side of reaction 9, 68.15 kcal-mol[-1] and 71.73 kcal-mol[-1], are quite large but nearly balancing, the difference being 3.58 kcal-mol[-1]. The BAC method thus supports the isodesmic reaction approach, in general. However, in CH_3NO , the nearest neighbor correction is worth 2.49 kcal-mol[-1] (due to g_{kij} multiplicative factor of 0.907). Furthermore, the spin correction is 1.37 kcal-mol[-1] larger on the right side. Without these additional corrections, the simple corrections based on equation (2) would have been 68.36 kcal-mol[-1] on the left vs. 68.08 kcal-mol[-1] on the right. Thus, the BAC method has improved on the isodesmic approach by including corrections for neighboring bonds and for spin corrections.

More importantly, since the BAC method includes the corrections, one no longer needs to restrict oneself in the type of reaction considered. One can break bonds and, indeed, calculate heats of atomization. Using the experimentally determined heats of formation of the atoms, one can therefore derive heats of formation for each molecule.

As a second example, we consider the isomers CH_3NO , CH_2NHO, and CH_2NOH. The BAC corrections are also given in Table III. The resulting net BAC corrections are 39.84 kcal-mol[-1] for CH_3NO , 48.06 kcal-mol[-1] for CH_2NHO, and

Table III. BAC Corrections for Example Problems

```
        CH3NO ONCH CIS NITROSO METHANE
            S**2 = 0.229          Espin = 2.29
2 1   R( CN ) = 1.4640  With Fmult = 0.907   Bac = 6.95  hr 1
3 2   R( NO ) = 1.1766  With Fmult = 0.907   Bac = 17.32
4 1   R( CH ) = 1.0819  With Fmult = 1.000   Bac = 4.44
5 1   R( CH ) = 1.0839  With Fmult = 1.000   Bac = 4.42
6 1   R( CH ) = 1.0839  With Fmult = 1.000   Bac = 4.42
MP4 = 61.17  Bac-MP4 = 21.33
             21.33   18.87    28.8  39.3  53.9  72.3  90.3
   18.88     1.38      u 2.29     -0.72 -0.43 1.06


        NH3
2 1   R( NH ) = 1.0025  With Fmult = 1.000   Bac = 9.44
3 1   R( NH ) = 1.0025  With Fmult = 1.000   Bac = 9.44
4 1   R( NH ) = 1.0025  With Fmult = 1.000   Bac = 9.44
MP4 = 19.08  Bac-MP4 = -9.23
             -9.23  -10.99    -3.9   3.7  14.8  28.9  43.0
  -10.98     1.00    0.00         0.01 -0.01  0.00


        HNO
            S**2 = 0.366          Espin = 3.66
2 1   R( NO ) = 1.1752  With Fmult = 1.000   Bac = 19.16
3 1   R( NH ) = 1.0317  With Fmult = 1.000   Bac = 8.90
MP4 = 55.66  Bac-MP4 = 23.94
             23.94   23.20    26.3  29.6  34.4  40.5  46.7
   23.20     1.32  -0.60  u 3.66     0.06 -0.44  0.19


        H3CNH2
2 1   R( CN ) = 1.4533  With Fmult = 1.000   Bac = 7.90  hr 1
3 1   R( CH ) = 1.0910  With Fmult = 1.000   Bac = 4.36
4 1   R( CH ) = 1.0839  With Fmult = 1.000   Bac = 4.42
5 1   R( CH ) = 1.0839  With Fmult = 1.000   Bac = 4.42
6 2   R( NH ) = 1.0014  With Fmult = 1.000   Bac = 9.46
7 2   R( NH ) = 1.0014  With Fmult = 1.000   Bac = 9.46
MP4 = 38.17  Bac-MP4 = -1.84
             -1.84   -5.55     7.7  22.1  42.8  68.6  94.1
   -5.53     1.03  -0.03         -0.10 -0.22 -0.12

        H2C=NOH FORMALDOXIME CNOH TRANS
            S**2 = 0.060          Espin = 0.60
2 1   R( CN ) = 1.2493  With Fmult = 0.886   Bac = 12.40  hr 3
3 2   R( NO ) = 1.3687  With Fmult = 0.886   Bac = 11.31  hr 1
4 3   R( OH ) = 0.9468  With Fmult = 1.000   Bac = 10.91
5 1   R( CH ) = 1.0775  With Fmult = 1.000   Bac = 4.48
6 1   R( CH ) = 1.0735  With Fmult = 1.000   Bac = 4.51
MP4 = 51.92  Bac-MP4 = 7.72
             7.72    5.11     15.5  26.4  41.7  60.9  79.9
    5.13     2.50     u 0.60     -1.63 -1.61  2.26
        H2CNHO
            S**2 = 0.425          Espin = 4.25
2 1   R( CN ) = 1.2692  With Fmult = 0.863   Bac = 11.42  hr 2
3 2   R( NO ) = 1.2542  With Fmult = 0.863   Bac = 14.01
4 1   R( CH ) = 1.0704  With Fmult = 1.000   Bac = 4.54
5 1   R( CH ) = 1.0703  With Fmult = 1.000   Bac = 4.54
6 2   R( NH ) = 1.0099  With Fmult = 1.000   Bac = 9.30
MP4 = 64.96  Bac-MP4 = 16.90
             16.90   14.09    24.8  36.3  52.4  72.6  92.7
   14.11     1.92     u 4.25     -0.43  1.33  1.94
```

44.20 kcal-mol-1 for CH_2NOH. The primary difference in these corrections results from the much smaller BAC correction for the CH bond compared to the NH and OH bonds and due to the large spin correction for CH_2NHO. Note that the NO bond in CH_2NHO is best described as a dative bond (i.e., the nitrogen is not trivalent). Thus, the types of bonds have changed quite significantly between CH_3NO , CH_2NHO, and CH_2NOH. But this does not matter in the BAC-MP4 method, since the BAC corrections only depend on the bond distance for a given pair of atomic elements. The BAC method does not require that one identify the bond type (e.g., a single, double, triple, dative, resonance, or some mixture). For instance, the CC bond distance in NCCN (1.398 Å) is similar in length to that in benzene (1.386 Å), indicative of significant delocalization (diradical resonance character).

It should be noted that for any given reaction, it is not necessary to know the experimental heats of formation of the atoms or of the element's standard state in order to determine the heat of reaction. This will become important as the BAC method is extended to heavier elements of the periodic table, for which the experimental heats of formation of the atoms are not well known. Indeed, the heats of reaction determined computationally will not be affected by changes in the heat of formation of the reference state, as has occurred for phosphorous. The important thing to remember is that the heats of formation on both sides of the reaction must be treated consistently. When one mixes theoretical and experimental heats of formation, potential inconsistencies can arise.

Conclusions

The bond additivity correction (BAC) procedure is a powerful method for extending the practical application of *ab-initio* methods for the prediction of thermochemical properties of molecular species. Comparison of the BAC-MP4 computations with experiment indicate the method works for both stable and radical species, and provides heats of formation with accuracies comparable to experiment. An error estimate provides an indicator of the reliability of a given computation and can be used as a screening procedure when a large body of data are being computed. The ultimate utility of the method lies in the promise of a computational economical method for computing large bodies of thermochemical information on systems where experimental data are unavailable. Such data are finding increased use as the feedstock to ever more sophisticated Computational Reacting Fluid Dynamics programs and are the ultimate motivation for much of the work.

Appendix

BAC-MP4   Run Procedure (Unix) For Gaussian94

```
set name=sch2
cat>"$name"_opt<<EOF
%chk=sch2.chk
#p HF/6-31G* FOPT=Z-MATRIX
GEOM=(DIHEDRAL)

CH2(1A1) hf/6-31g* optimization calculation

0,1
C
H 1 R21
H 1 R21 2 A312
```

R21   1.0970591
A312 103.02145745

EOF
g92 <"$name"_opt > /"$name"_opt.out
cp "$name".chk /"$name".chk
cat>"$name"_frq<<EOF
%chk=sch2.chk
#p HF/6-31G* FREQ GEOM=CHECK GUESS=READ

 CH2(1A1) HF/6-31g* frequencies

0,1

EOF
g92 <"$name"_frq > /"$name"_frq.out

cat>"$name"_mp4<<EOF
%chk=sch2.chk
#p MP4(SDTQ)/6-31g** GEOM=CHECK GUESS=READ

 CH2(1A1) MP4/6-31G** // HF/6-31g* structure

0,1

EOF
g92 <"$name"_mp4 > /"$name"_mp4.out

cat>"$name"_mp2<<EOF
%chk=sch2.chk
#p RHF/6-31G** STABLE=(OPT,RUHF) geom=check guess=read
SCFCON=5

 CH2(1A1) ump2/6-31G** // rhf/6-31g* structure

0,1

--link1--
%chk=sch2.chk
#P UMP2/6-31g** geom=check guess=read
 CH2(1A1) ump2/6-31G** // rhf/6-31g* structure

0,1

EOF
g92 <"$name"_mp2 > /"$name"_mp2.out
cp "$name".chk /"$name"_mp2.chk

Literature Cited

1. Benson, S. Thermochemical Kinetics John Wiley and Sons; New York   1976.
2. (a) Ho, P.; Coltrin, M.E.; Binkley, J.S.; Melius, C.F. *J. Phys. Chem.* **1985**, *89*, 4647.

(b) Ho, P.; Coltrin, M.E.; Binkley, J.S.; Melius, C.F. *J. Phys. Chem.* **1986**, *90*, 3399.
(c) Melius, C.F.; Binkley, J.S.*Symp. (Inter.) Combust.* **1986**, *21*, 1953.
(d) Ho, P., and Melius, C.F. *J. Phys. Chem.* **1990**, *94*, 5120.
(e) Melius, C.F.; Ho, P. *J. Phys. Chem.* **1991**, *95*, 1410.
(f) Ho, P.; Melius, C.F. *J. Phys. Chem.* **1995**, *99*, 2166.
(g) Melius, C.F.; Binkley, J.S.*Symp. (Inter.) Combust.* **1986**, *21*, 1953.
(h) Allendorf, M.D.; Melius, C.F. *J. Phys. Chem.* **1992**, *96*, 428.
(i) Allendorf, M.D.; Melius, C.F. *J. Phys. Chem.* **1993**, *97*, 720.
(j) Zachariah, M.R.; Tsang, W.*J. Phys. Chem..* **1995**, *99*, 5308.
(k) Zachariah, M.R.; Westmoreland, P.R., Burgess Jr., D.R, Tsang, W. Melius, C.F. *J. Phys. Chem.* **1996**, *100*, 8737.
(l) Zachariah, M.R.; Melius, C.F. *J. Phys. Chem.* **1997**, *101*, 913.
(m) Allendorf, M.D.; Melius, C.F., Ho, P., and Zachariah, M.R., *J. Phys. Chem.* **1995**, *99*, 15285.
3. Schlegel, H.B. *J.Chem. Phys.* **1986**, *84*, 4530.
4. Zachariah, M.R.; Westmoreland, P.R.; Burgess Jr., D.R.; Tsang, W.; Melius, C.F. *J. Phys. Chem* . **1996**, *100*, 8737.

Chapter 10

Computational Methods for Calculating Accurate Enthalpies of Formation, Ionization Potentials, and Electron Affinities

Larry A. Curtiss[1] and Krishnan Raghavachari[2]

[1]Chemical Technology Division, Argonne National Laboratory, 9700 South Cass Avenue, Argonne, IL 60439
[2]Bell Laboratories, Lucent Technologies, Murray Hill, NJ 07974

In this chapter we describe two methods being used in computational thermochemistry. The first is Gaussian-2 theory, which is based on a sequence of well-defined *ab initio* molecular orbital calculations. It has proven to be a useful technique for the calculation of accurate bond energies, enthalpies of formation, ionization potentials, electron affinities, and proton affinities. The second is density functional theory which is developing as a cost-effective procedure for studying molecular properties and energies.

Quantum chemical methods for the calculation of thermochemical data have developed beyond the level of just reproducing experimental data and can now make accurate predictions where the experimental data are unknown or uncertain. In this chapter we review two theoretical methods that are being used in computational thermochemistry. The more accurate of these methods is Gaussian-2 (G2) theory (*1*). It was the second in a series of methods termed Gaussian-n theories (*1-3*) based on *ab initio* molecular orbital theory for the calculation of energies of molecular systems containing the elements H-Cl. The other method is density functional theory (DFT) which is less accurate, but is a cost-effective method for including correlation effects that are needed for thermochemical calculations.

In this chapter we first review the elements of Gaussian-2 theory and the test set of 125 reaction energies that was developed to assess its reliability. This is followed by a description of several variants of G2 theory that have been proposed for saving computational time or improving accuracy and the extension of the method to include third-row non-transition metal elements. Then we provide a brief overview of density functional theory and several functionals currently being used in computational quantum chemistry. Recently, the original test set of 125 energies has been expanded to include larger and more diverse molecules. This new test set has been used to critically evaluate G2 theory, some of its variants, and several DFT methods. The results of this assessment are described. Finally, examples of several applications of G2 theory are discussed.

Theoretical Methods

Gaussian-2 Theory. Gaussian-2 theory is a composite technique in which a sequence of well-defined *ab initio* molecular orbital calculations is performed to arrive at a total energy of a given molecular species. Geometries are determined using second-order Møller-Plesset perturbation theory. Correlation level calculations are done using Møller-Plesset perturbation theory up to fourth-order and with quadratic configuration interaction. Large basis sets including multiple sets of polarization functions are used in the correlation calculations. A series of additivity approximations makes the technique fairly widely applicable. Unlike many other approaches, it is not dependent on calibration with experimental data for related species either through isodesmic reactions or in some other manner. It does have a *single molecule-independent* semi-empirical parameter which is chosen by fitting to a set of accurate experimental data. The principal steps in G2 theory are as follows.

1. An initial equilibrium structure is obtained at the Hartree-Fock (HF) level with the 6-31G(d) basis (*4*). Spin-restricted (RHF) theory is used for singlet states and spin-unrestricted Hartree-Fock theory (UHF) for others.

2. The HF/6-31G(d) equilibrium structure is used to calculate harmonic frequencies, which are then scaled by a factor of 0.8929 to take account of known deficiencies at this level (*5*). These frequencies give the zero-point energy Δ(ZPE) used to obtain E_0 in step 7.

3. The equilibrium geometry is refined at the MP2/6-31G(d) level [Møller-Plesset perturbation theory to second order with the 6-31G(d) basis set] using all electrons for the calculation of correlation energies. This is the final equilibrium geometry in the theory and is used for all single-point calculations at higher levels of theory in step 4. All of these subsequent calculations include only valence electrons in the treatment of electron correlation.

4. The first higher level calculation is at full fourth-order Møller-Plesset perturbation theory (*4*) with the 6-311G(d,p) basis set, i.e., MP4/6-311G(d,p). This energy is then modified by a series of corrections from additional calculations including (a) a correction for diffuse functions (*4*), ΔE(+); (b) a correction for higher polarization functions (*4*) on non-hydrogen atoms, ΔE(2df); (c) a correction for correlation effects beyond fourth-order perturbation theory using the method of quadratic configuration interaction (*6*), ΔE(QCI); and (d) a correction for larger basis set effects and for non-additivity caused by the assumption of separate basis set extensions for diffuse functions and higher polarization functions, ΔE(+3df,2p). The single-point energy calculations required for these corrections are described in Table I.

Table I. Energy corrections for G2, G2(MP2), and G2(MP2,SVP) theory

Method	Step	Corrections
G2	4(a)	$\Delta E(+) = E[MP4/6\text{-}311+G(d,p)]$ $- E[MP4/6\text{-}311G(d,p)]$
	4(b)	$\Delta E(2df) = E[MP4/6\text{-}311G(2df,p)]$ $- E[MP4/6\text{-}311G(d,p)]$
	4(c)	$\Delta E(QCI) = E[QCISD(T)/6\text{-}311G(d,p)]$ $- E[MP4/6\text{-}311G(d,p)]$
	4(d)	$\Delta E(+3df,2p) = E[MP2/6\text{-}311+G(3df,2p)]$ $-E[MP2/6\text{-}311G(2df,p)]$ $- E[MP2/6\text{-}311+G(d,p)]$ $+E[MP2/6\text{-}311G(d,p)]$
G2(MP2)	4'(d)	$\Delta E'(+3df,2p) = [E(MP2/6\text{-}311+G(3df,2p)] -$ $[E(MP2/6\text{-}311G(d,p)]$
G2(MP2,SVP)	4"(c)	$\Delta E"(QCI) = E[QCISD(T)/6\text{-}31G(d)]$ $- E[MP4/6\text{-}31G(d)]$
	4"(d)	$\Delta E"(+3df,2p) = [E(MP2/6\text{-}311+G(3df,2p)]$ $- [E(MP2/6\text{-}31G(d)]$

5. The MP4/6-311G(d,p) energy and the four corrections from step 4 are combined in an additive manner:

$$E(combined) = E[MP4/6\text{-}311G(d,p)] + \Delta E(+) + \Delta E(2df) + \Delta E(QCI) \qquad (1)$$
$$+ \Delta E(+3df,2p)$$

6. A "higher level correction" (HLC) is added to take into account remaining deficiencies

$$E_e(G2) = E(combined) + \Delta E(HLC). \qquad (2)$$

The HLC is equal to $-An_\beta - Bn_\alpha$, where the n_β and n_α are the number of β and α valence electrons, respectively, with $n_\alpha \geq n_\beta$. For G2 theory, A = 4.81 mhartrees and B = 0.19 mhartrees (equivalent to 5.00 mhartrees per electron pair). The B value was chosen so that E_e is exact for the hydrogen atom (2). The A value was determined to give a zero mean deviation from experiment for the atomization energies of 55 molecules having well-established experimental values. The higher level correction makes G2 theory "semi-empirical," though only a *single molecule-independent* parameter is used.

7. Finally, the total energy at 0 K is obtained by adding the zero-point correction, obtained from the frequencies of step 2 to the total energy:

$$E_0(G2) = E_e(G2) + \Delta(ZPE) \tag{3}$$

The energy E_0 is referred to as the "G2 energy."

Gaussian-2 theory was tested on a total of 125 reaction energies (55 atomization energies, 38 ionization energies, 25 electron affinities, and 7 proton affinities), chosen because they have well-established experimental values (*1-3*). The molecules in this test set contain elements from the first- and second-rows of the periodic chart. The mean absolute deviation of G2 theory for this test set from experiment is 1.21 kcal/mol, and the maximum deviation is for the atomization energy of SO_2 (-5.0 kcal/mol). The average absolute deviations of G2 theory from experiment for the different types of reaction energies in this test set are listed in Table II. The set of 125 reaction energies has since been used by others to test new quantum chemical methods and is often referred to as the "G2 test set."

Table II. Average absolute deviation (in kcal/mol) between theory and experiment for the original G2 test set of 125 reaction energies [a]

Method	All	Reaction Type[b]			
		AE	IP	EA	PA
G2	1.21	1.16	1.24	1.29	1.04
G2(MP2)	1.58	1.32	1.86	1.99	0.64
G2(MP2,SVP)	1.63	1.34	1.93	2.08	0.81
SVWN	25.96	39.59	15.70	17.21	5.85
BLYP	4.49	4.75	5.99	2.42	1.80
BPW91	4.48	5.32	5.29	2.22	1.53
BP86	7.31	10.51	5.34	4.91	1.51
B3LYP	2.81	2.40	3.88	2.49	1.29
B3PW91	2.89	2.59	3.96	2.43	1.17
B3P86	10.69	7.84	14.57	13.77	1.09

[a] G2 results from Ref. 1; G2(MP2) results from Ref. 9; G2(MP2,SVP) results from Ref. 11. The DFT results (unpublished work) are based on the 6-311+G(3df,2p) basis set, MP2(full)/6-31G(d) geometries, and scaled (0.893) HF/6-31G(d) frequencies.
[b] AE = atomization energy (55); IP = ionization potential (38); EA = electron affinity (25); PA = proton affinity (7); number of molecules in parentheses.

The final total energy is effectively at the QCISD(T)/6-311+G(3df,2p) level if the different additivity approximations work well. The validity of these additivity approximations was investigated by performing complete QCISD(T)/6-311+G(3df,2p) calculations on the set of 125 test reactions, and in most cases, the additivity approximations were found to work well (*7*). All of the calculations required for G2 theory are available in the quantum chemical computer program Gaussian 94 (*8*).

The steps in G1 theory (*2,3*), the predecessor to G2 theory, are the same as in G2 theory except step 4(d) is not included and the value of A in the HLC is 5.95 mhartrees instead of 4.81 mhartrees. The final energy in G1 theory is effectively at the QCISD(T)/6-311+G(2df,p) level and is less accurate than G2 theory (average absolute deviation of 1.58 kcal/mol for the 125 reaction energies).

Variants of G2 Theory. A number of variants of G2 theory have been proposed. The purpose of some of these has been to reduce the computational expense of the calculations while others have been aimed at improving the accuracy.

G2(MP2) Theory. A variation of G2 theory which uses reduced orders of Møller-Plesset perturbation theory is G2(MP2) theory (*9*). In this theory the basis set extension corrections of G2 theory in steps 4(a), 4(b), and 4(d) are replaced by a single correction obtained at the MP2 level with the 6-311+G(3df,2p) basis set, $\Delta E'(+3df,2p)$, as given by step 4'(d) in Table I. The total G2(MP2) energy is thus given by

$$E_0[G2(MP2)] = E[MP4/6\text{-}311G(d,p)] + \Delta E(QCI) + \Delta E'(+3df,2p) + \Delta E(HLC) \quad (4)$$
$$+ \Delta E(ZPE)$$

where the $\Delta E(QCI)$ and $\Delta E(HLC)$ terms are the same as in G2 theory.

The G2(MP2) energy requires only two single-point energy calculations, QCISD(T)/6-311G(d,p) and MP2/6-311+G(3df,2p), since the sum of the E[MP4/6-311G(d,p)] and $\Delta E(QCI)$ terms in equation (4) is equivalent to the QCISD(T)/6-311G(d,p) energy and the QCISD(T)/6-311G(d,p) calculation provides the MP2/6-311G(d,p) energy needed to evaluate $\Delta E'(+3df,2p)$. The absence of the MP4/6-311G(2df,p) calculation in G2(MP2) theory provides significant savings in computational time and disk storage such that larger systems can be handled than in G2 theory. The limiting calculation in G2(MP2) theory is the QCISD(T)/6-311G(d,p) calculation. G2(MP2) theory is somewhat less accurate than G2 theory, having an average absolute deviation of 1.58 kcal/mol for the 125 reaction energies used for validation of G2 theory (see Table II).

G2(MP2,SVP) Theory. A variation of G2 theory which uses reduced orders of Møller-Plesset perturbation theory in combination with a smaller basis set for the quadratic configuration correction is G2(MP2,SVP) theory (*10,11*). The "SVP" refers to the split-valence plus polarization basis, 6-31G(d), used in this QCISD(T) correction. In this theory the final energy is given by

$$E_0[G2(MP2,SVP)] = E[MP4/6\text{-}31G(d)] + \Delta E''(QCI) + \Delta E''(+3df,2p) \quad (5)$$
$$+ \Delta E(HLC) + \Delta E(ZPE)$$

where the $\Delta E''(QCI)$ and $\Delta E''(+3df,2p)$ correction terms are relative to the 6-31G(d) basis set (see Table I). The HLC for G2(MP2,SVP) theory is A = 5.13 mhartrees and

B = 0.19 mhartrees. The G2(MP2,SVP) energy requires only two single-point energy calculations, QCISD(T)/6-31G(d) and MP2/6-311+G(3df,2p), since the sum of the E[MP4/6-31G(d)] and $\Delta E''$(QCI) terms in equation (5) is equivalent to the QCISD(T)/6-31G(d) energy and the QCISD(T)/6-31G(d) calculation provides the MP2/6-31G(d) energy needed to evaluate $\Delta E''$(+3df,2p). The use of the 6-31G(d) basis set in the quadratic configuration interaction calculation instead of the 6-311G(d) basis, as in the G2 and G2(MP2) theories, reduces computational time and disk space. G2(MP2,SVP) theory has an average absolute deviation of 1.63 kcal/mol for the 125 reaction energies used for validation of G2 theory (see Table II). This is similar to that of G2(MP2) theory, but the method requires significantly less computer resources. The limiting calculation in G2(MP2,SVP) theory is the QCISD(T)/6-31G(d) calculation. A comparison of the different G2 theories is shown in Fig. 1. With the exception of the proton affinities, G2 theory performs the best of the three methods for the different types of reactions.

G2(COMPLETE) Theory. A variation of G2 theory in which the additivity assumptions are eliminated is referred to as G2(COMPLETE) theory (*7*). In this method the energy is calculated with the full 6-311+G(3df,2p) basis set using the quadratic configuration method:

$$E_0[G2(COMPLETE)] = E[QCISD(T)/6\text{-}311\text{+}G(3df,2p)] + \Delta E(HLC) \qquad (6)$$
$$+ \Delta E(ZPE)$$

The HLC for G2(COMPLETE) theory is A = 5.13 mhartrees and B = 0.19 mhartrees. Since the method uses the full basis set at the highest correlation level, it is applicable to only small molecules. The accuracy of G2(COMPLETE) is only slightly improved (average absolute deviation of 1.17 kcal/mol for the 125 molecule G2 test set) over that of G2 theory.

Improvement in Correlation Treatment, Geometries, and Zero-Point Energies. Other modifications of G2 theory have been investigated which use higher levels of theory for correlation effects, geometries, and zero-point energies (*12*). A higher level of correlation treatment was examined using Brueckner doubles [BD(T)] (*13,14*) and coupled cluster [CCSD(T)] (*15-17*) methods rather than quadratic configuration interaction [QCISD(T)]. These methods are referred to as G2(BD) and G2(CCSD), respectively. The use of geometries optimized at the QCISD level rather than the second-order Møller-Plesset level (MP2) and the use of scaled MP2 zero-point energies rather than scaled Hartree-Fock (HF) zero-point energies have also been examined. These methods are referred to as G2//QCI and G2(ZPE=MP2), respectively (*12*). The set of 125 reaction energies employed for validation of G2 theory was used to test these variations of G2 theory. Inclusion of higher levels of correlation treatment has little effect except in the cases of multiply-bonded systems. In these cases better agreement is obtained in some cases and poorer agreement in others so that there is no improvement in overall performance. The use of QCISD geometries yields significantly better agreement with experiment for several cases

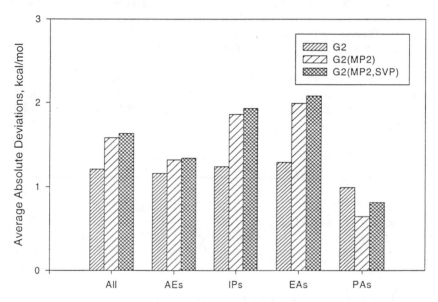

Figure 1. Average absolute deviations of atomization energies (AEs), ionization potentials (IPs), electron affinities (EAs), and proton affinities (PAs) calculated from G2, G2(MP2), and G2(MP2,SVP) theories for the 125 reaction energies in the G2 test set.

including the ionization potentials of CS and O_2, electron affinity of CN, and dissociation energies of N_2, O_2, CN, and SO_2. This leads to a slightly better agreement with experiment overall. The use of MP2 zero-point energies gives no overall improvement. These methods may be useful for specific systems.

Other Variants. Bauschlicher and Partridge (*18*) have proposed a modification of G2 theory which uses geometries and vibrational frequencies from density functional theory (DFT) methods. In this method, referred to as G2(B3LYP/MP2/CC), the QCISD(T) step in G2(MP2) theory is replaced by a CCSD(T) calculation. In addition, the HF/6-31G(d) zero-point energies and MP2/6-31G(d) geometries are replaced by zero-point energies and geometries obtained from density functional theory [B3LYP/6-31G(d)]. This modifcation does not improve the average absolute deviation, but does decrease the maximum errors compared with the G2(MP2) approach.

Mebel, Morokuma and Lin (*19,20*) have proposed a family of modified G2 schemes based on geometry optimization and vibrational frequency calculations using density functional theory [B3LYP/6-311G(d,p)] and electron correlation using coupled cluster methods. These schemes use spin-projected Møller-Plesset theory for radicals and triplets and may be more reliable for radicals with large spin contamination.

Extension to Third-Row Non-Transition Metal Elements. G2 theory has been extended to include molecules containing third-row non-transition elements Ga-Kr (*21*). Basis sets compatible with those used in G2 theory for first- and second-row molecules were derived for this extension. G2 theory for the third-row incorporates the following modifications:

1. The MP2 geometry optimizations and the HF vibrational frequency calculations use the 641(d) basis set of Binning and Curtiss (*22*) for Ga-Kr along with 6-31G(d) for first- and second-row atoms, referred to overall for simplicity as "6-31G(d)." The same scale factor (0.8929) is used for the zero-point energies.
2. The MP2, MP4 and QCISD(T) calculations (step 4 in Section 2.2.1) use the 6-311G basis and appropriate supplementary functions for first- and second-row atoms, and corresponding sets that were developed (*21*) for Ga-Kr, referred to overall again for simplicity as "6-311G."
3. The splitting factor of the d-polarization functions for the 3df basis set extension is three rather than the factor of four used for first- and second-row atoms. The 3d core orbitals and 1s virtual orbitals are frozen in the single-point correlation calculations.
4. First-order spin-orbit energy corrections, $\Delta E(SO)$, are included in the G2 energies for the third-row species that have first-order spin-orbit effects. This includes ^2P and ^3P atoms and $^2\Pi$ molecules. Values for these corrections are obtained from spin-orbit configuration interaction calculations (*23*).

The average absolute deviation from experiment of 40 test cases involving species containing Ga-Kr atoms (atomization energies, ionization energies, electron affinities, proton affinities) is 1.37 kcal/mol. This is only slightly greater than for the G2 treatment of first- and second-row molecules for the 125 reaction energies of small molecules, for which the average absolute deviation is 1.21 kcal/mol. The inclusion of first-order atomic and molecular spin-orbit corrections is important for attaining good agreeement with experiment. When the spin-orbit correction is not included, the deviation increases to 2.36 kcal/mol.

The G2(MP2) and G2(MP2,SVP) theories can also be used for the third-row. The formulation is analogous to the first- and second-rows with the exception of the additional spin-orbit correction term. The average absolute deviation for the third-row test set using G2(MP2) and G2(MP2,SVP) theories is slightly larger (1.92 kcal/mol for both methods).

Density Functional Methods. Over the past 30 years, density functional theory has been widely used by physicists to study the electronic structure of solids. More recently chemists have been using the Kohn-Sham version (*24*) of density funtional theory (DFT) as a cost-effective method to study properties of molecular systems. The density functional models currently being used by quantum chemists may be broadly divided into non-empirical and empirical types. The simplest non-empirical type is the local spin density functional, which treats the electronic environment of a given position in a molecule as if it were a uniform gas of the same density at that point. One of these is the SVWN functional that uses the Slater functional (*25*) for exchange and the uniform gas approximate correlation functional of Vosko, Wilk and Nusair (*26*). The more sophisticated functional BPW91 combines the 1988 exchange functional of Becke (*27*) with the correlation functional of Perdew and Wang (*28*). Both components involve local density gradients as well as densities. The Becke part involves a single parameter which fits the exchange functional to accurate computed atomic data. The BP86 functional is similar but uses the correlation functional of Perdew (*29*). The BLYP (*30*) functional also uses the Becke 1988 part for exchange, together with the correlation part of Lee, Yang and Parr (*31*). This LYP functional is based on a treatment of the helium atom and really only treats correlation between electrons of opposite spin.

A number of other functionals use parameters which are fitted to energies in the original G2 test set. These give a functional which is a linear combination of Hartree-Fock exchange, 1988 Becke exchange, and various correlation parts. This idea was introduced by Becke (*32*) who obtained parameters by fitting to the molecular data. This is the basis of the B3PW91 functional. The others (B3P86 and B3LYP) are constructed in a similar manner, although the parameters are the same as in B3PW91.

In several validation studies for molecular geometries and frequencies, DFT has given results of quality similar to that of MP2 theory (*33,34*). It has also been examined for use in calculation of thermochemical data (*32-38*). For example, Becke (*32*) found that the B3PW91 functional with a numerical basis set gave an average absolute deviation of 2.4 kcal/mol for the 55 atomization energies in the original G2 test set, about twice as large as G2 theory. He also found similar results for ionization potentials and proton affinities. Bauschlicher (*37*) has examined several DFT methods [BLYP, B3LYP, BP86, B3P86, BP] for the 55 atomization energies in the original G2 test set using the same 6-311+G(3df,2p) basis set. He found that B3LYP gave the best agreement with experiment (average absolute deviation of 2.20 kcal/mol).

The results for the 125 reaction energies in the original G2 test set are given in Table II for the seven density functionals (SVWN, BLYP, BPW91, BP86, B3LYP, B3PW91, B3P86) described above. The results are based on the 6-311+G(3df,2p) basis set and the geometries and zero-point energies used in the original G2 paper (*1*). The local density approximation has a very large average absolute deviation (26.2 kcal/mol) especially for the atomization energies due to overbinding. The gradient corrected functionals have average absolute deviations of 10 kcal/mol or less. The methods containing the Becke three-parameter functional perform better than the Becke 1988 functional for atomization energies, ionization energies, and proton affinities. For electron affinities the smallest average absolute deviation is found for the BPW91 functional. Overall, the B3LYP functional performs the best, with an average absolute deviation of 2.81 kcal/mol. The B3PW91 functional is only slightly worse with an average absolute deviation of 2.89 kcal/mol.

Assessment of Theoretical Methods for Computational Thermochemistry

Critical documentation and evaluation of theoretical models for calculating energies are essential if such methods are to become proper tools for chemical investigation. As mentioned earlier, G2 theory was tested on a total of 125 reaction energies, chosen because they have well-established experimental values (*1*). All of the molecules contained only one or two non-hydrogen atoms with two exceptions (CO_2 and SO_2). In recent work (*39*), the test set has been expanded to include larger, more diverse molecules with enthalpies of formation at 298 K being used for comparison between experiment and theory. This set, referred to as the "enlarged G2 neutral test set," includes the 55 molecules whose atomization energies were used to test G2 theory and 93 new molecules. The full set includes 29 radicals, 35 non-hydrogen systems, 22 hydrocarbons, 47 substituted hydrocarbons, and 15 inorganic hydrides. The set includes molecules containing up to six non-hydrogen atoms. In this section we critically evaluate G2 theory, some of its variants, and several density functional methods on this enlarged G2 test set. Before doing this, we describe the procedure employed to calculate enthalpies of formations from calculated energies.

Calculation of Enthalpies of Formation. One of the methods for calculating enthalpies of formation at 0 K is based on subtracting calculated atomization energies ΣD_0 from known enthalpies of formation of the isolated atoms. For any molecule, such as $A_x B_y H_z$, the enthalpy of formation at 0 K is given by

$$\Delta_f H^\circ (A_x B_y H_z, 0 \text{ K}) = x\Delta_f H^\circ(A, 0 \text{ K}) + y\Delta_f H^\circ (B, 0 \text{ K}) + z\Delta_f H^\circ (H, 0 \text{ K}) \quad (7)$$
$$- \Sigma D_0$$

Experimental values (*40*) for the atomic $\Delta_f H^\circ$ are used. The experimental enthalpies of formation of Si, Be, and Al have large uncertainties (2.0, 1.2 and 1.0 kcal/mol, respectively). This means that the calculated enthalpies of formation containing these atoms will have uncertainties due to the use of the atomic enthalpies in Eq. (7) and the theoretical methods. The other atomic enthalpies for first- and second-row elements are quite accurate (±0.2 kcal/mol or better).

Theoretical enthalpies of formation at 298 K are calculated by correction to $\Delta_f H^\circ$ (0 K) as follows:

$$\Delta_f H^\circ(A_x B_y H_z, 298 \text{ K}) = \Delta_f H^\circ (A_x B_y H_z, 0 \text{ K}) + \tag{8}$$
$$[H^\circ(A_x B_y H_z, 298 \text{ K}) - H^\circ(A_x B_y H_z, 0 \text{ K})]$$
$$- x[H^\circ(A, 298 \text{ K}) - H^\circ(A, 0 \text{ K})]_{st}$$
$$- y[H^\circ(B, 298 \text{ K}) - H^\circ(B, 0 \text{ K})]_{st}$$
$$- z[H^\circ(H, 298 \text{ K}) - H^\circ(H, 0 \text{ K})]_{st}$$

The enthalpy corrections (in square brackets) are treated differently for compounds and elements. The correction for the $A_x B_y H_z$ molecule is made using scaled HF/6-31G(d) frequencies for the vibrations in the harmonic approximation for vibrational energy (*41*), the classical approximation for translation (3RT/2) and rotation (3RT/2 for nonlinear molecules, RT for linear molecules) and the PV term. The harmonic approximation may not be appropriate for some low frequency torsional modes, although the error should be small in most cases. The elemental corrections are for the standard states of the elements [denoted as "st" in eqn (8)] and are taken directly from experimental compilations (*40*). The resulting values of $\Delta_f H^\circ$ (298 K) are often discussed as theoretical numbers, although they are based on some experimental data for monatomic and standard species. Enthalpies of formation of cations are calculated by combining enthalpies of formation and ionization potentials of the corresponding neutrals (*42*).

Gaussian-2 Theory and Variants. The enlarged G2 neutral test set of 148 molecules was used to assess the performance of G2, G2(MP2) and G2(MP2,SVP) theories (*39*). The mean absolute deviation between the theoretical and experimental enthalpies of formation at 298 K for the different methods is listed in Table III. The results indicate that G2 theory is the most reliable of the methods, with a mean absolute deviation of 1.58 kcal/mol for the 148 enthalpies. This is larger than for the atomization energies of the 55 small molecules in the original G2 test set, mainly due to the new molecules containing multiple halogens and molecules with unsaturated rings. The largest deviations between experiment and G2 theory (up to 8 kcal/mol) occur for molecules having multiple halogens. This leads to a poor performance of G2 theory on non-hydrogen systems (see Table III). The G2 enthalpies of formation for cyclic hydrocarbons with unsaturated rings deviate with experiment by 2-4 kcal/mol. The other hydrocarbons are generally in good agreement with experiment.

The modified versions of G2 theory, G2(MP2) and G2(MP2,SVP), have average absolute deviations of 2.04 and 1.93 kcal/mol, respectively. Both methods do poorly for non-hydrogens similar to G2 theory. G2(MP2,SVP) theory appears to be very good for hydrocarbons, radicals, and inorganic hydrides. Surprisingly, this approximation does better for hydrocarbons than G2 theory, especially cyclic systems, for which it has an average absolute deviation of 1.06 kcal/mol. Since G2(MP2,SVP) theory uses considerably less cpu time and disk storage than G2 theory, it may be a useful alternative for large hydrocarbons.

Table III. Average absolute deviation (in kcal/mol) between theory and experiment for 148 enthalpies of formation in the enlarged G2 neutral test set [a]

Method	All[c]	Type of Molecule[b]				
		A	B	C	D	E
G2	1.58(8.2)	2.53	1.29	1.48	1.16	0.95
G2(MP2)	2.04(10.1)	3.30	1.83	1.89	1.36	1.20
G2(MP2,SVP)	1.93(12.5)	3.57	0.77	2.04	1.20	0.91
SVWN	90.88(228.7)	73.65	133.71	124.40	54.64	33.80
BLYP	7.09(28.4)	10.30	8.09	6.10	6.09	3.13
BPW91	7.85(32.2)	12.25	4.85	7.99	6.48	4.21
BP86	20.19(49.7)	16.61	25.82	26.80	15.76	8.16
B3LYP	3.11(20.1)	5.35	2.76	2.10	2.98	1.84
B3PW91	3.51(21.8)	5.14	3.96	2.77	3.21	1.99
B3P86	17.97(49.2)	7.80	30.81	25.49	13.53	7.86

[a] From Ref. 39.
[b] A = non-hydrogen systems (35); B = hydrocarbons (22); C = substituted hydrocarbons (47); D = radicals (29); E = inorganic hydrides (15). Number of molecules in each type is listed in parentheses.
[c] Maximum absolute deviation in parentheses.

Gaussian-2 theory is based on non-relativistic energies, but for large molecules, such as those in the enlarged G2 neutral test set, spin-orbit effects become important in some cases. To take account of these effects a spin-orbit correction $\Delta E(SO)$ can be added to the G2 energy:

$$E_0[G2_{SO}] = E_0[G2] + \Delta E(SO) \qquad (9)$$

For first- and second-row molecules the spin-orbit effect is significant for the cases when it is a first-order effect, such as in ^2P and ^3P atoms and $^2\Pi$ molecules. It can be neglected for the other atoms and molecules of the first- and second-row for which it is not a first-order effect. The spin-orbit corrections can be derived from experimental data in Moore's tables (*43*) or, alternatively, they can be calculated quite accurately (*23,44*). The corrections are large enough for heavier atoms such as Cl (-1.34 mhartree per atom), S (-0.89 mhartree per atom) and F (-0.61 mhartree per atom) that in molecules containing several such atoms the effect on the enthalpy of formation will amount to several kcal/mol. For the 148 enthalpies of formation in the G2 neutral test set inclusion of spin-orbit effects reduces the average absolute deviation to 1.47 kcal/mol from 1.58 kcal/mol without spin-orbit corrections (*39*). The spin-orbit correction significantly improves the enthalpies for the chlorine substituted molecules, but little overall improvement is seen for the fluorine containing molecules.

Density Functional Theory Thermochemistry. The seven DFT methods described in the theory section have been assessed on the enlarged G2 test set (*39*). The DFT calculations were done with the 6-311+G(3df,2p) basis set and the geometries and zero-point energies used in the original G2 paper (*1*). The average absolute deviations for the G2 and DFT methods are illustrated in Fig. 2 and broken up into types of molecules in Table III. The DFT methods give a wide range of average absolute deviations (3.11 to 90.9 kcal/mol) for the G2 test set, all of which are larger than for the G2 methods. As expected, the local density method (SVWN) performs poorly with a deviation of 90.9 kcal/mol and overbinds all systems except Li$_2$. However, this method involves no parameterization, and application of empirical corrections as in other methods can significantly improve its performance. For the remaining gradient corrected functionals, the average absolute deviation ranges from 3.11 to 20.19 kcal/mol. The Becke three parameter functional performs better than the Becke exchange functional with all three correlation functionals. The B3LYP functional performs the best of the functionals tested with an average absolute deviation of 3.11 kcal/mol. The deviation for the B3PW91 functional is only slightly larger at 3.51 kcal/mol. As in the case of the G2 methods, the DFT methods do poorest for the systems containing multiple halogens as is seen from the large average absolute deviations for the non-hydrogen systems in Table III.

The maximum deviations of the DFT methods are significantly larger than those of the G2 methods (see Table III). For example, B3LYP has a maximum deviation of 20.1 kcal/mol compared to 8.2 kcal/mol for G2 theory. The largest errors for the B3LYP method occur for non-hydrogen systems (average absolute deviation of 5.35 kcal/mol) while hydrocarbons, substituted hydrocarbons, and radicals have smaller average absolute deviations (2 to 3 kcal/mol). The distribution of deviations for B3LYP is given in Fig. 3. About 50% of the B3LYP enthalpies fall within ±2 kcal/mol of the experimental values and 63% fall within ±3 kcal/mol. In comparison, over 70% of the G2 enthalpies fall within ±2 kcal/mol of the experimental values and 87% fall within ±3 kcal/mol. While the deviations for G2 theory are quite equally distributed (Fig. 3), the B3LYP method has more negative deviations (underbinding). The B3LYP distribution covers a much larger range (-20 to 8 kcal/mol) than G2 theory (-7 to 8 kcal/mol).

The performance measures discussed above have important consequences. The best performing B3LYP functional has an average absolute deviation (3.11 kcal/mol) almost twice that of G2 theory. Among the 148 molecules studied, only 5 have deviations of 5 kcal/mol or more with G2 theory, whereas 25 molecules have deviations of more than 5 kcal/mol with the B3LYP functional. These considerations may be important for assessing the thermochemistry of systems where there is disagreement between theory and experiment or for making predictions for systems where there are no experimental measurements. Typical timings of several DFT calculations for comparison with the G2 methods are shown for two molecules in

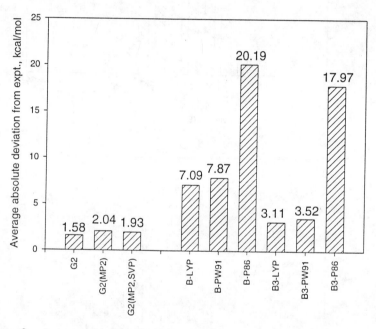

Figure 2. Average absolute deviations of G2 and DFT methods for the 148 molecule enlarged G2 test set of 148 enthalpies. (The SVWN density functional method has an average absolute deviation of 90.88 kcal/mol).

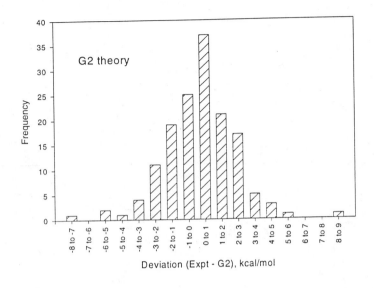

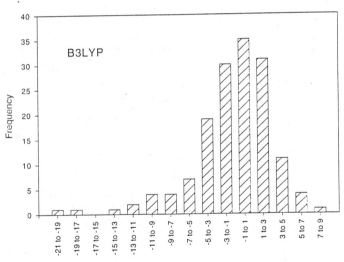

Figure 3. Histogram of G2 and B3LYP deviations for the enlarged G2 test set of 148 molecules. Each vertical bar represents one kcal/mol range for the G2 results and two kcal/mol range for the B3LYP results. (Reprinted with permission from Ref. 39. Copyright 1997 American Institute of Physics.)

Table IV. The DFT methods are about 80 times faster than G2 theory and 6 times faster than G2(MP2,SVP) theory for these molecules.

Table IV. Comparison of cpu times and disk storage used in G2 and DFT calculations   on benzene and t-butyl radical [a]

Method	Benzene (D_{6h})		t-Butyl Radical (C_{3v})	
	cpu time	disk storage	cpu time	disk storage
G2	851	1.4	868	1.1
G2(MP2)	170	0.6	226	0.6
G2(MP2,SVP)	58	0.4	64	0.4
BLYP	10	0.1	7	0.1
BPW91	6	0.1	10	0.1
B3LYP	10	0.1	10	0.1
B3PW91	11	0.1	9	0.1

[a] Using Gaussian 94 (Ref. 8) on a Cray YMP-C90. Time in minutes and maximum storage in Gb.

In the assessment of the DFT methods on the enlarged G2 test set MP2(full)/6-31G(d) optimized geometries and scaled HF/6-31G(d) vibrational frequencies were employed (*39*). The DFT geometries and zero-point energies for thermochemical calculations may be useful and have been examined in several studies (*37,45-47*). The results have indicated that DFT geometries and zero-point energies generally perform as well as those from *ab initio* methods as long as appropriate scale factors for the zero-point energies are used.

Applications

Gaussian-2 theory has been applied to many molecular systems and has in most cases been quite successful. It has been used to make predictions of bond dissociation energies, ionization energies, electron affinities, appearance energies, proton affinities, and enthalpies of formation. There have been several reviews of these applications (*48,49*). In this article we describe several typical examples of the use of G2 theory to obtain thermochemical data.

Atomization Energies. As discussed in the previous sections G2 theory was originally tested on 55 atomization energies (average absolute deviation of 1.19 kcal/mol) and in subsequent work it was tested on 148 enthalpies of formation at 298 K (average absolute deviation of 1.48 kcal/mol). It has been applied to numerous molecules to determine bond energies and enthalpies of formation. An example of this is the application to C_2H_n and Si_2H_n hydrides (*1,42*). For acetylene the experimental D_0(HCC-H) values have fallen in two groups, one around 126 kcal/mol and the other around 131-132 kcal/mol. The G2 value of 133.4 kcal/mol is in agreement with the larger experimental value. In a recent review of the experimental data, Berkowitz, Ellison and Gutman (50) give a value of 131.3 ± 0.7 kcal/mol, in agreement with the G2 results. For ethylene the experimental $D_0(H_2CCH-H)$ values have fallen in the range 98-117 kcal/mol. The G2 value is 110.1 kcal/mol. In their review of the experimental data, Berkowitz, Ellison and Gutman (50) give a value of 109.7 ± 0.7 kcal/mol, in agreement with the G2 results.

Ionization Energies. G2 theory was originally tested on 38 ionization potentials and found to be in good agreement with experiment (average absolute deviation of 1.24 kcal/mol). It has been used to calculate adiabatic ionization energies of numerous molecules. Some of these calculations (1,42,51-53) are summarized in Table V. The G2 results for the adiabatic ionization energies of Si_2H_n hydrides (42) n = 2, 4-6 are in agreement with measurements reported by Ruscic and Berkowitz (54,55) from a photoionization study. The G2 adiabatic ionization potential (51) of CH_3O (10.78 eV) is in agreement with the value of 10.726 ± 0.008 eV reported by Ruscic and Berkowitz (56) from a photoionization study. A previous photoelectron value (57) of 7.37 ± 0.03 eV appears to be incorrect. In a separate study (52) the ionization energy of C_2H_5O was calculated by G2 theory and is in good agreement with experiment (see Table V).

Electron Affinities. G2 theory was originally tested on 25 electron affinities and found to be in good agreement with experiment (average absolute deviation of 1.29 kcal/mol). It has since been used to calculate electron affinities of other molecules and clusters. For example, it was used to calculate the electron affinities of silicon clusters (Si_n) containing up to five silicon atoms (58). The theoretical electron affinities are in good agreement with the experimental data on n = 1-4.

Table V. Adiabatic ionization potentials (in eV) of selected molecules [a]

	G2 Theory	Expt.
$CH_3O \rightarrow CH_3O^+ + e^-$	10.78^b	10.726
$C_2H_5O \rightarrow C_2H_5O^+ + e^-$	10.32^c	10.29
$CH_3SH \rightarrow CH_3SH^+ + e^-$	9.46^d	9.44
$C_2H_2 \rightarrow C_2H_2^+ + e^-$	11.42^e	11.40
$C_2H_4 \rightarrow C_2H_4^+ + e^-$	10.32^e	10.29
$Si_2H_2 \rightarrow Si_2H_2^+ + e^-$	8.30^f	8.20
$Si_2H_4 \rightarrow Si_2H_4^+ + e^-$	8.11^f	8.09
$Si_2H_6 \rightarrow Si_2H_6^+ + e^-$	9.70^f	9.74

[a] References to experiment given in the theory references.
[b] Ref. 51.
[c] Ref. 52.
[d] Ref. 53.
[e] Ref. 1.
[f] Ref. 42.

Proton Affinities and Gas Phase Acidities. G2 theory was originally tested on seven proton affinities and found to be in good agreement with experiment (average absolute deviation of 1.04 kcal/mol). Smith and Radom (*59*) have used G2 theory to obtain proton affinities for 31 small molecules. The theoretical values are in good agreement for a range of bases with proton affinities spanning some 120 kcal/mol. Smith and Radom found a small number of discrepant cases, which can be reconciled if the currently accepted proton affinity of isobutene is lowered by 2-5 kcal/mol. Smith and Radom (*60*) also evaluated G2 theory in the calculation of gas phase acidities of 23 molecules. They found it to perform very well, with a mean error of 1.24 kcal/mol and consistently within 2 kcal/mol of reliable experimental data. In a small number of cases they found a larger discrepancy between experiment and theory and suggested that in these cases a re-examination of the experimental data may be warranted.

Conclusions

Gaussian-2 theory is a technique based on *ab initio* molecular orbital theory that has been widely used for the calculation of accurate thermochemical data. It is a composite method which is based on a well-defined sequence of calculations and has been tested on molecules having well-established experimental data. The original G2 test set contains 125 reaction energies including atomization energies, ionization potentials, electron affinities, and proton affinities. The average absolute deviation of G2 theory from experiment for this test set was 1.21 kcal/mol. In recent work, the test set has been expanded to include larger, more diverse molecules with comparison between theory and experiment being made using enthalpies of formation at 298 K. This set, referred to as the "enlarged G2 neutral test set," includes 148 molecules. The average absolute deviation of G2 theory from experiment for this new test set is 1.58 kcal/mol, slightly larger than the smaller test set. The largest deviations between experiment and G2 theory occur for molecules having multiple halogens and molecules having unsaturated rings. A number of variants of G2 theory such as G2(MP2) and G2(MP2,SVP) theory are available which save considerable computational time with some loss in accuracy. Gaussian-2 theory is the most accurate of the G2 methods, while G2(MP2) and G2(MP2,SVP) are cost-effective alternatives. In addition, Petersson et al. (*61-63*) have developed a series of related methods, referred to as complete basis set (CBS) methods, that are reviewed elsewhere in this book.

The other type of method that we have described in this chapter is density functional theory. These methods require much less computational resources than the G2 methods, but are not as accurate. The DFT methods have a wide range of average absolute deviations (3.11 to 90.9 kcal/mol) for the enthalpies in the enlarged G2 test set, all of which are larger than for the G2 methods. The best performing B3LYP functional has an average absolute deviation (3.11 kcal/mol) almost twice that of G2 theory and a much larger distribution of errors. The B3PW91 functional has an average absolute deviation of 3.51 kcal/mol, similar to that of B3LYP. The results of the assessment on the enlarged G2 test set indicate that B3LYP with a large basis set such as 6-311+G(3df,2p) is the preferred DFT method for thermochemical calculations at this time. The accuracy of the DFT methods can be significantly improved when used in schemes based on isodesmic bond separation reaction energies (*64*). Density functional theory is also useful for calculating geometries and vibrational frequencies for thermochemical calculations.

Acknowledgment

This work was supported by the U.S. Department of Energy, Division of Materials Sciences, under contract No. W-31-109-ENG-38.

Literature Cited

1. Curtiss, L. A.; Raghavachari, K.; Trucks, G.W.; Pople, J. A. *J. Chem. Phys.* **1991**, *94*, 7221.
2. Pople J.A.; Head-Gordon, M.; Fox, D.J.; Raghavachari, K.; Curtiss, L.A. *J. Chem. Phys.* **1989**, *90*, 5622.
3. Curtiss, L.A.; Jones, C.; Trucks, G. W.; Raghavachari, K.; Pople, J. A. *J. Chem. Phys.* **1990** *93*, 2537.
4. Hehre, W. J.; Radom, L.; Pople, J.A.; Schleyer, P.v.R. *Ab Initio Molecular Orbital Theory*; John Wiley: New York, 1987.
5. Pople, J. A.; Schlegel, H. B.; Krishnan, R.; Defrees, D. J.; Binkley, J. S.; Frisch, M. J.; Whiteside, R. A.; Hout, R. F.; Hehre, W.J. *Int. J. Quantum Chem. Symp.* **1981**, *15*, 269.
6. Pople, J. A.; Head-Gordon, M.; Raghavachari, K. *J. Chem. Phys.* **1987**, *87*, 5968.
7. Curtiss, L. A.; Carpenter, J. E.; Raghavachari, K.; Pople J. A. *J. Chem. Phys.* **1992**, *96*, 9030.
8. Frisch, M. J.; Trucks, G. W.; Schlegel, H. B.; Gill, P. M. W.; Johnson, B. G.; Robb, M. A.; Cheeseman, J. R.; Keith, T. A.; Petersson, G. A.; Montgomery, J. A.; Raghavachari, K.; Al-Laham, M. A.; Zakrzewski, V. G.; Ortiz, J. V.; Foresman, J. B.; Cioslowski J.; Stefanov, B. B.; Nanayakkara, A.; Challacombe, M.; Peng, C. Y.; Ayala, P. Y.; Chen, W.; Wong, M. W.; Andres, J. L.; Replogle, E. S.; Gomperts, R.; Martin, R. L.; Fox, D. J.; Binkley, J. S.; Defrees, D. J.; Baker, J.; Stewart, J. P.; Head-Gordon, M.; Gonzales, C.; Pople, J. A.; *Gaussian 94*, Gaussian, Inc. Pittsburgh, PA, 1995.
9. Curtiss, L. A.; Raghavachari, K.; Pople, J. A. *J. Chem. Phys.* **1993**, *98*, 1293.
10. Smith, B. J.; Radom, L. *J. Phys. Chem.* **1995**, *99*, 6468.
11. Curtiss, L. A.; Redfern, P. C.; Smith, B. J.; Radom, L. *J. Chem. Phys.* **1996**, *104*, 5148.
12. Curtiss, L. A.; Raghavachari, K.; Pople, J. A. *J. Chem. Phys.* **1995**, *103*, 4192.
13. Handy, N. C.; Pople, J. A.; Head-Gordon, M.; Raghavachari, K.; Trucks, G. W. *Chem. Phys. Lett.* **1989**, *164*, 185.
14. Raghavachari, K.; Pople, J. A.; Replogle, E. S.; Head-Gordon, M. *J. Phys. Chem.* **1990**, *94*, 5579.
15. Cizek, J. *J. Chem. Phys.* **1966**, *45*, 4256.

16. Pople, J. A.; Krishnan, R.; Schlegel, H. B.; Binkley, J.S. *Int. J. Quant. Chem.* **1978**, *14*, 561.
17. Bartlet, R. J.; Purvis, G. D. *Int. J. Quant. Chem.* **1978**, *14*, 561.
18. Bauschlicher, C. W.; Partridge, H. *J. Chem. Phys.* **1995**, *103*, 1788.
19. Mebel, A. M.; Morokuma, K.; Lin, M. C. *J. Chem. Phys.* **1994**, *101*, 3916.
20. Mebel, A. M.; Morokuma, K.; Lin, M. C. *J. Chem. Phys.* **1995**, *103*, 7414.
21. Curtiss, L. A.; McGrath, M. P.; Blaudeau, J.-P.; Davis, N. E.; Binning, R. C.; Radom, L. *J. Chem. Phys.* **1995**, *103*, 6104.
22. Binning, R. C.; Curtiss, L. A. *J. Comp. Chem.* **1990**, *11*, 1206.
23. Blaudeau, J.-P.; Curtiss, L.A. *Int. J. Quant. Chem.* **1997**, *61*, 943.
24. Kohn, W.; Sham, L. J. *Phys. Rev.* **1965**, *A140*, 1133.
25. Slater, J. C. *The Self-Consistent Field for Molecules and Solids: Quantum Theory of Molecules and Solids*; McGraw-Hill: New York, 1974 Vol. 4.
26. Vosko, S. H.; Wilk, L.; Nusair, M. *Can. J. Phys.* **1980**, *58*, 1200.
27. Becke, A. D. *Phys. Rev.* **1988**, *A38*, 3098.
28. Perdew, J. P.; Y. Wang, Y. *Phys. Rev.* **1992**, *B45*, 13244.
29. Perdew, J. P. *Phys. Rev.* **1986**, *B33*, 8822; **1986**, *B34*, 7406.
30. Gill, P. M. W.; Johnson, B. G.; Pople, J. A.; Frisch, M. J. *Int. J. Quant. Chem.* **1992**, *S26*, 319.
31. Lee, C.; Yang, W.; Parr, R. G. *Phys. Rev.* **1988**, *B37*, 785.
32. Becke, A. D. *J. Chem. Phys.* **1993**, *98*, 5648.
33. Andzelm, J.; Wimmer E. *J. Chem. Phys.* **1992**, *96*, 1280.
34. Johnson, B. G.; Gill, P. M. W.; Pople, J. A. *J. Chem. Phys.* **1993**, *98*, 5612.
35. Becke, A. D. *J. Chem. Phys.* **1992**, *96*, 2155.
36. Merrill, G. N.; Kass, S. R. *J. Phys. Chem.* **1996**, *100*, 17465.
37. Bauschlicher, C. W. *Chem. Phys. Let.* **1995**, *246*, 40.
38. Smith, B. J.; Radom, L. *Chem. Phys. Let.* **1995**, *231*, 345.
39. Curtiss, L. A.; Raghavachari, K.; Redfern, P. C.; Pople, J. A. *J. Chem. Phys.*, **1997**, *106*, 1063.
40. Chase, M. W., Jr.; Davies, C. A.; Downey, J. R., Jr.; Frurip, D. J.; McDonald, R. A.; Syverud, A.N. *J. Phys. Chem. Ref. Data* **1985**, *14*, Suppl. No. 1.
41. Lewis, G. N.; Randall, M. *Thermodynamics*, 2nd ed., revised by Pitzer, K. S.; Brewer, L.; McGraw-Hill: New York, 1961.
42. Curtiss, L. A.; Raghavachari, K.; Deutsch P. W.; Pople, J.A. *J. Chem. Phys.* **1991**, *95*, 2433.
43. Moore, C. Natl. Bur. Stand. (U.S.) Circ 467, 1952.
44. Wallace, N. M.; Blaudeau, J.-P.; Pitzer, R. M. *Int. J. Quant. Chem.* **1991**, *40*, 789.
45. Scott, A. P.; Radom, L. *J. Phys. Chem.* **1996**, *100*, 16502.
46. Wong, M. W. *Chem. Phys. Let.* **1996**, *256*, 391.
47. Curtiss, L. A.; Raghavachari, K.; Redfern, P. C.; Pople, J. A. *Chem. Phys. Let.*, submitted.
48. Curtiss, L. A.; Raghavachari, K. In *Quantum Mechanical Electronic Structure Calculations with Chemical Accuracy*, Langhoff S.R., Ed.; Kluwer Academic Press: Netherlands, 1995, p. 139-172.

49. Raghavachari, K.; Curtiss, L.A. In *Modern Electronic Structure Theory*, Yarkony D. R. Eds.; World Scientific: Singapore, 1995, p. 991-1021.

50. Berkowitz, J.; Ellison, G. B.; Gutman, D. *J. Phys. Chem.* **1994**, *78*, 2744.

51. Curtiss, L. A.; Kock, L. D.; Pople, J. A. *J. Chem. Phys.* **1991**, *95*, 4040.

52. Curtiss, L. A.; Lucas D. J.; Pople, J. A. *J. Chem. Phys.* **1995**, *102*, 3292.

53. Curtiss, L. A.; Nobes, R. H.; Pople, J. A.; Radom L. *J. Chem. Phys.* **1992**, *97*, 6766.

54. Ruscic, B.; Berkowitz, J. *J. Chem. Phys.* **1991**, *95*, 2407.

55. Ruscic, B.; Berkowitz, J. *J. Chem. Phys.* **1991**, *95*, 2416.

56. Ruscic, B.; Berkowitz, J. *J. Chem. Phys.* **1991**, *95*, 2407.

57. Dyke, J. M. *J. Chem. Soc. Faraday Trans. 2*, **1987**, *83*, 69.

58. Curtiss, L. A.; Deutsch, P. W.; Raghavachari K. *J. Chem. Phys.* **1992**, *96*, 6868.

59. Smith, B. J.; Radom, L. *J. Phys. Chem.* **1991**, *95*, 10549.

60. Smith, B. J.; Radom, L. *J. Am. Chem. Soc.* **1993**, *115*, 4885.

61. Petersson, G.A.; Tensfeldt, T.G.; Montgomery, J.A., Jr. *J. Chem. Phys.* **1991**, *94*, 6091.

62. Ochterski, J.W.; Petersson, G.A.; Wiberg, K. *J. Am. Chem. Soc.,* **1995**, *117*, 11299.

63. Ochterski, J.W.; Petersson, G.A.; Montgomery, J.A., Jr. *J. Chem. Phys.,* **1996**, *104*, 2598.

64. Ragahavachari, K.; Stefanov, B. B.; Curtiss, L. A. *Mol. Phys.*, in press.

Chapter 11

Calculating Bond Strengths
for Transition-Metal Complexes

Margareta R. A. Blomberg and Per E. M. Siegbahn

**Department of Physics, University of Stockholm, Box 6730,
S–113 85 Stockholm, Sweden**

The calculation of accurate bond strengths for transition metal
complexes is discussed. Emphasis is put on methods capable of
treating relatively large complexes. Results from parametrized
ab initio schemes (PCI-80) and hybrid density functional methods (B3LYP) are compared to accurate experiments for sequences of small transition metal complexes. In most cases both
methods give quite satisfactory results with the PCI-80 results
slightly better in general. However, for a few cases where the
correlation effects are unusually large, the PCI-80 scheme breaks
down. For large systems the B3LYP method is the most promising approach since it is very fast and sufficiently accurate for
most problems.

The calculation of accurate thermodynamical properties of transition metal
complexes is one of the most difficult problems in computational quantum chemistry. First, the calculation of accurate bond strengths for any molecule, even
for a small first row molecule, is difficult, requiring methods that account for
almost all of the valence correlation energy. As a simple rule of thumb, the relative error in the correlation effect on a bond strength will be the same as the
relative error in the total valence correlation energy. Secondly, transition metal
complexes are generally accepted to be more difficult to treat than molecules
composed of lighter atoms. From an *ab initio* viewpoint, there can be a very
strong coupling between near degeneracy effects and large dynamical correlation
effects for transition metal complexes. From a density functional theory (DFT)
viewpoint, one difficulty is the strong coupling between correlation effects and
exchange effects, at least for cases with unfilled d-shells. To these two difficulties
can be added the problem that even the smallest transition metal complexes

of chemical interest have a large number of valence electrons. It is therefore clear from the onset that the goal for the accuracy of a bond strength must be set lower for a transition metal complex than for most molecules mentioned in other chapters of this book.

Setting the goal for a useful accuracy of a calculated bond strength is an interesting question with many aspects. Conventionally, useful chemical accuracy has been set to 1.0 kcal/mol, which is a very high ambition. This goal has essentially been achieved as an average for a standard benchmark test consisting of 55 small first and second row molecules by the G2 method (1). However, if this goal is set for every molecule treated and not as an average, an accuracy of 2.0 kcal/mol is more realistic. This accuracy for individual cases is also almost achieved by the G2 method for the same benchmark test even if some cases exist with somewhat larger errors. Further improvements have recently been reached by modifications of the G2 method (2). Some aspects relating to the question of useful accuracy are clearly connected with what is known previously about the systems being investigated and also the types of questions that are asked and the chemical importance of the questions. From these aspects, transition metal chemistry is a chemically very important area with strong connections to catalysis and biochemistry and where even basic thermodynamical information of moderate accuracy is missing. Two recent examples from biochemical applications can be given. In one case a recently suggested model for water oxidation in photosystem II relies entirely on the O-H bond strengths of water coordinated to manganese (3). No direct information on these bond strengths is available from experiments and suggestions given vary by up to 20 kcal/mol. Clearly, to aim at an accuracy of 1-2 kcal/mol would in this case be too ambitious and not necessary in order to support or invalidate the water oxidation model. A second example relates to the methane monooxygenase (4) and ribonucleotide reductase (5) enzymes, where a suggested hydrogen abstraction mechanism relies on the O-H bond strengths for hydroxyl ligands in certain iron dimer complexes, where there is also a nearly complete lack of direct experimental information. These examples indicate that an accuracy of a calculated bond strength in the case of a transition metal complex, rather than the usual 1-2 kcal/mol, can be set to 4-6 kcal/mol in individual cases and a somewhat higher accuracy as an average. This is a realistic goal as will be demonstrated in the present chapter.

In the present chapter on bond strengths in transition metal complexes, the results from mainly two different methods will be discussed. The first one of these is based on standard quantum chemical *ab initio* methods. Since a couple of years ago it has been evident that small deviations from an entirely pure *ab initio* theory are necessary if useful results for chemically interesting systems are to be generated. A small number of carefully chosen empirical parameters therefore have to be incorporated into the computational scheme. These empirical parameters should be based on a comparison to a selection, which is as broad as possible, of accurate experimental results. One of the first realizations of this type of *ab initio* based empirical method is the G2 method (1) mentioned above. In this method two empirical parameters (in practice only one parameter) are used to obtain a correction which is additive and proportional to the number of strongly correlated electron pairs (the number of closed shells). Large basis sets are used requiring long computational times even for rather small systems. The G2 method is therefore not very useful at present for transition metal complexes and will not be discussed further here. Instead, another parametrized *ab initio* based scheme will be the focus of most comparisons made here. This is the parametrized configuration interaction method with parameter 80 (PCI-80) scheme (*6, 7*). This scheme has a single multiplicative empirical parameter based on the fraction of the correlation energy that is calculated using a particular method and basis set. To compensate for the arbitrariness introduced by using a basis set dependent parameter, the PCI-80 scheme should normally be used only with one particular basis set. This basis set is rather small, of double zeta plus polarization (DZP) type, allowing applications for larger systems than using the G2 method.

The second main method which is used in the applications discussed below is the DFT method suggested by Becke, termed B3LYP (*8-10*), which uses a hybrid functional including Hartree-Fock exchange and contains three empirical parameters. This is probably the most promising method for future applications on transition metal complexes, at least for larger systems. Detailed comparisons between the results using this method and the results both from the PCI-80 scheme and from experiments will be made, in particular for systems containing first row transition metals. Apart from this, comparisons are mainly made to experimental results. For comparisons to other theoretical results see the individual papers referred to in the results section.

Computational Details

The *ab initio* type calculations in the present study were performed using the recently developed PCI-80 scheme (*6, 7*). This parametrized scheme is based on calculations performed using the modified coupled pair functional (MCPF) method (*11*), which is a standard quantum chemical, size-consistent, single reference state method. The zeroth order wave-functions were determined at the SCF level and all valence electrons were correlated. If standard double zeta plus polarization (DZP) basis sets are used it has been shown that about 80% of the correlation effects on bond strengths are obtained irrespective of the system studied. A good estimate of a bond strength is obtained by simply adding 20% of the full correlation effects. In the PCI-80 scheme (*6, 7*) this is done by adding 25% of the calculated correlation effect (the difference between the MCPF and Hartree-Fock results). The parameter 80 is thus an empirical parameter chosen to give agreement with experiment. It was later found (*7*) that the value 80 of the parameter also yields good agreement with experiment for a standard benchmark test consisting of 32 first row molecules (*12*). For several first row systems it was shown (*6, 7*) that a Hartree-Fock limit correction is also strictly needed in the PCI-80 scheme. However, this correction has been shown to usually be small for transition metal systems and a useful procedure is to consider these effects together with basis set superposition errors as included in the parameterization. This procedure has been used for the transition metal systems discussed here. The use of a single scaling parameter for the entire periodic table is the key feature of the PCI-80 scheme. Other scaling schemes exist, such as those due to Truhlar et al (*13-16*). In those schemes different parameters are used for different systems. In the PCI-80 calculations relativistic effects were added using perturbation theory for the mass-velocity and Darwin terms (*17, 18*). The PCI-80 calculations are performed using the STOCKHOLM set of programs (*19*).

The present DFT calculations were made using the empirically parametrized B3LYP method (*8-10*). The B3LYP functional can be written as

$$F^{B3LYP} = (1 - A) * F_x^{Slater} + A * F_x^{HF} + B * F_x^{Becke} + C * F_c^{LYP} + (1 - C)F_c^{VWN}$$

where F_x^{Slater} is the Slater exchange, F_x^{HF} is the Hartree-Fock exchange, F_x^{Becke} is the exchange functional of Becke (*8*), F_c^{LYP} is the correlation functional of Lee, Yang and Parr (*20*) and F_c^{VWN} is the correlation functional of Vosko, Wilk and Nusair (*21*). A, B and C are the coefficients determined by Becke (*8-10*) using a fit to experimental heats of formation for the same benchmark test mentioned above but in this case consisting of 55 first and second row molecules. However, it should be noted that Becke did not use F_c^{LYP} in the expression above when the coefficients were determined, but used the correlation functional of Perdew and Wang instead (*22*). Since empirical parameters are used there are thus clear similarities between the PCI-80 scheme and the B3LYP method in this respect. In the B3LYP molecular results, relativistic effects taken from the MCPF calculations are included. The B3LYP calculations were carried out using the GAUSSIAN 94 package (*23*).

In the MCPF calculations essentially valence double zeta plus polarization (DZP) basis sets were used for all atoms. In the B3LYP energy calculations the large 6-311+G(2d,2p) basis sets (*23*) were used. This basis set has diffuse functions and two polarization functions on all atoms including two f-functions on the metal. The geometries were optimized at the Hartree-Fock or B3LYP levels using double zeta basis sets, or taken from other calculations. Zero-point vibrational effects were taken from other calculations or calculated at the Hartree-Fock or B3LYP level. It should also be mentioned that binding energies relative to naked metal atoms or cations are calculated relative to the lowest s[1] state, and experimental data for the atomic splittings are used to refer to the ground state, if necessary. For further details about the calculations see each reference.

The choice of the basis set, correlation method and in particular the parameter X = 80 might seem a bit arbitrary and therefore a set of investigations were performed on a set of first row benchmark molecules (*7*). Different one-particle basis sets and different correlation methods were used to calculate the atomization energies and the resulting average errors for the first row benchmark test molecules are summarized in Table I. In these results the parameter X has been optimized for each set of calculations to minimize the average error in the atomization energies. The conclusion from this investigation is that the PCI-80(MCPF) procedure using a DZP one-particle basis set gives very good agreement with experiment for these small first row molecules and the results can not easily be improved by changing the one-particle basis set or the correlation method.

Table I. Mean absolute deviations Δ (kcal/mol) obtained after scaling the correlation error using different basis sets and methods for the first row benchmark test (*7*)

Method	Basis set	Unscaled Δ	X^a	Δ
MCPF	DZ	40.0	64.1	15.8
MP2	DZP	25.1	74.1	7.9
MCPF	DZP	22.2	77.8	2.4
MCPF	VDZ	23.0	77.6	2.4
CCSD	VDZ	21.0	76.8	2.2
CCSD(T)	VDZ	17.6	80.0	2.6
MCPF	W321	23.1	76.8	4.0
MCPF	W432	19.5	81.4	2.8
MCPF	W5521	12.7	87.3	2.9
CCSD	W6532	10.6	88.2	3.2
CCSD(T)	W6532	5.4	94.1	1.2

[a] X is the optimized scale parameter.

SOURCE: Reprinted with permission from reference 7. Copyright 1995 American Institute of Physics.

The PCI-X method was originally designed for the treatment of transition metal systems. Therefore this procedure has to be tested for this kind of systems too. The first test of the PCI-80 scheme on transition metal systems is performed on the atomic spectra, which is the only set of very accurate experimental data available. The average deviation from experimental results for certain groups of atomic excitations for first (*24*) and second (*6*) row metals are collected in Table II. The main conclusion from this table is that the PCI-80 procedure leads to an improved description of atomic excitation energies. For the second row transition metal atoms (*6*) the scaling decreases the average error from about 5 kcal/mol at the MCPF level to about 2.5 kcal/mol, which has to be considered as quite satisfactory. The first row transition metal atoms are more difficult to describe due to the tighter valence d-orbitals and the more different radii of the valence d-and s-orbitals. Therefore the average MCPF error in the splitting is larger for these metals, around 10 kcal/mol (*24*) as compared to around 5 kcal/mol for the second row metals. The scaling of the correlation energy decreases the average error in the atomic splittings for the first row down to around 5 kcal/mol. However, core correlation effects are *not* included for the first row metals but *are* included for the second row metals. If the core correlation effects, as taken from ref. (*25*), are added to the results for the neutral first row metals the error after scaling decreases down to 2.3 kcal/mol, which is very similar to the results for the second row metals. In conclusion, if the core correlation effects are taken into account, the PCI-80 scheme gives an average absolute deviation from experiment for transition metal atomic splittings of the same size as for the atomization energies of the benchmark first row molecule test, about 2.5 kcal/mol.

Table II. Mean absolute deviations from experimental results in kcal/mol for atomic excitation energies for transition metal atoms and cations

Systems	First row TM[a]		Second row TM[b]	
	MCPF	PCI-80	MCPF	PCI-80
$M(s^1 \rightarrow s^0)$			5.8	2.5
$M(s^1 \rightarrow s^2)$	11.3	4.8[c]	4.8	2.8
$M^+(s^1 \rightarrow s^0)$	10.1	5.5		

[a] Ref. (*24*). [b] Ref. (*6*). [c] With core correlation effects taken from ref. (*25*) the value decreases to 2.3 kcal/mol.

Results

In this section we discuss some results for transition metal bond strengths obtained by the PCI-80 scheme. For some of the systems also B3LYP results have been obtained. In the first two subsections calculated M-H and M-C single bond energies for first row transition metal cations are compared to experimental values (*24*). The corresponding second transition row cationic systems have larger uncertainties on the experimental side and maybe also on the theoretical side, and they are therefore not included here. In the next subsection the calculated M=C double bond energies of the cationic carbenes, MCH_2^+, for both the first (*24*) and the second (*6*) row transition metals are compared to experiment. In the fourth subsection some second transition row diatomics selected from Huber and Herzberg (*26*) are treated (*6*) and in the final subsection some comparisons are made between calculated potential energy surfaces and gas phase experiments for neutral second row transition metal atoms and small hydrocarbons (*27*).

MH$^+$ Cations. The MH$^+$ cations are the simplest possible transition metal molecules that can be used to test computational methods. In these systems only a single bond, mainly involving the valence s orbital, is formed. The calculated binding energies for MH$^+$ for the first transition row metals are given in Table III (*24*). As seen in this table, already the MCPF results without scaling using a DZP basis are reasonable with an average deviation compared to experiments of only 6.0 kcal/mol. Applying a scaling improves the results to an average deviation of only 1.9 kcal/mol at the PCI-80 level. This agreement between theory and experiment is remarkable in at least two ways. First, the average uncertainty in the experimental results is 2.1 kcal/mol for the MH^+ systems (*28*). Secondly, the average error in the splittings at the PCI-80 level for the M^+ cations is 5.5 kcal/mol, which is substantially larger than the deviation for MH^+. This means that the procedure of dissociating to the s^1 state and thereby partly including core correlation effects is very successful. Results obtained with the B3LYP method are also given in Table III. In this case the average deviation from experiment is 4.7 kcal/mol. Even this result is quite acceptable for transition metal containing systems, although it is not as accurate as the PCI-80 result.

Table III. Metal-ligand bond strengths (D_0) in first row transition metal MH^+ systems (kcal/mol) (*24*)

M	State	B3LYP	err	MCPF	err	PCI-80	err	Exp.[a]
Sc	$^2\Delta$	60.0	+3.7	50.7	-5.6	55.9	-0.4	56.3
Ti	$^3\Phi$	56.3	+3.0	48.1	-5.2	55.3	+2.0	53.3
V	$^4\Delta$	54.1	+6.8	42.7	-4.6	49.6	+2.3	47.3
Cr	$^5\Sigma^+$	36.5	+4.9	27.4	-4.2	30.6	-1.0	31.6
Mn	$^6\Sigma^+$	48.6	+1.1	37.7	-9.8	44.7	-2.8	47.5
Fe	$^5\Delta$	57.4	+8.5	45.0	-3.9	52.2	+3.3	48.9
Co	$^4\Phi$	48.7	+3.0	37.0	-8.7	46.4	+0.7	45.7
Ni	$^3\Delta$	45.2	+6.5	32.6	-6.1	41.6	+2.9	38.7
Δ^b			4.7		6.0		1.9	

[a] Ref. (*28*). [b] Δ is the average absolute deviation compared to experiment.

MCH_3^+ Cations. Another set of simple test systems are the MCH_3^+ cations. The main difference between these systems and the MH^+ cations is that for MCH_3^+ there is direct repulsion between the CH-bond electrons of the methyl group and the valence d shell. The results for the M-CH_3^+ binding energies for the first transition row metals are given in Table IV (*24*). The increased difficulty to describe the metal methyl cations can be seen in the average deviation at the MCPF level, which increases from 6.0 kcal/mol for MH^+ to 8.9 kcal/mol for MCH_3^+. This size error is probably representative of what can be expected for a single covalent M-X bond at the MCPF level using DZP basis sets. The improvement obtained by scaling at the PCI-80 level for MCH_3^+ is even more surprising than the results for the MH^+ systems. The average deviation compared to experiments at the PCI-80 level is only 1.8 kcal/mol, which is only slightly larger than the average uncertainty in the experimental results of 1.3 kcal/mol (28). It is clear that these PCI-80 results must be regarded as fortuitous to some extent. Nevertheless, the agreement between theory and experiment is not only a support for the scaling scheme but can also be used as a confirmation of the validity of the experimental techniques and interpretations (28). The results obtained at the B3LYP level are also given in Table IV. These results must also be regarded as satisfactory with an average deviation of 5.5 kcal/mol.

Table IV. Metal-ligand bond strengths (D_0) in first row transition metal MCH_3^+ systems (kcal/mol) (24)

M	State[b]	B3LYP	err	MCPF	err	PCI-80	err	Exp.[a]
Sc	$^2A'$	60.8	+5.0	48.0	-7.8	55.5	-0.3	55.8
Ti	$^3A'$	58.8	+7.7	43.3	-7.8	53.3	+2.2	51.1
V	$^4A'$	49.9	+3.9	36.9	-9.1	45.0	-1.0	46.0
Cr	$^5A'$	35.6	+9.4	19.9	-6.3	29.2	+3.0	26.2
Mn	$^6A'$	50.2	+1.1	37.6	-11.5	46.2	-2.9	49.1
Fe	$^5A'$	62.3	+7.7	47.3	-7.3	56.4	+1.8	54.6
Co	$^4A'$	53.4	+4.9	39.7	-8.8	48.8	+0.3	48.5
Ni	$^3A'$	49.2	+4.5	32.1	-12.6	41.8	-2.9	44.7
Δ^c			5.5		8.9		1.8	

[a] From ref. (28). [b] Titanium and cobalt have A" ground states in the B3LYP calculations. [c] Δ is the average absolute deviation compared to experiment.

SOURCE: Reprinted with permission from reference 24. Copyright 1996 American Institute of Physics.

MCH_2^+ **Cations.** The transition metal carbene cations are significantly more difficult to describe than the systems discussed in the previous subsections. When a metal cation binds to methylene two additional closed shells are in general created for the product. For MH^+ and MCH_3^+ only one more closed shell is created when the bond is formed. For this reason alone, the errors in the computed ligand bond strengths for MCH_2^+ can be expected to be twice as large as for MCH_3^+. In addition to this increased difficulty to describe the carbenes, the near degeneracy effects are also more severe for these systems. These systems are on the borderline of what is reasonable to attack using a single reference method like MCPF. The results from the calculations are given in Table V (24) for the first row metals and in Table VI (6) for the second row metals. The binding energies obtained at the MCPF level confirm the above expectations of larger errors for the carbenes. The average deviation at this level is 20.2 kcal/mol for the first row metals and 20.4 kcal/mol for the second row metals. When scaling is applied the results are substantially improved but not quite as far as for MH^+ and MCH_3^+. The average deviation at the PCI-80 level is 3.6 kcal/mol for the first row metals and 4.0 kcal/mol for the second row metals. In Tables V and VI it can be noted that the results are highly satisfactory for all metals except for vanadium, chromium and niobium. $CrCH_2^+$ with an error as large as 14.4 kcal/mol can thus be concluded to be one of a few cases where the scaling scheme based on MCPF does not work. Considering the fact that there are MCSCF coefficients as large as 0.40 for this system, this failure can not be considered surprising. It is in fact more surprising that the other carbenes, which have equally large MCSCF coefficients, work so well. The PCI-80 errors for vanadium and chromium can be understood from the contribution to the wave-function of a configuration having two d-bonds, see further ref. (*24*). For the niobium case it can be noted that good agreement is obtained with the theoretical value of 89 kcal/mol obtained in ref. (*29*), where very large basis sets were used in combination with multireference techniques. For the first row metals, B3LYP results are also included in Table V. The average deviation for the MCH_2^+ systems at the B3LYP level is 3.6 kcal/mol, which is the same as the deviation found at the PCI-80 level. However, the largest error at the B3LYP level is only 6.1 kcal/mol (for iron) compared to 14.4 kcal/mol (for chromium) at the PCI-80 level. The results obtained for MCH_2^+ at the B3LYP level can therefore be considered quite satisfactory.

Table V. Metal-ligand bond strengths (D_0) in first row transition metal MCH_2^+ systems (kcal/mol) (*24*)

M	State	B3LYP	err	MCPF	err	PCI-80	err	Exp.[a]
Sc	$^1A_1$	83.8	-5.0	70.7	-18.1	87.6	-1.2	88.8[b]
Ti	$^2A_1$	84.9	-6.0	71.3	-19.6	92.1	+1.2	90.9
V	$^3B_2$	76.1	-1.7	61.0	-16.8	84.7	+6.9	77.8
Cr	$^4B_1$	57.1	+5.4	38.1	-13.6	66.1	+14.4	51.7
Mn	$^5B_1$	68.6	-0.3	43.6	-24.7	70.6	+2.3	68.3
Fe	$^4B_2$	75.4	-6.1	55.1	-26.4	79.1	-2.4	81.5
Co	$^3A_2$	77.7	+1.8	54.1	-21.8	76.1	+0.2	75.9.
Ni	$^2A_1$	71.0	-2.1	52.1	-21.0	73.7	+0.6	73.1
Δ^c			3.6		20.2		3.6	

[a] From ref. (*28*). [b] Corrected for involvement of an excited state of Sc^+ (*28*).
[c] Δ is the average absolute deviation compared to experiment.

Some Second Row Diatomics. The PCI-80 method has also been tested on bond strengths for most of the second row transition metal systems listed by Huber and Herzberg (*26*). The results for the individual systems are given in ref. (6). One problem in this context is that several of the systems listed in Ref. (26) happen to be unusually difficult. Another problem of testing the present method against experiments is that the experiments can be very uncertain. In fact, it is our belief that if all the experimental bond strengths listed in Ref. (26) were exchanged with the present theoretical values, the overall accuracy would probably not be lower. For the 18 comparisons of neutral diatomic molecules discussed in ref. (6) the average absolute deviation from experiments is 4.2 kcal/mol at the PCI-80 level. This is a large improvement, of the same size as the ones discussed in the previous subsections, compared to the unparametrized MCPF deviation of 22.4 kcal/mol. The average absolute deviation at the SCF level of 88.3 kcal/mol for these systems also illustrates the difficulty to treat these systems theoretically. A PCI-80 discrepancy of less than 5 kcal/mol must be considered as quite acceptable. For a few systems the discrepancy is larger than this.

Table VI. Metal-ligand bond strengths (D_0) in second row transition metal MCH_2^+ systems (kcal/mol) (6)

System	State	MCPF	err	PCI-80	err	Exp
YCH_2^+	$^1A_1$	76.6	16.2	89.9	2.9	92.8^a
$ZrCH_2^+$	$^2A_1$	86.4		101.4		-
$NbCH_2^+$	$^3B_2$	78.0	28.1	95.0	11.1	106.1^b
$MoCH_2^+$	$^4B_1$	62.1		81.6		-
$TcCH_2^+$	$^5B_1$	65.9		84.4		-
$RuCH_2^+$	$^2A_2$	59.5	22.8	80.9	1.4	82.3^b
$RhCH_2^+$	$^1A_1$	64.3	23.3	87.1	0.5	$87.6^c(85.1^d)$
$PdCH_2^+$	$^2A_1$	56.6	11.4	71.9	3.9	68.0^b
Δ^e			20.4		4.0	

[a] Ref. (28). [b] Ref. (30). [c] Ref. (31), adjusted to 0K. [d] Ref. (32). [e] Δ is the average absolute deviation compared to experiment.

Gas Phase Reactivity of Some Second Row Systems. It has been shown above that for sequences of transition metal bond strengths, the PCI-80 scaling scheme works very well in most cases. However, so far no examples have been discussed concerning the applicability of the scaling scheme for studies of transition states and chemical reactivity. In the present subsection the results from this scheme will therefore be compared to results obtained without scaling for a set of chemical reactions studied experimentally. Recently the reactivity of the neutral atoms of the second transition row metals Y, Zr, Nb, Mo, Rh and Pd with some small hydrocarbons (ethane, ethylene and cyclo-propane) has been investigated experimentally and some of the resulting rate constants are summarized in Table VII (27). Reactions of small hydrocarbons with naked metal atoms fall in two general categories: termolecular stabilization of long-lived M(hydrocarbon) complexes and bimolecular elimination of H_2. Calculated relative energies and vibrational frequencies can be used in a statistical unimolecular rate theory (RRKM theory) to model the lifetimes of M(hydrocarbon) complexes. In this way the plausibility of a saturated termolecular mechanism can be assessed and compared to the bimolecular reaction mechanism. Therefore PCI-80 calculations were performed on the reactions between all second row metal atoms, yttrium to palladium, with ethane, ethylene and cyclopropane (*27*). The bimolecular elimination of H_2 must be initiated by C-H activation. Therefore the C-H activation barriers are calculated for all systems and some of the results are summarized in Table VII. For ethane and cyclopropane C-C insertion presents alternative reaction paths and the calculated C-C activation barriers are also collected in Table VII. In this table we have given both the MCPF and the PCI-80 results, to evaluate the importance of the scaling procedure when quantitative results are needed.

Table VII. Comparison of MCPF and PCI-80 barrier heights (kcal/mol) for the insertion of second row metal atoms into a C-H or a C-C bond in some small hydrocarbons (*27*). Effective bimolecular rate constants k_1 (10^{-12} cm^3 s^{-1}) for reaction of the metal atoms with the hydrocarbons at 0.80 ± 0.05 Torr of He and 300 ± 5 K (*27*) are also given.

| Systems | C-H activation | | C-C activation | | |
	MCPF	PCI-80	MCPF	PCI-80	k_1 (exp)
Ethane activation[a]:					
Y	31.6	20.7	33.3	28.5	NR
Zr	27.2	16.9	43.0	35.8	NR
Mo	49.0	37.8	82.0	41.2	NR
Rh	11.0	0	35.0	21.8	3.7
Ethylene activation:					
Y	13.0	1.9			8.2
Zr	11.6	1.8			59
Nb	9.4	0			314
Mo	30.3	18.2			NR
Cyclopropane activation:					
Y	22.9	11.3	14.4	9.6	0.70
Zr	18.1	7.2	23.3	14.4	0.66
Nb	15.1	4.3	19.5	8.1	3.04
Rh	7.7	0	0	0	14.2
Pd	12.0	0.5	0	0	58.9

[a] C-H activation barriers are taken from methane activation (*27*).

Rhodium is the only one of the metals listed in Table VII that reacts with ethane. The C-H activation barrier calculated at the MCPF level is 11.0 kcal/mol, which is too high to allow an H_2 elimination process. However, when the scaling procedure is applied this barrier on the low-spin potential surface disappears, and agreement with the experimental observation is obtained. The PCI-80 results for the rest of the potential surface for the H_2 elimination process (27) are also consistent with an interpretation of the experimental data as a bimolecular H_2 elimination process. For yttrium, zirconium and molybdenum no reaction with ethane is observed, which can be explained by the calculated C-H activation barriers, which are too high to allow C-H insertion both at the MCPF and the PCI-80 level. For all the metals the calculated C-C activation barriers are too high to allow a C-C insertion reaction to occur.

Ethylene reacts with all the investigated metal atoms except molybdenum. Niobium is the most reactive atom. The detailed PCI-80 investigation of the quartet Nb + C_2H_4 potential surface shows that there is no barrier to exothermic H_2 elimination on that low-spin surface (27), which is consistent with the high reactivity observed for niobium. On the contrary, as can be seen in Table VII, the C-H activation barrier calculated at the MCPF level, 9.4 kcal/mol, is too high to explain the experimentally observed reaction rate of niobium. For Y and Zr the reaction rates with ethylene are substantially smaller than for Nb. For both Y and Zr reactions with ethylene the PCI-80 calculations give small barriers of 1.9 and 1.8 kcal/mol, respectively, to C-H insertion. These insertion barriers should limit the ethylene reaction rates. Once C-H insertion occurs, the extensive PCI-80 calculations along the entire reaction path for Zr + ethylene find no further impediment to H_2 elimination (27), and the same is expected to hold for Y + ethylene. Again, the C-H activation barriers calculated at the MCPF level for Y and Zr are too high to explain the experimentally observed reaction rates, see Table VII. For Mo, the PCI-80 energy at the transition state to C-H insertion on the low-spin surface is clearly too high (18.2 kcal/mol) to be overcome at 300 K, in agreement with the experimental observation that no reaction is observed.

All the atoms listed in Table VII react with cyclopropane, and the PCI-80 results permit either C-C or C-H activation. For the most reactive atoms, rhodium and palladium, both C-C and C-H insertion are clearly energetically feasible. Termolecular stabilization into the deep metallacyclobutane well (C-C insertion) can explain the observed reactions of Y, Zr, Nb, Rh and Pd with cyclopropane. For Y, Zr, Nb and Rh collisional stabilization of C-H insertion complexes can not be ruled out. The PCI-80 results are thus in good agreement with the experimentally observed reactivity of these metal atoms with cyclopropane. On the other hand, as can be seen from Table VII, the activation barriers calculated at the MCPF level are too high for both C-H and C-C insertion for Y, Zr and Nb, and the PCI-80 scaling procedure is necessary for the interpretation of the experimental results.

In summary, it was shown in ref. (27) that the combination of PCI-80 bound state energies and RRKM modeling of lifetimes works extremely well for explaining the experimental results for the second row neutral M + hydrocarbon systems. On the other hand, it was indicated above that for several of the activation barriers involved in these reactions, results at the MCPF level without scaling were not accurate enough to allow a reasonable interpretation of the experimental observations.

Conclusions

In the present chapter, a survey of results has been given showing present possibilities for obtaining accurate bond strengths for transition metal complexes. Examples from the use of methods and schemes which have the capability of treating reasonably large systems have been discussed. For *ab initio* based schemes, this means that only medium size basis sets can be afforded. For many of the systems discussed above where the metal complex is quite small, for example for the hydride cations, it is certainly possible to use much more extended basis sets and thus obtain possibly even higher accuracy for these systems, although the accuracy demonstrated above is already very high. For the first transition row MH^+ and MCH_3^+ systems, the average accuracy using the PCI-80 scheme is about the same as the one obtained experimentally. For a few of the other sequences of complexes, individual cases have been found where the error of the PCI-80 scheme is much larger. These cases include $CrCH_2^+$, $NiHCH_3$ and $Co(C_5H_5)(CO)HCH_3$. The common factor for these systems is that the correlation effect on the binding energy is unusually large compared to similar systems. The origin of the large correlation effect is a near degeneracy of valence bond type, which unfortunately does not always reveal itself by the presence of large coefficients in the wave-function. Instead, these cases have to be identified by a comparison to results for other methods and also to comparisons to similar systems.

The usefulness of the PCI-80 scheme is mainly restricted by the size of system that can be treated. At present, and also for the near future, the reaction between $Ir(C_5H_5)(CO)$ and methane, which was recently treated, is at the limit of what can be handled. For a less expensive treatment of this system and for the treatment of larger systems, the hybrid density functional B3LYP method is the most useful method. This method is substantially faster than the PCI-80 scheme and has the additional advantage that there appear to be fewer cases of severe breakdown. In fact, for the systems presently studied where there is a large discrepancy between the results of the PCI-80 scheme and of the B3LYP method, the B3LYP result tends to be most reliable. On the other hand, in cases where these methods give reasonable agreement, the PCI-80 result is generally most accurate.

Literature Cited

1. Curtiss L. A.; Raghavachari K.; Trucks G. W.; Pople J. A. *J. Chem. Phys.* **1991**, *94*, 7221.
2. Bauschlicher, C. W., Jr.; Partridge, H. *J. Chem. Phys.* **1995**, *103*, 1788.

3. Blomberg, M. R. A.; Siegbahn, P. E. M.; Styring, S.; Babcock, G. T.; Åkermark, B.; Korall, P., *J. Am. Chem. Soc.*, in press.

4. Siegbahn, P. E. M.; Crabtree, R. H. *J. Am. Chem. Soc.*, in press.

5. Siegbahn, P. E. M.; Blomberg, M. R. A.; Crabtree, R. H. *Theor. Chem. Acc.*, in press.

6. Siegbahn P. E. M.; Blomberg M. R. A.; Svensson M. *Chem. Phys. Lett.* **1994**, *223*, 35.

7. Siegbahn P. E. M.; Svensson M.; Boussard P. J. E. *J. Chem. Phys.* **1995**, *102*, 5377.

8. Becke A. D. *Phys. Rev.* **1988**, *A38*, 3098.

9. Becke A. D. *J. Chem. Phys.* **1993**, *98*, 1372.

10. Becke A. D. *J. Chem. Phys.* **1993**, *98*, 5648.

11. Chong D. P.; Langhoff S. R. *J. Chem. Phys.* **1986**, *84*, 5606.

12. Pople J. A.; Head-Gordon M.; Fox D. J.; Raghavachari K.; Curtiss L. A. *J. Chem. Phys.* **1989**, *90*, 5622.

13. Brown F. B.; Truhlar D. G. *Chem. Phys. Letters* **1985**, *117*, 307.

14. Gordon M. S.; Truhlar D. G. *J. Am. Chem. Soc.* **1986**, *108*, 5412.

15. Gordon M. S.; Truhlar D. G. *Int. J. Quantum Chem.* **1987**, *31*, 81.

16. Gordon M. S.; Nguyen K.A.; Truhlar D.G. *J. Phys. Chem.* **1989**, *93*, 7356.

17. Martin, R. L. *J. Phys. Chem.* **1983**, *87*, 750.

18. Cowan, R. D.; Griffin, D. C. *J. Opt. Soc. Am.* **1976**, *66*, 1010.

19. STOCKHOLM is a general purpose quantum chemical set of programs written by Siegbahn, P. E. M.; Blomberg, M. R. A.; Pettersson, L. G. M.; Roos, B. O.; Almlöf, J.

20. Lee, C.; Yang, W.; Parr, R.G. *Phys. Rev.* **1988**, *B37*, 785.

21. Vosko, S. H.; Wilk, L.; Nusair, M. *Can. J. Phys.* **1980**, *58* 1200.

22. Perdew J. P.; Wang, Y. *Phys. Rev. B* **1992**, *45*, 13244.

23. GAUSSIAN 94 (Revision A.1), Frisch, M. J.; Trucks, G. W.; Schlegel, H. B.; Gill, P. M. W; Johnson, B. G.; Robb, M. A.; Cheeseman, J. R.; Keith, T. A.; Petersson, G. A.; Montgomery, J. A.; Raghavachari, K.; Al-Laham, M. A.; Zakrzewski, V. G.; Ortiz, J. V.; Foresman, J. B.; Cioslowski, J.; Stefanov, B. R.; Nanayakkara, A.; Challacombe, M.; Peng, C. Y.; Ayala, P. Y.; Chen, W.; Wong, M. W.; Andres, J. L.; Replogle, E. S.; Gomperts, R.; Martin, R. L.; Fox, D. J.; Binkley, J. S.; Defrees, D. J.; Baker, J.; Stewart, J. P.; Head-Gordon, M.; Gonzales C.; Pople, J. A.; Gaussian, Inc., Pittsburgh PA, 1995.

24. Blomberg M. R. A.; Siegbahn P. E. M.; Svensson M. *J. Chem. Phys.* **1996**, *104*, 9546.

25. Bauschlicher, C. W., Jr.; Walch S. P.; Partridge H. *J. Chem. Phys.* **1982**, *76*, 1033.

26. Huber, K. P.; Herzberg, G. *Molecular Spectra and Molecular Structure*; van Nostrand-Reinhold: New York, 1979; Vol. IV.

27. Carroll J. J.; Weisshaar J. C.; Blomberg M. R. A.; Siegbahn P. E. M.; Svensson M. *J. Phys. Chem.* **1995**, *99*, 13955.

28. Armentrout P. B.; Kickel, B. L. In *Organometallic Ion Chemistry*; Freiser B.S., Ed.; Kluwer: Dordrecht, 1995; p 1.

29. Bauschlicher, C. W., Jr.; Partridge, H.; Sheehy, J. A.; Langhoff, S. R.; Rosi, M. *J. Phys. Chem.* **1992**, *96*, 6969.

30. Armentrout, P. B., personal communication.

31. Hettich, R. L.; Freiser, B. S. *J. Am. Chem. Soc.* **1987**, *109*, 3543.

32. Chen. M.; Armentrout, P. B. *J. Phys. Chem.* **1995**, *99*, 10775.

Chapter 12

Calibration Study of Atomization Energies of Small Polyatomics

Jan M. L. Martin

Department of Organic Chemistry, Weizmann Institute of Science, IL–76100 Rehovot, Israel

A number of aspects involved in computing thermochemical properties beyond mere 'chemical accuracy' are discussed. This includes the major issues of n-particle space (electron correlation treatment) and 1-particle basis set, as well as subsidiary issues such as core correlation, atomic spin-orbit coupling, anharmonicity in the zero-point energy, and the quality of the reference geometry. Through stepwise basis set extrapolation at the CCSD(T) level and involving basis sets of up to $[7s6p5d4f3g2h]$ quality, total atomization energies accurate to 0.1 kcal/mol, on average, can be obtained. A somewhat more modest accuracy goal of 0.25 kcal/mol can be met using $[6s5p4d3f2g]$ basis sets and a 3-parameter empirical correction.

In the context of this volume, the importance of having accurate thermochemical data available requires no explanation. The total atomization energy (TAE, ΣD_0), from a computational chemistry point of view, is the most fundamental such quantity and, using the experimental heats of formation of the atoms in the gas phase, can be directly related to the gas-phase heat of formation.

In the present chapter, we will focus on methods for obtaining 'calibration quality' TAEs in two senses of the word.

In the first sense 'calibration quality', or 'benchmark quality', means a result that is not merely of 'chemical accuracy', but one that would be acceptable as a reference benchmark value. The target accuracy, therefore, is not 1 kcal/mol, but (ideally) on the order of 1 kJ/mol (0.24 kcal/mol) or less.

In the second sense, the use of the word 'calibration' implies that, rather than merely calculating a number using one level of theory and saying 'that's it', a systematic assessment is carried out of computed values at successively higher levels of theory, which will allow, at the very least, to approximately quantify any further errors that might be involved.

In the first Section, we will discuss the major issues arising in the quality of a calculation — basis set and electron correlation treatment. In the next Section, an overview of some of the minor — but quite nonnegligible — ones will be given. Finally, some practical approaches will be reviewed.

Major Issues

Zero-Order Description. The exact wave function can be expanded as

$$\psi = \psi_0 + \sum_{ia} C_{ia}\psi_{i\to a} + \sum_{ijab} C_{ijab}\psi_{ij\to ab} + \dots$$

$$\equiv (1 + \hat{C}_1 + \hat{C}_2 + \dots + \hat{C}_n)\psi_0 \tag{1}$$

that is, as a linear combination of some zero-order wave function ψ_0 and all possible Slater determinants that can be generated by exciting electrons from occupied into unoccupied orbitals within that basis set. The $C_{ij\dots ab\dots}$ are the configuration interaction coefficients, and the excitation operators $\hat{C}$, labeled by excitation level (i.e. the number of electrons moved from occupied to virtual orbitals), permit a compact notation of the wave function.

In most cases, ψ_0 will be the Hartree-Fock (SCF) wave function, since for most systems (other than exotica such as the Cr_2 molecule (*1*)), the exact wave function will be dominated by a single reference determinant. Even for first-row compounds this is by no means always the case, however. For example (*2–5*), the $a\ ^1\Sigma^+$ state of BN has a low-lying excited state $\dots(3\sigma)^2(4\sigma)^0(1\pi)^4(5\sigma)^2$ of the same symmetry and spin as the ground state $\dots(3\sigma)^2(4\sigma)^2(1\pi)^4$, and in the actual wave function the first excited state contributes about 30%. It goes without saying that under such circumstances, a single-determinant SCF wave function is generally *not* an adequate zero-order wave function.

The effect on the energy and wave function of such occurences is known as *nondynamical correlation* (also known as near-degeneracy correlation).

The alternative, when such effects are important, is then to use a multiconfigurational zero-order reference wave function such as the result from a CASSCF (complete active space SCF) calculation (*6*), i.e. a linear combination of all the determinants that can be generated by rearranging electrons within an 'active space' of orbitals, which may range from two nearly degenerate orbitals to the full valence space. Since the computational cost of full-valence CASSCF references rapidly becomes intractable beyond 12 orbitals (i.e. 3 heavy atoms, or 2 heavy atoms and 4 hydrogens), and dynamical correlation methods based on multiconfigurational zero-order wave functions are in rather shorter supply than those based in single-configuration zero-order wave functions (so-called "single-reference methods"), multireference methods are generally only used as a last resort, particularly since the advent of efficient and accurate coupled-cluster methods (see below).

n-Particle Calibration: Quality of the Electron Correlation Method. In cases where the Hartree-Fock approximation does not break down, it generally recovers about 99% of the total energy. However, the remaining 1% is on the same order of magnitude as the quantity we are interested in, namely the atomization energy.

Since the molecular SCF energy converges relatively quickly with the 1-particle basis set, we can make fairly accurate estimates of the energies at the SCF limit. (Or alternatively we may carry out a numerical Hartree-Fock calculation.) By comparing these with the actual atomization energies, we find that for molecules like H_2, CH_4, and C_2H_2, SCF recovers 76.6%, 78.9%, and 73.9%, respectively, of the TAE. For more polar molecules like H_2O and CO_2, this drops to 68.7 and 66.2%, respectively. But only about half (52.4%) of the binding energy of N_2 is accounted for at the SCF level, and merely a third (35.1%, to be precise) for N_2O. Most spectacularly, for F_2 the binding energy is actually found to be negative at the SCF level, and in error by 179.6%. It should be clear from these examples that the importance of an adequate treatment of electron correlation (that is, of an adequate n-particle calibration) can hardly be overestimated.

Full configuration interaction by definition gives the exact n-particle energy within the given basis set. (Because it is the exact solution, this happens irrespective of the quality of the zero-order wave function.) Because its computational requirements ascend factorially with the size of the system, application to practical systems using one-particle basis sets of useful size will be essentially impossible for the foreseeable future. Even using the fastest available computational hardware and parallelized codes, an FCI calculation on H_2O in a double-zeta plus polarization basis set is about the state of the art at present (7, 8).

Nevertheless, such calculations (and the heroic earlier work of the NASA Ames group (9)) were of inestimable value for another purpose: evaluating the performance of more computationally efficient correlation methods.

Limited configuration interaction (CI) is perhaps the oldest approximate correlation method, being discussed by Slater (10) as early as 1929. The most common approximation is to truncate the CI expansion at single and double excitations, which is known as CISD (configuration interaction with all single and double excitations). It corresponds to a wave function of the form $\psi = (1 + \hat{C}_1 + \hat{C}_2)\psi_0$. The main problem with this method (or any truncated CI expansion) is that it is not size-extensive: that is, the energy does not scale properly with the size of the system, and the error becomes more important as the system grows larger. Various approximate correction formulas for this behavior have been introduced (see Ref. (11) for a review) — and one of them, the familiar Davidson correction (12), actually was rated a Citation Classic in the Science Citation Index. However, none of these formulas meets the exacting standards required for the present purpose.

An alternative is to take the sum of the one-electron Fock operators as the zero-order Hamiltonian and and treat the difference between this and the exact Hamiltonian as a perturbation. This is known as many-body perturbation theory (MBPT), or, more pedantically, the Møller-Plesset (13) (MP) partition thereof. The zero-order energy is the sum of the orbital energies: the sum of zero-order and first-order energies is actually the SCF energy. The second-order term recovers a large percentage of the correlation energy for many molecules, and can be evaluated very rapidly. Moreover, it can be shown (14) using an algebraic technique known as diagrammatic perturbation theory that MPn is size-extensive at all orders n.

So what is wrong with it? Principally, that products of differences between orbital energies appear in the denominator. Now if there are no low-lying excited electronic states to speak of, all these differences will be fairly large, and the perturbation series will converge rapidly. If low-lying excited states *are* present, however, they will slow down the convergence of the MP series to such an extent that low-order MP energies are essentially unusable. And not only the computational cost, but the algebraic complexity involved in deriving the equations, increases so rapidly (see e.g. (*15–17*)) that MP4 (i.e. fourth-order MBPT) is essentially the highest order available for routine use.

However, it is posssible to iteratively sum certain classes of excitations to infinite order, thus yielding a set of methods which are size extensive but are more robust than finite-order MBPT. Such methods are known as *coupled cluster* methods, and are equivalent to CI wave functions in which the coefficients of higher excitations are taken as products of those of lower excitations. Such a wave function can be formally written as

$$\psi = \exp(1 + \hat{T}_1 + \hat{T}_2 + \ldots)\psi_0 \qquad (2)$$

where the T_n operators are known as cluster operators and the coefficients $t_{ij\ldots ab\ldots}$ they contain are cluster amplitudes.

Coupled cluster theory has been reviewed in great detail in Refs. (*18–22*). The beauty of the coupled cluster wave function is that the expansion converges quite rapidly: even CCSD (coupled cluster with all single and double substitutions (*23*), i.e. $\psi = \exp(1 + \hat{T}_1 + \hat{T}_2)\psi_0$) already consistently recovers a very large percentage of the correlation energy. A method (*24*) known as CCSD(T), in which the effect of $\hat{T}_3$ is estimated using a perturbation theory expression but with the converged CCSD wave function substituted for the first-order wave function (which causes certain higher-order terms to be implicitly accounted for), has proven to be particularly successful and is now widely considered the electron correlation method of choice for accurate but affordable work.

The QCISD(T) (quadratic configuration interaction) method (*25*) that is implemented in the Gaussian 94 program (*26*) and widely employed by its users, is a simplification of CCSD(T) in which higher-order terms in $\hat{T}_1$ have been deleted (see pp.179–181 of (*19*) for discussion and further references) — despite having been derived on different grounds. In most situations its performance is comparable to that of CCSD(T), but with a fast coupled cluster code like those in ACES II (*27*), MOLPRO 96 (*28*), or TITAN (*29*), there is absolutely no speed advantage over CCSD(T) anymore.

One situation in which the performance of all single-reference correlation methods will suffer to a smaller or larger extent is strong nondynamical correlation, that is, when several other individual determinants beside ψ_0 are prominently figuring in the expansion (generally because of low-lying excited electronic states). Lee and Taylor (*30*) proposed a simple diagnostic known as $\mathcal{T}_1$, defined in the closed-shell case as

$$\mathcal{T}_1 = \sqrt{\frac{\sum_{ia} t_{ia}^2}{N}} \qquad (3)$$

where N is the number of electrons being correlated. (In the open-shell case, some double-counting needs to be avoided: see Ref. (*31*) for details.) Experience has taught (*21*) that, while CCSD(T) will produce acceptable results for $\mathcal{T}_1$ values as high as 0.055, MP2 results are essentially unusable for $\mathcal{T}_1$ values as low as 0.02. QCISD(T) breaks down for lower T_1 values than CCSD(T), because of the omission of the higher-order terms in $\hat{T}_1$.

For all of the molecules considered in the present paper, the T_1 diagnostic is in a regime where CCSD(T) is known to yield correlation energies close to full CI. This then essentially solves our n-particle calibration problem.

For certain radical or outlandish molecules with low-lying excited states, CCSD(T) will result in significant errors (although it might still be quite usable for assessing the basis set extension effects) and the only practical option may be multireference configuration-interaction (MRCI) (9), or the related (and approximately size-extensive) MRACPF (multireference averaged coupled pair functional (32)) method. These methods employ a CASSCF wavefunction as reference (i.e. CASSCF/CI), or the subset thereof with expansion coefficients greater than a particular threshold (e.g. MRCI(0.01) or MRCI(0.025)). Such calculations, when carried out with great care, will yield excellent results for even the most problematic diatomic molecules but the computational cost involved precludes routine application to even small polyatomics. Two more economical methods that are starting to be commonly used are IC-MRCI (internally contracted MRCI (33–35)) and CAS-PT2 (MP2 with a CASSCF reference (36)). Representative recent examples of benchmark calculations carried out using the IC-MRCI method are a series of papers by Dunning and coworkers (37–39).

Note that MRCI is still not size extensive. However, with an adequate reference space the extensivity error is sufficiently small that correction using a multireference analog (40) of the Davidson correction is generally adequate.

1-Particle Calibration: Quality of the Finite Basis Set. A comprehensive review of basis sets is beyond the scope of this chapter: perhaps the most recent and comprehensive one is that by Helgaker and Taylor (41). We will only mention a few salient points here.

For an atomic calculation at the SCF level, a basis set can be of 'minimal' quality and still recover essentially the exact SCF energy, as long as the individual functions closely mimic true Hartree-Fock orbitals. In a molecular calculation, flexibility is required — which requires splitting up the valence functions — as well as the ability to accommodate polarization of the atomic charge cloud in the molecular environment, which is done by adding higher angular momentum $(d, f, \ldots)$ basis functions (so-called *polarization functions*). Nevertheless, the basis set convergence of the SCF energy is fairly rapid compared to the correlation energy.

In a correlated calculation on an atom, the basis set must accommodate two important kinds of dynamical correlation effects. The first, *radial correlation* (or "in-out correlation"), involves the tendency of one electron to be near the nucleus when the other is near the periphery, or conversely. It is accommodated by permitting basis functions with extra *radial* nodes to mix into the wave function, i.e. by uncontracting s and p functions.

The second, *angular correlation* (or left-right correlation), involves the tendency of one electron to be on a different side of the atom as the other. This is accommodated by permitting basis functions with extra angular nodes to mix into the wave function, i.e. by adding $d, f, g, \ldots$ functions. The convergence of this effect in particular is quite slow.

The basis set in a correlated molecular calculation, then, will have to accommodate polarization, radial correlation, and angular correlation, with the additional complication for the latter two of a nonspherical environment.

We will now briefly review the two families of basis sets most commonly used in calibration studies.

Both of them begin by trying to find a wave function that gives the best compromise between compactness and description of the atomic correlation energy. Both are moreover based on general contractions, (42) i.e. all primitive Gaussians can contribute to all contracted functions.

In the atomic natural orbital (ANO) basis sets of Almlöf and Taylor (43), an atomic CISD calculation is carried out in a very large primitive basis set, and the natural orbitals and their corresponding occupation numbers obtained by diagonalization of the first-order reduced density matrix. The natural orbitals with the highest occupation numbers are then selected as basis functions. It was found that these always occur in groups of almost equal occupation numbers: e.g., the first f, second d, and third p function have similar natural orbital occupations. This systematically leads, for first-row elements, to contractions like $[4s3p2d1f]$, $[5s4p3d2f1g]$, $[6s5p4d3f2g1h]$, and so forth. (Corresponding contractions for second-row elements are $5s4p2d1f]$, $[6s5p3d2f1f]$, $7s6p4d3f2g1h]$, and the like.)

In the second family, the correlation consistent (cc) basis sets of Dunning (44) and coworkers, relatively compact atomic basis sets are subjected to energy optimization, and the energy gain from adding different kinds of primitives is considered. It is then found that these energy contributions likewise occur in groups: thus, the energy gain from adding the first f, second d, or third p function is similar. Again this suggests adding them in shells, which again leads to the same typical contraction patterns as for their ANO counterparts. Based on the number of different functions available for the valence orbitals, these basis sets are known as cc-pVnZ (correlation consistent valence n-tuple zeta), where n=D for double (a [3s2p1d] contraction), T for triple (a [4s3p2d1f] contraction), Q for quadruple (a [5s4p3d2f1g] contraction), and 5 for quintuple zeta (a [6s5p4d3f2g1h] contraction).

Martin (45) carried out a detailed comparison of computed TAE values with equivalent ANO and cc basis sets. The results were found to be nearly identical, while the integral evaluation time for the cc basis sets was considerably shorter due to their more compact primitive size. Therefore cc basis sets are more commonly used, although for certain other applications (like weak molecular interaction or electrical properties) ANO basis sets and particularly the "averaged ANO" variant (46) may be preferable.

For the computation of electron affinities and calculations on anions in general, special low exponent s and p functions (so-called 'soft' or 'diffuse' functions) are required at the SCF level (e.g. (47)). At correlated levels, the regular s and p functions are adequate as radial correlation functions thereto, but angular correlation in the tail range requires the addition of 'soft' d, f, ... functions. Kendall et al. (48) proposed the aug-cc-pVnZ basis sets, in which the cc-pVnZ basis set is 'augmented' with one 'soft' (or low-exponent) basis set of each angular momentum. It was subsequently found (e.g. (49, 50)) that these functions are indispensable for calculating properties such as geometries and harmonic frequencies of highly polar *neutral* molecules as well. It is also noteworthy (*49*) that including just the soft s and p functions only recovers about half the effect.

Del Bene (*51*) noted that, except in such compounds as LiH in which hydrogen has a significant negative partial charge, omission of the diffuse functions on H generally does not affect results. This practice is denoted by the acronym aug'-cc-pVnZ.

Some authors (e.g. (*52*)) have obtained excellent results for the first-row hydrides using basis sets of only *spdf* quality, combined with *sp* bond functions (i.e. basis functions centered around the bond midpoint). However, as the extension of bond function basis sets to multiple bonds will require d bond functions, which in turn will require (*53*) atom basis sets of up to *spdfg* quality to keep to keep down basis set superposition error to an acceptable level, the usefulness of bond functions for our purpose appears to be somewhat limited.

1-Particle Calibration versus n-Particle Treatment. In many situations, it will simply not be feasible to directly carry out calculations with the desired (1-particle) basis set using the desired n-particle treatment (correlation method). A common workaround is to carry out separate 1-particle and n-particle calibrations. That is, one selects a basis set A in which the more accurate correlation method M is still feasible, as well as a more economical correlation treatment M′ which permits a larger basis set B to be used. One then carries out the following calculations: M′/A, M/A, and M′/B, and assumes that the following additivity approximation holds:

$$E(M/B) \approx E(M'/A) + [E(M/A) - E(M'/A)] + [E(M'/B) - E(M'/A)]$$
$$= E(M'/B) + E(M/A) - E(M'/A) \qquad (4)$$

The first step is then termed the n-particle calibration, and the second step the 1-particle calibration.

The underlying assumption, of course, is that the effects of improving the electron correlation method and of enlarging the 1-particle basis set are only weakly coupled. How important 1-particle/n-particle space coupling will actually be will depend on the quality of both M′ and A, as well as on the system under study.

For example, in a system where SCF provides a relatively poor zero-order wave function, any method M′ based on low-order single-reference perturbation theory will probably yield dubious results at best. Conversely, if your system needs a particular addition to the basis set for even semiquantitative results (e.g., diffuse functions for an anion), using a reference basis set A that does not contain such functions will likely result in erratic 1-particle calibration behavior.

In some cases, the process can be taken one level deeper, involving an extra even lower-level correlation method M″ for assessing the effect of extension from the large basis set B to an even larger basis set C. G2 theory (see the contribution on the subject elsewhere in this volume) is a good example: here M, M′, and M″ are the QCISD(T), MP4, and MP2 methods, respectively.

Also, if the difference between basis sets A and B comprises two or more sets of basis functions that do not overlap appreciably, and cover quite different effects and/or regions of the wave function (e.g. diffuse functions and core correlation functions), additivity approximations may be invoked. E.g. if B is the set union of B_1 and B_2, both of which are supersets of A, then

$$E(M/B) \approx E(M/B_1) + E(M/B_2) - E(M/A) \tag{5}$$

In general, for all but the smallest systems, even the largest basis set one can afford in the 1-particle calibration step will not be sufficient to reach the 1-particle basis set limit. The available alternatives will then be either extrapolation to the basis set limit, empirical correction, or a hybrid scheme. All of these will be discussed in detail in the relevant Sections.

Another alternative would be to invoke a thermochemical cycle involving a calculated reaction energy for which convergence behavior is more rapid than for the total atomization energy. As an example (*54*), the convergence behavior of the total atomization energy of HCN, and of the ΔE_e of the reaction $2HCN - C_2H_2 + N_2$ is compared in Table I.

Table I: Comparison of the Convergence Behavior of ΣD_e(HCN) and ΔE_e for the Reaction Relating it to ΣD_e(C_2H_2) and D_e(N_2). All Values Are in kcal/mol.

	$N_2 \rightarrow 2 N$	$2 HCN \rightarrow C_2H_2 + N_2$
CCSD(T)/cc-pVTZ	301.36	-7.00
CCSD(T)/cc-pVQZ	307.67	-6.97
CCSD(T)/cc-pV5Z	309.67	-7.10
CCSD(T)/aug′-cc-pVTZ	302.67	-6.90
CCSD(T)/aug′-cc-pVQZ	308.30	-7.15
CCSD(T)/aug′-cc-pV5Z	309.94	-7.21
estimated limit	311.60	-7.27
core correlation	+1.67	+0.05

It is seen there that a thermochemically useful value for the reaction — which conserves the number of bonds of each bond order and the hybridization of each atom even though it is not, strictly speaking, an isodesmic reaction — can be obtained with basis sets as 'small' as cc-pVTZ, which fall some 10 kcal/mol short of the valence correlation basis set limit for the directly computed property. The computationally intensive core correlation contribution (*vide infra*) essentially cancels out for the isodesmic reaction and can hence be neglected.

Extensive applications of the accelerated convergence properties of isodesmic and isogyric reactions can be found in the work of Lee and coworkers (e.g., (*55,56*) and references therein). The main limitation to their usage, of course, is the availability of accurate experimental or *ab initio* data for the other reactions in the thermochemical cycle.

Secondary Issues

Quality of the Zero-Point Energy. Here one has to contend both with a difference between the computed and experimental harmonic frequencies, and with the difference between the harmonic and anharmonic zero-point energy. As pointed out e.g. by Grev et al. (*58*) the harmonic zero-point energy is generally too high, while one-half the sum of the fundamentals is generally too low. They suggest tackling the issue by introducing a scaling factor that is somewhat closer to unity than those generally used (e.g. Ref. (*66*)) to generate approximate fundamentals from computed harmonics. This issue is discussed in great detail by Scott and Radom (*67*) (see also Ref. (*68*)), who propose different scale factors for obtaining estimated fundamentals as well as zero-point energies for a variety of density functional and conventional ab initio methods.

How important can these effects be? For our sample of 14 reference molecules (Table II), accurate zero-point energies are available from experimental or high-level ab initio anharmonic force fields. We find that the procedure used to estimate the zero-point energy in G2 theory (HF/6-31G* scaled by 0.8929) results in mean and maximum absolute errors for our sample of 0.26 and 0.72 kcal/mol, respectively. Such errors (1 and 3 kJ/mol) are already significant for accurate work, and as the size of the molecules under study grows they will only become more important.

For work of the highest accuracy, and if no full anharmonic force field is available experimentally, the only option may be to compute an accurate anharmonic force field ab initio (e.g. (*69–72*)).

Quality of the Reference Geometry. The error from using a reference geometry that is not optimum for the level of theory at which the energy calculation is carried out can of course be roughly quantified. For example, the experimental (*73*) NN stretching force constant in the NNO molecule is 18.251 aJ/Å^2, corresponding to an r(NN) coefficient in the energy expansion of about 2 hartree/Å^2. Consequently, an error of about 0.01 Å in the NN bond distance will result in an error on the order of 0.1 millihartree, or 0.06 kcal/mol, in the total energy. If all bond lengths in a molecule are overestimated by similar amounts, the errors may become significant, while in cases where the geometry is qualitatively incorrect (e.g. the MP2/6-31G* geometry of NNO), errors on the order of 1 kcal/mol or more may result. For example, the CCSD(T)/cc-pVTZ TAE of NNO is computed to be 252.00 kcal/mol at the MP2/6-31G* geometry, compared to 253.44 kcal/mol at the CCSD(T)/cc-pVDZ geometry and 253.76 kcal/mol at the optimum geometry.

Morokuma and coworkers (*74*) and Bauschlicher and Partridge (*77*) both suggest modifications of G2 theory using reference geometries computed at the B3LYP (*75,76*) level, which appear to improve overall performance. It was found before (e.g. (*78*)) that B3LYP geometries are quite close to experiment, particularly with a basis set of *spdf* quality (mean signed error 0.003 Å in bond distances with the cc-pVTZ basis set).

As noted by Martin (*79*), the parametrization in empirical correction methods should largely absorb any residual error resulting from approximate reference geometries of reasonable quality. For example, while the mean absolute error for Martin-corrected CCSD(T)/cc-pVTZ energies is 1.08 kcal/mol using MP2/6-31G* reference geometries and this drops to 0.95 kcal/mol using CCSD(T)/cc-pVDZ geometries, the latter value is virtually identical to that using optimum (i.e., CCSD(T)/cc-pVTZ) geometries (0.93 kcal/mol).

The bottom line appears to be that, as long as the reference geometry used is of reasonable quality, one should not have any problems. Given how cost-effective B3LYP

Table II: Experimental ΣD_0 Values, Zero Point Energies, ΣD_e Values, and Computed Core Correlation Contributions. All Units Are kcal/mol.

	$\Sigma D_0{}^a$	ZPE^b	ΣD_e	w/o spin-orbit^c	core corr.^{d}
C_2H_2	$388.90(24)^e$	16.46	405.36(24)	405.53(24)	2.44
CH_4	392.51(14)	27.6^f	420.15(14)	420.23(14)	1.25
CO	256.16(12)	3.11	259.27(12)	259.58(12)	0.96
CO_2	381.91(6)	7.24	389.15(6)	389.68(6)	1.78
H_2	103.27(0)	6.21	109.48(0)	109.48(0)	0.00
H_2O	219.35(2)	13.25	232.60(2)	232.83(2)	0.38
HCN	$303.10(25)^g$	10.09^h	313.19(25)1	313.27(25)	1.67
HF	$135.33(17)^i$	5.85	141.18(17)	141.57(17)	0.18
NH_3	276.73(10)	21.33^j	298.06(10)	298.06(10)	0.66
N_2	225.06(3)	3.36	228.42(3)	228.42(3)	0.85
H_2CO	$357.25(16)^k$	16.53^l	373.78(16)	374.09(16)	1.32
F_2	36.94(10)	1.30	38.24(10)	39.01(10)	-0.07
HNO	$196.85(6)^m$	8.56^n	205.41(6)	205.64(6)	0.48
N_2O	263.61(10)	6.77	270.38(10)	270.60(10)	1.26

^{a} All data taken from the JANAF tables (*57*) unless indicated otherwise
^{b} Taken from the compilation in Ref. (*58*) unless indicated otherwise
^{c} All relevant spin-orbit energy levels taken from Ref. (*57*)
^{d} taken from Ref. (*59*), except for HNO, Ref. (*60*)
^{e} The thermodynamic functions of C_2H_2 in Ref. (*57*) are in error for all temperatures other than 298.15 K; see Ervin, K. M.; S. Gronert, Barlow, S. E.; Gilles, M. K.; Harrison, A. G.; Bierbaum, V. M.; DePuy, C. H.; Lineberger, W. C.; Ellison, G. B. *J. Am. Chem. Soc.* **1990**, *112*, 5750, Table 1, footnote e. Present data taken from Wagman, D. D.; Evans, W. H.; Parker, V. B.; Schumm, R. H.; Halow, I.; Bailey, S. M.; Churney, K. L.; Nuttall, R. L.; *J. Phys. Chem. Ref. Data* **1982**, *11*, Supplement 2.
^{f} from accurate ab initio quartic force field in Lee, T. J.; Martin, J. M. L.; Taylor, P. R.; *J. Chem. Phys.* **1995**, *102*, 254.
^{g} Ref. (*54*). The commonly cited experimental value from the JANAF tables (*57*) is based on experiments from the previous century (*61,62*).
^{h} variational anharmonic zero point energy from CCSD(T)/[4s3p2d1f/4s2p1d] force field in Ref. (*63*) corrected for difference between computed and experimental (*64*) harmonic frequencies.
^{i} Huber, K. P.; Herzberg, G. *Constants of diatomic molecules*; Van Nostrand Reinhold: New York, 1979.
^{j} Martin, J. M. L.; Lee, T. J.; Taylor, P. R. *J. Chem. Phys.* **1992**, *97*, 8361.
^{k} Baulch, D. L.; Cox, R. A.; Crutzen, P. J.; Hampson, R. F., Jr.; Kerr, J. A.; Troe, J.; Watson, R. T. *J. Phys. Chem. Ref. Data* **1982**, *11*, 327.
^{l} Martin, J. M. L.; Lee, T. J.; Taylor, P. R.; *J. Mol. Spectrosc.* **1993**, *160*, 105.
^{m} Ref. (*65*).
^{n} best value from Dateo, C. E.; Lee, T. J.; Schwenke, D. W.; *J. Chem. Phys.* **1994**, *101*, 5853.

calculations even with fairly large basis sets are these days, B3LYP/cc-pVDZ or (for accurate work) B3LYP/cc-pVTZ geometry optimizations appear to be especially attractive.

Core Correlation. Conventional wisdom would have it that core correlation effects will not be important for first-row compounds. Recent experience, however, has shown this to be a half-truth at best.

Explicit consideration of core correlation requires the use of special basis sets that accommodate inner-shell correlation effects by the addition of extra radial nodes in the s functions (in practice, uncontracting the innermost s function a little will do the job) as well as extra 'hard' (high-exponent) p and d functions. Basis sets designed explicitly for assessing core correlation effects have been proposed by Bauschlicher and Partridge (*80, 81*), by Martin and Taylor (*49*), and by Woon and Dunning (*82*). Of those, both the Martin-Taylor basis set and the Woon-Dunning cc-pCVQZ basis set will recover essentially the whole effect at reasonable computational cost. It cannot be stressed enough that including core correlation in basis sets not designed to handle core correlation (such as the regular cc-pVnZ basis sets or the Pople basis sets (*83*), which are only minimal in the core) will generally yield erratic core correlation contributions, and therefore is simply a waste of CPU time. (All electronic structure programs presently allow for correlated calculations with frozen core electrons.)

Case studies of core correlation effects on binding energies were made, inter alia, by Grev and Schaefer (*84*) for CH_n and SiH_n, by Bauschlicher and Partridge (*85*) for C_2H_2 and CH_4, and by a succession of groups for N_2 (see (*81*) and references therein). A systematic study was carried out by Martin (*59*) at the CCSD(T) level with the Martin-Taylor basis set. (Results for molecules studied previously (*81, 85*) were in essentially complete agreement with those papers.)

Core correlation contributions as high as 2.44 kcal/mol (for C_2H_2) and 1.78 kcal/mol (for CO_2) were found. Predictably, larger contributions were seen for multiple than for single bonds. While some empirical correction schemes might be able to absorb most of this effect in their parametrization, any extrapolation method would obviously have to account for the core correlation contribution.

Since however the basis set additions required to deal with core correlation are quite different from those required to recover more of the valence correlation energy, the use of an additivity approximation appears warranted.

Finally, it is worth mentioning that the core correlation contributions to TAEs are relatively sensitive towards the electron correlation method (as found by Bauschlicher and Partridge (*81*) for N_2) and that including core correlation at, say, the MP2 level may not lead to very accurate results.

Thermal Contributions. Except for floppy molecules, thermal contributions at room temperature can be quite accurately evaluated using the familiar rigid rotor-harmonic oscillator (RRHO) approach. If data at high temperatures are required, this approach is no longer sufficient, and an anharmonic force field and analysis, combined with a procedure for obtaining the rotation-vibration partition function therefrom, are required. Two practical procedures have been proposed. The first one, due to Martin and coworkers (*86, 87*) is based on asymptotic expansions for the nonrigid rotor partition function inside an explicit loop over vibration. It yields excellent results in the medium temperature range but suffers from vibrational level series collapse above 2000 K or more. A representative application (to FNO and ClNO) is found in Ref. (*50*).

The second method, due to Topper and coworkers (*88*), is based on Feynman path integrals, and works best in the high temperature limit. Therefore the two methods are complementary.

Basis Set Superposition Error. When one carries out a calculation on the AB diatomic using a basis set for A and B that is incomplete (as all finite basis sets by definition are), the atomic energy of A in AB, and of B within AB, will be slightly overestimated (in absolute value) due to the fact that the basis functions on the other atom have become available. (It is easily verified that basis functions on B can be expanded as a series of higher angular momentum functions around A.) This phenomenon is known as *basis set superposition error* (BSSE). The standard estimate is using the Boys-Bernardi (*89*) counterpoise method:

$$\text{BSSE} \approx E[A(B)] + E[B(A)] - E[A] - E[B] \tag{6}$$

where $E[A(B)]$ represents the energy of A with the basis set of B present on a 'ghost atom', and conversely for $E[B(A)]$.

While the counterpoise correction is commonly used as a correction term for interaction energies in weak molecular complexes, virtually no authors apply it to the calculation of total atomization energies, for the simple reason that it invariably produces *worse* results. In addition, the extension of the counterpoise correction to systems with more than two fragments is not uniquely defined (*90–92*).

The anomaly that neglecting BSSE would yield better results is only an apparent one: after all, BSSE is a measure of basis set incompleteness — which is precisely what we are trying to get rid of — but the correction has the opposite sign. For sufficiently large basis sets (say, of $spdf g$ quality), the NASA Ames group actually found that 150% of the counterpoise BSSE is a fair estimate of the remaining basis set incompleteness (*93*). However, given the complications for systems larger than diatomics, the present author prefers the use of extrapolation to the infinite-basis limit above such methods. (It goes without saying that the BSSE goes to zero at the infinite-basis set limit. Therefore, a sufficiently reliable extrapolation to the infinite-basis set limits effectively obviates the issue.)

Quality of Experimental Reference Values

For this purpose, it is very important to have a set of very accurate reference values available. Most values in the "G2 set" (*94*) are of chemical accuracy (error bars on the order of 1 kcal/mol), which is appropriate if that is the target accuracy of one's method but is not very useful for the present purpose — assessing calibration methods with a target accuracy of 0.25 kcal/mol or less — since many of the values therein have larger experimental error bars than our target calculated accuracy.

We have selected the 14 molecules given in Table II, all of which are closed-shell species for which the experimental ΣD_0 is known with an accuracy on the order of 0.1

kcal/mol. This set includes two very recent values, namely for HNO (*65, 95*) and for HCN (*54*), the latter substantially revising the JANAF value (*57*) which is based on experiments from the previous century (*61, 62*).

In cases where no experimental anharmonic zero point energy is available, literature values from accurate ab initio quartic force fields have been taken (see Table II for detailed references).

In addition, it should be kept in mind that the present calculations are all nonrelativistic. For the total atomization energies of first-row molecules, the main consequence is that the spin-orbit components of $B(^2P)$, $C(^3P)$, $O(^3P)$, and $F(^2P)$ are all degenerate, which of course is not the case in Nature. In order not to "compare apples with oranges", therefore, the effect of spin-orbit splitting in the atoms has been removed from the experimental TAEs. For example, for every oxygen atom present, TAE should be increased by $[E(^3P_0) + 3E(^3P_1) + 5E(^3P_2)]/9 - E(^3P_0)$ (see e.g. (*96*)). The largest two effects of atomic spin-orbit splitting in the present work are 0.8 kcal/mol (for F_2) and 0.6 kcal/mol (for CO_2), clearly on the order of magnitude of the accuracy we are trying to achieve.

Empirical Correction Methods

G2 Theory. In G2 theory (*94*), an approximate QCISD(T)/6-311+G(3df,2p) energy is derived using a basis set additivity scheme. Further basis set incompleteness is then corrected for using an empirical "high level correction" (HLC)

$$\Delta E(\text{HLC}) = c_1 n_\alpha + c_2 n_\beta = (c_1 + c_2)n_{\text{pair}} + c_1(n_\alpha - n_{\text{pair}}) \qquad (7)$$

where n_α and n_β stand for the numbers of spin-up and spin-down valence electrons, respectively, and it is assumed by convention that $n_\alpha > n_\beta$.

In the original version, G1 theory (*97*), c_1 and c_2 were determined by insisting on exact energies for $H(^2S)$ and H_2. In G2 theory, the parameters are chosen to minimize the mean absolute error for the thermochemical properties of a fairly large set of reference compounds (the "G2 set").

It should be noted that many of the values in that reference set actually have error bars equal to or greater than the target accuracy. For our reference sample (Table III), the mean absolute error is 1.50 kcal/mol for the ΣD_0 values, and 1.32 kcal/mol for the ΣD_e values. (The latter is not qualitatively different from the 1.21 kcal/mol for the G2 set.) Note that the difference suggests that the handling of zero-point energy in G2 theory (HF/6-31G* harmonic frequencies scaled by 0.8929) could bear some improvement.

Several modifications and improvements of G2 theory have been suggested — for a detailed discussion, see the relevant chapter of the present volume. Some are aimed at reducing computational requirements (at the expense of some accuracy), others at extension to third-row or heavier elements, yet others at improving the accuracy. The latter are the ones relevant for the purpose of the present chapter.

While Curtiss et al. (*98, 99*) came to the conclusion that such improvements as eliminating the basis set additivity approximations, using better zero point energies, using CCSD(T) rather than the approximation QCISD(T) thereto, all yielded negligible overall improvements, these authors (like Durant and Rohlfing (*100*)) did come to

Table III: Performance of G2 Theory and CBS Schemes for our Reference Set. All Values in kcal/mol

	Expt. value	Errors			
		G2	G2M	CBS-Q	CBS-QCI/APNO
C_2H_2	405.53(24)	0.06	-2.09	-1.88	-1.07
CH_4	402.23(14)	2.37	0.64	-0.55	0.52
CO	259.58(12)	1.84	0.99	0.44	-0.15
CO_2	389.68(6)	3.12	-0.22	1.65	0.17
H_2	109.48(0)	1.14	1.44	1.07	0.38
H_2O	232.83(2)	0.57	-0.06	-0.06	-0.15
HF	141.57(17)	0.58	-0.49	0.28	-0.29
NH_3	298.06(10)	0.96	0.55	-0.96	-0.29
N_2	228.42(3)	-0.77	-0.02	-1.74	-0.98
H_2CO	374.09(16)	3.33	1.13	0.84	0.47
F_2	39.01(10)	-0.54	-1.40	0.05	-0.33
HNO	205.64(6)	2.17	1.55	0.69	-0.06
N_2O	270.60(10)	0.32	-0.88	-0.10	0.56
HCN	313.27(25)	0.75	-0.07	-1.15	-0.89
mean abs. error		1.32	0.82	0.82	0.45
max. abs. error		3.33	2.09	1.88	1.07

the conclusion that using higher-level reference geometries than MP2/6-31G* does improve results for radicals. Bauschlicher and Partridge (*77*) and Morokuma and coworkers (*74*) independently proposed G2-modified (G2M) theories based on B3LYP reference geometries and zero-point energies, which also seem to represent small improvements. The scheme denoted G2M(RCC) by Morokuma and coworkers — which employs B3LYP reference geometries and harmonic frequencies, CCSD(T) instead of QCISD(T) for the n-particle calibration step, and spin-projected MP4 for the one-particle calibration step — was found to represent a significant improvement (*74*): for our reference sample, it yields a mean absolute error of 0.82 kcal/mol. Given the fairly small extra computational expense over regular G2 theory, it is probably to be recommended over the latter.

Martin's 3-Parameter Correction (3PC). In a detailed convergence study using both ANO and correlation consistent basis sets, Martin (*45*) found that bonds of different bond order display clearly different convergence characteristics. Specifically, while σ bonds appear to approach convergence for $spdfg$ basis sets, π bonds appear to be still some distance removed from this. This suggests expanding the pair correction formula used in G2 theory to

$$\Delta E = a_\sigma n_\sigma + b_\pi n_\pi + c_{\text{pair}} n_{\text{pair}} \tag{8}$$

where n_σ and n_π stand for the number of σ and π bonds, respectively, $n_{\text{pair}} = n_\beta$ as before, and the coefficients $a_\sigma, b_\pi, c_{\text{pair}}$ are specific to the basis set, electron correlation method, and level of theory for the reference geometry. The most up-to-date collection of parameters is given in Ref. (*101*), where the parameters were determined by fitting to our set of 14 reference values. These parameters are summarized for easy reference in Table IV. Values obtained at the CCSD(T)/BASIS level with the 3-parameter correction will be denoted here as 3PC/BASIS.

Table IV: Summary of Error Statistics and Parameters for the 3-Parameter Correction (3PC). All Values in kcal/mol

	a_σ	b_π	c_{pair}	mean abs err.	max. abs. err.
		Core correlation absorbed in parameters			
cc-pVDZ	0.620	0.946	9.083	2.64	5.42
cc-pVTZ	-0.838	0.937	4.026	0.99	2.09
cc-pVQZ	-0.251	0.756	1.558	0.45	0.98
cc-pV5Z	0.121	0.860	0.547	0.31	0.78
aug'-cc-pVDZ	2.426	5.155	5.101	1.50	3.78
aug'-cc-pVTZ	-0.134	1.871	2.528	0.57	1.40
aug'-cc-pVQZ	0.177	1.155	0.777	0.32	0.78
aug'-cc-pV5Z	0.307	0.958	0.246	0.28	0.66
		Core correlation included explicitly			
cc-pVDZ	-0.094	0.074	9.612	2.56	5.42
cc-pVTZ	-1.551	0.065	4.555	0.91	1.91
cc-pVQZ	-0.965	-0.116	2.087	0.38	1.24
cc-pV5Z	-0.593	-0.013	1.076	0.24	1.03
aug'-cc-pVDZ	1.713	4.282	5.630	1.42	3.31
aug'-cc-pVTZ	-0.848	0.999	3.057	0.49	1.23
aug'-cc-pVQZ	-0.537	0.283	1.306	0.24	0.63
aug'-cc-pV5Z	-0.407	0.086	0.775	0.20	0.64

While 3PC/cc-pVTZ yields results (Table IV) only slightly better than G2 quality (mean absolute error 0.99 kcal/mol), this improves to 0.45 kcal/mol for 3PC/cc-pVQZ, and for the (admittedly very costly) 3PC/cc-pV5Z level to 0.31 kcal/mol. Using augmented basis sets results in a notable improvement: even 3PC/aug'-cc-pVDZ values, at 1.50 kcal/mol mean absolute error, could be quite useful for larger systems. For 3PC/aug'-cc-pVnZ (n=T,Q,5), respectively, these numbers improve to 0.57, 0.32, and 0.28 kcal/mol. It should be noted that 3PC/aug'-cc-pVnZ seem to be almost of the same quality as 3PC/cc-pV$(n+1)$Z, and particularly 3PC/aug'-cc-pVQZ stands out.

In addition, is was previously found (72) that proton affinities are seriously in error with 3PC/cc-pVQZ, while no such problem exists with 3PC/aug'-cc-pVQZ.

In all these schemes, core correlation is being absorbed into the parameters. While this appears to work very satisfactorily, treating core correlation explicitly turns out to be worthwhile for the larger basis sets. Thus, 3PC+core/cc-pVQZ and 3PC+core/cc-pV5Z result in mean absolute errors of only 0.37 and 0.23 kcal/mol, respectively, while these drop further to 0.23 and 0.20 kcal/mol, respectively, for 3PC+core/aug'-cc-pVQZ and 3PC+core/aug'-cc-pV5Z. The very small improvement from aug'-cc-pVQZ to aug'-cc-pV5Z, which corresponds to a vast increase in computational expense, clearly suggests that the point of diminishing returns has been reached. The 3PC+core/aug'-cc-pVQZ scheme appears to be especially attractive, yielding a mean absolute error matching our 1 kJ/mol target at high, but for small systems still easily manageable, computational expense.

Hybrid Schemes

The main representative of this category is a family of methods known as CBS (complete basis set [extrapolation]) models and developed by Petersson and coworkers (*102*). These are intricate additivity schemes, whose main distinguishing feature is the inclusion of a procedure to extrapolate pair correlation energies to the infinite-basis set limit. Since the numerical factor involved in the extrapolation is determined empirically, the method is classified as a hybrid method here. For a detailed discussion, see the chapter by Petersson in the present volume.

The main schemes of concern in the family are (a) the one that is labeled CBS-Q in the Gaussian 94 package, and is a variant of the scheme denoted CBS2/[6s6p3d2f,4s2p1d]//QCI/6-311G** in Ref. (*103*), and (b) the one labeled CBS-APNO in the Gaussian 94 package, and CBS-QCI/APNO in Ref. (*102*).

For our set of reference molecules, CBS-Q yields a mean absolute error on the ΣD_e values of 0.82 kcal/mol, which is on a par with the 0.82 kcal/mol of the G2M scheme and clearly better than the 0.99 kcal/mol of the 3PC/cc-pVTZ scheme. (The results can be inspected and compared in Table III.) 3PC/aug'-cc-pVTZ, however, which uses a basis set of similar size as the largest one occuring in the CBS-Q scheme, decidedly outperforms it with a mean absolute error of 0.57 kcal/mol. The comparatively low computational cost and ease of use of CBS-Q make it an attractive proposition for routine work, and it is surprising that it has not been applied more extensively.

The CBS-QCI/APNO scheme involves QCISD(T)/6-311++G(2df,p) calculations, which are fairly time-consuming with Gaussian 94 but (if the system has any symmetry at all) can be carried out quite rapidly using a fast coupled cluster program such as that in MOLPRO 96. It also approximately includes core correlation. The results are quite respectable: the mean absolute error for our reference set is 0.45 kcal/mol, on a par with 3PC/cc-pVQZ (0.45 kcal/mol) or 3PC+core/cc-pVQZ (0.48 kcal/mol). For those who have fast coupled cluster codes available, however, and require that kind of accuracy, 3PC/aug'-cc-pVQZ (0.32 kcal/mol) or even 3PC+core/aug'-cc-pVQZ (0.23 kcal/mol) might be more attractive.

Extrapolation Methods

Feller's Geometric Extrapolation. Dunning (*44*), in his seminal work on correlation consistent basis sets, made a number of important observations, including that the energy gain from adding extra functions of a given angular momentum, as well as that from adding the first function of the next higher angular momentum, roughly followed a geometric series. In addition he observed (parallelling a similar observation about the atomic natural orbital occupations by Almlöf and Taylor (*43*)) that adding the first function of angular momentum l, the second of $l - 1$, the third of $l - 2$, ... all have similar energy contributions, which led to his suggestion of the cc-pVnZ basis sets in the first place.

Feller (*104*) then observed that total energies for molecules calculated with successive cc-pVnZ basis set themselves roughly followed geometric series. The use of an expression of the form

$$E(n) = E_\infty + A\exp(-Bn) \tag{9}$$

for extrapolation to the infinite-basis limit then follows naturally.

Such a procedure, evidently, will only recover the valence correlation energy, such that a core correlation contribution will still have to be added in. Lee and coworkers (e.g. (95, 105) and references therein) have carried out a number of studies in which exactly this was done.

The great attraction of Feller's formula, besides its simplicity, is that no empiricism of any kind is involved. It has been used extensively by Dunning and coworkers (e.g. (37, 106)). For applications to other properties than energies, which may not converge monotonically, Martin and Taylor (70) proposed the following generalization of the Feller formula:

$$E(\mathring{n}) = E_\infty + AB^{-n} \tag{10}$$

For Eq.(10) to be defined, at least three points are required. We now introduce the notation Feller(lmn) for the result of Eq.(10) from CCSD(T)/cc-pVlZ, cc-pVmZ, and cc-pVnZ calculations, and Feller(lmn)+core for the same plus an additive correction for core correlation. The use of aug'-cc-pVnZ instead of regular cc-pVnZ basis sets can then be denoted using the aug'-Feller(lmn) and aug'-Feller(lmn)+core notations.

For our sample, the results are somewhat disappointing: Feller(DTQ)+core yields a mean absolute error of 0.72 kcal/mol, while Feller(TQ5)+core yields essentially no improvement at 0.70 kcal/mol. aug'-Feller(DTQ)+core and aug'-Feller(TQ5)+core yield even more lackluster results — 0.66 and 0.73 kcal/mol, respectively — given that the latter involves a [$7s6p5d4f3g2h/5s4p3d2f1g$] basis set which comes to no fewer than 127 basis functions per first-row atom! Evidently alternatives need to be sought.

Extrapolation Based on Schwartz's Expression for Limiting Convergence Behavior. Following the pioneering contribution (107) of Schwartz, it has been shown (108–110) that the contributions to the correlation energy of a two-electron atom converge in terms of the angular momentum l as

$$\Delta E(l) = A/(l+1/2)^4 + B/(l+1/2)^5 + O(l^{-6}) \tag{11}$$

If so, the error for a calculation in a basis set truncated at angular momentum L is given by

$$E_\infty - E(L) = \sum_{l=L+1}^{\infty}\left[\frac{A}{(l+1/2)^4} + \frac{B}{(l+1/2)^5} + \dots\right] \tag{12}$$

$$= \frac{A\psi^{(3)}(L+3/2)}{6} + \frac{B\psi^{(4)}(L+3/2)}{24} + \dots \tag{13}$$

where $\psi^{(n)}(x)$ represents the polygamma function (111) of order n. Its asymptotic expansion has the leading terms

$$\psi^{(n)}(x) = (-1)^{n-1}[\frac{(n-1)!}{x^n} + \frac{n!}{2x^{n+1}} + O(x^{-n-2})]$$

$$= (-1)^{n-1}\frac{(n-1)!}{(x-1/2)^n} + O(x^{-n-2}) \tag{14}$$

Hence

$$E(L) = E_\infty - \frac{A(L+1)^{-3}}{3} + \frac{B(L+1)^{-4}}{4} + O(L^{-5}) \qquad (15)$$

If we have a sequence of 'correlation consistent' basis sets with increasing maximum angular momentum l, it would be natural to at least attempt the use of an expression based on this behavior as an extrapolation formula. (One might remark that ideally the basis set should be saturated at every angular momentum present: however, the next best solution, which is to use basis sets which are balanced in their radial and angular quality, is computationally much more efficient. Both ANO and cc basis sets were designed with this property in mind.) The most obvious one would be $A + B/(L+1)^{-3}$, but as pointed out by the present author (*60*), this term is not the dominant one (yet) for practical values of L. Moreover, since hydrogen basis sets commonly are truncated one angular momentum below the corresponding basis sets for the elements B–F, some compromise between L and $L + 1$ is in order. After some numerical experimentation the following three-point extrapolation formulas were suggested (*60*):

$$E(l) = E_\infty + A/(l+1/2)^4 + B/(l+1/2)^6 \qquad (16)$$
$$E(l) = E_\infty + A/(l+1/2)^\alpha \qquad (17)$$

The results for E_∞ obtained from these extrapolations are denoted Schwartz6(kmn) and Schwartzα(kmn), respectively, when derived from cc-pVkZ, cc-pVmZ, and cc-pVnZ energies. In addition, Schwartz4(mn) denotes a 2-point extrapolation using

$$E(l) = E_\infty + A/(l+1/2)^4 \qquad (18)$$

The prefix "aug'-" and the suffix "+core" can be added to this notation to denote the use of augmented basis sets and the inclusion of core correlation by additivity approximation, respectively.

Table V: Summary of Errors (kcal/mol) in Extrapolated CCSD(T) Values after Correction for Core Correlation

	cc-pVnZ		aug'-cc-pVnZ	
	mean	maximum	mean	maximum
	absolute error		absolute error	
Feller(DTQ)	0.72	1.86	0.66	1.50
Feller(TQ5)	0.70	1.87	0.73	1.89
Schwartz4(TQ)	0.46	1.27	0.35	0.69
Schwartzα(TQ5)	0.32	0.72	0.36	1.18
with triple bond correction	0.22	0.64	0.23	0.78
Schwartz4(Q5)	0.37	0.90	0.31	0.90
with triple bond correction	0.26	0.83	0.22	0.69
Schwartz6(TQ5)	0.35	0.81	0.33	0.94
with triple bond correction	0.24	0.67	0.22	0.68

It was found (*60*) that Schwartz4(Q5), Schwartz6(TQ5), and Schwartzα(TQ5) all yielded very similar results (Table V), all of which were a dramatic improvement over the Feller geometric extrapolation. The mean absolute error for Schwartzα(TQ5)+core over our reference sample was found to be only 0.32 kcal/mol: interestingly, the mean absolute error for the aug$'$-Schwartzα(TQ5)+core results is slightly less good (0.35 kcal/mol). However, in the latter all the major errors remaining are for triple bonds or cumulated double bonds: errors for such molecules as H_2O (0.05 kcal/mol), HF (-0.04 kcal/mol), CH_4 (0.02 kcal/mol) and H_2 (0.00 kcal/mol) can only be described as impressive. Schwartzα(TQ5)+core gives equally excellent results for nonpolar molecules but, because of the lack of diffuse functions, has some trouble handling such polar systems as H_2O and HF.

It of course would stand to reason that basis set convergence for a single bond would be more rapid than for a double and particularly a triple bond or cumulated pair of double bonds. Inspection of the convergence behavior for N_2 up to cc-pV6Z basis sets revealed that basis set expansion beyond cc-pV5Z would increase the bond energy by a further 0.3 or 0.4 kcal/mol, depending on the extrapolation used. If we now add these numbers as a "fudge term" to the Schwartz-extrapolated results, we can bring the mean absolute error for Schwartzα(TQ5)+core back to 0.22 kcal/mol, which is less than 1 kJ/mol and can definitely be considered a result of calibration quality. At first sight, the aug$'$-Schwartzα(TQ5)+core results do not appear to be any better. However, about half of the average error of 0.23 kcal/mol is due to two molecules: N_2O and F_2. Both of these molecules exhibit some degree of nondynamical correlation: for N_2O, T_1=0.020, while for F_2, the deceptively low T_1=0.013 does not reflect the fact that the double excitation $\sigma \rightarrow \sigma^*$ has a coefficient of about 0.20 in the wave function. The latter suggests that connected quadruple excitations (which are completely neglected at the CCSD(T) level) may be of some importance. (It is recalled here that only about a third of the N_2O binding energy is recovered at the SCF level, while F_2 is actually found to be *endo*energetic at that level.) If we eliminate F_2 and N_2O from consideration, the mean absolute error drops to 0.14 kcal/mol, which is comparable to the error bars on the experimental results! We can therefore consider these to be results of calibration quality by any definition: however, these numbers can still be improved slightly further at no extra computational cost (the latter admittedly being quite Gargantuan).

It can rightly be argued that perhaps it would be worthwhile to fit SCF and correlation energies separately, since the two will have different limiting behavior in the very large basis set limit. Specifically, increments to the SCF energy will go $\propto l^{-6}$ (cf. the SCF extrapolation used in the CBS method (*102*)) and those to the correlation energy $\propto l^{-4}$. Now since the SCF binding energy is essentially converged for a basis set the size of aug$'$-cc-pV5Z, one would expect this not to make a big difference: yet in the accuracy range we are dealing with it might have a perceptible effect. The use of different exponents for the SCF and correlation energies will be indicated by the notation Schwartz$\alpha\beta$, and so forth.

For aug$'$-Schwartz$\alpha\beta$(TQ5)+core, we find (Table VI) that the mean absolute error without any triple bond correction is now 0.25 kcal/mol; adding a triple bond correction further decreases this value to 0.20 kcal/mol. It should be noted that the nature of the extrapolation procedure for the SCF binding energy — whether one uses $A + B/(l + 1/2)^5$, $A + B.C^{-l}$, or even the unextrapolated aug$'$-cc-pV5Z energy — does not affect the results by more than a few hundredths of a kcal/mol.

Table VI: Performance of the aug′-Schwartz$\alpha\beta$(TQ5) Extrapolation in Conjunction with the CCSD(T) Electron Correlation Method and Separate Treatment of Core Correlation. All Units are kcal/mol

	Calculated (a)	Calculated (b)	Experiment	Error (a)	Error (b)
HNO	205.30	205.67	205.64(6)	0.34	-0.03
CO_2	389.75	389.75	389.68(6)	-0.07	-0.09
CO	259.56	259.56	259.58(12)	0.02	0.02
N_2	228.16	228.53	228.42(3)	0.26	-0.11
N_2O	269.73	270.23	270.60(10)	0.87	0.37
C_2H_2	405.04	405.04	405.53(24)	0.49	0.49
CH_4	420.18	420.18	420.23(14)	0.05	0.05
H_2CO	374.33	374.33	374.09(16)	-0.24	-0.24
H_2O	232.83	232.83	232.83(2)	0.00	0.00
H_2	109.48	109.48	109.48(0)	0.00	0.00
HCN	312.96	313.33	313.29(25)	0.33	-0.04
HF	141.54	141.54	141.47(17)	-0.07	-0.07
NH_3	297.77	298.15	298.06(10)	0.29	-0.09
C_2H_4	563.77	563.77	563.68(29)	0.09	0.09
Average abs. error				0.22	0.12
Maximum abs. error				0.87	0.49

[a] without correction term for triple bonds
[b] adding 0.126 kcal/mol per bond order unit for bonds involving N

Interestingly, while the triple bond/cumulated double bond correction does improve overall agreement with experiment slightly, in some cases (CO and CO_2) it does more harm than good. Upon close inspection, we see that the largest error without triple bond correction are almost invariably in N-containing compounds — and it has been known for quite some time (e.g. (*112,113*)) that basis set convergence for nitrogen compounds is particularly slow.

Now if we introduce a small correction of 0.126 kcal/mol (an average over the reference molecules) per bond order unit involving an N atom (e.g. 3 such units in N_2, 4 in N_2O, 3 in HNO, ...), we obtain the lowest mean absolute error yet — 0.12 kcal/mol, on a par with the experimental error bars. The results in Table VI speak for themselves.

For aug′-Schwartz$\alpha\beta$(DTQ)+core and Schwartz46(DTQ)+core, we do obtain results of slightly better quality than with one joint extrapolation: however, the results are still inferior to those obtained using the corresponding 3PC+core/aug′-cc-pVQZ method.

We therefore conclude that aug′-Schwartz$\alpha\beta$(TQ5)+core — or, for nonpolar compounds, Schwartz$\alpha\beta$(TQ5)+core — may be the best overall procedure for work of the utmost accuracy on small systems.

A very recent application (*114*) to C_2H_4 may well represent the limit of what is presently computationally feasible. Schwartz$\alpha\beta$(TQ5)+core yielded a ΣD_e of 563.77 kcal/mol. Together with the best available zero-point energy (from a CCSD(T)/cc-pVTZ complete quartic force field (69) and benchmark values (70) for the harmonic frequencies), we obtain ΣD_0=532.17 kcal/mol, in perfect agreement with the experimental value (57) of ΣD_0=532.08±0.29 kcal/mol after subtracting out the spin-orbit coupling in the atoms. The largest CCSD(T) calculation in this study involved 402 basis functions, of which fully 400 were employed in the correlated part.

Conclusions

At the end of the day, a calibration calculation on a given molecule — assuming it does not exhibit extraordinary nondynamical correlation effects — will involve a compromise between accuracy and computational feasability from the following recommendations:

- For utmost accuracy, aug'-Schwartz$\alpha\beta$(TQ5)+core or, for nonpolar systems, Schwartz$\alpha\beta$(TQ5)+core. Your obtainable accuracy may well lie in the 0.10 kcal/mol range if no triple bonds are involved. Largest required basis set: 55 functions per hydrogen atom, 91 per first-row atom in nonpolar compounds, 127 in polar compounds.

- If this is not a practical option, 3PC+core/aug'-cc-pVQZ will permit a mean absolute error of only 0.23 kcal/mol, using only 30 basis functions per hydrogen atom and 80 basis functions per heavy atom.

- The next lower step in accuracy would be a choice between 3PC+core/aug'-cc-pVTZ, 3PC/cc-pVQZ, or CBS-QCI/APNO, all of which yield mean absolute errors around 0.45 kcal/mol. Which method is to be preferred will depend to some extent on the available software — a CCSD(T)/cc-pVQZ calculation using Gaussian 94 will be very costly, but the same calculation using a fast coupled-cluster code such as those in ACES II, MOLPRO 96, and TITAN may actually run faster than the CBS-QCI/APNO setup using Gaussian 94 exclusively, in many cases.

- If the system is too large for even those calculations, both G2M and CBS-Q will yield mean absolute errors on the order of 0.85 kcal/mol. If Gaussian 94 is used exclusively, CBS-Q will normally be the faster of the two methods, but if the CCSD(T) calculations involved in G2M are carried out using a faster coupled cluster code the tables may actually be turned.

Acknowledgments

JM is a Yigal Allon Fellow of the Israel Academy of Sciences and the Humanities as well as a Honorary Research Associate ("Onderzoeksleider in eremandaat") of the National Science Foundation of Belgium (NFWO/FNRS). He thanks many colleagues, but especially Dr. Timothy J. Lee (NASA Ames Research Center) and Prof. Peter R. Taylor (San Diego Supercomputer Center and University of California, San Diego), for enlightening discussions.

This research was supported by the University of Antwerp "Geconcerteerde actie" program, and by generous computer time grants from the Institute of Chemistry (Hebrew University of Jerusalem, Israel) and San Diego Supercomputer Center.

Literature Cited

1. Andersson, K. *Chem. Phys. Lett.* **1995**, *237*, 212 and references therein.
2. Martin, J. M. L.; Lee, T. J.; Scuseria, G. E.; Taylor, P. R. *J. Chem. Phys.* **1992**, *97*, 6549.
3. Peterson, K. A. *J. Chem. Phys.* **1995**, *102*, 262.
4. Bauschlicher, C. W.; Partridge, H. *Chem. Phys. Lett.* **1996**, *257*, 601.
5. Lorenz, M.; Agreiter, J.; Smith, A. M.; Bondybey, V. E. *J. Chem. Phys.* **1996**, *104*, 3143.
6. Roos, B. O. *Adv. Chem. Phys.* **1987**, *69*, 399.
7. Olsen, J.; Jorgensen, F.; Koch, H.; Balkova, A.; Bartlett, R. J. *J. Chem. Phys.* **1996**, *104*, 8007.
8. Olsen, J.; Jørgensen, P.; Simons, J. *Chem. Phys. Lett.* **1990**, *169*, 463.
9. Bauschlicher, C. W., Jr.; Langhoff, S. R.; Taylor, P. R. *Adv. Chem. Phys.* **1990**, *77*, 103.
10. Slater, J. C. *Phys. Rev.* **1929**, *34*, 1293.
11. Martin, J. M. L.; François, J. P.; Gijbels, R. *Chem. Phys. Lett.* **1990**, *172*, 346.
12. Langhoff, S. R.; Davidson, E. R. *Int.J. Quantum Chem.* **1974**, *8*, 61.
13. Møller, C.; Plesset, M. S. *Phys. Rev.* **1934**, *46*, 618.
14. Goldstone, J. *Proc. Royal Soc. (London) A* **1957**, *239*, 267.
15. Kucharski, S.; Bartlett, R. J. *Adv. Chem. Phys.* **1986**, *18*, 281.
16. Cremer, D.; He, Z. *J. Phys. Chem.* **1996**, *100*, 6173.
17. Cremer, D.; He, Z. *Int.J. Quantum Chem.* **1996**, *59*, 15, 31, 57, 71.
18. Bartlett, R. J. *J. Phys. Chem.* **1989**, *93*, 1697.
19. Taylor, P. R. In *Lecture Notes in Quantum Chemistry II*; Roos, B. O. , Ed.; Lecture Notes in Chemistry 64; Springer: Berlin, 1994, pp 125–202.
20. Bartlett, R. J.; Stanton, J. F. In *Reviews in Computational Chemistry, Vol. V*; Lipkowitz, K. B.; Boyd, D. B. , Eds.; VCH: New York, 1994, pp 65–169.
21. Lee, T. J.; Scuseria, G. E. In *Quantum mechanical electronic structure calculations with chemical accuracy*; Langhoff, S. R. , Ed.; Kluwer Academic Publishers: Dordrecht, The Netherlands, 1995, pp 47–108.
22. Bartlett, R. J. In *Modern Electronic Structure Theory*; Yarkony, D. R. , Ed.; World Scientific Publishing: Singapore, 1995, pp 1047–1131.
23. Purvis, G. D., III; Bartlett, R. J. *J. Chem. Phys.* **1982**, *76*, 1910.
24. Raghavachari, K.; Trucks, G. W.; Pople, J. A.; Head-Gordon, M. *Chem. Phys. Lett.* **1989**, *157*, 479.
25. Pople, J. A.; Head-Gordon, M.; Raghavachari, K. *J. Chem. Phys.* **1987**, *87*, 5968.
26. Frisch, M. J.; Trucks, G. W.; Schlegel, H. B.; Gill, P. M. W.; Johnson, B. G.; Robb, M. A.; Cheeseman, J. R.; Keith, T.; Petersson, G. A.; J. A. Montgomery, Raghavachari, K.; Al-Laham, M. A.; Zakrzewski, V. G.; Ortiz, J. V.; Foresman, J. B.; Cioslowski, J.; Stefanov, B. B.; Nanayakkara, A.; Challacombe, M.; Peng, C. Y.; Ayala, P. Y.; Chen, W.; Wong, M. W.; Andres, J. L.; Replogle, E. S.; Gomperts,

R.; Martin, R. L.; Fox, D. J.; Binkley J. S.; DeFrees, D. J.; Baker, J.; Stewart, J. P.; Head-Gordon, M.; Gonzalez, C.; Pople, J. A. *GAUSSIAN 94 Revision D.4*; Gaussian, Inc.: Pittsburgh, 1995.

27. ACES II is an ab initio program system written by: Stanton, J. F.; Gauss, J.; Watts, J. D.; Lauderdale, W. J.; Bartlett, R. J., incorporating the MOLECULE molecular integral program by: Almlöf, J.; Taylor, P. R.;,and a modified version of the ABACUS integral derivative package by: Helgaker, T.; Jensen, H. J. Aa.; Jørgensen, P.; Olsen, J.; Taylor, P. R.

28. MOLPRO 96 is an ab initio MO package by Werner, H. J.; Knowles, P. J. , with contributions from: Almlöf, J.; Amos, R. D.; Deegan, M. J. O.; Elbert, S. T.; Hampel, C.; Meyer, W.; Peterson, K. A.; Pitzer, R. M.; Stone, A. J.; Taylor, P. R.; Lindh, R.

29. TITAN is a set of electronic structure programs written by: Lee, T. J.; Rendell, A. P.; Rice, J. E.

30. Lee, T. J.; Taylor, P. R. *Int. J. Quantum Chem. Symp.* **1989**, *23*, 199.

31. Jayatilaka, D.; Lee, T. J. *J. Chem. Phys.* **1993**, *98*, 9734.

32. Gdanitz, R.; Ahlrichs, R. *Chem. Phys. Lett.* **1988**, *143*, 413.

33. Werner, H. J.; Knowles, P. J. *J. Chem. Phys.* **1988**, *89*, 5803.

34. Knowles, P. J.; Werner, H. J. *Chem. Phys. Lett.* **1988**, *145*, 514.

35. For internally contracted MRACPF see: Werner, H. J.; Knowles, P. J. *Theor. Chim. Acta* **1990**, *78*, 145.

36. Andersson, K.; Roos, B. O. In *Modern Electronic Structure Theory*; Yarkony, D. R., Ed.; World Scientific Publishing: Singapore, 1995, pp 55–109.

37. Woon, D. E.; Dunning, T. H., Jr. *J. Chem. Phys.* **1993**, *99*, 1914.

38. Peterson, K. A.; Kendall, R. A.; Dunning, T. H., Jr. *J. Chem. Phys.* **1993**, *99*, 1930.

39. Peterson, K. A.; Kendall, R. A.; Dunning, T. H., Jr. *J. Chem. Phys.* **1993**, *99*, 9790.

40. Blomberg, M. R. A.; Siegbahn, P. E. M. *J. Chem. Phys.* **1983**, *78*, 5682.

41. Helgaker, T.; Taylor, P. R. In *Modern Electronic Structure Theory*; Yarkony, D. R., Ed.; World Scientific Publishing: Singapore, 1995, pp 725–856.

42. Raffenetti, R. C. *J. Chem. Phys.* **1973**, *58*, 4452.

43. Almlöf, J.; Taylor, P. R. *J. Chem. Phys.* **1987**, *86*, 4070.

44. Dunning, T. H., Jr. *J. Chem. Phys.* **1989**, *90*, 1007.

45. Martin, J. M. L. *J. Chem. Phys.* **1992**, *97*, 5012.

46. Widmark, P. O.; Malmqvist, P. Å.; Roos, B. O. *Theor. Chim. Acta* **1990**, *77*, 291.

47. Clark, T.; Chandrasekar, J.; Spitznagel, G. W.; von Ragué Schleyer, P. *J. Comp. Chem.* **1983**, *4*, 294.

48. Kendall, R. A.; Dunning, T. H.; Harrison, R. J. *J. Chem. Phys.* **1992**, *96*, 6796.

49. Martin, J. M. L.; Taylor, P. R. *Chem. Phys. Lett.* **1994**, *225*, 473.

50. Martin, J. M. L.; François, J. P.; Gijbels, R. *J. Phys. Chem.* **1994**, *98*, 11394.

51. Del Bene, J. E. *J. Phys. Chem.* **1993**, *97*, 107.

52. Martin, J. M. L.; François, J. P.; Gijbels, R. *Chem. Phys. Lett.* **1989**, *163*, 387 and references therein.

53. Martin, J. M. L.; François, J. P.; Gijbels, R. *J. Comput. Chem.* **1989**, *10*, 875.

54. Martin, J. M. L. *Chem. Phys. Lett.* **1996**, *259*, 679.
55. Lee, T. J. *J. Phys. Chem.* **1995**, *99*, 1943.
56. Lee, T. J.; Rice, J. E.; Dateo, C. E.; *Mol. Phys.* **1996**, *89*, 1359.
57. Chase, M. W., Jr.; Davies, C. A.; Downey, J. R., Jr.; Frurip, D. J.; McDonald, R. A.; Syverud, A. N. *JANAF Thermochemical Tables, 3rd Edition, J. Phys. Chem. Ref. Data* **1985**, *14*, Supplement 1.
58. Grev, R. S.; Janssen, C. L.; Schaefer, H. F., III *J. Chem. Phys.* **1991**, *95*, 5128.
59. Martin, J. M. L. *Chem. Phys. Lett.* **1995**, *242*, 343.
60. Martin, J. M. L. *Chem. Phys. Lett.* **1996**, *259*, 669.
61. Berthelot, J. *Ann. Chim. Phys.* **1881**, *23*, 252.
62. Thomsen, J. *Thermochemische Untersuchungen*; Barth: Leipzig, 1886.
63. Bentley, J. A.; Bowman, J. M.; Gazdy, B.; Lee, T. J.; Dateo, C. E. *Chem. Phys. Lett.* **1992**, *198*, 563.
64. Smith, A. M.; Coy, S. L.; Klemperer, W.; Lehmann, K. K. *J. Mol. Spectrosc.* **1989**, *134*, 134.
65. Dixon, R. N. *J. Chem. Phys.* **1996**, *104*, 6905.
66. Pople, J. A.; Schlegel, H. B.; Krishnan, R.; DeFrees, D. J.; Binkley, J. S.; Frisch, M. J.; Whiteside, R. A.; Hout, R. F.; Hehre, W. J. *Int.J. Quantum Chem. Symp.* **1981**, *15*, 269.
67. Scott, A. P.; Radom, L. *J. Phys. Chem.* **1996**, *100*, 16502.
68. Pople, J. A.; Scott, A. P.; Wong, M. W.; Radom, L. *Israel J. Chem.* **1993**, *33*, 345.
69. Martin, J. M. L.; Lee, T. J.; Taylor, P. R.; François, J. P. *J. Chem. Phys.* **1995**, *103*, 2589.
70. Martin, J. M. L.; Taylor, P. R. *Chem. Phys. Lett.* **1996**, *248*, 336.
71. Martin, J. M. L.; Lee, T. J. *Chem. Phys. Lett.* **1996**, *258*, 129.
72. Martin, J. M. L.; Lee, T. J. *Chem. Phys. Lett.* **1996**, *258*, 136.
73. Teffo, J.; Chédin, A. *J. Mol. Spectrosc.* **1989**, *135*, 389.
74. Mebel, A. M.; Morokuma, K.; Lin, M. C. *J. Chem. Phys.* **1995**, *103*, 7414.
75. Becke, A. D. *J. Chem. Phys.* **1993**, *98*, 5648.
76. Lee, C.; Yang, W.; Parr, R. G. *Phys. Rev. B* **1988**, *37*, 785.
77. Bauschlicher, C. W., Jr.; Partridge, H. *J. Chem. Phys.* **1995**, *103*, 1788; *erratum* **1996**, *105*, 4398
78. Martin, J. M. L.; El-Yazal, J.; François, J. P. *Mol. Phys.* **1995**, *86*, 1437.
79. Martin, J. M. L. *J. Chem. Phys.* **1994**, *100*, 8186.
80. Pradhan, A.; Partridge, H.; Bauschlicher, C. W., Jr. *J. Chem. Phys.* **1994**, *101*, 3857.
81. Bauschlicher, C. W., Jr.; Partridge, H. *J. Chem. Phys.* **1994**, *100*, 4329.
82. Woon, D. E.; Dunning, T. H., Jr. , *J. Chem. Phys.* **1995**, *103*, 4572.
83. Hehre, W. J.; Radom, L.; von Ragué Schleyer, P.; Pople, J. A. *Ab Initio Molecular Orbital Theory* ; J. Wiley and Sons: New York, 1986.
84. Grev, R. S.; Schaefer, H. F., III *J. Chem. Phys.* **1992**, *97*, 8389.
85. Bauschlicher, C. W., Jr.; Partridge, H. *J. Chem. Phys.* **1995**, *103*, 10589.
86. Martin, J. M. L.; François, J. P.; Gijbels, R. *J. Chem. Phys.* **1991**, *95*, 8374.
87. Martin, J. M. L.; François, J. P.; Gijbels, R. *J. Chem. Phys.* **1992**, *96*, 7633.
88. Topper, R. Q.; Zhang, Q.; Liu, Y. P.; Truhlar, D. G. *J. Chem. Phys.* **1993**, *98*, 4991 and references therein.

89. Boys, S. F.; Bernardi, F. *Mol. Phys.* **1970**, *19*, 553.
90. Wells, B. H.; Wilson, S. *Chem. Phys. Lett.* **1983**, *101*, 429.
91. Martin, J. M. L.; François, J. P.; Gijbels, R. *Theor. Chim. Acta* **1989**, *76*, 195.
92. Parasuk, V.; Almlöf, J.; DeLeeuw, B. *Chem. Phys. Lett.* **1991**, *176*, 1.
93. Taylor, P. R. In *Lecture Notes in Quantum Chemistry*; Roos, B. O. , Ed.; Lecture Notes in Chemistry 58; Springer: Berlin, 1992, pp 325–415.
94. Curtiss, L. A.; Raghavachari, K.; Trucks, G. W.; Pople, J. A. *J. Chem. Phys.* **1991**, *94*, 7221.
95. Lee, T. J.; Dateo, C. E. *J. Chem. Phys.* **1995**, *103*, 9110.
96. Lee, T. J.; Martin, J. M. L.; Dateo, C. E.; Taylor, P. R. *J. Phys. Chem.* **1995**, *99*, 15858.
97. Pople, J. A.; Head-Gordon, M.; Fox, D. J.; Raghavachari, K.; Curtiss, L. A. *J. Chem. Phys.* **1989**, *90*, 5622.
98. Curtiss, L. A.; Raghavachari, K.; Pople, J. A. *J. Chem. Phys.* **1995**, *103*, 4192.
99. see also Curtiss, L. A.; Carpenter, J. E.; Raghavachari, K.; Pople, J. A. *J. Chem. Phys.* **1992**, *96*, 9030.
100. Durant, J. L., Jr.; Rohlfing, C. M. *J. Chem. Phys.* **1993**, *98*, 8031.
101. Martin, J. M. L. *J. Mol. Struct. (THEOCHEM)* **1997**, in press. (WATOC '96 special issue).
102. Montgomery, J. A., Jr.; Ochterski, J. W.; Petersson, G. A.; *J. Chem. Phys.* **1994**, *101*, 5900 and references therein.
103. Petersson, G. A.; Tensfeldt, T. G.; Montgomery, J. A., Jr. *J. Chem. Phys.* **1991**, *94*, 6091.
104. Feller, D. *J. Chem. Phys.* **1992**, *96*, 6104.
105. Lee, T.J.; Martin, J. M. L.; Dateo, C. E.; Taylor, P. R.; *J. Phys. Chem.* **1995**, *99*, 15858.
106. Peterson, K. A.; Dunning, T. H., Jr. *J. Phys. Chem.* **1995**, *99*, 3898.
107. Schwartz, C. In *Methods in Computational Physics 2*; Alder, B. J., Ed.; Academic Press: New York, 1963.
108. Kutzelnigg, W. *Theor. Chim. Acta* **1985**, *68*, 445.
109. Kutzelnigg, W.; Morgan, J. D., III *J. Chem. Phys.* **1992**, *96*, 4484; *erratum* **1992**, *97*, 8821.
110. Hill, R. N. *J. Chem. Phys.* **1985**, *83*, 1173.
111. Abramowitz, M.; Stegun, I. A. *Handbook of Mathematical Functions*; Dover: New York, 1972.
112. Frisch, M. J.; Pople, J. A.; Binkley, J. S. *J. Chem. Phys.* **1984**, *80*, 3265.
113. Almlöf, J.; DeLeeuw, B. J.; Taylor, P. R.; Bauschlicher, C. W., Jr.; Siegbahn, P. E. M. *Int.J. Quantum Chem. Symp.* **1989**, *23*, 345.
114. Martin, J. M. L.; Taylor, P. R. *J. Chem. Phys.*, in press.

Chapter 13

Complete Basis-Set Thermochemistry and Kinetics

George A. Petersson

Hall-Atwater Laboratories of Chemistry, Wesleyan University,
Middletown, CT 06459–0180

The major source of error in most *ab initio* calculations of molecular energies is the truncation of the one-electron basis set. Extrapolation to the complete basis set (CBS) limit using the N^{-1} asymptotic convergence of N-configuration pair natural orbital (PNO) expansions has been combined with the use of relatively small basis sets for the higher-order (*i.e.* MP3, MP4, and QCI) correlation energy to develop cost effective computational models, denoted CBS-4, CBS-Q, and CBS-QCI/APNO. The RMS errors for the 125 chemical energy differences of the G2 test set are 2.5, 1.3, and 0.7 kcal/mol respectively. In favorable circumstances, bond additivity corrections (BAC) can reduce these errors by an order-of-magnitude. An "IRCMax" extension of these models to the characterization of transition states (TS) for chemical reactions gives RMS errors in the barrier heights for ten atom exchange reactions of 1.3, 0.6, and 0.3 kcal/mol respectively. The rate constants for five hydrogen abstraction reactions with barrier heights from 1.3 kcal/mol ($H_2 + F$) to 20.6 kcal/mol ($H_2O + H$) over temperatures from 250 K to 2500 K range from 10^{-18} to 10^{-10} cm^3 molecule^{-1} sec^{-1}. If we include variations of the zero-point energy along the reaction path, (*i.e.* variational transition state theory) the calculated APNO absolute rates are all within the uncertainty of the experiments.

Advances in computational methods and computer hardware have made possible the accurate *ab initio* calculation of molecular energies including electron correlation for small and medium-size molecules. The accurate description of molecular wave functions requires the convergence of both the one-particle expansion (basis set) and the *n*-particle expansion (correlation energy). Currently available configuration interaction and coupled-cluster methods allow inclusion of the most important terms in the n-particle expansion, leaving basis set truncation as the primary source of error in the accurate calculation of molecular energies.

Background

The slow convergence of the correlation energy with the one-electron basis set expansion has provided the motivation for several attempts to extrapolate to the complete basis set limit (1-8). Such extrapolations require a well defined sequence of basis sets, and a model for the convergence of the resulting sequence of approximations to the correlation energy. The various extrapolation schemes that have been proposed differ in the method used to obtain a well defined sequence of one-electron basis sets. The complete basis set (CBS) extrapolations described in this chapter employ the asymptotic convergence of pair natural orbital (PNO) expansions.

Complete Basis Set Extrapolations. In a series of papers (3-5), we demonstrated the N^{-1} asymptotic convergence of second-order pair correlation energies (9) calculated with the leading N pair natural orbitals (10):

$$\lim_{N \to \infty} e_{ij}^{(2)}(N) = e_{ij}^{(2)}(\infty) + (25 / 512) |S|_{ij}^2 (N + \delta_{ij})^{-1} \tag{1}$$

and the modification necessary to extrapolate infinite-order (e.g. QCISD) pair energies:

$$\lim_{N \to \infty} e_{ij}^{(\infty)}(N) = e_{ij}^{(\infty)}(\infty) + \left(\sum_{\mu=1}^{N} C_{\mu_{ij}} \right)^2 \left[e_{ij}^{(2)}(N) - e_{ij}^{(2)}(\infty) \right] \tag{2}$$

where the $C_{\mu_{ij}}$'s are obtained from the first-order wave function, $\psi^{(1)}$, after diagonalization of the coefficient matrix over virtual orbital pairs, $C_{ab}^{(1)}$, for each occupied pair ij:

$$\psi_{ij}^{(1)}(N) = C_0 \Phi_0 + \sum_{a=a'=2}^{N} C_{aa',ij}^{(1)} \Phi_{ij}^{aa'} \tag{3}$$

The interference factor, $\left(\Sigma\, C_\mu \right)^2$, is the square of the trace of the first-order wave function. Since we have published (3-5) the derivation and quantitative verification of equation 2, we shall just explain the qualitative significance here. The interference factor relating the second-order basis set truncation error to the infinite-order basis set truncation error describes the effects of HOMO - LUMO near degeneracies. For example, the two configuration $1\,\sigma_g^2$ - $1\,\sigma_u^2$ description of H_2 is exact at infinite bond length. Thus, $e_{ij}^{(\infty)}(\infty)$ should be equal to $e_{ij}^{(\infty)}(2)$ in this case, even though

$e_{ij}^{(2)}(\infty)$ is not equal to $e_{ij}^{(2)}(2)$. Since the coefficients of the $1\,\sigma_g^2$ and $1\,\sigma_u^2$ configurations are equal but opposite in sign, equation 2 satisfies this condition. This interference effect accounts for the rapid convergence with basis set associated with the slow convergence with order of perturbation theory in species such as a beryllium atom that have HOMO - LUMO near degeneracies (5).

Subsequently (11-15), we used these asymptotic results to develop practical methods for extrapolating finite basis set calculations to obtain estimates of the complete basis set (CBS) limit. The essential idea is conveyed by a graph of $e_{ij}(N)$ as a function of $(\Sigma C_\mu)^2(N+\delta_{ij})^{-1}$. As $N\to\infty$, $(N+\delta_{ij})^{-1}$ approaches zero, so the intercept is the CBS limit, $e_{ij}(\infty)$. Note that only certain closed-shell sets of pair natural orbitals (denoted by filled circles in Figure 1) are useful for the extrapolation. The precise CBS extrapolation algorithm that has been implemented in the commercially available computer program Gaussian 94 has been described in detail elsewhere (16-18). The extrapolated CBS energy is size-consistent and will approach the exact energy as the basis set employed in the numerical calculations is systematically improved.

Compound Model Chemistries. A *theoretical model chemistry* is a complete algorithm for the calculation of the energy of any molecular system (19). It cannot involve subjective decisions in its application. It must be size consistent so that the energy of every molecular species is uniquely defined. A model chemistry is useful if for some class of molecules it is the most accurate calculation we can afford to do. A simple model chemistry employs the same basis set for each of the energy components (*i.e.* geometry, zero-point vibrational energy, and each component of the electronic energy: SCF, MP2, MP3, *etc.*) (9, 20, 21). A compound model chemistry employs different basis sets for each of the energy components. The earliest examples use a low level of correlation energy and a small basis set for the geometry and ZPE, since these calculations require multiple gradient and curvature calculations (22-28). The electronic energy is then determined with a higher level of correlation energy and a larger basis set: Energy[Method(1)]//Geometry[Method(2)]. A popular example is the MP2/6-31G**//UHF/3-21G* compound model (29).

The CBS-n Models

The accurate calculation of molecular energies requires convergence of both the one-particle (basis set) expansion and the n-particle (CI, perturbation, or coupled-cluster) expansion. However, the order-by-order contributions to chemical energies, and thus the number of significant figures required, generally decrease with increasing order of perturbation theory. For example, an accurate estimation of the bond dissociation energy for O_2 requires that the calculation of the SCF energy be converged to six significant figures, the MP2 component to three significant figures, and the higher-order correlation energy contribution to two significant figures (17). The MP4(SDQ) contribution (beyond the correlation energy already recovered in MP2) is usually a bit more important than the contributions from triple excitations and still higher-orders of perturbation theory [*i.e.* QCISD(T)]. The contributions from the correlation energy with the core electrons and from the zero-point energy need only be determined to one significant figure (*17*). Concomitant with the decreasing contributions, the computational demands increase rapidly for the higher-orders of perturbation theory (*30*). These two complementary trends combine to dictate that efficient computational models should employ smaller basis sets for the higher-orders of perturbation theory.

Once the treatment of the higher-order terms has been selected, we require procedures for the geometry, zero-point energy, SCF energy, and MP2 energy that are compatible with the higher-order treatment in both accuracy and speed. The components of the CBS-n and G2 models (*17, 31-34*) are indicated in Figure 2. The general approach for both is to first determine the geometry and ZPE at a low level of theory, and then perform a high level single point electronic energy calculation at this geometry using large basis sets for the SCF calculation, medium basis sets for the MP2 calculation, and small basis sets for the higher-order calculations. The components of each model have been selected to be balanced so that no single component dominates either the computer time or the error. The CBS-n models employ the above asymptotic extrapolation to reduce the error from truncation of the basis sets used to calculate the correlation energy. The compound model single point energy is evaluated at a geometry determined at a lower level of theory (*e.g.* CBS-4//UHF/3-21G or G2//MP2/6-31G*) (*17, 31-34*), which we shall reexamine when we consider transition states.

The guiding principle for the design of the fourth-order CBS-4 model was to attain the maximum speed consistent with a meaningful implementation of complete basis set pair natural orbital extrapolations. The minimum basis set that can give reliable results including polar molecules and negative ions is 6-31+G††. The CBS-4 model has a total of three empirical corrections. Since the UHF wåve function is not an eigenfunction of $<S^2>$, the resulting spin contamination can result in slow convergence of the perturbation series. Quadratic CI completely removes the contribution of the leading spin contaminant to the energy of the CBS-Q and CBS-QCI/APNO models. However, the fourth-order CBS-4 model requires an empirical correction proportional to the error in $<S^2>$. Two additional empirical corrections, a one-electron correction and an intraorbital pair two-electron correction, further reduce systematic errors in this model (*17*). The collective effect of these empirical corrections is to reduce the RMS error for the 125 chemical energy differences in the G2 test set from 5.73 kcal/mol to 2.51 kcal/mol.

Each component of the CBS-Q model was selected to employ the minimum effort necessary to achieve an accuracy of 1 mE$_h$ (0.6275 kcal/mol) *for that component* of chemical energy differences. The components of the CBS-Q model represent a compromise between the accuracy of the CBS-QCI/APNO model and the speed of the CBS-4 model. Although we have reduced the error contribution from each of the six components (*i.e.* geometry, ZPE, E$_{SCF}$, E$_{MP2}$, E$_{MP3,4}$, and E$_{QCISD(T)}$) of the CBS-Q model below 1 mE$_h$, the cumulative effect of all six errors is still several kcal/mol. Fortunately, this residual error is quite systematic and thus easily predicted by simple empirical models. The combination of two empirical corrections reduces the RMS error

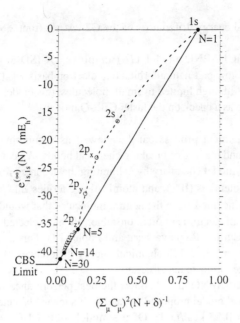

Figure 1. Only completed shells of helium PNO's (e.g. N=1, 5, 14, or 30) indicated by filled circles are useful for extrapolations (solid line) to the complete basis set (CBS) limit. The dashed curves demonstrate the typical behavior of partially completed shells (open circles). Closed shells can be recognized by a cusp in the graph. (Adapted from Ref. 17)

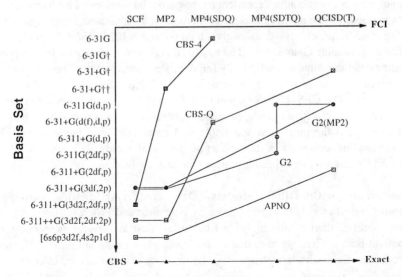

Figure 2. The components of the CBS-n and G2 models. (Adapted from Ref. 17)

for the 125 chemical energy differences of the G2 test set from 2.80 kcal/mol to 1.26 kcal/mol.

The CBS-QCI/APNO model (*16*) employs QCISD/6-311G** geometry optimization, and atomic pair natural orbital one-electron basis sets (*11-13*) for the SCF and MP2 energies. Although limited to small molecules, this model (RMS error = 0.68 kcal/mol) is valuable as a check on the faster CBS-Q model.

Accuracy. Any chemical process can be viewed as a combination of adding or removing electrons and making or breaking chemical bonds. We can therefore evaluate the reliability of any model chemistry by calibrating the accuracy for electron affinities (EA's), ionization potentials (IP's), and atomization energies Σ D_0's. Proton affinities (PA's) must also be included since the neutral molecules that would give a protonated species after loss of an electron are often unstable. We have elected to use the "G2 test set" introduced by Curtiss, Raghavachari, and Pople (*34*). This test set of 147 atoms, molecules, and ions provides 125 chemical energy differences of all four types and includes a wide range of molecular structure.

The CBS-QCI/APNO model shows the best performance with this test set of any general theoretical model proposed to date, with a mean absolute deviation (MAD) from experiment of 0.53 kcal/mol. Of the models defined for both the first- and second-row elements, the CBS-Q model gives the greatest accuracy (MAD = 1.01 kcal/mol), followed by G2 theory (MAD = 1.21 kcal/mol), G2(MP2) theory (MAD = 1.59 kcal/mol), and finally the CBS-4 model (MAD = 1.98 kcal/mol). All are at or under the accuracy of ~2 kcal/mol required for meaningful thermochemical predictions.

Range of Applicability. Each of the CBS models has a range of molecular size for which it is the most accurate computational model currently available. Absolute computer times are of course dependent on the particular versions of the hardware and software used, but the relative times are of more general significance. Every calculation in Figure 3 requires less than twelve hours on the IBM RS/6000 model 590 workstation running Gaussian 94. The newly emerging massively parallel computers should extend the range of applicability for all models considerably. Thus the abscissa of the figure should be scaled by a factor of 1.5 to 2 depending on the hardware one has available. The CBS-4, CBS-Q, and CBS-QCI/APNO models form a convenient sequence of cost-effective models, each successive member of which reduces both the RMS error and the maximum accessible molecular size by a factor-of-two. We recommend this sequence as a standard set of tools for the study of chemical problems, using the highest level model that is practical for each application.

Comparison with Other Models. Data comparing a wide variety of thermochemical models have been given by Foresman and Frisch (*30*). We repeat a small subset of their results in Table I below to illustrate several elementary, but important points. First, we note that *ab initio* calculations are not necessarily *good* calculations. Even with large basis sets, Hartree-Fock calculations are inferior to the better semiempirical calculations (*35*). Electron correlation is important for thermochemistry.

Using the same level of theory for the geometry, zero-point vibrational energy, and electronic energy, as in MP2/6-311+G(2d,p), provides a model that is inferior to the fourth-order CBS-4 compound thermochemical model in *both* speed and accuracy. To develop faster models, we are inclined to employ gradient corrected density functional methods (*36*) such as Becke's three parameter exchange functional (*37*) with the Lee, Yang, and Parr correlation functional (*38*). The 6-31+G(d,p) basis set gives good results at modest cost, and continues our CBS-QCI/APNO, CBS-Q, CBS-4 sequence with an additional doubling of the error while extending the range (with HF/3-21G* geometries and ZPEs) to molecules twice as large. Such calculations probably represent the limiting speed for useful thermochemical predictions. Local DFT methods as presently formulated (*e.g.* SVWN5) (*39*) offer no advantage in accuracy over semiempirical methods.

Table I. Summary of error measurements (kcal/mol) for the G2 test set of 125 chemical energy differences

Type	Model Chemistry	MAD[a]	Max. Error
ab initio	CBS-QCI/APNO	0.5	1.5
	CBS-Q	1.0	3.8
	G2	1.2	5.1
	G2(MP2)	1.6	6.2
	CBS-4	2.0	7.0
	MP2/6-311+G(2d,p)	8.9	39.2
	UHF/6-311+G(2d,p)	46.1	173.8
DFT	Becke3LYP/6-311+G(3df,2df,2p)	2.7	12.5
	Becke3LYP/6-31+G(d,p)//HF/3-21G*	3.9	33.9
	SVWN5/6-311+G(2d,p)	18.1	81.0
Semiempirical	AM1	18.8	95.5

[a] Mean absolute deviation from experiment.

"Random" and Systematic Errors. Ideally, a computational model chemistry should not only provide chemical energy differences, but it should also provide a realistic estimate of the uncertainty. The 125 examples in the G2 test set have been used to determine the approximate shapes of the distribution functions for the errors from the CBS-4, CBS-Q, G2(MP2), G2, and CBS-QCI/APNO models. The results (*17*) are

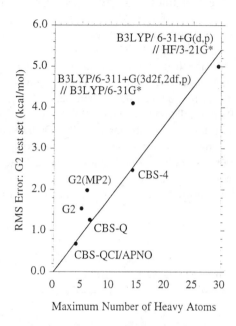

Figure 3. Both the range of applicability (time < 12 hrs. on a single CPU) and the RMS error increase as more approximate methods are incorporated in a model chemistry. (Adapted from Ref. 17)

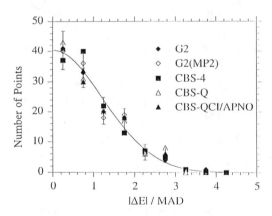

Figure 4. The error distribution for each of the models follows a Gaussian distribution function (curve) to within the variations expected from the square-root of the number of points. The errors from each model have been rescaled by the mean absolute deviation for that model. (Adapted from Ref. 17)

quite striking - none of the six models shows a statistically significant deviation from a Gaussian distribution function (Figure 4), indicating that the RMS errors for the G2 test set (or $\sqrt{\pi/2}$ times the MAD in Table I) can be used in the same way that experimental uncertainties are used to assess reliability. Nevertheless, the possibility of relatively large errors in cases with high symmetry (or in large molecules) should be recognized for any computational model chemistry. The largest errors uncovered to date (*17*) for the CBS-4, G2(MP2), and G2 models are all more than four standard deviations measured with the G2 test set. On the other hand errors in individual bond dissociation energies are generally smaller than errors in atomization energies (*40*).

Within an isoelectronic sequence, errors can be sufficiently constant to permit empirical corrections. For example, the CBS-4 model consistently overestimates singlet-triplet gaps in carbenes by ~1 kcal/mol (Figure 5).

Isodesmic Bond Additivity Corrections

An isodesmic reaction conserves the number of bonds of each order between each pair of atom types. For example, the reaction:

$$H_2C=C=CH_2 + CH_4 \rightarrow 2\ H_2C=CH_2 \tag{4}$$

preserves two carbon-carbon double bonds and eight C-H bonds. If the error in the calculated bond energy were constant for each type of bond and these errors were exactly additive, then the calculated enthalpy change for an isodesmic reaction would be exact. In our example, if the heats of formation of ethylene and methane are known more accurately than the heat of formation of allene, then the calculated enthalpy change for equation (4) can be used to determine the heat of formation of allene.

Alternatively, we can determine the error, BAC_{ij}, in each bond ij and add these *bond additivity corrections* to obtain the total correction to the heat of formation:

$$\Delta_f H\,(\,BAC\,) \;=\; \sum_{ij} N_{ij} BAC_{ij} \tag{5}$$

Such isodesmic bond additivity corrections have already been used to advantage in correcting G2 and CBS-n heats of formation for halocarbons (*41*) (see also chapter by R. J. Berry, M. Schwartz, and P. Marshall), and Melius has made detailed studies (*42, 43*) of the behavior of more general BACs (see chapter by M. R. Zachariah and C. F. Melius). Corrections for such systematic errors will become more important as it becomes possible to apply *ab initio* models to larger and larger molecules. Table II illustrates the potential power of such methods in an ideal case. To achieve comparable results in general we will have to understand the effects on BACs of such variables as hybridization, bond order, and charge, which did not vary significantly for methane, ethane, and propane in our example.

Table II. Convergence of the calculated heat of formation of propane
(an ideal example) with and without isodesmic bond additivity
corrections [a]

Method	$\Delta_f H^\circ{}_0$ (kcal/mol)	
	Without BAC	With BAC
UHF/3-21G	233.5	-18.4
MP3/6-31G*	56.8	-19.2
B3LYP/6-311+G(2df,p)	-14.2	-18.2
CBS-4	-22.8	-19.2
G2(MP2)	-18.7	-19.3
G2	-19.2	-19.4
CBS-Q	-20.1	-19.4
CBS-QCI/APNO	-22.1	-19.2
Experiment	-19.5 ±0.1	

[a] Using BAC_{C-H} from CH_4 and BAC_{C-C} from C_2H_6.

Transition States for Chemical Reactions

The geometry and energy of a stable molecule can often be measured experimentally to greater accuracy than is currently available computationally. However, transition states cannot generally be isolated for experimental study, and thus are obvious targets for computational studies. The development of analytical gradient and curvature methods (22-28) has made possible the rigorous determination of transition states within a given level of correlation energy and basis set (e.g. MP2/6-31G*) (29).

Transition States and Reaction Paths. The potential energy surface (PES) for a typical bimolecular chemical reaction such as (Figure 6):

$$H_2 + OH \rightarrow H + H_2O \tag{6}$$

includes valleys (leading to the reactants and products) connected at the transition state (TS), which is a first-order saddle point (i.e. a stationary point with exactly one negative force constant). The reaction path or intrinsic reaction coordinate (IRC) is defined (44, 45) as the path beginning in the direction of negative curvature away from the TS and following the gradient of the PES to the reactants and products.

If we move in a direction perpendicular to the reaction path, we find a potential energy curve (or surface) corresponding to a stable reactant or product molecule if we are far from the TS. Even around the TS, the variation of the PES perpendicular to the IRC is very similar to the PES for a stable molecule. Transition States differ from stable

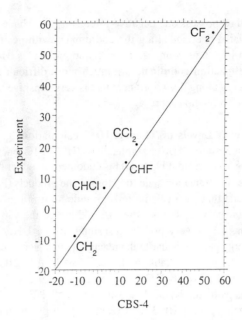

Figure 5. The CBS-4 model gives a constant error for the singlet - triplet splitting in carbenes, $E(^3B_1) - E(^1A_1)$.

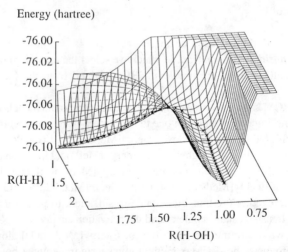

Figure 6. The potential energy surface for the reaction: $H_2 + OH \rightarrow H + H_2O$. The reaction path or intrinsic reaction coordinate (IRC) begins in the direction of negative curvature from the transition state (TS) and follows the gradient through the valleys to the reactants and products.

molecules in that they possess one negative force constant which defines the reaction coordinate. Calculated energies along the coordinates with positive force constants behave very much like their counterparts in stable molecules. However, the energy changes along the reaction coordinate are much more difficult to predict. It is the variation of the energy along this coordinate that is very sensitive to (and thus requires the inclusion of) the correlation energy.

Combining Different Levels of Theory. UHF calculations give notoriously bad results for transition states. For example, the UHF/3-21G energy profile for the transfer of a hydrogen atom from H_2 to OH is endothermic rather than exothermic and consequently places the transition state too close to the products (Figure 7). One might erroneously conclude that such calculations provide no useful information about the reaction path. Fortunately, this is not the case. Although the variation of the energy along the reaction path is very poorly described by the UHF/3-21G method, the variation of the energy perpendicular to the reaction path is very nicely described by the UHF/3-21G method, just as in stable molecules. The UHF/3-21G reaction path very closely approximates the MP2/6-31G* reaction path (Figure 8). However, the energy variation along this path, and hence the position of the UHF/3-21G transition state, is incorrect. Nevertheless, the UHF/3-21G reaction path passes through (or near) the MP2/6-31G* transition state. Hence, if we calculate the MP2/6-31G* energy along the UHF/3-21G reaction path, we obtain an energy profile that differs trivially from the MP2/6-31G* energy profile along the MP2/6-31G* reaction path (Figure 9). This is rather remarkable for the H_2 + OH reaction, given the very poor UHF/3-21G//UHF/3-21G energy profile.

The IRCMax Method. Based on these observations, we have proposed (*46*) the *"IRCMax" transition state method*, in which we select the maximum of the high-level Energy[Method(1)] along the low-level IRC obtained from the Geom[Method(2)] calculations.

The IRCMax transition state extension of the compound models takes advantage of the enormous improvement (from one to two orders-of-magnitude) in computational speed (*30*) achieved by using low-level, Geom[Method(2)], IRC calculations. We then perform several single point higher level, Energy[Method(1)], calculations along the Geom[Method(2)] reaction path to locate the Energy[Method(1)] transition state, that is, the maximum of Energy[Method(1)] along the Geom[Method(2)] IRC. Calculations at three points bracketing the transition state are sufficient to permit a parabolic fit to determine the transition state geometry and the activation energy.

Since we determine the maximum of Energy[Method(1)] along a path from reactants to products, the IRCMax method gives a rigorous upper bound to the high-level Method(1) transition state energy. In addition, when applied to the compound CBS and G2 models, the IRCMax method reduces to the normal treatment of bimolecular reactants and products, Energy[Method(1)]//Geom[Method(2)]. Thus the IRCMax method can be viewed as an extension of these compound models to transition states. The results below demonstrate the numerical superiority of this IRCMax

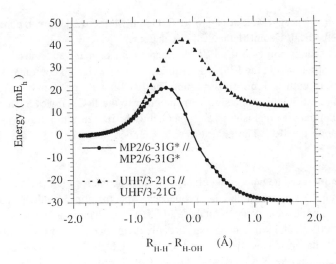

Figure 7. The energy profile along the reaction path for the reaction: $H_2 + OH \rightarrow H + H_2O$. The UHF/3-21G method indicates an endothermic reaction with a high barrier and a late transition state. The inclusion of electron correlation (MP2/6-31G*) makes the reaction exothermic with a modest barrier and an early transition state.

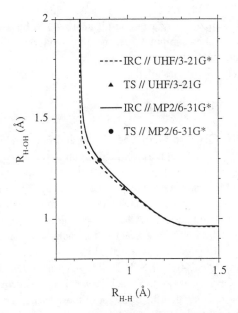

Figure 8. The IRC for the reaction: $H_2 + OH \rightarrow H + H_2O$, changes very little when we include electron correlation. However, the position of the TS along the IRC shows a large shift towards the reactants.

method, Max{Energy[Method(1)]}// IRC{Geom[Method(2)]}, over conventional Energy[Method(1)]//Geom[Method(2)] calculations.

The saddle point on the PES is independent of mass and thus invariant to isotopic substitution, but the IRC is determined in mass weighted coordinates and thus varies slightly with isotopic substitution. The IRCMax TS geometry will therefore show a small artificial isotope shift perpendicular to the IRC. However, this effect is much smaller than the real isotope shift parallel to the IRC resulting from variations in the ZPE along the IRC. The latter effect is included in the IRCMax search for the variational TS.

Geometry. The determination of molecular geometries is the first step in any computational study of molecular properties. Spectroscopic studies of diatomic molecules provide a convenient database (47-50) of molecular geometries known to within ±0.0001 Å, and thus can be used to calibrate the accuracy of our computational methods. We have used the diatomic species H_2, CH, NH, OH, FH, CN, and N_2, and the polyatomic species H_2O, HCN, CH_3, and CH_4 for this geometry calibration (46). The RMS errors in the UHF/3-21G, MP2/6-31G*, and MP2/6-31G† geometries are 0.012, 0.016, and 0.015 Å respectively. The CBS-4 method is comparable in accuracy (±0.0048 Å) to (but much faster than) the QCISD/6-311G** method (±0.0052 Å). The G2(MP2), G2, and CBS-Q methods are very close in accuracy (±0.0036, 0.0038, and 0.0031 Å respectively). The compound methods (*i.e.* CBS-4, G2(MP2), G2, CBS-Q, and CBS-QCI/APNO) are clearly superior to the simple methods. In fact, the CBS-QCI/APNO method is sufficiently accurate (±0.0013 Å) to be used to calibrate the accuracy of any of the other computational methods, and was therefore used as the reference for transition states (46).

The calculated geometries of the transition states for six degenerate atom exchange reactions:

$$\text{A-B} + \text{A} \rightarrow \text{A--B--A} \rightarrow \text{A} + \text{B-A} \qquad (7)$$

where A = H, F, CN, OH, and CH_3, and B = H and F (with A = H) provide an indirect test of the IRCMax concept. Since the position of the transition state along the IRC is determined by symmetry, the accuracy of the calculated geometries depends only on the behavior of the calculated potential energy surface perpendicular to the IRC. Thus, the accuracy of these calculated transition state geometries should be comparable to the accuracy of calculated geometries for stable molecules. In fact, the RMS error (relative to the CBS-QCI/APNO TS geometry) for each of the models (±0.035 Å for UHF/3-21G, ±0.012 Å for MP2/6-31G*, ±0.013 Å for MP2/6-31G†, ±0.007 Å for QCISD/6-311G**, ±0.009 Å for CBS-4, ±0.0023 Å for G2(MP2), ±0.0033 Å for G2, and ±0.0032 for CBS-Q) is not too much larger than the corresponding error for stable molecules. We conclude that the behavior of these surfaces perpendicular to the IRC's is similar to that of stable molecules, but with somewhat smaller force constants accounting for the modest reduction in the accuracy of calculated geometries. The accuracy of calculated transition state energies should be unaffected, since the smaller force constants make the accuracy of the transition state geometry less critical.

The calculated geometries of the transition states for four exothermic hydrogen abstraction reactions:

$$R\text{-}H + R' \rightarrow R\text{-}\text{-}H\text{-}\text{-}R' \rightarrow R + H\text{-}R' \tag{8}$$

with R = H and R' = F, OH, or CN, and R = CH$_3$ with R' = OH provided a direct test of the IRCMax method. We first compared the UHF/3-21G and Max{MP2/6-31G*}//IRC{UHF/3-21G} optimized transition state geometries with the fully optimized MP2/6-31G* geometries, which were the presumed goal of these IRCMax calculations. The relative RMS errors indicated that the IRCMax method reduces geometry errors by a factor of five compared to the UHF/3-21G geometries. The residual relative RMS error (±0.027 Å) was similar to the error in UHF/3-21G geometries for the symmetric transition states for degenerate reactions (±0.035).

We next compared the MP2/6-31G* and Max{QCISD/6-311G**}//IRC{MP2/6-31G*} optimized transition state geometries with the fully optimized QCISD/6-311G** geometries, which are the goal for these IRCMax calculations. The relative RMS errors indicated that this time the IRCMax method reduced geometry errors by a factor of four compared to the MP2/6-31G* geometries. The residual relative RMS error here (±0.013 Å) was similar to the error in MP2/6-31G* geometries for the symmetric transition states for degenerate reactions (±0.012 Å).

Our third test compared the QCISD/6-311G** and Max{CBS-QCI/APNO}//IRC {QCISD/6-311G**} optimized transition state geometries with the fully optimized CBS-QCI/APNO geometries, which were the goal for these IRCMax calculations. The relative RMS errors indicate that this time the IRCMax method reduced geometry errors by a factor of five compared to the QCISD/6-311G** geometries. The residual relative RMS error here (±0.016 Å) is more than twice as large as the error in QCISD/6-311G** geometries for the symmetric transition states for degenerate reactions (±0.007 Å). These strongly exothermic reactions have "early" transition states in which the reactants are still quite far apart and only weakly interacting. This makes the transition states less well defined and difficult to locate to better than 0.01 Å. Nevertheless, all three tests of the IRCMax method gave substantial improvement over the low level saddle points and demonstrated that the major error in calculated transition state geometries is the location along the IRC as we anticipated.

Barrier Height. The enthalpies of reaction, ΔH°_{298}, for the four exothermic reactions are known from experiment (*47, 50*) and thus could be used to calibrate the reliability of each of the computational methods for the energy changes along these reaction paths (*46*). The RMS errors were: UHF/3-21G* = 25.4, MP2/6-31G* = 16.3, QCISD/6-311G** = 5.3, CBS-4 = 1.41, G2(MP2) = 1.05, G2 = 1.20, CBS-Q = 0.79, and CBS-QCI/APNO = 0.70 kcal/mol. Given the small size of the current sample set, the agreement with the larger calibration studies (*17, 34*) is good.

Since the location of the transition state along the reaction path is completely determined by symmetry for the six degenerate reactions, the standard Energy[Method(1)]//Geom[Method(2)] procedure is equivalent to the IRCMax method for these cases. They thus provided a benchmark for the accuracy attainable when the location of the transition state is not in question. In comparison sets of three calculations each (Table III), we first calculated the barrier height using the simple geometry method, then the higher-level single point energy evaluated at the transition state geometry obtained from the simple geometry method, and finally the energy of the higher-level method evaluated at the transition state geometry obtained from the higher-level method. The relative RMS error was evaluated with respect to the higher-level//higher-level calculation, and the absolute RMS error was evaluated with respect to our best estimate of $\Delta E_e^{\ddagger}$, the CBS-QCI/APNO model (Table III).

Table III. Relative and absolute errors in calculated barrier heights

| Method/Basis Set | RMS Error (kcal/mol) | | | |
| | Degenerate | | Nondegenerate | |
	Rel.	Abs.	Rel.	Abs.
UHF/3-21G*	6.37	9.49	13.19	20.87
MP2/6-31G* // UHF/3-21G*	3.38	9.75	6.48	3.70
Max{MP2/6-31G*} // IRC{UHF/3-21G*}			0.59	7.57
MP2/6-31G*	-	7.12	-	7.12
MP2/6-31G*	4.23	7.12	2.93	7.12
QCISD/6-311G**// MP2/6-31G*	1.63	5.13	0.40	4.72
Max{QCISD/6-311G**}// IRC{MP2/6-31G*}			0.14	4.88
QCISD/6-311G**	-	4.07	-	4.79
QCISD/6-311G**	3.93	4.07	4.64	4.79
CBS-QCI/APNO // QCISD/6-311G**	0.09	0.19	0.49	0.37
Max{CBS-QCI/APNO} // IRC{QCISD/6-311G**}			0.24	0.37
CBS-QCI/APNO	-	0.15	-	0.15

The calculated barrier heights for the four nondegenerate hydrogen abstraction reactions given in three sets of four calculations each (Table III) provided a direct test of the IRCMax method. In each set of four calculations, the first row is the calculated barrier height using the simple geometry method, the second calculation is the more elaborate single point energy evaluated at the transition state geometry obtained from the simple geometry method, the third calculation is the maximum energy of the more elaborate method evaluated along the IRC obtained from the simple geometry method, and the fourth calculation is the energy of the more elaborate method evaluated at the transition state geometry obtained from the more elaborate method. The fourth calculation was used as the reference to evaluate the relative errors for the first three calculations.

The first set of four calculations illustrates the behavior of the IRCMax method very clearly. This set compares the UHF/3-21G* method with the MP2/6-31G* energies calculated at UHF/3-21G* and MP2/6-31G* geometries. The UHF barriers are all much too high. This is an extreme example of the tendency of calculated barrier heights to converge to the exact value from above. The MP2//UHF barrier heights are all lower than the MP2//MP2 values. The UHF transition state geometry is at the wrong position along the reaction path and thus gives a value for the MP2//UHF transition state energy that is too low. The extreme case is the H-H-F barrier for which the MP2//UHF value is *negative*. In less extreme cases, this negative error can cancel the positive error (from the use of limited correlation/basis sets) to reduce the absolute RMS error. However, the results are erratic with no clear convergence pattern as the basis set and level of correlation treatment are improved. The IRCMax method gives an upper bound to the MP2 energy of the transition state. The Max{MP2/6-31G*}//IRC{UHF/3-21G*} procedure reduces the relative RMS error by a factor-of-ten (from 6.48 to 0.59 kcal/mol). The difference between the IRCMax barriers and the MP2//MP2 barriers is minor, and we retain the inherent convergence pattern of calculated barrier heights.

The calculated barrier heights of all ten test reactions are summarized in Table IV below for the simple UHF, MP2, and QCI methods and for the five recommended IRCMax methods. The eight sets of calculations are arranged in order of decreasing RMS errors. For each reaction, the sequence of improving calculations tends to converge to the exact barrier height from above. The CBS-QCI/APNO barrier height for the $H_2 + H$ reaction has converged to within 0.15 kcal/mol of the exact barrier obtained by numerical solution of the Schrödinger equation (*51*). Having verified the accuracy of the CBS-QCI/APNO method in the one case for which the exact barrier height is known, we have justified use the CBS-QCI/APNO// CBS-QCI/APNO values for the nine remaining reactions to determine the errors in the less computationally demanding methods.

A graph of the barrier heights calculated with the less demanding methods vs the CBS-QCI/APNO barrier heights demonstrates the systematic tendency to overestimate barrier heights (Figure 10). Linear least-squares fits give lines with unit slope and nonzero intercept for all but the simple UHF/3-21G* calculations. Each of these computational methods consistently overestimates all ten barrier heights by about the same error. This suggests a simple empirical correction to the calculated barrier heights (*46*).

Recall that these compound model chemistries already include two empirical parameters, a one-electron correction and a two-electron correction (*17, 34*). In the case of the original G1 method (*31*), the one-electron correction [Const$_1$ x ($n_\alpha + n_\beta$)] was selected to give the exact energy for the hydrogen atom. The two-electron correction [Const$_2$ x ($n_\beta = n_{pairs}$)] was selected to give the exact energy for the hydrogen molecule (*31*). We therefore proposed (*46*) a third empirical correction selected to give the exact energy (*51*) for the H_3 transition state [Const$_3$ x (number of imaginary frequencies)].

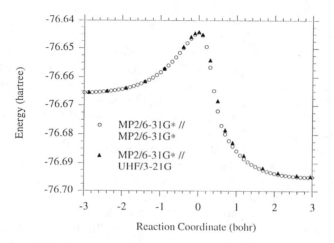

Figure 9. The MP2/6-31G* energy profile calculated along the IRC{UHF/3-21G} is almost indistinguishable from the MP2/6-31G* energy profile calculated along the IRC{MP2/6-31G*} for the reaction: $H_2 + OH \rightarrow H + H_2O$. Compare this result to the UHF/3-21G energy profile in Figure 7.

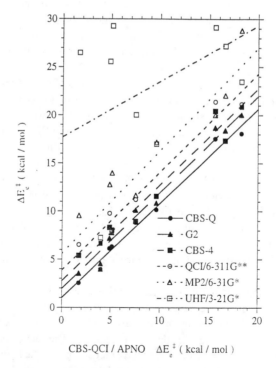

Figure 10. The barrier heights, $\Delta E_e^{\ddagger}$, for the ten chemical reactions in Table IV calculated with the seven more approximate methods show a strong linear correlation with the more accurate CBS-QCI/APNO results.

Table IV. Calculated barrier heights, $\Delta E_e^{\ddagger}$ (kcal/mol), including empirical corrections

Method/Basis Set	H$_2$ +F	H$_2$ +CN	H$_2$ +OH	CH$_4$ +OH	H$_2$O +OH	H$_2$ +H	HF +F	CH$_4$ +CH$_3$	HCN +CN	HF +H	Ave Error	RMS Error
UHF/3-21G*	19.11	-0.14	18.18	21.88	12.64	9.61	21.75	19.81	16.09	37.16	5.08	9.40
MP2/6-31G*	1.97	-3.59	5.21	6.40	4.08	9.61	12.45	14.47	21.24	42.93	-1.06	3.07
QCISD/6-311G**	4.82	5.10	8.06	9.39	9.51	9.61	19.67	18.18	19.46	45.91	2.44	2.80
IRCMax												
CBS-4 [a]	3.47	4.18	6.36	6.09	7.01	9.61	18.48	15.39	18.05	43.33	0.66	1.34
G2(MP2)	2.51	3.44	6.02	6.57	9.11	9.61	17.93	17.22	18.89	42.92	0.89	1.18
G2	2.30	3.31	5.96	6.53	8.87	9.61	17.45	17.17	18.80	42.75	0.74	1.02
CBS-Q [b]	1.97	3.86	5.57	5.77	8.27	9.61	16.98	16.79	18.38	41.79	0.37	0.62
CBS-QCI/APNO	2.04	3.95	4.95	5.45	7.59	9.61	15.75	16.50	18.26	42.14	0.10	0.17
CBS-QCI/APNO -.15	1.70	3.96	4.87	5.11	7.51	9.61	15.58	16.60	18.28	42.17	-	-

[a] The CBS-4 barriers were increased by 3.63 kcal/mol for each CN group.
[b] The CBS-Q barriers were increased by 0.43 kcal/mol for each CN group.

The value for this empirical constant decreases with increasing quality of the computational model: UHF/3-21G* = -7.39, MP2/6-31G* = -7.57, QCISD/6-311G** = -1.72, CBS-4{IRCMax} = -1.95, G2(MP2){IRCMax} = -1.24, G2{IRCMax} = -1.24, CBS-Q{IRCMax} = -0.57, CBS-QCI/APNO = -0.15 kcal/mol.

All nine methods now give the exact barrier height for H_3 (Table IV). The average errors are now generally small compared to the RMS errors, demonstrating that the dominant systematic errors have been removed. The correction based on H_3 gives a general improvement for all these atom exchange reactions. An additional empirical correction for the spin contaminated cyano group significantly improved the CBS-4 and CBS-Q barriers. This suggests that a general bond additivity correction (BAC) (*42, 43*) might also be useful for calculations of activation energies.

Based on our experience (*17*) that the CBS-Q errors are half as large as the CBS-4 errors and the CBS-QCI/APNO errors are half as large as the CBS-Q errors, we estimate the uncertainty in the CBS-QCI/APNO barrier heights to be ±0.3 kcal/mol. This is consistent with the CBS-QCI/APNO overestimation of the H_3 barrier height by 0.15 kcal/mol before the empirical correction was applied. If we use the nonlinear CBS-QCI/APNO optimized geometry for the H_2 + F transition state, our empirically corrected CBS-QCI/APNO barrier height (1.70 ±0.3 kcal/mol) for this highly exothermic reaction is in agreement with the best CCSD(T) (1.7 ±0.2 kcal/mol) (*52*) and MRCI calculations (1.53 ±0.15 kcal/mol) (*53*).

All five compound models give barrier heights with RMS errors sufficiently small to give absolute reaction rates accurate to within a factor of ten at room temperature (*i.e.* RMS error < 1.4 kcal/mol). The RMS errors are consistently about half as large as those for the same models applied to dissociation energies, ionization potentials, and electron affinities (*17*). However, this accuracy depends upon an empirical correction (obtained from the H_3 TS) which may not be applicable to more general reaction mechanisms.

Chemical Reaction Rates

The absolute rates of chemical reactions present a formidable challenge to theoretical predictions. An elementary bimolecular chemical reaction:

$$R_1 + R_2 \rightarrow P_1 + P_2 \tag{9}$$

proceeds at a rate that is proportional to the concentrations of the reactants and the specific rate constant, $k_{rate}(T)$:

$$\frac{d[P_1]}{dt} = \frac{d[P_2]}{dt} = -\frac{d[R_1]}{dt} = -\frac{d[R_2]}{dt} = k_{rate}(T)[R_1][R_2] \tag{10}$$

It is now over one hundred years since Arrhenius published the seminal article (*54*) on the variation of the rate constant with absolute temperature:

$$k_{rate}(T) = A \ e^{-\Delta \ E^{\ddagger}/ \ RT} \tag{11}$$

The principal difficulty for theoretical predictions lies in the extreme sensitivity of $k_{rate}(T)$ to small errors in the activation energy, $\Delta E^{\ddagger}$, which can be interpreted as the difference in energy between the "transition state" and the reactants. An error of only 1.4 kcal/mol in $\Delta E^{\ddagger}$ leads to an error of an order-of-magnitude in $k_{rate}(T)$ at room temperature.

Absolute Rate Theory. More than sixty years ago, Eyring proposed (55) the use of statistical mechanics to evaluate the preexponential factor, A, in equation 11 using partition functions, Q, and the "collision velocity", $k_B T/h$:

$$A \ (T) = \frac{Q^{\ddagger}(T)}{Q_{R_1}(T) \ Q_{R_2}(T)} \bullet \frac{k_B \ T}{h} \tag{12}$$

thereby accounting for the small temperature variations in A. The evaluation of the translational, vibrational, and rotational partition functions:

$$Q_{translation} = V \ \left(2 \ \pi M \ k_B T\right)^{3/2} \Big/ h^3 \tag{13}$$

$$Q_{rotation} = 2 \ \pi \left(4 \ \pi/\sigma\right) \prod_j \left(2 \ \pi I_j \ k_B T\right)^{1/2} \Big/ h \tag{14}$$

$$Q_{vibration} = \prod_j \left(1 - e^{-h \ v_j/k_B T}\right)^{-1} \tag{15}$$

$$Q_{electronic} = \sum_j e^{-\Delta E_j / k_B T} \tag{16}$$

$$Q_{total} = Q_{electronic} \ Q_{vibration} \ Q_{rotation} \ Q_{translation} \tag{17}$$

requires only a knowledge of the mass (for $Q_{translation}$), geometry (for the moments of inertia, I_j, in $Q_{rotation}$), vibrational frequencies (v_j for $Q_{vibration}$), and ΔE of any low-lying electronic states for the reactants and transition state ($Q_{electronic}$).

The electronic, vibrational, and rotational energy levels can often be determined from experiment for the reactants, but not for the transition state. The use of *ab initio* quantum mechanical methods to evaluate the required electronic, vibrational, and rotational energy levels is therefore of considerable practical importance. We shall first be concerned with establishing the validity of transition state theory for quantitative absolute rate predictions if we use state of the art computational quantum methods. We shall then address the question of cost-effective compromises necessary for a widely applicable method.

Eyring presumed that motion along the path from reactants to products could be treated classically. However, if the reduced mass for this motion is finite, the quantum mechanical wave packet can tunnel through the barrier, $V(R)$, rather than climb over the top (Figure 11). The probability of reactants with energy, E, tunneling through such a barrier is determined by the initial and final wave amplitudes:

$$\kappa(E) = \left| \frac{\psi[R_f(E), E]}{\psi[R_i(E), E]} \right|^2 \tag{18}$$

Eckart introduced (56) a simple mathematical form for the potential energy function, $V(R)$:

$$V(R) = \frac{\Delta E_{reaction}}{e^{-2\pi R/L} + 1} + \frac{\left(\sqrt{\Delta E_{forward}^{\ddagger}} + \sqrt{\Delta E_{reverse}^{\ddagger}}\right)^2}{e^{2\pi R/L} + 2 + e^{-2\pi R/L}} \tag{19}$$

for which he determined the exact transmission probability (56):

$$\kappa(E) = 1 - \frac{\cosh(\alpha - \beta) + \cosh \delta}{\cosh(\alpha + \beta) + \cosh \delta} \tag{20}$$

in terms of the intermediate quantities:

$$\alpha = \frac{L}{\hbar}\sqrt{2\mu E}, \qquad\qquad \beta = \frac{L}{\hbar}\sqrt{2\mu(E - \Delta E_{reaction})},$$

and: $\tag{21}$

$$\delta = \frac{L}{\hbar}\sqrt{2\mu\left(\sqrt{\Delta E_{forward}^{\ddagger}} + \sqrt{\Delta E_{reverse}^{\ddagger}}\right)^2 - \frac{h^2}{4L^2}}.$$

This transmission probability must then be weighted by the Boltzmann distribution function. The final expression for the rate constant is therefore:

$$k_{rate}(T) = \frac{Q^{\ddagger}(T)}{Q_{R_1}(T)\,Q_{R_2}(T)} \cdot \frac{k_B T}{h} \int_0^{\infty} \kappa(E)\, e^{-E/k_B T}\, dE \tag{22}$$

Truhlar and Kuppermann introduced (44) the correct definition of the reaction path, R, in mass weighted coordinates and were the first to recognize the importance of including in $V(R)$, the quantum mechanical zero-point vibrational energy for all normal modes that are orthogonal to R. The challenge we face is to accurately determine all the quantities required to evaluate equation 22.

Vibrational Frequencies. Vibrational energy levels include harmonic components linear in the quantum number, n, and anharmonic corrections:

$$E_n = \left(n + \tfrac{1}{2}\right)hc\omega_e - \left(n + \tfrac{1}{2}\right)^2 hc\omega_e x_e + \cdots \qquad (23)$$

The harmonic term generally dominates the low-lying levels, especially for the stiff normal modes that dominate zero-point energies. For example, the values of ω_e and $\omega_e x_e$ for H_2 are 4401.23 cm^{-1} and 121.34 cm^{-1} respectively (47). We therefore employ the harmonic approximation for all calculated frequencies and zero-point energies. The convergence of the harmonic frequencies obtained with the computational methods used for the IRC calculations is illustrated by the results for H_2 and the $H_2 + H$ transition state in Table V (the standard CBS nand G2 methods are in bold). Note the extremely

Table V. Convergence of calculated harmonic frequencies (cm^{-1})

Method/Basis Set	H_2	H_3		
	Stretch	Stretch	Bend(2)	Rxn Path
UHF/3-21G	4657.1	2054.8	1121.4	2292.2 i
MP2/6-31G*	4533.6	2087.8	1164.6	2325.5 i
MP2/6-31G**	4609.2	2173.1	1037.7	2108.4 i
MP2/6-311G**	4533.2	2152.7	973.7	1994.0 i
QCISD/6-311G**	4420.1	2054.2	958.2	1589.1 i
QCISD(T)/6-311+G(2df,2pd)	4404.1	2050.3	917.6	1506.4 i
CBS-QCI/APNO				1412.3 i
Experiment	4401.23			

slow convergence of the imaginary frequency for the H--H--H transition state. Also note that both p-polarization functions and a triple-ζ valence basis on the central hydrogen are necessary to obtain an accurate bending frequency. Without them, we would overestimate the ZPE of the transition state by 0.71 kcal/mol. To minimize such errors, we shall employ QCISD(T)/6-311+G(2df, 2pd) frequencies for our initial study of absolute reaction rates.

The change in the total ZPE at the H_3 transition state decreases the barrier height, $\Delta E^{\ddagger}$ in equation 19, by 0.74 kcal/mol, and thus accelerates the reaction by more than a factor-of-three at room temperature. The variation of the H_3 stretch and bend as

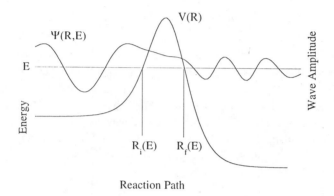

Figure 11. Tunneling through an Eckart barrier.

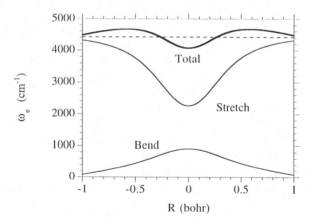

Figure 12. The vibrational frequencies for the reaction: $H_2 + H \rightarrow H + H_2$ vary along the reaction path. The TS is located at R=0.

we move along the reaction path increases $V(R)$ at $R = \pm 0.5$ bohr (1 bohr = 0.529177 Å) to either side of the maximum (Figure 12), and thus broadens the Eckart barrier, equation 19, as illustrated for $H_2 + D \rightarrow H + HD$ in Figure 13. This increase in the Eckart width parameter, L in equation 19, reduces the tunneling through the barrier and hence the calculated reaction rate by a factor-of-ten at low temperatures.

Changes in zero-point energies along the reaction paths for the exothermic reactions can increase the barrier height (by 0.6 kcal/mol for $H_2 + OH$) or decrease the barrier height (by 0.4 kcal/mol for $H_2 + F$) depending on the stiffness of the bending force constant (Figure 12). Accurate rate constants can only be obtained if we include the ZPE in our determination of the IRCMax transition state geometry and energy. Our final algorithm is thus an adaptation of Truhlar's "zero curvature variational transition state theory" (ZC-VTST) (57) to our CBS models (16, 17) through use of the IRCMax technique (46).

Absolute Reaction Rates. We are finally ready to evaluate all of the terms in equation 22. We have selected five hydrogen abstraction reactions for the initial test of our methodology. The barrier heights for these reactions range from 1.3 kcal/mol ($H_2 + F$) to 20.6 kcal/mol ($H_2O + H$). We include temperatures from 250 K to 2500 K. The rate constants range from 10^{-18} up to 10^{-10} cm^3 / molecule sec. All absolute rate constants obtained from our CBS-QCI/APNO model are within the uncertainty of the experiments (58) (Figure 14). The dashed curves and open symbols for $H_2 + OH$, $H_2 + D$, and $D_2 + H$ represent the least-squares fits of smooth curves to large experimental data sets in an attempt to reduce the noise level in the experimental data (58). The close agreement with theory suggests that this attempt was successful.

The problem of predicting absolute rates for a wide range of gas phase chemical reactions has in principle been solved. We have made contact with experiment, demonstrating that in favorable circumstances a sufficiently accurate evaluation of equation 22 can provide absolute rate constants within the uncertainty of experimental rates. We must now determine the precise limitations of this model. Reactions lacking a well defined transition state pose an obvious challenge.

The use of the IRCMax method with the G2 and CBS-Q computational models is giving immediate improvements in our ability to model flame chemistry (see chapter by R. J. Berry, M. Schwartz, and P. Marshall). However, much remains to be done. We must now systematically determine the minimum level of calculation necessary to maintain the accuracy achieved above for each part of these calculations. It is only through such a systematic study that we will be able to extend this very high accuracy to the much larger species generally required for solution of practical problems in chemistry.

We can partition the evaluation of equation 22 into several steps:

1. Find the transition state at a low-level of theory.
2. Determine a portion of the IRC at this low-level.
   Determine the harmonic ZPE, $\Delta E_{ZPE}(\mathbf{R})$, at this low-level along this path.

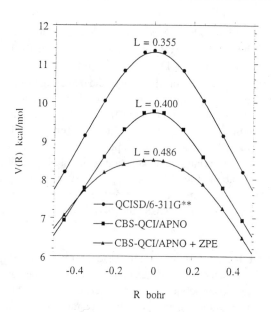

Figure 13. The energy profile along the reaction path for $H_2 + D \rightarrow H + HD$ varies with the method of calculation.

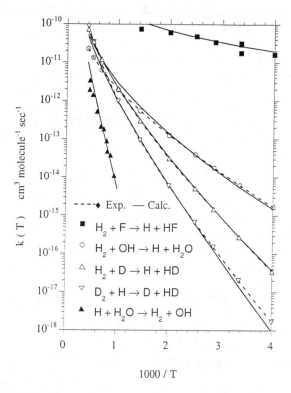

Figure 14. Calculated and experimental absolute rate constants

3. Determine the high-level electronic energy, $E_e(\mathbf{R})$, along a portion of the low-level IRC. The maximum gives the IRCMax geometry, $\mathbf{R}_e^{\ddagger}$, and energy, $E_e^{\ddagger}$.

4. Add $\Delta E_{ZPE}(\mathbf{R})$ to $E_e(\mathbf{R})$ giving $E_0(\mathbf{R})$ along the path. The maximum of $E_0(\mathbf{R})$ gives the zero curvature variational transition state theory (ZC-VTST) (57) IRCMax geometry, $\mathbf{R}_0^{\ddagger}$, and energy, $E_0^{\ddagger}$.

5. The harmonic frequencies at this geometry, $\{\omega_e(\mathbf{R}_0^{\ddagger})\}$, are used to evaluate the vibrational partition functions.

6. The rotational partition functions are evaluated at $\mathbf{R}_0^{\ddagger}$.

7. An Eckart function, $V_0(\mathbf{R})$ in equation 19, is least-squares fit to the potential function, $E_0(\mathbf{R})$. This Eckart function is then used to evaluate the transmission probability, $\kappa(E)$, in equation 20.

8. The absolute rate constant is then evaluated by numerical integration of equation 22.

Each of these steps must be converged to the desired accuracy (± 1.3 kcal/mol for CBS-4, ± 0.6 kcal/mol for CBS-Q, and ± 0.3 kcal/mol for CBS-QCI/APNO) with respect to both the one-electron expansion and the many-electron expansion. In addition, we must examine the convergence of the partition functions, and we must check for convergence with respect to the anharmonic expansion of the vibrational energy levels in equation 23, and complications such as centrifugal distortion.

Summary and Conclusions

The slow convergence of the correlation energy with the one-electron basis set expansion has provided the motivation for the complete basis set (CBS) extrapolations described in this chapter. The CBS method employs the N^{-1} asymptotic convergence of pair natural orbital (PNO) expansions to estimate basis set truncation errors in the correlation energy.

These CBS extrapolations have been incorporated into a hierarchy of computational thermodynamics models ranging from the very fast CBS-4 model applicable to up to twenty nonhydrogen atoms and giving RMS errors of 2.5 kcal/mol, through the CBS-Q model applicable to up to ten nonhydrogen atoms giving RMS errors of 1.3 kcal/mol, and finally to the most accurate CBS-QCI/APNO model limited to only five nonhydrogen atoms, but giving RMS errors of 0.7 kcal/mol.

Since the errors in the CBS thermochemical models fit Gaussian distribution functions, the RMS errors for the G2 test set can be used in the same way that experimental RMS deviations are used to assess the uncertainty of measured values. However, we should also be alert to the possibility of systematic errors for which empirical corrections such as bond additivity corrections can be useful.

Rate constants describe how fast individual chemical reactions occur, absolutely and in comparison to other reactions. The electronic energy of the transition state is of primary importance and has thus been the focus of our work to date. The IRCMax method in combination with our CBS extrapolations again provides a hierarchy of models achieving chemically useful accuracy for the calculated barrier heights (± 1.3 kcal/mol for CBS-4, ± 0.6 kcal/mol for CBS-Q, and ± 0.3 kcal/mol for CBS-QCI/APNO). The zero point energy (ZPE) contributions must be calculated to still higher accuracy (± 0.6 kcal/mol for CBS-4, ± 0.3 kcal/mol for CBS-Q, and ± 0.15 kcal/mol for CBS-QCI/APNO) so that we do not compromise these hard won gains.

The energy of the transition state is of primary importance, and we have developed a new computational method to improve the accuracy of these calculated energies. Also of great importance is the quantum mechanical tunneling, which depends on the width of the potential barrier. This barrier width controls the rates at low temperatures where comparisons with experiment are more easily made. If variations in the zero-point vibrational energy along the reaction path are taken into account, zero curvature variational transition state theory (57) quantitatively accounts for tunneling effects. By working with simple reactions, we demonstrated that our new methods give excellent agreement with experimental results for a wide range of temperatures and reaction rates.

The tunneling calculations presented in this chapter did not include "bob sled" or "corner cutting" effects (59, 60). The close agreement with experiment must therefore be the result of a cancellation of this omission internally or with some other error. If we are to have confidence in our final method, it is imperative that we identify the source of this effect, and adjust our methodology according to our findings.

The problem of predicting absolute rates for a wide range of gas phase chemical reactions has been solved. However, much remains to be done. We must systematically determine the minimum level of calculation necessary to maintain the accuracy of each part of these calculations. Nevertheless, the use of the IRCMax method with the G2 and CBS-Q computational models is giving immediate improvements in our ability to solve practical problems (see chapter by R. J. Berry, M. Schwartz, and P. Marshall).

Acknowledgments

The complete basis set extrapolation was developed in collaboration with Marc R. Nyden, Thomas G. Tensfeldt, and Mohammed A. Al-Laham. The CBS-4, CBS-Q, and CBS-QCI/APNO models were developed in collaboration with Joseph W. Ochterski and John A. Montgomery, Jr. The IRCMax method was developed in collaboration with John A. Montgomery, Jr. and David Malick. The absolute rate calculations were a collaborative effort with Paul Marshall. We are grateful to the Air Force Office of Scientific Research, the Petroleum Research Fund, and Gaussian. Inc. for their continued support of this research.

Literature Cited

1. Schwartz, C. *Phys. Rev.* **1962**, *126*, 1015.
2. Schwartz, C. In *Methods in Computational Physics*; Alder, B.; Fernbach, S.; Rotenberg, M., Eds., Academic: New York, 1963, Vol. 2.

3. Nyden, M. R.; Petersson, G. A. *J. Chem. Phys.* **1981**, *75*, 1843.
4. Petersson, G. A.; Nyden, M. R. *J. Chem. Phys.* **1981**, *75*, 3423.
5. Petersson, G. A.; Licht, S. L. *J. Chem. Phys.* **1981**, *75*, 4556.
6. Brown, F. B.; Truhlar, D. G. *Chem. Phys. Lett.* **1985**, *117*, 307.
7. Dunning, T. H., Jr. *J. Chem. Phys.* **1989**, *90*, 1007.
8. Siegbahn, P. E. M.; Blomberg, M. R. A.; Svensson, M. *Chem. Phys. Lett.* **1994**, *223*, 35.
9. Møller, C.; Plesset, M. S. *Phys. Rev.* **1934**, *46*, 618.
10. Löwdin, P. O. *Phys. Rev.* **1955**, *97*, 1474.
11. Petersson, G. A.; Yee, A. K.; Bennett, A. *J. Chem. Phys.* **1985**, *83*, 5105.
12. Petersson, G. A.; Braunstein, M. *J. Chem. Phys.* **1985**, *83*, 5129.
13. Petersson, G. A.; Bennett, A; Tensfeldt, T. A.; Al-Laham, M. A.; Shirley, W. A.; Mantzaris, J. *J. Chem. Phys.* **1988**, *89*, 2193.
14. Petersson, G. A.; Al-Laham, M. A.; *J. Chem. Phys.* **1991**, *94*, 6081.
15. Petersson, G. A.; Tensfeldt, T. G.; Montgomery, J. A., Jr. *J. Chem. Phys.* **1991**, *94*, 6091.
16. Montgomery, J. A., Jr.; Ochterski, J. W.; Petersson, G. A. *J. Chem. Phys.* **1994**, *101*, 5900.
17. Ochterski, J. W.; Petersson, G. A.; Montgomery, J. A., Jr. *J. Chem. Phys.* **1996**, *104*, 2598.
18. Frisch, M. J.; Trucks, G. W.; Schlegel, H. B.; Gill, P. M. W.; Johnson, B. G.; Robb, M. A.; Cheeseman, J. R.; Keith, T.; Petersson, G. A.; Montgomery, J. A., Jr., Raghavachari, K.; Al-Laham, M. A.; Zakrzewski, V. G.; Ortiz, J. V.; Foresman, J. B.; Peng, C. Y.; Ayala, P. Y.; Chen, W.; Wong, M. W.; Andres, J. L.; Replogle, E. S.; Gomperts, R.; Martin, R. L.; Fox, D. J.; Binkley, J. S.; Defrees, D. J.; Baker, J.; Stewart, J. P.; Head-Gordon, M.; Gonzalez, C.; Pople, J. A. *Gaussian94*, Revision B.3: Gaussian, Inc.: Pittsburgh PA, **1995**.
19. Pople, J. A. In *Energy, Structure and Reactivity*, Smith, D. W; McRae, W. B., Eds.; Wiley: New York, 1973; p 51.
20. Roothaan, C. C. J. *Rev. Mod. Phys.* **1951**, *23*, 69.
21. Pople, J. A.; Nesbet, R. K. *J. Chem. Phys.* **1959**, *22*, 571.
22. Peng, C.; Schlegel, H. B. *Israel J. Chem.* **1993**, *33*, 449.
23. Handy, N. C.; Schaefer, H. F., III, *J. Chem. Phys.* **1984**, *81*, 5031.
24. Pople, J. A.; Krishnan, R.; Schlegel, H. B.; Binkley, J. S. *Int. J. Quant. Chem. Symp.* **1979**, *13*, 325.
25. Schlegel, H. B. *J. Comput. Chem.* **1982**, *3*, 214.
26. Head-Gordon, M.; Head-Gordon, T. *Chem. Phys. Lett.* **1994**, *220*, 122.
27. Frisch, M. J.; Head-Gordon, M.; Pople, J. A. *Chem. Phys. Lett.* **1990**, *166*, 275.
28. Frisch, M. J.; Head-Gordon, M.; Pople, J. A. *Chem. Phys. Lett.* **1990**, *166*, 281.
29. Hehre, W. J.; Radom, L.; Schleyer, P. v. R.; Pople, J. A. *Ab Initio Molecular Orbital Theory*; John Wiley & Sons: New York, 1986.
30. Foresman, J. B.; Frisch, A. *Exploring Chemistry with Electronic Structure Methods*, 2nd ed.; Gaussian, Inc.: Pittsburgh, PA., 1996.

31. Pople, J. A.; Head-Gordon, M.; Fox, D. J.; Raghavachari, K.; Curtiss, L. A. *J. Chem. Phys.* **1989**, *90*, 5622.

32. Curtiss, L. A.; Jones, C.; Trucks, G. W.; Raghavachari, K.; Pople, J. A. *J. Chem. Phys.* **1990**, *93*, 2537.

33. Curtiss, L. A.; Raghavachari, K.; Trucks, G. W.; Pople, J. A. *J. Chem. Phys.* **1991**, *94*, 7221.

34. Curtiss, L. A.; Raghavachari, K.; Pople, J. A. *J. Chem. Phys.* **1993**, *98*, 1293.

35. Dewar, M. J. S.; Zoebisch, E. G.; Healy, E. F. *J. Am. Chem. Soc.* **1985**, *107*, 3902.

36. Becke, A. D. *J. Chem. Phys.* **1993**, *98*, 5648.

37. Becke, A. D. *Phys. Rev. A* **1988** *38*, 3098.

38. Lee, C; Yang, W.; Parr, R. G. *Phys. Rev. B* **1988**, *37*, 785.

39. Vosko, S. H.; Wilk, L.; Nusair, M. *Can. J. Phys.* **1980**, *58*, 1200.

40. Ochterski, J. W.; Petersson, G. A.; Wiberg, K. B. *J. Am. Chem. Soc.* **1995**, *117*, 11299.

41. Berry, R. J.; Burgess, D. R. F., Jr.; Nyden, M. R.; Zachariah, M. R.; Schwartz, M. *J. Phys. Chem.* **1995**, *99*, 17145.

42. Ho, P.; Melius, C. F. *J. Phys. Chem.* **1990**, *94*, 5120.

43. Allendorf, M. D.; Melius, C. F. *J. Phys. Chem.* **1993**, *97*, 72.

44. Truhlar, D. G.; Kuppermann, A. *J. Am. Chem. Soc.* **1971**, *93*, 1840.

45. Gonzalez, C.; Schlegel, H. B. *J. Phys. Chem.* **1989**, *90*, 2154.

46. Malick, D. K.; Petersson, G. A.; Montgomery, J. A., Jr. manuscript in preparation.

47. Huber, K. P.; Herzberg, G. *Constants of Diatomic Molecules*; Van Nostrand Reinhold: New York, 1979.

48. Gray, D. L.; Robiette, A. G. *Mol. Phys.* **1979**, *37*, 1901.

49. Hirota, E. *J. Mol. Spectrosc.* **1979**, *77*, 213.

50. Chase, M. W.; Davies, C. A.; Downey J. R.; Frurip, D. J.; Mc Donnell, R. A.; Syverud, A. N. *J. Phys. Chem. Ref. Data* **1985**, *14, Suppl. 1*.

51. Diedrich, D. L.; Anderson, J. B. *Science* **1992**, *258*, 786.

52. Scuseria, G. E. *J. Chem. Phys.* **1991**, *95*, 7426.

53. Knowles, P. J.; Stark, P.; Werner, H. *Chem. Phys. Lett.* **1991**, *185*, 555.

54. Arrhenius, S. *Z. Phys. Chem.* **1889**, *4*, 226.

55. Eyring, H. *J. Chem. Phys.* **1935**, *3*, 107.

56. Eckart, C. *Phys. Rev.* **1930**, *35*, 1303.

57. Truhlar, D. G. *J. Chem. Phys.* **1970**, *53*, 2041.

58. NIST Chemical Kinetics Database, version 6.0.

59. Marcus, R. A. *J. Chem. Phys.* **1966**, *45*, 4493.

60. Skodje, R. T.; Truhlar, D. G.; Garrett, B. C. *J. Chem. Phys.* **1982**, *77*, 5955.

Chapter 14

Computational Thermochemistry
and Transition States

Joesph L. Durant

**Combustion Research Facility, Sandia National Laboratories,
Livermore, CA 94551-0969**

Recent advances in ab initio methods and computer hardware have made
the accurate calculation of properties of transition states possible. We
review some terminology necessary for our discussion, and proceed to
examine multireference, single reference and density functional
techniques for characterizing transition states. The results of
calculations carried out using the three different methods are compared
to gain insight into the overall accuracy of property prediction.

The study of chemical transformations is at the heart of chemistry. And in
understanding chemical transformations we must understand the "transition states"
which a molecule goes through as it moves from reactant to product. These "transition
states" have but a fleeting existence, and are only now beginning to yield up a few of
their secrets to the experimentalist. In the meantime the characterization of transition
states remains a task for the theoretician.

The inability to directly measure the energies, geometries and frequencies of
transition states presents a problem: how do we know when we have the right
answer? The information which is experimentally accessible, reaction rates as
functions of temperature and pressure, as well as kinetic isotope effects, do not yield
unique values for transition state properties, since the problem of inverting rate data to
obtain molecular properties is highly underspecified. Instead, what is typically done is
to estimate transition state properties, often with reference to Benson's
thermochemical kinetic method (*1*), and then modify the properties to obtain good
agreement with experiment. While this method provides a physically-based
parameterization of the rate coefficient data it does not produce a unique
parameterization, and it cannot be used to obtain definitive transition state properties.
Conversely, it is sometimes necessary to go to great lengths to calculate accurate
transition state theory rate coefficients even when the transition state properties are

known, since some reactions are quite sensitive to treatment of hindered rotors, tunnelling contributions and variational effects (_2_). Reproducing experimental rate data is a necessary, but not sufficient, condition for evaluating correctness of transition state properties.

In the absence of clear experimental guidelines we will use consensus to determine what the "right" answer is. In recent years it has become possible to carry out very high level calculations using different methodologies on small reactive systems. Thankfully these calculations, which are claimed to·be accurate and precise by their progenitors, often do agree to within quoted error limits! In instances where these calculations have been used as input to sophisticated rate constant calculations predictions have been in generally good agreement with experimental observations.

In the remainder of this chapter we will first briefly touch on some of the terminology we will use, and also spend some time discussing Arrhenius activation energies and energetic barriers, highlighting why they are not the same. This will be followed by a brief discussion of the problems specific to calculation of properties of transition states, followed by a survey of methods used for calculating transition state properties, and a comparison between some of the results of these calculations.

Our purpose in this chapter is to paint a picture in broad strokes; there is not space for an encyclopedic survey of this field. In particular we do not have space to cover contributions from experimental kinetics, or the field of direct dynamics calculations, either of which could be the subject of a review of considerable length. We have attempted to touch on important problems and important methodologies, but have left some problems and some methods out. Because of the aforementioned problems with knowing the "right" answer, we have emphasized work which compares, contrasts and attempts to refine theoretical characterization of transition state properties.

Terminology

Before we launch into the bulk of this discussion we want to clarify our use of the terms "stationary points", "equilibrium structures, states or species" and "transition states."

First, we will identify certain portions of the super-molecule's potential energy surface as "stationary points." A stationary point is simply a place on the surface where all the first derivatives of the potential energy are zero. A molecule dropped onto a stationary point does not feel any forces, and is (meta)stable there.

If all of the second derivatives at the stationary point are positive we have located a minimum on the surface. We will call these "equilibrium structures" or "equilibrium states", regardless of whether they correspond to stable molecules or reactive radicals, and we will refer to these species as "equilibrium species."

If the stationary point has exactly one negative second derivative we call it a "transition state." In contrast, transition state theory generally defines a transition state as a minimum-flux surface dividing reactants and products. These two definitions will often apply to the same place in configuration space, but they are not required to, and will not for certain classes of reactions. Most often this is encountered when the barrier to reaction is small; the asymmetric change of vibrational frequencies across the reaction coordinate leads to the minimum flux (maximum free energy) dividing surface being displaced to the high frequency side of the calculated stationary point. Rate coefficients for reactions of this type are typically calculated using a variational version of transition state theory, with the position of the transition state being varied along the reaction coordinate until a minimum in the reactive flux is found.

Barrier Heights and Activation Energies

Barrier heights are not equal to activation energies. Activation energies are not equal to barrier heights.

The activation energy, or more precisely the Arrhenius activation energy, is obtained from the derivative of the logarithm of the reaction rate coefficient with respect to the inverse temperature ($\mathcal{3}$),

$$E_a = -R \, d[\ln(k)]/d(1/T).$$

As experimentalists have become more precise in their measurements of rate coefficients over ever broader temperature ranges, it has become apparent that most Arrhenius plots are curved. This necessitates amending the above definition by requiring specification of the temperature at which the derivative is evaluated, or using a modified Arrhenius equation of the form

$$k = A \, T^n \exp(-E_a/RT)$$

to fit the data. In this latter case, n and E_a are generally highly correlated, and often become little more than fitting parameters.

The direct result of an ab initio electronic structure calculation is the electronic energy of a species. Taking the difference between the electronic energy of a transition state and the electronic energy of the reactants we can calculate the classical barrier height for a reaction. This quantity is equivalent to Johnston's potential energy of activation ($\mathcal{4}$). To obtain the barrier height generally used in RRKM (and other) calculations we must correct for zero point effects, adding the zero point energy of the transition state and subtracting the zero point energy of the reactants. This zero point corrected barrier height is sometimes termed the "activation energy at absolute zero ($\mathcal{5}$)."

These various energies are all different, and should not be used interchangeably. In particular, neither the classical barrier height or the zero point corrected barrier is equal to the Arrhenius activation energy; typically they will differ by several kJ/mol ($\mathcal{4}$, $\mathcal{6}$, $\mathcal{7}$).

Problematic Characteristics of Transition States

Transition states present a number of challenges to the computational chemist. The most obvious problem relates to locating the transition state. Locating a minimum on a potential energy surface is relatively straightforward; determine which direction is downhill and go that direction. However, a transition state is defined as a maximum in one dimension, and a minimum in the other n-1 degrees of freedom. Thus, it is not simply a question of walking uphill, but rather one of walking in the correct uphill direction. From the bottom of the well it is not always obvious what the correct direction is. Sometimes a good guess can be made of the transition state's geometry. If the guess is good enough the reaction coordinate can be identified by its negative eigenvalue; then the correct direction is known and the same types of methods which are used to locate minima can be used to refine the transition state's geometry.

However, getting "close enough" to the correct transition state geometry on the first try is difficult. One solution is to perform a potential energy surface scan, slowly varying a coordinate, hopefully one close to the true reaction coordinate, performing a constrained energy minimization at each step, and looking for the maximum energy along the path. This approach works reasonably well for simple bond fissions and metathesis reactions, while it is considerably more difficult to implement for multicenter eliminations or intramolecular atom transfers. Because many reaction barriers are the result of avoided crossings between different electronic surfaces care must be taken to ensure that the surface scan stays on the lowest lying electronic surface. Otherwise it is possible to follow the incorrect electronic surface quite far beyond the location of the true transition state, before the SCF solution falls through to the (now) lower electronic surface. SCF stability calculations are useful in this regard since they will, presumably, detect the presence of lower-lying SCF solutions; they are also frequently necessary to correct very bad initial SCF guesses. Initial guesses of SCF orbital occupancies are generally good for equilibrium species, but bonding in transition states can be considerably different, and frequently the initial SCF guess for a species far from equilibrium will lead to an unstable SCF solution, or even a lack of SCF convergence. Fortunately, SCF convergence problems can normally be overcome by judicious use of non-default parameters and methods in the electronic structure code.

Considerable effort has been devoted to implementing various "synchronous transit" algorithms. These algorithms attempt to use information about the reactant and product geometries to locate the transition state. The concept is straightforward; the transition state is assumed to be on (or at least near) a path interpolated between reactants and products. Linear and quadratic interpolation schemes have been proposed, allowing the interpolated path to be searched for a maximum, which is assumed to be close to the transition state. Linear synchronous and quadratic synchronous transit algorithms were introduced by Halgren and Lipscomb (8), and their use further examined and refined by Schlegel and coworkers (9, 10). They have been implemented in the Gaussian suite of programs (11), but the author's experience in using them to locate transition states of three and four center eliminations and intermolecular H-atom transfers demonstrates that they are still not robust algorithms.

Another approach is to find the correct uphill direction and follow it (*12-17*). These methods attempt to follow a minimum energy path uphill to the approximate location of the transition state, after which the transition state geometry can be optimized using conventional techniques. A disadvantage of the basic method is that it follows the "softest" mode uphill, and thus will only locate one of the transition states associated with a minimum. For multi-atom systems following the softest mode will often lead to transition states for isomerization; work still needs to be done to incorporate constraints which will allow other transition states to be located (*17*).

Once an optimized structure has been found, it should be checked for the presence of one and only one imaginary frequency. Additionally, it is often useful to follow the intrinsic reaction coordinate, IRC, from the structure to reactants and products, thus ensuring that the structure corresponds to the desired reaction.

Once a transition state has been located and confirmed, generally at a low level of theory, the next task is to refine the geometry at higher levels of theory. As we will see below, many of the most accurate quantum chemical methods are very computationally expensive, and often require costly numerical evaluation of energy gradients. Because the dimensionality of a geometry optimization problem grows as $3N-6$ it is clear that the number of single point energies required to calculate a gradient will rapidly grow with the size of the problem, as will the size of the space which must be explored to find the optimum geometry. This has led to the common practice of using geometries calculated at low levels of theory together with single point energies calculated at high levels of theory. The success of this approach hinges on the accuracy of the geometries calculated at the lower levels of theory. There is considerable experience in this for equilibrium species; it is generally considered that approaches such as MP2/6-31G(d) yield good geometries for the vast majority of stable species (*18*). However, there is not a similar body of experience with respect to transition states. Experience in developing the G2Q methodology suggests that MP2/6-31G(d) geometries are not sufficiently accurate for transition states (*19*). There are instances of G2 single point energies varying by as much as 57 kJ/mol between MP2/6-31G(d) and QCISD/6-311G(d,p) geometries (*20*).

The problem of evaluating geometries and energies at different levels of theory is particularly a problem with small reaction barriers, such as are often encountered for association reactions. Page and Soto have carried out a careful study of this phenomenon for the association reaction between H and HNO (*21*). They have documented the expected behavior: as the level of theory is increased, the barrier height decreases and the barrier moves toward the higher energy asymptote. In situations like this high level single-point energy calculations carried out at lower level geometries often give energies below the product asymptote, despite the fact that calculations carried out at higher level geometries still show an energetic barrier.

Computational Approaches

Multi-reference Methods. The electronic Schroedinger equation can be solved exactly using configuration interaction (CI) methods provided we employ complete basis sets and include all configurations in the CI calculation (*22*). Unfortunately, this calculation has very unfavorable scaling with system size, scaling roughly as the factorial of the system size (*23*), and is not presently feasible except for very small systems. Calculations using all configurations, so called "full configuration interaction" calculations, have been performed for some small systems with limited basis sets, and many CI studies include an attempt to extrapolate to the complete basis set limit. Experience from these studies leads us to conclude that properly performed multi-reference CI (MRCI) calculations can yield accurate single point energies (*22*).

"Properly performed" is an important caveat, since poor choices for the active space, as well as poor choices in the configurations to be included in the CI, can seriously compromise the calculations. MRCI calculations are also size-inconsistent, which can lead to errors in calculating the energies of separated fragments unless care is taken. Computer resources often limit the size of the active space and the number of configurations which can be included in the CI. Lack of raw computer power makes older calculations suspect; however, small systems studied in the last 10 years seem to have had the resources necessary to achieve reasonable accuracy.

Often problems with MRCI calculations become evident in examining asymptotic behavior as the super-molecule separates into reactants or products. Often the exothermicity of one or more product channels will disagree with experimental values. Situations where the number and type of bonding interactions changes in going from reactants to products are often most difficult to treat. Thus a first step in assessing an MRCI study is to carefully examine the overall thermochemistry of the channels, as well as insuring accurate properties for the stable species treated.

MRCI methods lack analytic gradients for the electronic energy with respect to nuclear positions, making geometry optimization at high levels of theory impossible except for very small systems. A related problem arises when we attempt to evaluate frequencies using MRCI: we now need numerically evaluated second derivatives, and even more costly calculations. A common solution to this problem is to use complete active space SCF (CASSCF) geometries and frequencies.

Despite these limitations there is a growing number of systems for which MRCI calculations have been used at a uniformly high level to characterize transition states. Systems include $H + H_2$ (*24*), $H + N_2$ (*25*), $H + F_2$ (*26, 27*), $O(^3P) + HCl$ (*28*), HNO (*29*), $H + C_2H_4$ (*30*), $CN + H_2$ (*31*), and $Cl + H_2$ (*32*), to mention only a few.

HF and MPn. Single reference methods use a single Slater determinant wavefunction as the basis for calculating electronic energies. The simplest single reference method is Hartree Fock (HF), also known as Self Consistent Field (SCF), which does not include any correlation energy. The HF method has long been known to be highly inaccurate for transition states. A standard part of any book on electronic structure theory is a discussion of the fact that HF theory leads to asymptotically incorrect behavior for dissociation of closed shell species (*33-35*). Some of this poor behavior can be corrected by using the spin-unrestricted form of HF, but it still systematically overpredicts energetic barriers, and is also known to produce geometries of limited accuracy, even for equilibrium species (*35*).

However, HF theory does have its uses. Numerous studies have documented the extremely good performance of scaled HF/6-31G(d) calculations in reproducing frequencies of equilibrium species (*36-39*). Indeed, scaled HF/6-31G(d) outperforms other post-SCF methods examined until one uses QCISD, which is considerably more computationally expensive (*39*). Unfortunately this performance is not carried over to transition state species. Recent work has documented the relatively poor performance of HF, as well as MP2, in predicting vibrational frequencies for transition states (*19*). The errors in calculated frequencies are random, and cannot be significantly improved by judicious choice of a scaling factor, as is the case for HF prediction of vibrational frequencies for equilbrium species.

HF calculations are useful for ruling out the presence of barriers, since UHF systematically overpredicts barrier heights. If a UHF surface scan does not reveal the presence of an energetic barrier then higher level calculations will not reveal one. This is particularly useful when examining association reactions. Absence of a HF barrier is strong evidence for the reaction proceeding without a barrier; presence of a barrier means that more work must be done to establish the presence or absence of a barrier at higher levels of theory.

Second-order Møller-Plesset perturbation theory (MP2), with a variety of basis sets, has been used to obtain "better" transition state properties. MP2 corrects many of the most egregious errors of HF calculations, but still seems to systematically overpredict energy barriers. Geometries, while much improved over HF geometries, still have significant errors. While these errors are perhaps not bad enough to seriously degrade the already only qualitative accuracy of MP2 energies, they do affect accuracy of more accurate single point energy calculations. Thus, when MP2/6-31G(d) geometries are used for G2 single point calculations on transition states the accuracy of the combined procedure is degraded from that observed when performing G2 calculations of equilibrium structures (*19*). This is because an accurate energy is being calculated for a place on the potential energy surface which is not the transition state. Use of better geometries, such as QCISD geometries, restores the performance of the G2 method for transition states (vide infra). As noted above, MP2/6-31G(d) frequencies for transition states are worse than HF/6-31G(d) frequencies.

HF and MPn methods also are quite vulnerable to spin contamination. Transition states are notorious for having geometry-dependent spin contamination; this undoubtedly adds to the problems encountered in using these methods on transition states. Jensen has found that even small amounts of spin contamination can cause significant errors in calculated vibrational frequencies at the unrestricted MP2 (UMP2) level (*40*). Many authors have made use of projected UMP energies, PUMPn (*41*). While many production codes allow easy calculation of these energies they do not allow geometry optimizations or frequency calculations to be carried out based on the

PUMPn energetics. Thus one is faced with what may be relatively good single-point energies at bad geometries with bad vibrational frequencies.

Quadratic Configuration Interation (QCI) and Coupled Cluster(CC) methods are closely related techniques which are quite efficient at capturing most of the electron correlation energy (*42*). They both have implementations including single and double substitutions, QCISD and CCSD, as well as including triple substitutions, QCISD(T) and CCSD(T) (*43-46*). The methods have been extensively tested on "difficult" molecules, such as ozone, and shown to perform exceptionally well (*47-50*). QCISD (or CCSD) seems to be the minimum level of single reference theory necessary to accurately predict transition state geometries and frequencies (*19*); inclusion of triple excitations seems to be necessary for accurate treatment of energetics. QCISD and CCSD methods do not suffer from the degree of spin-contamination encountered in HF and MPn calculations. Geometries and frequencies calculated at these levels normally are close to those calculated in high level MRCI calculations (*19, 30*). Unfortunately these methods are computationally very expensive; CCSD has a formal scaling of N^6 and CCSD(T) has a formal scaling of N^7 (*23*).

G2/G2Q. The recent development of the Gaussian 2 (G2) method (*51-53*) as an accurate, fast prescription for undertaking ab initio electronic structure calculations has made it possible for non-specialists to carry out high quality characterizations of molecular species. The method was developed for use on molecules in their equilibrium geometries, and offers impressive performance for these species. Energies for the 150 species in the G2 benchmark set were reproduced with an average absolute deviation of 4.6 kJ/mol (*53*). G2 is a composite method, approximating a QCISD(T)/6-311+G(3df,2p) calculation by a series of smaller calculations. A QCISD(T)/6-311G(d,p) single point energy is used as the base for the calculation, with effects of modifying the basis set evaluated by a series of MP2 and MP4 calculations. (This interpretation of the method is somewhat different from, but mathematically identical to, the one originally proposed.) An empirical "High Level Correction" term based on the number of α and β electrons is added to yield the G2 energy. It should be noted that this correction term cancels out when evaluating barrier heights, removing empiricism from G2 calculations of barrier heights. The accuracy of the additivity assumptions used in G2 have been confirmed by Pople and coworkers (*54*) by comparing G2 and QCISD(T)/6-311+G(3df,2p) energies. Because of the ability of the QCISD(T) method to capture electron correlation energy (*43*), we expect that G2 will provide an accurate source of single point energies.

Durant and Rohlfing have examined the performance of G2 for transition states (*19*). They identified 12 transition states, mainly atom + diatom reactions, which had been "accurately" characterized by a variety of other techniques. These included high quality multi-reference CI calculations as well as lower quality theoretical calculations calibrated using high quality reaction rate calculations and extensive experimental data. In order to separate out errors in vibrational frequency calculations they compared classical barrier heights, and then further compared calculated vibrational frequencies and geometries.

They determined that the unmodified G2 method performed reasonably well for energies, although the vibrational frequencies had large errors, and the geometries had small errors. Because accurate vibrational frequencies are necessary for accurate rate constant calculation further attention was devoted to find accurate ways to calculate these frequencies. For the transition states they studied there was little improvement in going to MP2/6-31G(d) frequencies. This observation is consistent with observations of the relative performance of MP2/6-31G(d) and HF/6-31G(d) in predicting frequencies for equilibrium species (*39*).

Their solution was to use QCISD/6-311G(d,p) to calculate geometries and frequencies for the transition states. This approach led to much improved prediction of vibrational frequencies, as well as improved geometries and energies. Based on this, they proposed a modification of the G2 method, which they called G2Q, substituting QCISD/6-311G(d,p) geometries and frequencies for the scaled HF/6-31G(d) frequencies and MP2/6-31G(d) geometries of the original G2 method.

G2Q has been used for a number of other systems, both in NHx systems (*55-57*) and in chlorination reactions (*20*). Several of the NHx systems were studied concurrently by MRCI methods (*58-62*), and the results are in generally good agreement. One difference is found in comparing results for the reaction of NH with NO. The G2Q results were systematically displaced from the MRCI results by approximately 12 kJ/mol (*55*). The source of the offset was traced to the MRCI calculations, which failed to correctly reproduce the experimental exothermicity of the non-isogyric reaction $NH + NO \rightarrow N_2O + H$. The exothermicity of the isogyric $H + N_2O \rightarrow N_2 + OH$ reaction was well reproduced in the MRCI calculations, and when the transition state energies were evaluated relative to the $H + N_2O$ asymptote instead of the NH + NO asymptote the discrepancy with the G2Q results vanished. This demonstrates an important feature of G2 and G2Q calculations: their relatively uniform treatment of species, regardless of change in number of bonds or spin.

The study of the alkyl + Cl_2 reactions performed by Tirtowidjojo and coworkers was an instance in which the small differences in geometry between G2's MP2/6-31G(d) and G2Q's QCISD/6-311G(d,p) geometries made a huge difference in calculated barriers; the G2Q barrier for ethyl + Cl_2 is 57 kJ/mol lower than the G2 value. Comparison of calculated and measured rates for ethyl + Cl_2 suggest that the G2Q value is closer to reality than the G2 value (*20*).

In its unmodified form G2 has been used by a number of workers to characterize transition states (*63-67*). The results have been generally good, although many of the studies have been more qualitative. In a combined experimental/theoretical study Marshall and coworkers found it necessary to further refine the MP2/6-31G(d) frequencies (both imaginary and real) in their study of the reaction of H with SiH_4 (*66*).

A number of other modifications to the basic G2 method have been proposed and tested for equilibrium species (*68-72*). In general, they have been directed at replacing some of the most computationally expensive steps with less expensive calculations. They generally offer faster calculations at the price of less accuracy for equilibrium species, with generally untested performance for transition states.

BACMP4. BACMP4, or Bond Additivity Corrected MP4, is a method which has been applied to calculation of thermochemistry of both equilibrium and transition state species (for more information see the article by Zachariah and Melius in this volume). BACMP4 uses HF/6-31G(d) geometries, upon which MP4/6-31G(d) single point energies are calculated. Frequencies are calculated using scaled HF/6-31G(d) frequencies. The energy is corrected using a rather complicated parameterization, which takes into account bond lengths and types, as well as atom types. The parameters are determined by fitting a corpus of known thermochemistry. BACMP4 is a relatively inexpensive computational tool, and has been widely used, both on small systems and on larger systems.

Because of BACMP4's use of HF geometries we expect that it will have troubles producing accurate energies because of the inaccuracy of the underlying geometries. However, because it is a highly parameterized model, it can be thought of as basically an engineering approach to the prediction of transition state properties. Hartree Fock barriers are systematically high, so we expect that the transition states will be systematically displaced toward the lower energy reaction asymptote. As long as the errors remain systematic they can be corrected for in such an empirical scheme.

Density Functional Methods. Recent years have seen a renewed interest in the use of density functional methods for thermochemical calculations on gas phase species (*73, 74*). Density functional methods have very attractive scaling behavior with molecular size, behaving like HF methods (*75*). Additionally, they are much less affected by spin contamination (*76*), even for transition states (*77*). Best of all, they are reasonably accurate, with average absolute errors as small as 8 kJ/mol on the G2 benchmark set (*78*).

Hohenberg and Kohn have shown that there exists a functional which relates the electron density to the exact energy for the ground state of any multi-electron system (*79*). Unfortunately, their proof offers no clues as to how to find the functional, or even how to approach its performance in a systematic fashion. This last fact has presented the computational chemist with a problem. Typically calculations are carried out at increasingly high levels of theory, and the results carefully examined for evidence that the calculation has converged. Since the present functionals are not easily grouped into such a hierarchy such an approach is not possible.

Despite, or perhaps because, of this lack of a systematic path to better functionals, the development of new functionals has been the focus of considerable effort over the past few years (*78, 80-84*).

As was the case for G2, it is also necessary to examine the performance of the various density functionals as applied to transition states. There have been a number of studies which have used density functional theory (DFT) to characterize transition states (*85-98*), but only a few to systematically examine the performance of various functionals and basis sets when applied to transition states (*77, 99-102*). The various exchange functionals which do not include gradient corrections (eg. Xα, LSDA) have been largely abandoned outside of the solid-state community because of their poor performance in predicting gas phase thermochemistry (*74*). Studies utilizing gradient-corrected functionals have shown them to have mixed performance for transition states. Both Baker and coworkers (*101, 102*) and Durant (*77*) documented the failure of pure DFT functionals to predict even substantial barriers. Inclusion of Hartree-Fock exchange improves this situation, but does not completely solve it. Indeed, Durant found that the B3LYP, B3P86 and B3PW91 functionals all systematically underpredicted barrier heights (*77*); Baker et al. found a similar underprediction for the ACM functional (*102*). Durant did find that the Gaussian implementation of the Becke half-and-half/Lee, Yang and Parr (BH&HLYP) functional did not systematically under (or over)predict transition state barrier heights. Truong and coworkers examined a series of proton transfer reactions using DFT, MP2, MP4 and CCSD(T) (*100*). They found that, of the DFT methods they examined, only the Gaussian implementation of BH&HLYP was capable of predicting transition state structures and energetics with an accuracy comparable to MP2. This finding has been confirmed in later work by Truong and coworkers (*92*), who examined a number of functionals for use in their direct ab initio dynamics calculations. They again found that the BH&HLYP functional gave them the most uniformly accurate predictions of transition state properties.

In addition to the effects of varying functionals, we should also inquire into the effect of various basis sets on DFT calculations. Durant has examined performance of 6-31G(d), 6-311G(d,p), 6-311++G(d,p) and 6-311G(3df,2p) basis sets (*77*). He found that there was essentially no trend in calculated barrier heights, vibrational frequencies or geometries as the basis set was improved. This behavior has been seen with a study of DFT on equilibrium species; its origin has been ascribed to the different response of DFT to basis set incompleteness (*103*). Traditional electronic structure methods need an extensive virtual orbital space in order to calculate correlation energies; density functional methods only need the basis set to be flexible enough to represent the electron density accurately. Calculations of electron density seem to converge for basis sets of double-zeta quality or better (*104, 105*), suggesting that this size basis function may be large enough to give good quality DFT results, at least for neutral species.

The future looks even brighter for DFT. Becke has derived a new dynamic correlation functional and demonstrated excellent performance on the G2 benchmark set (*78*). This functional has not yet been widely implemented and tested, but it should offer improved performance for transition states. Other functional developments are continuing in the search for the functional which will allow exact calculations to be performed. And perhaps most interesting is the development of DFT methods which will scale linearly with problem size (*106-109*). These methods will allow DFT to be used on systems of hundreds of atoms, allowing calculations of transition states in biological molecules and in systems with explicit inclusion of solvent molecules.

Convergence of Results?

While not as satisfying as an experiment, the existence of independent calculations, using different techniques with different sources of error, can be used to judge the approach to a converged answer. And this convergence has in fact begun to occur. Thus we see for a number of triatomic systems there are multireference calculations, single reference calculations (G2Q, QCISD(T) or CCSD(T)) and density functional methods for which the error bars overlap (*19, 30, 77*).

In particular, a number of transition states have been systematically studied using all three of these methods. In Table I we have listed values for the classical energy barriers for several of these transition states. We have chosen to use classical barrier heights in order to uncouple errors made in calculations of frequencies (and hence zero point energies) from errors in the electronic energy evaluation. It is gratifying to note the generally good agreement between the different methods; BH&HLYP appears to have the most difference from MRCI and G2Q. Because G2 is often used for transition states we have included it in Table I also. For this small set of transition states it performs acceptably well.

Table I. Classical barrier heights (kJ/mol)[a]

	MRCI	G2	G2Q	BH&HLYP/ 6-311G(d,p)
$H+H_2$	40	45	45	27
$H+N_2$	64	60	61	48
$H+NO$ $(^3A")$	17	12	17	
$NH+O$ $(^5\Pi)$	23	22	23	
$NH+O$ $(^3\Pi)$	49	54	50	
$H+F_2$	9	8	12	13
$O(^3P)+H_2$	53	64	62	
$O+HCl$	36	44	44	52
$H+N_2O \rightarrow NN\text{-}OH$	61	76	61	57
$H+N_2O \rightarrow H\text{-}NNO$	40	41	37	33

[a]Data from references (*19*), (*77*) and references contained therein.

Table II presents frequencies for a number of transition states which have been studied using various theoretical approaches. In an effort to distill out some of the information in these frequencies we have calculated %errors ($\pm 1\sigma$), relative to the QCISD and the MRCI calculations. Again we see reasonably good agreement between the MRCI and QCISD/6-311G(d,p) results. The BH&HLYP/6-311G(d,p) frequencies are somewhat degraded, and the other methods tabulated are degraded even further.

Table III presents the imaginary frequencies calculated by each method. The imaginary frequency gives us a second-order picture of the shape of the barrier to reaction; the higher the frequency, the narrower the barrier, and the more important

Table II. Transition state frequencies (cm^{-1})^a

	CASSCF	MRCI	scaled HF/ 6-31G(d)	scaled MP2/ 6-31G(d)	QCISD/ 6-311G(d,p)	BH&HLYP/ 6-311G(d,p)
H-H$_2$		2062	1805	1970	2062	2097
		908	1004	1096	952	951
H-N$_2$		2072	1907	2741	2104	2305
		771	798	786	774	733
NH-O ($^5\Pi$)		855	523	669	804	
		600	470	392	566	
F-H$_2$		3768	1524	2498	3118	3249
O(^{3}P)-H$_2$			1292	1605	1589	
			642	595	489	
O(^{3}P)-HCl			393	1151	1310	1212
			368	255	345	326
NN-OH	2035		1765	2987	2113	2141
	1587		1280	2282	1665	1799
	890		805	969	1015	1062
	467		471	851	665	620
	948		958	1101	919	945
H-NNO	2033		1830	2611	2173	2304
	1259		1173	1380	1288	1361
	585		759	830	764	746
	464		375	465	411	413
	604		590	771	619	658
H-C$_2$H$_4$	469	404	370	472	374	
	501	428	397	497	406	
	838	832	787	808	828	
	868	882	830	930	908	
	1024	973	916	1043	983	
	1055	1031	927	1061	1037	
	1327	1256	1180	1209	1242	
	1344	1342	1196	1306	1346	
	1577	1481	1429	1440	1484	
	1686	1633	1512	1581	1632	
	3312	3224	2971	3044	3159	
	3323	3225	2981	3054	3173	
	3402	3308	3042	3126	3243	
	3426	3327	3064	3149	3268	
%error ± std. dev.						
QCISD	-1±12	-2±5	11±17	-5±17		-1±5
MRCI	-5±5		11±14	1±16	2±5	0±10

^{a}Data from references (*19*), (*77*), (*30*) and references contained therein. HF/6-31G(d) and MP2/6-31G(d) frequencies scaled by 0.8929 and 0.9427, respectively (*38*).

tunnelling can potentially become (assuming the reaction involves H-atom transfer). Alternatively, the lower the imaginary frequency, the broader the barrier, and the more likely that variational effects will become important. Examination of the imaginary frequencies is not too gratifying; even the values for MRCI and QCISD/6-311G(d,p) are somewhat different. Calculating the curvature along the reaction path at the barrier appears to be a test of ab initio methods for which we cannot yet claim to have converged answers.

Conclusions

In this brief chapter we have attempted to give the reader a flavor of the problems which characterizing transition states pose, as well as approaches which have been followed to surmount them. The problem of knowing what the "right" answer is still remains, although diverse theoretical approaches are beginning to converge on common

Table III. Imaginary frequencies (cm^{-1})^a

	CASSCF	MRCI	scaled HF/ 6-31G(d)	scaled MP2/ 6-31G(d)	QCISD/ 6-311G(d,p)	BH&HLYP/ 6-311G(d,p)
H-H$_2$		1496i	2138i	2188i	1566i	1160i
H-N$_2$		1662i	1711i	2180i	1615i	1437i
NH-O ($^5\Pi$)		1175i	3100i	2332i	1513i	
F-H$_2$		605i	2896i	1669i	1041i	799i
O(^{3}P)-HCl			2976i	1752i	1930i	1933i
NN-OH	1934i		2421i	2067i	2035i	2116i
H-NNO	1724i		1286i	2056i	1401i	1221i
H-C$_2$H$_4$	1188i	900i	588i	1143i	822i	

^{a}Data from references (*19*), (*77*), (*30*) and references contained therein. HF/6-31G(d) and MP2/6-31G(d) frequencies scaled by 0.8929 and 0.9427, respectively (*38*).

answers. MRCI, QCISD(T), CCSD(T), G2Q and DFT methods all appear to be able to make good predictions of energetics, geometries and real vibrational frequencies of transition states. The methods do not appear to have converged yet relative to the imaginary frequency.

Recent advances in both the speed of the computers and the efficiency of the algorithms should continue into the future, allowing more precise calculations to be performed on ever larger systems. New approaches to precise calculation of molecular properties, such as integrated molecular orbital plus molecular orbital (IMOMO) (*110, 111*) and complete basis set extrapolation (CBS), (*112*) continue to be developed. Of considerable promise are the twin developments in density functionals and in linear scaling algorithms for carrying out DFT calculations. With these tools the quantum chemist should be able to approach whole new classes of real world problems by the turn of the century.

Acknowledgments

This work was funded by the U.S. Dept. of Energy, Office of Basic Energy Sciences, Division of Chemical Sciences and by Sandia through its Laboratory Directed Research and Development program.

Literature Cited

1.   Benson, S. W. *Thermochemical Kinetics*, 2nd ed. John Wiley & Sons: New York, 1976.
2.   Truhlar, D. G.; Garrett, B. C.; Klippenstein, S. J. *J. Phys. Chem.* **1996**, *100*, 12771.
3.   Arrhenius, S. *Z. Phys. Chem.* **1889**, *4*, 226.
4.   Johnston, H. *Gas Phase Reaction Rate Theory*, Ronald Press Company: New York, 1966.
5.   Glasstone, S.; Laidler, K. J.; Eyring, H. *The Theory of Rate Processes*, McGraw-Hill Book Co.: New York, 1941.
6.   Robinson, P. J.; Holbrook, K. A. *Unimolecular Reactions*, Wiley-Interscience: New York, 1972.
7.   Gilbert, R. G.; Smith, S. C. *Theory of Unimolecular and Recombination Reactions*, Blackwell Scientific Publications: Oxford, 1990.
8.   Halgren, T. A.; Lipscomb, W. N. *Chem. Phys. Lett.* **1977**, *49*, 225.
9.   Peng, C.; Schlegel, H. B. *Isr. J. Chem.* **1993**, *33*, 449.
10. Peng, C.; Ayala, P. Y.; Schlegel, H. B.; Frisch, M. J. *J. Comp. Chem.* **1996**, *17*, 49.
11. Frisch, M. J.; Trucks, G. W.; Schlegel, H. B.; Gill, P. M. W.; Johnson, B. G.; Robb, M. A.; Cheeseman, J. R.; Keith, T.; Petersson, G. A.; Montgomery, J. A.; Raghavachari, K.; Al-Laham, M. A.; Zakrzewski, V. G.; Oritz, J. V.; Foresman, J. B.; Cioslowski, J.; Stefanov, B. B.; Nanayakkara, A.; Challacombe, M.; Peng, C. Y.; Ayala, P. Y.; Chen, W.; Wong, M. W.; Andres, J. L.; Replogle, E. S.; Gomperts, R.; Martin, R. L.; Fox, D. J.; Binkley, J. S.; Defrees, D. J.; Baker, J.; Stewart, J. P.; Head-Gordon, M.; Gonzalez, C.; Pople, J. A. *Gaussian 94, Revision B.1.* 1995, Gaussian, Inc: Pittsburgh, PA.
12. Cerjan, C. J.; Miller, W. H. *J. Chem. Phys.* **1981**, *75*, 2800.
13. Simons, J. P.; Jorgensen, P.; Taylor, H.; Ozment, J. *J. Phys. Chem.* **1983**, *87*, 2745.
14. Banerjee, A.; Adams, N.; Simons, J.; Shepard, R. *J. Phys. Chem.* **1985**, *89*, 52.
15. Baker, J. *J. Comp. Chem.* **1986**, *7*, 385.
16. Jorgensen, P.; Jensen, H. J. A.; Helgaker, T. *Theor. Chim. Acta* **1988**, *73*, 55.
17. Abashkin, Y.; Russo, N. *J. Chem. Phys.* **1994**, *100*, 4477.
18. DeFrees, D. J.; Levi, B. A.; Pollack, S. K.; Hehre, W. J.; Binkley, J. S.; Pople, J. A. *J. Am. Chem. Soc.* **1979**, *101*, 4085.
19. Durant, J. L.; Rohlfing, C. M. *J. Chem. Phys.* **1993**, *98*, 8031.

20. Tirtowidjojo, M. M.; Colegrove, B. T.; Durant, J. L. *I&EC Research* **1995**, *34*, 4202.

21. Page, M.; Soto, M. R. *J. Chem. Phys.* **1993**, *99*, 7709.

22. Bauschlicher, C. W.; Langhoff, S. R. *Science* **1991**, *254*, 394.

23. Head-Gordon, M. *J. Phys. Chem.* **1996**, *100*, 13213.

24. Boothroyd, A. I.; Keogh, W. J.; Martin, P. G.; Peterson, M. R. *J. Chem. Phys.* **1991**, *95*, 4343.

25. Walch, S. P. *J. Chem. Phys.* **1990**, *93*, 2384.

26. Bauschlicher, C. W.; Walch, S. P.; Langhoff, S. R.; Taylor, P. R.; Jaffe, R. L. *J. Chem. Phys.* **1988**, *88*, 1743.

27. Stark, K.; Werner, H.-J. *J. Chem. Phys.* **1996**, *104*, 6515.

28. Koizumi, H.; Schatz, G.; Gordon, M. S. *J. Chem. Phys.* **1991**, *95*, 6421.

29. Guadagnini, R.; Schatz, G. C.; Walch, S. P. *J. Chem. Phys.* **1995**, *102*, 774.

30. Hase, W. L.; Schlegel, H. B.; Balbyshev, V.; Page, M. *J. Phys. Chem.* **1996**, *100*, 5354.

31. ter Horst, M. A.; Schatz, G. C.; Harding, L. B. *J. Chem. Phys.* **1996**, *105*, 558.

32. Allison, T. C.; Lynch, G. C.; Truhlar, D. G.; Gordon, M. S. *J. Phys. Chem.* **1996**, *100*, 13575.

33. Hehre, W. J.; Radom, L.; Schleyer, P. v. R.; Pople, J. A. *Ab Initio Molecular Orbital Theory*, Wiley-Interscience: New York, 1986.

34. Hirst, D. M. *A Computational Approach to Chemistry*, Blackwell Scientific Publications: Oxford, 1990.

35. Bartlett, R. J.; Stanton, J. F. In *Reviews in Computational Chemistry*, Lipkowitz, K.B.; Boyd, D.B., Eds.; VCH Publishers: New York, 1994; Vol. 5; p. 65.

36. Hout, R. F.; Levi, B. A.; Hehre, W. J. *J. Comp. Chem.* **1982**, *3*, 234.

37. DeFrees, D. J.; McLean, A. D. *J. Chem. Phys.* **1985**, *82*, 333.

38. Pople, J. A.; Scott, A. P.; Wong, M. W.; Radom, L. *Isr. J. Chem.* **1993**, *33*, 345.

39. Scott, A.; Radom, L. *J. Phys. Chem.* **1996**, *100*, 16502.

40. Jensen, F. *Chem. Phys. Lett.* **1990**, *169*, 519.

41. Schlegel, H. B. *J. Chem. Phys.* **1986**, *84*, 4530.

42. Raghavachari, K. *Ann. Rev. Phys. Chem.* **1991**, *42*, 615.

43. Pople, J. A.; Head-Gordon, M.; Raghavachari, K. *J. Chem. Phys.* **1987**, *87*, 5968.

44. Purvis, G. D.; Bartlett, R. J. *J. Chem. Phys.* **1982**, *76*, 1910.

45. Urban, M.; Noga, J.; Cole, S. J.; Bartlett, R. J. *J. Chem. Phys.* **1985**, *83*, 4041.

46. Raghavachari, K. *J. Chem. Phys.* **1985**, *82*, 4607.

47. Raghavachari, K.; Trucks, G. W.; Pople, J. A.; Replogle, E. *Chem. Phys. Lett.* **1989**, *158*, 207.

48. Raghavachari, K.; Trucks, G. W. *Chem. Phys. Lett.* **1989**, *162*, 511.

49. Lee, T. J.; Scuseria, G. E. *J. Chem. Phys.* **1990**, *93*, 489.

50. Watts, J. D.; Stanton, J. F.; Bartlett, R. J. *Chem. Phys. Lett.* **1991**, *178*, 471.

51.   Pople, J. A.; Head-Gordon, M.; Fox, D. J.; Raghavachari, K.; Curtiss, L. A. *J. Chem. Phys.* **1989**, *90*, 5622.

52.   Curtiss, L. A.; Jones, C.; Trucks, G. W.; Raghavachari, K.; Pople, J. A. *J. Chem. Phys.* **1990**, *93*, 2537.

53.   Curtiss, L. A.; Raghavachari, K.; Trucks, G. W.; Pople, J. A. *J. Chem. Phys.* **1991**, *94*, 7221.

54.   Curtiss, L. A.; Carpenter, J. E.; Raghavachari, K.; Pople, J. A. *J. Chem. Phys.* **1992**, *96*, 9030.

55.   Durant, J. L. *J. Phys. Chem.* **1994**, *98*, 518.

56.   Yang, D. L.; Koszykowski, M. L.; Durant, J. L., Jr. *J. Chem. Phys.* **1994**, *101*, 1361.

57.   Wolf, M.; Yang, D.; Durant, J. L. *manuscript in preparation* **1996**, .

58.   Walch, S. P. *J. Chem. Phys.* **1993**, *98*, 1170.

59.   Walch, S. P. *J. Chem. Phys.* **1993**, *99*, 3804.

60.   Page, M.; Soto, M. R. *J. Chem. Phys.* **1993**, *100*, 7709.

61.   Soto, M. R.; Page, M.; McKee, M. L. *Chem. Phys. Lett.* **1991**, *187*, 335.

62.   Walch, S. P. *J. Chem. Phys.* **1993**, *99*, 5295.

63.   Smith, B. J.; Nguyen, M. T.; Bouma, W. J.; Radom, L. *J. Am. Chem. Soc.* **1991**, *113*, 6452.

64.   Ma, N. L.; Smith, B. J.; Pople, J. A.; Radom, L. *J. Am. Chem. Soc.* **1991**, *113*, 7903.

65.   Ding, L.; Marshall, P. *J. Chem. Phys.* **1993**, *98*, 8545.

66.   Goumri, A.; Yuan, W.-J.; Ding, L.; Shi, Y.; Marshall, P. *Chem. Phys.* **1993**, *177*, 233.

67.   Goumri, A.; Laakso, D.; Rocha, J. R.; Smith, C. E.; Marshall, P. *J. Chem. Phys.* **1995**, *102*, 161.

68.   Curtiss, L. A.; Raghavachari, K.; Pople, J. A. *J. Chem. Phys.* **1993**, *98*, 1293.

69.   Curtiss, L. A.; Raghavachari, K.; Pople, J. A. *J. Chem. Phys.* **1995**, *103*, 4192.

70.   Bauschlicher, C. W.; Partridge, H. *J. Chem. Phys.* **1995**, *103*, 1788.

71.   Mebel, A. M.; Morokuma, K.; Lin, M. C. *J. Chem. Phys.* **1995**, *103*, 7414.

72.   Curtiss, L. A.; Redfern, P. C.; Smith, B. J.; Radom, L. *J. Chem. Phys.* **1995**, *104*, 5148.

73.   Parr, R. G.; Yang, W. *Density-Functional Theory of Atoms and Molecules*, Oxford University Press: New York, 1989.

74.   Kohn, W.; Becke, A. D.; Parr, R. G. *J. Phys. Chem.* **1996**, *100*, 12974.

75.   Johnson, B. G.; Gill, P. W. M.; Pople, J. A. *J. Chem. Phys.* **1992**, *97*, 7846.

76.   Baker, J.; Scheiner, A.; Andzelm, J. *Chem. Phys. Lett.* **1993**, *216*, 380.

77.   Durant, J. L. *Chem. Phys. Lett.* **1996**, *256*, 595.

78.   Becke, A. D. *J. Chem. Phys.* **1996**, *104*, 1040.

79.   Hohenberg, P.; Kohn, W. *Phys. Rev.* **1964**, *3B*, 864.

80.   Perdew, J. P. *Phys. Rev. B* **1986**, *45*, 8822.

81.   Becke, A. D. *Phys. Rev. A* **1988**, *38*, 3098.

82.   Lee, C.; Yang, W.; Parr, R. G. *Phys. Rev. B* **1988**, *37*, 785.

83.   Perdew, J. P.; Wang, Y. *Phys. Rev. B* **1992**, *45*, 13244.

84.   Becke, A. D. *J. Chem. Phys.* **1993**, *98*, 5648.
85.   Fan, L.; Ziegler, T. *J. Chem. Phys.* **1990**, *92*, 3645.
86.   Fan, L.; Ziegler, T. *J. Am. Chem. Soc.* **1992**, *114*, 10890.
87.   Sosa, C.; Lee, C. *J. Chem. Phys.* **1993**, *98*, 8004.
88.   Deng, L.; Ziegler, T.; Fan, L. *J. Chem. Phys.* **1993**, *99*, 3823.
89.   Andzelm, J.; Sosa, C.; Eades, R. A. *J. Phys. Chem.* **1993**, *97*, 4664.
90.   Stanton, R. V.; Merz, K. M. *J. Chem. Phys.* **1994**, *100*, 434.
91.   Sosa, C.; Andzelm, J.; Lee, C.; Blake, J. F.; Chenard, B. L.; Butler, T. W. *Int. J. Quant. Chem.* **1994**, *49*, 511.
92.   Truong, T. N.; Duncan, W. *J. Chem. Phys.* **1994**, *101*, 7408.
93.   Deng, L.; Branchadell, V.; Ziegler, T. *J. Am. Chem. Soc.* **1994**, *116*, 10645.
94.   Deng, L.; Ziegler, T. *J. Phys. Chem.* **1995**, *99*, 612.
95.   Wiest, O.; Houk, K. N.; Black, K. A.; Thomas, B. *J. Am. Chem. Soc.* **1995**, *117*, 8594.
96.   Porezag, D.; Pederson, M. R. *J. Chem. Phys.* **1995**, *102*, 9345.
97.   Pai, S. V.; Chabalowski, C. F.; Rice, B. M. *J. Phys. Chem.* **1996**, *100*, 15368.
98.   Goldstein, E.; Beno, B.; Houk, K. N. *J. Am. Chem. Soc.* **1996**, *118*, 6036.
99.   Johnson, B. G.; Gozales, C. A.; Gill, P. M. W.; Pople, J. A. *Chem. Phys. Lett.* **1994**, *221*, 100.
100. Zhang, Q.; Bell, R.; Truong, T. N. *J. Phys. Chem.* **1995**, *99*, 592.
101. Baker, J.; Muir, M.; Andzelm, J. *J. Chem. Phys.* **1995**, *102*, 2063.
102. Baker, J.; Andzelm, J.; Muir, M.; Taylor, P. R. *Chem. Phys. Lett.* **1995**, *237*, 53.
103. Hertwig, R. H.; Kock, W. *J. Comp. Chem.* **1995**, *16*, 576.
104. Cioslowski, J. *J. Am. Chem. Soc.* **1989**, *111*, 8333.
105. Bachrach, S. M. In *Reviews in Computational Chemistry*; Lipkowitz, K.B.; Boyd, D.B., Eds.; VCH Publishers, Inc.: New York, 1994; Vol. 5; p. 171.
106. Yang, W. *Phys. Rev. Lett.* **1991**, *66*, 1438.
107. Zhou, Z. *Chem. Phys. Lett.* **1993**, *203*, 396.
108. White, C. A.; Johnson, B. G.; Gill, P. M. W.; Head-Gordon, M. *Chem. Phys. Lett.* **1994**, *230*, 8.
109. White, C. A.; Johnson, B. G.; Gill, P. M. W.; Head-Gordon, M. *Chem. Phys. Lett.* **1996**, *253*, 268.
110. Humbel, S.; Sieber, S.; Morokuma, K. *J. Chem. Phys.* **1996**, *105*, 1959.
111. Svensson, M.; Humbel, S.; Morokuma, K. *J. Chem. Phys.* **1996**, *105*, 3654.
112. Ochterski, J. W.; Petersson, G. A.; Montgomery, J. A. *J. Chem. Phys.* **1996**, *104*, 2598.

Chapter 15

Modeling Free Energies of Solvation and Transfer

David J. Giesen, Candee C. Chambers, Gregory D. Hawkins,
Christopher J. Cramer[1], and Donald G. Truhlar[1]

Department of Chemistry and Supercomputer Institute, University of Minnesota,
Minneapolis, MN 55455-0431

The free energy of transfer of a solute from one medium to another,
which is the free energy of solvation if the first medium is the gas
phase and the second is a liquid-phase solution, controls all solvation
and partitioning phenomena. The SM5.4 quantum mechanical
solvation model allows for the calculation of (i) partitioning free
energies between the gas phase and a solvent (i.e., free energies of
solvation) or (ii) partitioning free energies between two solvents. The
model provides a framework for interpreting the factors responsible
for differential solvation effects and can be used to predict solvation
effects on chemical equilibria and kinetics—examples in this chapter
include partitioning of the nucleic acid bases between water and
chloroform, solvation effects on anomeric conformational equilibria,
and solvation effects on the rate of the Claisen rearrangement.

A collection of solute molecules will partition between two (or more) phases so as to
equalize the concentration-dependent chemical potential of the solute in all phases *(1)*.
This behavior has far reaching consequences in chemistry. For instance, a molecule
might have very high specific activity against a target enzyme, but if its concentration
in aqueous biophases or fatty tissues is extremely low it may be inefficacious as an
oral drug (since it must be carried by the bloodstream and, if the enzyme is located
intracellularly, it must pass through a lipid bilayer membrane). A second example of
medium effects on equilibria is that the fraction of molecules in any one of several
potentially accessible conformations can depend significantly on solvent, and this
structural equilibrium may have significant effects on molecular properties,
recognition, and reactivity. As a final example, medium effects on reaction rates can
be viewed as changes in the relative equilibrium concentrations of reactants and
activated complexes on going from one environment to the next. These examples

[1]Corresponding authors

illustrate the practical importance of being able to predict concentrations for solutes in different media.

In an ideal solution, the chemical potential μ for a solute in a given medium can be expressed as (1)

$$\mu = \mu^o + RT\ln\left(\frac{X}{X^o}\right) \qquad (1)$$

where μ^o and X^o are the standard state chemical potential and concentration, respectively, for the solute in the medium, and X is the concentration of the solute. For a solute in equilibrium between two phases i and j, equality of the chemical potentials implies that

$$\mu_i^o - \mu_j^o = RT\ln\left(\frac{X_i^o}{X_j^o}\right) - RT\ln\left(\frac{X_i}{X_j}\right). \qquad (2)$$

In practice, the standard state concentrations are typically chosen to be the same in both phases so that the first term on the right-hand-side is zero, in which case one recovers the well known

$$\Delta G_{j \to i}^o = - RT\ln\left(\frac{X_i}{X_j}\right). \qquad (3)$$

That is, the standard state free energy of transfer from phase j to phase i is proportional to the logarithm of the ratio of concentrations of the solute in the two phases (sometimes also written $\Delta G_{i/j}^o$; the equilibrium constant for the transfer is called the "partition coefficient"). From an experimental standpoint, this means that the standard state free energy of transfer can be determined by measuring the concentration of substrate in each of the two phases at equilibrium under ideal-solution conditions (which typically requires low concentrations in both phases). From the theoretical standpoint, a model that predicts the standard state free energy of transfer also permits the prediction of differential substrate concentrations between the two phases.

The SMx series of quantum mechanical solvation models are designed specifically to calculate standard state free energies of transfer between the gas phase and solution $(2\text{-}15)$. By appropriate use of thermodynamic cycles as illustrated in Figure 1, this also allows for the calculation of free energies of transfer between two *different* solutions.

This chapter focuses on a particular member of the SMx family, namely the SM5.4 model. This model has specific parameters for water (12), chloroform (15), benzene and toluene (16), and a general set of parameters that can be used for any other organic solvent (14). The SM5 part of the name refers to the solute-geometry-

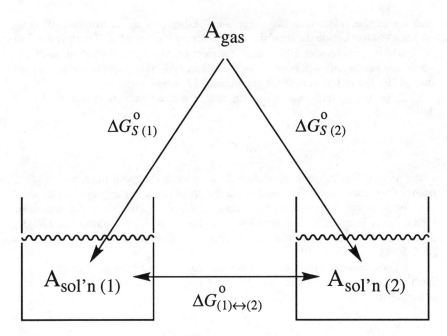

Figure 1. Free energy cycle for transfer of A from the gas phase into either of two solvents or between the two solvents. Since the complete cycle must sum to zero, knowing any two sides of the triangle permits calculation of the remaining side.

dependence assumed for first-solvation-shell effects, and the ".4" in the model name indicates that the electrostatic portion of the solvation free energy is calculated using Class IV atomic partial charges *(6,17)* (vide infra). In this chapter we use the water, chloroform, benzene, and toluene parameters wherever applicable and the general organic parameters otherwise.

Solvation Model 5.4

The functional forms for the SM5.4 model have been presented in full detail in the elsewhere *(12,14,15)*. In this section we review those features of the models that are important to understanding the rest of this chapter.

Framework of Model. The free energy of solvation is partitioned into two terms *(2,5,18-20)*

$$\Delta G_S^0 = \Delta G_{ENP} + G_{CDS} \tag{4}$$

where ΔG_{ENP} includes the change in the electronic and nuclear internal energy of the solute and the electric polarization free energy of the solute-solvent system upon insertion of the solute into the solvent, and G_{CDS} is the contribution of first-solvation-shell effects to the standard-state free energy of transfer. Note that the S in ΔG_S^0 stands for solvation, ENP stands for electronic-nuclear-polarization, and CDS stands for cavitation–dispersion–solvent-structure. All SMx models use a standard state of 298 K and 1 M in both the gas phase and solution.

The ΔG_{ENP} term, which is often called the electrostatic term, can be further decomposed as

$$\Delta G_{ENP} = \Delta E_{EN} + G_P \tag{5}$$

where ΔE_{EN} is the change in the electronic and nuclear energy of the solute in going from the gas phase to solution, and G_P is the polarization free energy. The terms in equation 6 are calculated based on neglect of diatomic differential overlap molecular orbital theory using either the Austin Model 1 *(21-23)* (AM1) or Parameterized Model 3 *(24)* (PM3) Hamiltonian; these choices are distinguished by the notations SM5.4/AM1 and SM5.4/PM3, respectively.

G_P is calculated from Class IV atomic charges q using the AM1 or PM3 wave functions and Charge Model 1A *(17)* or Charge Model 1P *(17)*, respectively, according to

$$G_P = -\frac{1}{2}\left(1 - \frac{1}{\varepsilon}\right)\sum_{k,k'} q_k q_{k'} \gamma_{kk'} \tag{6}$$

where ε is the solvent dielectric constant, k and k' label atoms, and $\gamma_{kk'}$ is a Coulomb integral that accounts for either the self-energy of a charge in a dielectric medium $(k=k')$ or the screened Coulomb interaction of two charges $(k \neq k')$. Computation of $\gamma_{kk'}$ involves the use of atomic Coulomb radii ρ_k; these Coulomb radii are parameters originally optimized for water *(12)* and the same values are used in all SM5.4 parameterizations discussed here. Although the accuracy of the SM5.4 model is fairly insensitive to the atomic Coulomb radii, the magnitudes of the ENP and CDS components of the free energy of solvation are more sensitive. Final radii were selected based on careful inspection of various classes of solutes to ensure physically meaningful ENP values. These intrinsic radii are modified in molecular calculations by an algorithm that accounts for descreening of solute atoms (i.e., displacement of the dielectric screening of the surrounding medium) by the molecular volume of the solute *(2,11,25)*.

The first-solvation-shell term has the general form:

$$G_{CDS} = \sum_k \sigma_k A_k\left(R_S^{CD}\right) + \sigma^{CS}\sum_k A_k\left(R_S^{CS}\right) \tag{7}$$

where k denotes an atom, $A_k(R_S^{CX})$ is the solvent-accessible surface area *(11,26)* of atom k as calculated for a rolling solvent ball having radius R_S^{CX}, and the various σ are surface tensions (having units of energy per unit area). The atomic surface tensions σ_k typically depend upon the local geometry of the solute and one or more surface tension coefficients $\hat{\sigma}^i_{Z_k}$ that are themselves dependent on the atomic number Z of atom k and are parameters of the model, i.e.,

$$\sigma_k = f_{Z_k}\left(\mathbf{R}, \left\{\hat{\sigma}^i_{Z_k}\right\}\right) \tag{8}$$

where $\mathbf{R}$ denotes the solute geometry and the set of functions f_k that define the SM5 set of models has been described in detail *(12-15)*. The molecular surface tension σ^{CS} is a constant for a given solvent.

We note that in the parameterization of *all* SM5.4 models, solute geometries and electronic wave functions were allowed to fully relax in response to surrounding solvent. All surface tension coefficients are different (separately optimized) for SM5.4/AM1 and SM5.4/PM3, but other parameters are the same for the /AM1 and /PM3 cases.

Parameterization for water *(12)*. The dielectric constant employed in equation 6 is 78.3. In the SM5.4-water model, the solvent radii R_S^{CD} and R_S^{CS} are both taken to be 1.7 Å, so that equation 8 simplifies to

$$G_{CDS} = \sum_k \sigma_k A_k\left(R_S\right). \tag{9}$$

Thus, σ^{CS} is not a separately optimized parameter in the SM5-water model, but is absorbed into the optimization of the atomic surface tension coefficients contributing to each σ_k. The Coulomb radii and surface tension coefficients were optimized against a set of 34 ionic and 215 neutral experimental free energies of aqueous solvation. Table I provides information on the accuracy of the model.

Parameterization for general organic solvents *(14)*. The hallmark of the general SM5.4 model is that it is designed to be employed for *all* organic solvents. In order to accomplish this, the molecular surface tension appearing in equation 7 and the atomic surface tension coefficients appearing in equation 8 are taken to be functions of experimental *solvent* properties. In particular,

$$\sigma^{CS} = \sigma^{CS,n}n + \sigma^{CS,\gamma}\gamma \tag{10}$$

$$\hat{\sigma}^i_{Z_k} = \hat{\sigma}^{i,n}_{Z_k}n + \hat{\sigma}^{i,\alpha}_{Z_k}\alpha + \hat{\sigma}^{i,\beta}_{Z_k}\beta \tag{11}$$

where n is the solvent index of refraction, γ is the macroscopic solvent surface tension, and α and β are Abraham's *(27-29)* indices of solvent hydrogen bonding acidity and basicity, respectively (more specifically, these are the $\Sigma\alpha_2^H$ and $\Sigma\beta_2^H$ values for a solvent molecule were it to be taken as a *solute* in the Abraham model). The solvent radii R_S^{CD} and R_S^{CS} are taken to be 1.7 and 3.4 Å, respectively. Two radii appear to be required in order to accurately capture both short-range (cavitation-dispersion) and intermediate-range (cavitation-solvent-structure) effects—for the

Table I. Calculated mean signed errors (MSE), mean unsigned errors (MUE), and root-mean-square errors (RMS) for solvent/gas molar free energies of transfer using SM5.4 solvation models (kcal).

Solvents	Solutes	AM1			PM3		
		MSE	MUE	RMS	MSE	MUE	RMS
Water	ions[a]	0.0	4.1	5.6	0.1	4.2	5.6
	neutrals[b]	0.0	0.5	0.8	0.0	0.5	0.6
Chloroform	neutrals[c]	0.1	0.5	0.7	0.0	0.4	0.5
Benzene	neutrals[d]	0.2	0.5	0.8	0.2	0.4	0.5
Toluene	neutrals[e]	0.1	0.3	0.4	0.2	0.3	0.4
General organic	neutrals[f]	-0.1	0.5	0.6	-0.1	0.4	0.6

[a] 34 data points for both AM1 and PM3. Fitted values range from -55 to -107 kcal. [b] 215 data points for AM1, 214 for PM3. Fitted values range from 4 to -11 kcal. [c] 88 data points for both AM1 and PM3. Fitted values range from 0 to -13 kcal. [d] 60 data points for AM1, 59 for PM3. Fitted values range from 0 to -8 kcal. [e] 45 data points for both AM1 and PM3. Fitted values range from 0 to -8 kcal. [f] 1599 data points for AM1, 1597 for PM3. Fitted values range from 2 to -15 kcal.

alkane models, the magnitudes of the CD and CS terms are consistent with available data for cavitation and dispersion energies *(9)*.

The set of surface tension coefficients $\hat{\sigma}_{Z_k}^{i,j}$ together with $\hat{\sigma}^{CS,\gamma}$ were optimized against a set of 1786 experimental free energies of transfer from the gas phase into various organic solvents (either measured directly or derived from solvent/solvent partition coefficients and solvent/gas solvation free energies according to the scheme illustrated in Figure 1). These data spanned 206 different solutes and 90 different organic solvents (including chloroform, benzene, and toluene). Table I provides information on the accuracy of the model. The errors in Table I for this parameterization do not include the chloroform, benzene, and toluene data points; inclusion of these points affects the errors by less than 0.1 kcal, but we want to emphasize that for calculations in these solvents the specific parameterizations are nearly always to be preferred.

Parameterization for chloroform *(15)*. The dielectric constant employed in equation 6 is 4.2. The solvent radii R_S^{CD} and R_S^{CS} are taken to be the same as in the general organic model. The atomic surface tension coefficients and the molecular surface tension were optimized against a set of 88 experimental free energies of transfer from the gas phase to chloroform (either measured directly or derived from chloroform/water partition coefficients and gas/water solvation free energies according to the scheme illustrated in Figure 1). In addition, the optimization was restrained by (i) including data for 123 free energies of solvation into solvents other than chloroform for molecules containing functional groups poorly represented in the chloroform data and (ii) including data for 26 chloroform water partition coefficients where the free energies of aqueous solvation were calculated using SM5.4-water. The restraint is based on the solvent dependencies of the general organic parameterization discussed above. Table I provides information on the accuracy of the model.

Parameterizations for benzene and toluene *(16).* The dielectric constants employed in equation 6 were standard values *(30).* The solvent radii R_S^{CD} and R_S^{CS} are taken to be the same as in the general organic model. The atomic surface tension coefficients were taken to be the same as in the general organic model with the exception of $\hat{\sigma}^{CS,\gamma}$, the magnitude of which was reduced by 34% and 23% for benzene and toluene, respectively; these factors were optimized against sets of experimental free energies of transfer from the gas phase to the aromatic hydrocarbon solvent (free energies either measured directly or derived from solvent/water partition coefficients and gas/water solvation free energies according to the scheme illustrated in Figure 1). Table I provides information on the number of data and the accuracy of the models.

Interpreting ENP and CDS components. An important aspect of the various SM5.4 parameterizations is that considerable care went into ensuring that the various parameter values and the individual terms in equation 4 are physically meaningful. This involved a certain degree of chemical intuition, since the ENP and CDS components are not state functions like their sum *(20).* In addition, special care was paid to subsets of the parameterization data organized by solute functional group(s) to attempt to minimize any systematic bias in the models, and trends in parameter values were examined to determine if they were consistent with expected trends based on chemical behavior. For example, one would expect SM5.4 surface tension coefficients $\hat{\sigma}_{Z_k}^{l,n}$ associated with the index of refraction (a measure of solvent polarizability) to become increasingly negative with increasing atomic polarizability (i.e., dispersion interactions will become more favorable for these atoms), and indeed that trend is observed *(14).* Similar trends over groups having varying degrees of hydrogen bond donating and accepting capabilities are also observed *(14).* So, while the separation of ENP and CDS contributions is necessarily ambiguous, we believe that the parameter sets yield physically meaningful terms and provide for an additional level of detail that may be analyzed from an SM5 calculation.

Partition Coefficients

For many interesting solutes, lack of volatility makes the measurement of gas/solvent transfer free energies difficult. One example of biological interest is that of the methylated nucleic acid bases (Figure 2). For such solutes, experimental data are more typically available in the form of solvent/solvent partition coefficients. In the case of the methylated nucleic acid bases, chloroform/water partition coefficients are available for six cases *(31).* Table II is a compilaton of experimental values (when available) and predicted values from two models. In particular, the SM5.4 results are compared to published work of Orozco et al. *(32)* using the Miertus-Scrocco-Tomasi *(33-36)* algorithm for electrostatics plus semiempirical surface tensions parameterized specifically for chloroform by Luque et al. *(36).* We note that the mean unsigned

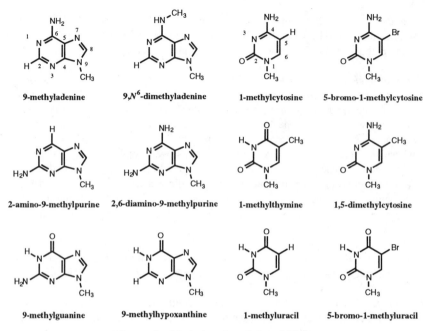

Figure 2. Methylated nucleic acid bases.

Table II. Chloroform/water partition coefficients ($\log_{10}$ units) for methylated nucleic acid bases.[a]

Solute	SM5.4	MST/ST	Experiment
9-Methyladenine	–1.7	–0.3	–0.8
9,N^6-Dimethyladenine	–0.3		
2-Amino-9-methylpurine	–1.9		
2,6-Diamino-9-methylpurine	–2.4		
9-Methylguanine	–4.1	–4.8	–3.5
9-Methylhypoxanthine	–3.6	–1.4	–2.5
1-Methylcytosine	–4.2	–3.4	–3.0
5-Bromo-1-methylcytosine	–2.4		
1-Methylthymine	–0.3	–0.4	–0.5
1,5-Dimethylcytosine	–3.1		
1-Methyluracil	–1.2	–1.0	–1.2
5-Bromo-1-methyluracil	–0.3		
Mean unsigned error:	*0.6*	*0.6*	

[a] All calculations used the AM1 Hamiltonian.

error comparing the SM5.4 partition coefficients to experiment is only 0.6 log units. [Applying the general organic parameters for chloroform (as opposed to the chloroform-specific ones) makes every chloroform free energy of solvation slightly more negative and decreases the mean unsigned error to 0.5 log units, which is also very respectable and comparable to the results obtained with the MST/ST specific chloroform/water models. The difference seems to be primarily a consequence of a less positive σ^{CS} value for chloroform in the general organic model.]

One advantage of the generalized Born approach, compared to other continuum models, is that solvation free energies (and partition coefficients) can be rather simply decomposed into contributions from individual molecular fragments, e.g., functional groups. This permits an analysis of such issues as the transferability of group contributions to partition coefficients for different chemical environments, and we have examined this elsewhere for the specific case of the methylated nucleic acid bases discussed above (37).

Solvent Effects on an Anomeric Equilibrium

Stereoelectronic effects on conformational equilibria are another example of a thermochemical phenomenon that may have a large medium effect. A classic example is the anomeric effect. The term "anomeric effect" refers to the unexpectedly low energies of pyranose and pyranoside structures with axial substituents at the anomeric, i.e., C(2), position (Figure 3) (38-44); this stability is unexpected insofar as axial substituents on six-membered rings typically experience strong destabilizing steric interactions with other axial C–X bonds. The stabilization, unique among six-membered rings to the pyranoses, pyranosides, and analogs, has been rationalized as arising from decreased dipole-dipole repulsion between C–O bonds in the axial anomer (45-47) and/or from greater hyperconjugative stabilization of the pyranose oxygen lone pair when delocalized into the empty axial σ^{*}_{CO} orbital (48-51). The degree to which these two phenomena contribute to the free energy in any given system (or the degree to which it is profitable to single out either of these two explanations for the single physical effect) depends on the molecular structure (42,52-61) and, importantly, also the surrounding medium (42,44,54,61-74).

In this section, we compare predictions of the SM5.4 model (using the PM3 Hamiltonian and fully relaxed geometries—AM1 results are quantitatively very similar) to experiment for the effect of solvation on the anomeric equilibrium of

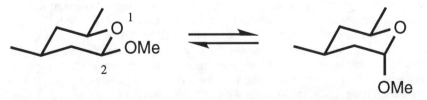

Figure 3.   The   anomeric   equilibrium   for   2-methoxy-4,6-dimethyltetrahydropyran.

Table III. Solvent effect (kcal) on the molar free energy of anomerization of 2-methoxy-4,6-dimethyltetrahydropyran.

Solvent	OPLS[a]	SM5.4/PM3	Expt.[b]	Other expt.[c]
benzene		0.4	0.2	0.0–0.2
CCl4	0.1	0.4	0.3	0.1–0.3
n–butyl ether		0.5	0.3	
DMSO		0.8	0.5	0.3–0.4
acetone		0.7	0.5	0.3–0.4
pyridine		0.7	0.6[d]	
methanol		0.7	0.7	0.5–0.7
acetonitrile	1.6	0.7	0.7	0.5–0.7
water	2.1	0.6		0.9

[a] Reference *(72)*. [b] Reference *(71)*. [c] Reference *(58)*. [d] Change in enthalpy, not free energy.

2-methoxy-4,6-dimethyltetrahydropyran. Wiberg and Marquez have performed thermochemical measurements of the solvation effect on this anomeric equilibrium *(71)*, and those results are provided in Table III. Also included is a summary of solvation effect data compiled by Tvaroska and Carver *(58)* for 2-methoxytetrahydropyran, 2-methoxy-6-methyltetrahydropyran, 2-methoxy-4-methyltetrahydropyran, and 2-ethoxytetrahydropyran. Finally, Jorgensen et al. have modeled solvation effects on the anomeric equilibrium of 2-methoxytetrahydropyran using a Monte Carlo statistical model with explicit solvent and the OPLS force field *(72)*, and their predictions are also provided in Table III. All of the values in Table III are positive free energies, implying a preferential solvation of the equatorial anomer.

An inspection of the SM5.4 results suggests that there is a systematic overestimation of the solvation effect (i.e., the equatorial anomer is oversolvated relative to the axial) by about 0.2 kcal in most organic solvents. In water, on the other hand, the SM5.4 prediction is 0.3 kcal smaller than that found experimentally *(64)*. Nevertheless, the overall agreement between the SM5.4 results and experiment is encouraging, especially when one considers alternative models. The much more expensive OPLS simulations of Jorgensen et al. *(72)* appear to strongly oversolvate the equatorial anomer in acetonitrile and water. Similarly, using a continuum solvent model that considered *only* electrostatic effects, Montagnani and Tomasi calculated the equatorial anomer to be better solvated than the axial by 0.6 kcal in CCl4 and 1.4 kcal in water *(65)*. One might infer from these latter results, which overestimate the effect of solvent on the anomeric equilibrium, that non-electrostatic effects oppose electrostatic effects in their influence. With the SM5.4 models, however, we do not find a significant difference between the CDS components of the free energies of solvation for the two anomers. The sensitivity of ΔG_{ENP} to geometry relaxation for each anomer, on the other hand, is particularly noteworthy. For instance, with frozen PM3 gas-phase structures, the calculated solvation effect on the anomeric equilibrium is doubled compared to using relaxed structures; this difference is found entirely in ΔG_{ENP}. When MP2/cc-pVDZ gas-phase geometries were used, the quality of the predictions was also degraded by about the same magnitude. This example illustrates how difficult it can be to assess the quality of a given theoretical approach in the absence of experimental data against which to validate different approximations, e.g., choice of geometry, choice of solvation model, etc., particularly when the range of solvent effects being compared spans less than 1.0 kcal/mol.

Solvent Effects on the Rate of the Claisen Rearrangement

The [3,3]-sigmatropic shift of an allyl vinyl ether to produce a 4-pentenal is called the Claisen rearrangement *(75)*. This reaction has attracted considerable attention primarily because in a number of organisms it is the mechanism by which chorismate rearranges to prephenate in the committed step for the biosynthesis of aromatic amino acids *(76,77)*. In an effort to better understand the function of the enzyme involved much work has appeared in the area of molecular biology *(78-80)*, and, in addition, experimental studies have investigated the character of the transition state by examination of substituent and solvent effects on the rate of the rearrangement *(81-86)*. Based on rate accelerations observed in polar solvents and sensitivity to substituent positioning, there is consensus *(81-86)* that the transition state has greater charge separation than the reactant, and that negative charge concentrates on the oxygen atom, so that hydrogen-bonding solvents accelerate the reaction. Theory is entirely in concert with this analysis, with studies having been done with both continuum *(6,87-92)* and explicit solvent models *(88,93-98)*, as well as simulations including the enzyme catalyst *(98)*. The SM5.4 model, being quantum mechanical, is well suited to calculations on transition states, and it is interesting to apply it to this problem.

We have previously examined *(6)* aqueous acceleration of the Claisen rearrangement for allyl vinyl ether using a SM4 water model with specific range parameters (i.e., parameters chosen for a subset of molecules bearing some similarities to those involved in the rearrangement, in this case aldehydes, ethers, and hydrocarbons). That study demonstrated how sensitive the rate acceleration is to the looseness of the transition state, and we found that using a transition state structure calculated at the multiconfigurational SCF level *(99)* gives good agreement between theory and the experimental rate acceleration in water, which is estimated to be about 1×10^3 *(93)*. Here we examine the same rearrangement using the same reactant and transition state structures as before, but with SM5.4, which allows us to compare to the study of Brandes et al. *(85)*, which examined the effect of solvation over a broad range of solvents by comparing the reaction rates for two different substituted Claisen substrates (see Table IV). In solvents where both reactants were soluble, the rates were identical to within a factor of two, allowing a "ladder" to be constructed from cyclohexane to water. Those results, together with the SM5.4 predictions, are presented in Table IV.

The predicted accelerations relative to cyclohexane using SM5.4 are good for benzene and trifluoroethanol. The aqueous acceleration is underestimated, although a factor of 4 represents an error of only 0.8 kcal/mol. On the other hand, the rates in

Table IV. Relative rates for Claisen rearrangements in different solvents.

Solvent	SM5.4/AM1[a]	Experiment[b]
cyclohexane	1.0	1.0
benzene	1.3	2.0
methanol	57.9	8.6
trifluoroethanol	56.6	56.0
water	53.7	214.3

[a] 298 K. [b] R = $(CH_2)CO_2Me$ for organic solvents and $(CH_2)CO_2Na$ for water. All results for 333 K.

methanol and trifluoroethanol are predicted to be essentially the same, whereas experimentally they differ by a factor of 7. There are a number of possible explanations for this discrepancy, but, based on results from molecular simulations *(88,93)*, the most likely explanation is that specific solvation in the form of hydrogen bonding to the oxygen atom can modify the overall transition state structure in a way not well accounted for in the MCSCF structure. In the general case, this situation does not pose a special problem for a continuum model like SM5.4 in one can include the special solvent molecule(s) and consider the continuum solvation of the supersolute. However, for the present case, reoptimization of the transition state structure at the semiempirical level is not an option since AM1 does not provide an adequate description of the gas-phase transition state structure. It will be interesting to revisit this system when the SM*x* models are extended to ab initio levels of theory, in particular density functional theory, since Houk and co-workers have found this level of theory to give excellent Claisen transition state structures based on analysis of kinetic isotope effects *(100)*.

Conclusions

The SM5.4 solvation model provides a robust general method for the quantum mechanical calculation of organic and aqueous solvation effects on a variety of chemical phenomena. Partition coefficients can be predicted, and the total partition coefficient can be decomposed into atomic or group contributions—this should prove particularly useful in the area of molecular design, e.g., for pharmaceutical purposes or to enhance nonlinear optical effects. Computation of differential solvation effects for isomers (conformational, tautomeric, or structural) and for reactants vs transition state structures permits prediction of solvent-induced changes in equilibrium and rate constants, respectively. The physical separation of electrostatics from first-solvation-shell effects allows chemical interpretation of these phenomena, although care is warranted when the effects become small so that the model is not overinterpreted.

Acknowledgments

This work was supported in part by the National Science Foundation and the Army Research Office. We thank Profs. Bill Jorgensen and Modesto Orozco for stimulating discussions, and we are grateful for high-performance vector and parallel computing resources made available by the Minnesota Supercomputer Institute and the University of Minnesota-IBM Shared Research Project, respectively.

Additional Information.

Further details about the SM*x* solvation models and software implementing them may be found at http://amsol.chem.umn.edu/~amsol.

Literature Cited

(1)   Ben-Naim, A. *Statistical Thermodynamics for Chemists and Biochemists*; Plenum: New York, 1992, p. 421.

(2)   Cramer, C. J.; Truhlar, D. G. *J. Am. Chem. Soc.* **1991**, *113*, 8305 and 9901(E).

(3)   Cramer, C. J.; Truhlar, D. G. *Science* **1992**, *256*, 213.

(4)   Cramer, C. J.; Truhlar, D. G. *J. Comp. Chem.* **1992**, *13*, 1089.

(5)   Cramer, C. J.; Truhlar, D. G. *J. Comput.-Aid. Mol. Des.* **1992**, *6*, 629.

(6)   Storer, J. W.; Giesen, D. J.; Hawkins, G. D.; Lynch, G. C.; Cramer, C. J.; Truhlar, D. G.; Liotard, D. A. In *Structure and Reactivity in Aqueous Solution, ACS Symposium Series 568*; C. J. Cramer and D. G. Truhlar, Eds.; American Chemical Society: Washington, DC, 1994; p. 24.

(7)   Barrows, S. E.; Dulles, F. J.; Cramer, C. J.; Truhlar, D. G.; French, A. D. *Carbohydr. Res.* **1995**, *276*, 219.

(8)   Hawkins, G. D.; Cramer, C. J.; Truhlar, D. G. *Chem. Phys. Lett.* **1995**, *246*, 122.

(9)   Giesen, D. J.; Storer, J. W.; Cramer, C. J.; Truhlar, D. G. *J. Am. Chem. Soc.* **1995**, *117*, 1057.

(10) Giesen, D. J.; Cramer, C. J.; Truhlar, D. G. *J. Phys. Chem.* **1995**, *99*, 7137.

(11) Liotard, D. A.; Hawkins, G. D.; Lynch, G. C.; Cramer, C. J.; Truhlar, D. G. *J. Comp. Chem.* **1995**, *16*, 422.

(12) Chambers, C. C.; Hawkins, G. D.; Cramer, C. J.; Truhlar, D. G. *J. Phys. Chem.* **1996**, *100*, 16385.

(13) Hawkins, G. D.; Cramer, C. J.; Truhlar, D. G. *J. Phys. Chem.* **1996**, *100*, 19824.

(14) Giesen, D. J.; Gu, M. Z.; Cramer, C. J.; Truhlar, D. G. *J. Org. Chem.* **1996**, *61*, 8720.

(15) Giesen, D. J.; Chambers, C. C.; Cramer, C. J.; Truhlar, D. G. *J. Phys. Chem. B* **1997**, *101*, 2061.
(16) Giesen, D. J.; Cramer, C. J.; Truhlar, D. G. *Theor. Chem. Acc.* in press.
(17) Storer, J. W.; Giesen, D. J.; Cramer, C. J.; Truhlar, D. G. *J. Comput.-Aid. Mol. Des.* **1995**, *9*, 87.
(18) Cramer, C. J.; Truhlar, D. G. In *Quantitative Treatments of Solute/Solvent Interactions, Theoretical and Computational Chemistry*; P. Politzer and J. S. Murray, Eds.; Elsevier: Amsterdam, 1994; Vol. 1; p. 9.
(19) Cramer, C. J.; Truhlar, D. G. In *Reviews in Computational Chemistry*; K. B. Lipkowitz and D. B. Boyd, Eds.; VCH: New York, 1995; Vol. 6; p. 1.
(20) Cramer, C. J.; Truhlar, D. G. In *Solvent Effects and Chemical Reactivity*; O. Tapia and J. Bertrán, Eds.; Kluwer: Dordrecht, 1996; p. 1.
(21) Dewar, M. J. S.; Zoebisch, E. G.; Healy, E. F.; Stewart, J. J. P. *J. Am. Chem. Soc.* **1985**, *107*, 3902.
(22) Dewar, M. J. S.; Zoebisch, E. G. *J. Mol. Struct. (Theochem)* **1988**, *180*, 1.
(23) Dewar, M. J. S.; Yate-Ching, Y. *Inorg. Chem.* **1990**, *29*, 3881.
(24) Stewart, J. J. P. *J. Comp. Chem.* **1989**, *10*, 209.
(25) Still, W. C.; Tempczyk, A.; Hawley, R. C.; Hendrickson, T. *J. Am. Chem. Soc.* **1990**, *112*, 6127.
(26) Lee, B.; Richards, F. M. *J. Mol. Biol.* **1971**, *55*, 379.
(27) Abraham, M. H.; Grellier, P. L.; Prior, D. V.; Duce, P. P.; Morris, J. J.; Taylor, P. *J. Chem. Soc., Perkin Trans. 2* **1989**, 699.
(28) Abraham, M. H. *Chem. Soc. Rev.* **1993**, 73.
(29) Abraham, M. H. In *Quantitative Treatments of Solute/Solvent Interactions*; P. Politzer and J. S. Murray, Eds.; Elsevier: Amsterdam, 1994; p. 83.
(30) *CRC Handbook of Chemistry and Physics*; 75th ed.; Lide, D. R., Ed.; CRC Press: Boca Raton, FL, 1995.
(31) Cullis, P. M.; Wolfenden, R. *Biochemistry* **1981**, *20*, 3024.
(32) Orozco, M.; Colominas, C.; Luque, F. J. *Chem. Phys.* **1996**, *209*, 19.
(33) Miertus, S.; Scrocco, E.; Tomasi, J. *Chem. Phys.* **1981**, *55*, 117.
(34) Tomasi, J. In *Structure and Reactivity in Aqueous Solution, ACS Symposium Series 568*; C. J. Cramer and D. G. Truhlar, Eds.; American Chemical Society: Washington, DC, 1994; p. 10.
(35) Luque, F. J.; Bachs, M.; Orozco, M. *J. Comp. Chem.* **1994**, *15*, 847.
(36) Luque, F. J.; Zhang, Y.; Alemán, C.; Bachs, M.; Gao, J.; Orozco, M. *J. Phys. Chem.* **1996**, *100*, 4269.
(37) Giesen, D. J.; Chambers, C. C.; Cramer, C. J.; Truhlar, D. G. *J. Phys. Chem. B* **1997**, *101*, 5084.
(38) Jungius, C. L. *Z. Phys. Chem.* **1905**, *52*, 97.
(39) Lemieux, R. U. In *Molecular Rearrangements*; P. de Mayo, Ed.; Interscience: New York, 1964; p. 709.
(40) Pearson, C. G.; Rumquist, O. *J. Org. Chem.* **1968**, *33*, 2572.
(41) Kirby, A. J. *The Anomeric Effect and Related Stereoelectronic Effects at Oxygen*; Springer-Verlag: Berlin, 1983.
(42) Tvaroska, I.; Bleha, T. *Adv. Carbohydr. Chem. Biochem.* **1989**, *47*, 45.
(43) *The Anomeric Effect and Associated Stereoelectronic Effects*; Thatcher, G. R. J., Ed.; American Chemical Society: Washington DC, 1993.

(44) Graczyk, P. P.; Mikolajczyk, M. In *Topics in Stereochemistry*; E. L. Eliel and S. H. Wilen, Eds.; John Wiley & Sons: New York, 1994; Vol. 21; p. 159.
(45) Edward, J. T. *Chem. Ind. (London)* **1955**, 1102.
(46) Anderson, C. B.; Sepp, D. T. *Tetrahedron* **1968**, *24*, 1707.
(47) Box, V. G. S. *Heterocycles* **1990**, *31*, 1157.
(48) Romers, C.; Altona, C.; Buys, H. R.; Havinga, E. *Top. Stereochem.* **1969**, *4*, 39.
(49) Jeffrey, G. A.; Pople, J. A.; Radom, L. *Carbohydr. Res.* **1972**, *25*, 117.
(50) Wolfe, S.; Whangbo, M.-H.; Mitchell, D. J. *Carbohydr. Res.* **1979**, *69*, 1.
(51) Jeffrey, G. A.; Yates, J. H. *Carbohydr. Res.* **1981**, *96*, 205.
(52) Wiberg, K. B.; Murcko, M. A. *J. Am. Chem. Soc.* **1989**, *111*, 4821.
(53) Krol, M. C.; Huige, C. J. M.; Altona, C. *J. Comp. Chem.* **1990**, *11*, 765.
(54) Cramer, C. J. *J. Org. Chem.* **1992**, *57*, 7034.
(55) Juaristi, E.; Cuevas, G. *Tetrahedron* **1992**, *48*, 5019.
(56) Petillo, P. A.; Lerner, L. E. In *The Anomeric Effect and Related Stereoelectronic Effects, ACS Symposium Series 539*; G. R. J. Thatcher, Ed.; American Chemical Society: Washington, DC, 1993; p. 156.
(57) Salzner, U.; Schleyer, P. v. R. *J. Am. Chem. Soc.* **1993**, *115*, 10231.
(58) Tvaroska, I.; Carver, J. P. *J. Phys. Chem.* **1994**, *98*, 9477.
(59) Salzner, U.; Schleyer, P. v. R. *J. Org. Chem.* **1994**, *59*, 2138.
(60) Kneisler, J. R.; Allinger, N. L. *J. Comp. Chem.* **1996**, *17*, 757.
(61) Cramer, C. J.; Truhlar, D. G.; French, A. D. *Carbohydr. Res.* **1997**, *298*, 1.
(62) Tvaroska, I.; Kozár, T. *J. Am. Chem. Soc.* **1980**, *102*, 6929.
(63) Tvaroska, I.; Kozár, T. *Int. J. Quant. Chem.* **1983**, *23*, 765.
(64) Praly, J.-P.; Lemieux, R. U. *Can. J. Chem.* **1987**, *65*, 213.
(65) Montagnani, R.; Tomasi, J. *Int. J. Quant. Chem.* **1991**, *39*, 851.
(66) Ha, S.; Gao, J.; Tidor, B.; Brady, J. W.; Karplus, M. *J. Am. Chem. Soc.* **1991**, *113*, 1553.
(67) Kysel, O.; Mach, P. *J. Mol. Struct. (Theochem)* **1991**, *227*, 285.
(68) Cramer, C. J.; Truhlar, D. G. *J. Am. Chem. Soc.* **1993**, *115*, 5745.
(69) van Eijck, B. P.; Hooft, R. W. W.; Kroon, J. *J. Phys. Chem.* **1993**, *97*, 12093.
(70) Perrin, C. L.; Armstrong, K. B. *J. Am. Chem. Soc.* **1993**, *115*, 6825.
(71) Wiberg, K. B.; Marquez, M. *J. Am. Chem. Soc.* **1994**, *116*, 2197.
(72) Jorgensen, W. L.; Detirado, P. I. M.; Severance, D. L. *J. Am. Chem. Soc.* **1994**, *116*, 2199.
(73) Perrin, C. L.; Armstrong, K. B.; Fabian, M. A. *J. Am. Chem. Soc.* **1994**, *116*, 715.
(74) Marcos, E. S.; Pappalardo, R. R.; Chiara, J. L.; Domene, M. C.; Martínez, J. M.; Parrondo, R. M. *J. Mol. Struct. (Theochem)* **1996**, *371*, 245.
(75) Ziegler, F. E. *Chem. Rev.* **1988**, *88*, 1423.
(76) Walsh, C. T.; Liu, J.; Rusnak, F.; Sakaitani, M. *Chem. Rev.* **1990**, *90*, 1105.
(77) Anderson, K. S.; Johnson, K. A. *Chem. Rev.* **1990**, *90*, 1131.
(78) Hilvert, D.; Carpenter, S. H.; Nared, K. D.; Auditor, M.-T. M. *Proc. Natl. Acad. Sci., USA* **1988**, *85*, 4953.
(79) Haynes, M. R.; Stura, E. A.; Hilvert, D.; Wilson, I. A. *Science* **1994**, *263*, 646.

(80) Kast, P.; Hartgerink, J. D.; Asif-Ullah, M.; Hilvert, D. *J. Am. Chem. Soc.* **1996**, *118*, 3069.

(81) Burrows, C. J.; Carpenter, B. K. *J. Am. Chem. Soc.* **1981**, *103*, 6983.

(82) Burrows, C. J.; Carpenter, B. K. *J. Am. Chem. Soc.* **1981**, *103*, 6984.

(83) Coates, R. M.; Rogers, B. D.; Hobbs, S. J.; Peck, D. R.; Curran, D. P. *J. Am. Chem. Soc.* **1987**, *109*, 1160.

(84) Gajewski, J. J.; Jurayj, J.; Kimbrough, D. R.; Gande, M. E.; Ganem, B.; Carpenter, B. K. *J. Am. Chem. Soc.* **1987**, *109*, 1170.

(85) Brandes, E.; Grieco, P. A.; Gajewski, J. J. *J. Org. Chem.* **1989**, *54*, 515.

(86) Grieco, P. A. *Aldrichim. Acta* **1991**, *24*, 59.

(87) Cramer, C. J.; Truhlar, D. G. *J. Am. Chem. Soc.* **1992**, *114*, 8794.

(88) Severance, D. L.; Jorgensen, W. L. In *Structure and Reactivity in Aqueous Solution, ACS Symposium Series*; C. J. Cramer and D. G. Truhlar, Eds.; American Chemical Society: Washington, DC, 1994; Vol. 568; p. 243.

(89) Davidson, J. M.; Hillier, I. H. *Chem. Phys. Lett.* **1994**, *225*, 293.

(90) Davidson, M. M.; Hillier, I. H. *J. Chem. Soc., Perkin Trans. 2* **1994**, 1415.

(91) Davidson, M. M.; Hillier, I. H.; Hall, R. J.; Burton, N. A. *J. Am. Chem. Soc.* **1994**, *116*, 9294.

(92) Davidson, M. M.; Hillier, I. H.; Vincent, M. A. *Chem. Phys. Lett.* **1995**, *246*, 536.

(93) Severance, D. L.; Jorgensen, W. L. *J. Am. Chem. Soc.* **1992**, *114*, 10966.

(94) Gao, J. *J. Am. Chem. Soc.* **1994**, *116*, 1563.

(95) Jorgensen, W. L.; Blake, J. F.; Lim, D.; Severance, D. L. *J. Chem. Soc., Faraday Trans.* **1994**, *90*, 1727.

(96) Sehgal, A.; Shao, L.; Gao, J. *J. Am. Chem. Soc.* **1995**, *117*, 11337.

(97) Gao, J. L. *Acc. Chem. Res.* **1996**, *29*, 298.

(98) Carlson, H. A.; Jorgensen, W. L. *J. Am. Chem. Soc.* **1996**, *118*, 8475.

(99) Yoo, H. Y.; Houk, K. N. *J. Am. Chem. Soc.* **1994**, *116*, 12047.

(100) Wiest, O.; Black, K. A.; Houk, K. N. *J. Am. Chem. Soc.* **1994**, *116*, 10336.

APPLICATIONS

Chapter 16

Practical Chemistry Modeling Applied to Process Design Studies

J. T. Golab and M. R. Green

Amoco Research Center, Amoco Corporation, 150 West, Warrenville Road, Naperville, IL 60566–7011

Chemistry modeling is capable of accurately predicting the physical, material, and performance properties of both established and new materials and processes. A relatively new technology, it is gaining rapid acceptance as a way to understand practical industrial chemical problems in a structured and detailed manner. Based on quantum chemistry, chemistry modeling is a novel science that uses theoretical models and computer software instead of experimental observation to measure the properties of molecules. This paper discusses how chemistry modeling assists engineers in the design, pilot, and production stages of chemical processes. Several current examples taken from the modern chemical industry are reported. Finally, the promise of chemistry modeling as a modern business tool is addressed.

Modeling by computational methods is a general purpose, less expensive alternative to the traditional experimental methods of problem solving. At the macroscopic level, engineers are applying software solutions to a myriad of critical industrial problems successfully (*1*). Computational chemistry is a newer technology that attempts to better understand chemistry at the microscopic level. It is an enabling technology that, together with engineering programs, has the potential to significantly alter the manner in which research and development is conducted. In fact, by coupling these simulation technologies, the necessary parts to take a product from initial conception to plant production, using calculational methodologies, are in place (*2*).

Computational chemistry modeling, as an independent approach to the solution of industrial problems, provides a systematic methodology to formulate well-defined, structurally detailed, computer models of molecules or processes at the atomistic level. The results of these simulations produce explicit information which augments, complements and, in a growing number of cases, directs or replaces experimental research. Like the experimental chemist, who uses chemistry to verify a model, or hypothesis, of a particular process, the computational chemist uses models to gain insight into the chemistry of a particular process.

Background

Briefly defined, computational chemistry is the science relating molecular properties to the motion and interaction of electrons and nuclei (*3*). It attempts to quantitatively model all aspects of chemical behavior on a computer using the formalism of theoretical, rather than experimental, chemistry. Its theoretical roots are grounded in quantum mechanics derived by Schroedinger and others, as summarized in the well known equation (*4,5*),

$$H_e(R,r)\Psi_e(R,r) = E_e(R)\Psi_e(R,r). \tag{1}$$

Mainstream computational chemistry is focused on solutions to the time-independent, nonrelativistic, fixed nuclei version of this equation, in which the electronic Hamiltonian, $H_e(R,r)$, and the electronic wavefunction, $\Psi_e(R,r)$, are functions of the nuclear coordinates, R, and the electronic coordinates, r, of the molecule. The solutions to equation 1, the electronic energy levels, $E_e(R)$, and the nuclear repulsion energy of the system, define a potential energy surface, $V(R)$, on which the molecule can move,

$$V(R) = E_e(R) + \Sigma_n \Sigma_{n>m} Z_n Z_m/D_{nm} \tag{2}$$

where Z_n is the atomic number of nucleus n, and D_{nm} is the distance between nuclei n and m. It should be noted that numerous different methods exist in the literature (*3-12*) to solve equation 1 and they are all more or less appropriate depending on the chemical system under study and the level of accuracy desired. Great strides are being taken to improve the efficiency of each technique and/or algorithm, even though in what follows, we are speaking with a quantum chemistry bias.

The potential energy defined by equation 2 is a function of 3N-6 coordinates for nonlinear molecules (3N-5 for linear systems), the other six (or five) coordinates split between translational and rotational motions. $V(R)$ is quite formidable to obtain in detail (*13*). For example, the potential energy surface of methane, a five atom molecule, depends on nine independent coordinates. Neglecting symmetry, creating a potential energy surface composed of ten points for each coordinate, would require one billion separate calculations!

Luckily, all the points on a potential energy surface are not of equal interest. For example, only stationary points provide the details of the molecule's structure, conformational preference, and bond strengths. By definition, all stationary points on a potential energy surface have a first derivative of energy with respect to nuclear motion that is equal to zero (*3-5*). These points can be further classified into two groups by inspecting the molecule's diagonalized second derivative matrix, or Hessian. If all these eigenvalues are positive, then the stationary point represents a minimum on the potential energy surface. This minimum could correspond to a reactant, a product, or an intermediate. On the other hand, if all the eigenvalues are positive except one (and *only* one) then we have found a transition state. The transition state is the saddle point that connects the reactants and products on the potential energy surface (*14,15*). The motion along this path is described by the eigenvector specified by the negative eigenvalue and is commonly called the reaction coordinate at the saddle point. Thus, knowledge of the transition state is a critically important piece of kinetic information that cannot be obtained easily from experiment since the transition state exists so briefly. Many chemical reactions of interest occur on potential energy surfaces which can be described adequately with fewer than fifty reactants, products, intermediates, and transition states, well within the grasp of leading edge computer technology.

To illustrate how this applies to reaction thermodynamics, consider the energy diagram for an exothermic reaction (Figure 1). ΔE_{rxn} represents the difference in electronic energies of the reactants (E_r) and products (E_p) as obtained directly from equation 1,

$$\Delta E_{rxn} = \Sigma\, E_p - \Sigma\, E_r. \tag{3}$$

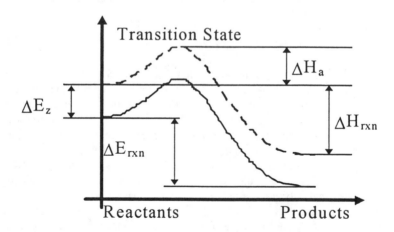

Figure 1. Energy Diagram of a General Exothermic Chemical Reaction. The solid line depicts a simple reaction path from reactants to products; the dashed line depicts the same path corrected by the zero point energy of reactants and products.

E_z, the zero point energy, is a correction that accounts for the vibrational motion of the molecule at the absolute zero of temperature. Assuming that the molecular bonds of the (nonlinear) chemical species can be represented by harmonic oscillators,

$$E_z = (1/2) \Sigma_i h\nu_i, \quad \text{for } i = 1, 2, ..., 3N\text{-}6 \quad (4)$$

where h is Plank's constant (*3*). The frequencies (ν_i) of the system are obtained from the diagonalization of the 3N x 3N dimensional mass-weighted Hessian matrix (*16*). When E_z is added to the electronic energies, the entire energy diagram is shifted up (exaggerated for emphasis), as shown by the dotted line, and now the energy difference between reactants and products corresponds to the enthalpy of reaction ΔH_{rxn} (at 0K),

$$\Delta H_{rxn} = \Delta E_{rxn} + (\Sigma E_{zp} - \Sigma E_{zr}) \quad (5)$$

where E_{zp} is the zero point energy of the products and E_{zr} is the zero point energy of the reactants. Similarly, the activation energy of reaction, ΔH_a, can be determined from the energy of the transition state and reactants (at 0K)

$$\Delta H_a = E_{ts} + E_{zts} - \Sigma (E_r + E_{zr}) \quad (6)$$

where E_{ts} and E_{zts} are the electronic and zero point energies of the transition state, respectively. Thus, the "simple" reaction coordinate diagram shown in Figure 1 defines the geometries of reactants, products, intermediates, and transition state(s), provides the overall reaction thermochemistry, and qualitatively indicates the energy barrier to reaction.

To design materials, a chemist needs a detailed understanding of the factors controlling the molecular interactions of a reaction. These properties, such as electronic and vibrational energy levels, zero point and activation energies, as well as ionization potentials, electron affinities, and IR, Raman, and NMR frequencies and intensities, can be calculated for any chemical moiety from knowledge of the molecule's quantum mechanical energies, wavefunctions, and/or potential energy surface (*12,13*). Furthermore, the application of statistical mechanics provides us with a methodology to calculate a wide variety of thermochemical properties such as the enthalpy, entropy, heat capacities, Gibb's free energy, and various kinetic parameters (*3,17*).

In order to calculate thermochemical properties as a function of temperature, such as enthalpy, $\Delta H(T)$, and absolute entropy, $S^\circ(T)$, it is necessary to employ the equations of statistical mechanics, assuming ideal gas behavior. These quantities are sums of translational, rotational, vibrational, and electronic terms which depend on temperature and characteristics of the molecule, e.g. molecular mass and the moments of inertia (*17*). Obviously, experimental spectroscopic methods can determine thermodynamic properties if the moments of inertia (for example, microwave and x-ray) and vibrational frequencies (for example, IR/Raman) can be measured. Even then, computationally generated spectra can assist in the interpretation of experimental spectra that are complex and overlapped. And more importantly, for systems whose spectra might be very hard to measure, such as transition states, calculational methods provide the only practical means of determining a thermochemical property value.

At this point, all the information that is needed to calculate rates of reaction is available to us. Arrhenius observed that rate constants depend upon temperature according to the second law of thermodynamics ($14,15$),

$$k_f(T) = A(T)\exp(-\Delta H_a(T)/kT) \tag{7}$$

where $k_f(T)$ is the forward reaction rate constant, $A(T)$ is the Arrhenius preexponential factor, k is the Boltzmann constant.

In addition to the calculation of forward (and reverse) reaction rates, we can also determine the equilibrium constant of a chemical process by using the Gibb's free energy function. By thermodynamic definition (17), the Gibb's free energy for a reaction is written as,

$$\Delta G_{rxn}(T) = \Delta H_{rxn}(T) - T\Delta S_{rxn}(T) \tag{8}$$

and at equilibrium,

$$\Delta G_{rxn}(T) = -(RT)\ln K_p \tag{9}$$

where K_p is the equilibrium constant of reaction.

Current advances in software and hardware technologies have made the use of computational chemistry techniques practical for industrial applications. Indeed, the recent emergence of high performance scientific workstations and vector or parallel supercomputers has brought about an explosion of user friendly programs which can be directly brought to bear on significant problems ($10-12$). In addition, very fast reduced instruction set chip (RISC) based workstations have become a viable alternative to the traditional central supercomputer. RISC based workstations are extremely cost effective, typically delivering more throughput than the usually small fractional share of most supercomputers. As a result, these technologies can be placed easily at the fingertips of the developers, practitioners, and implementors of chemistry modeling. This type of multidiscipline, cross functional collaboration will serve to accelerate the amount of innovation in this area.

Applications

In this section, a few highlights taken from Amoco Corporation chemistry modeling projects are discussed. The first two examples show that computational chemistry techniques yield similar or improved results relative to the group contribution methods that have been used more commonly by the engineering community to predict thermodynamic properties ($15,18,19$). The third example shows that simulations can be used to determine thermochemical properties with experimental accuracy. The final case demonstrates how chemistry modeling can assist in the understanding of a chemical process.

Group Contribution Methods. In a very early example, a model reaction, the dehydration of acetic acid to acetic anhydride and water at 312K,

$$2CH_3CO_2H \cdots \!\!\!> (CH_3CO)_2O + H_2O \qquad (10)$$

was studied using data taken from the DIPPR database (*18*), using the group contribution method, CHETAH (*19*), and ab initio computational chemistry results, calculated by the Hartree-Fock (HF) (*3-5*) method (see Table I). Group contribution methods empirically estimate thermodynamic properties assuming that the property is composed of piecewise additive contributions from groups within the molecule. A group is defined as a "polyvalent atom in a molecule together with all its ligands." (*15*)

Table I. Thermochemistry at 312K for Reaction 10, kcal/mol

Method	ΔG_{rxn}	ΔH_{rxn}	$T\Delta S_{rxn}$
DIPPR	11.3±6.1	12.3±3.5	1.0±2.7
CHETAH	11.0	12.2	1.2
HF/3-21G	14.1	13.6	-0.5

As expected, the data obtained from CHETAH compares quite favorably to the DIPPR results. The ab initio results are also well within the uncertainty values listed in DIPPR. Considering that the calculation used a small (3-21G) basis set and that the HF method does not account for electron correlation, the agreement obtained is very good. This example demonstrates good agreement between experimental and computational data for a practical chemical system. If more accurate agreement is required in the computed data for a chemical system for which highly accurate experimental data exist, then higher levels of calculations can be performed.

Amoco uses chemistry modeling as a tool that provides additional information on which to base technical business decisions. For example, the temperature dependence of the thermodynamics for the hydrogenation of 4-carboxybenzaldehyde, (4-CBA), to hydroxymethylbenzoic acid,

$$C_8O_3H_6 + H_2 \cdots \!\!\!> C_8O_3H_8 \qquad (11)$$

is relevant to our purified terephthalic acid process in which removal of 4-CBA is a step in the purification process.

Shown in Table II is the thermochemistry at 555K, obtained from CHETAH, by two independent users, denoted A and B, compared to values obtained from MOPAC using the PM3 parameter set (*8*).

Table II. Thermochemistry at 555K for Reaction 11, kcal/mol

Method	K_{eq}	ΔG_{rxn}	ΔH_{rxn}	$T\Delta S_{rxn}$
PM3	0.6	0.6	-15.9	-16.5
CHETAH-A	2.1	-0.8	-18.5	-17.7
CHETAH-B	0.4	0.9	-16.7	-17.7

User B computed the thermodynamic values using different and fewer groups than User A. User B employed base groups that were closer to the target molecule, wherein CHETAH is expected to yield more accurate results.

As shown, all the results agree that the reaction is exothermic and leads to a decrease in entropy. However, CHETAH-A predicts a decrease in free energy, owing to the larger exothermicity of the reaction, and as a result, a larger value of the equilibrium constant than either CHETAH-B or PM3.

Reaction Thermodynamics. One of the greatest strengths of computational chemistry methods is that one can study any molecule or chemical reaction. This ability profoundly impacts the acquisition of detailed, pertinent chemical information on compounds that have not been experimentally characterized or even synthesized. In fact, computational experiments are limited only by the imagination of the scientist and, of course, the computer resources that are available.

The catalytic conversion of methane to methanol is a topic of significant industrial interest. Methane activation, reaction 12, is known to occur in the presence of several transition metals. Conversion of a bound methyl group to a bound methoxy group has also been shown. Subsequent methanol formation from supported methoxy-metal species can occur via thermal hydrolysis or thermal hydrogenolysis, reaction 13, i.e.

$$M\text{-}H + CH_4 \;\cdots\!\!\succ\; M\text{-}CH_3 + H_2 \qquad\qquad (12)$$

$$M\text{-}OCH_3 + H_2 \;\cdots\!\!\succ\; M\text{-}H + CH_3OH \qquad\qquad (13)$$

where M is a metal hydroxide and/or oxo fragment, e.g. -Ta(OH)$_2$.

Experimental work (20) done under contract for Amoco's Natural Gas Program found that, for tantalum species, the catalytic conversion of methane to methanol is not appreciable; while methane activation does occur, subsequent hydrogenolysis does not occur. To better understand this behavior and to identify more promising transition metals, oxidation states, and coordination geometries for experimental study, chemistry modeling calculations were performed to obtain the thermochemistry of the above reactions substituting third row transition metal species, tantalum, tungsten, and rhenium for M.

The diverse electronic structure of transition metal containing systems leads to their particularly rich and commercially important chemistries. However, the theoretical study of reactions involving transition metal atoms presents three fundamental difficulties. First, transition metals have many electrons (relative to carbon, oxygen, and hydrogen) and hence naturally lead to large calculations. The computer resources required to perform a computational chemistry calculation are roughly proportional to the fourth power of the number of functions used to describe the molecule's electron orbitals. Second, the third row transition metals investigated in this study require the inclusion of relativistic effects in the computational method. Relativistic effects are important in transition metal atoms because the relativistic increase of mass with velocity of their electrons contracts or shrinks the s and p orbitals. These orbitals in turn shield or screen the nucleus more effectively leading to an expansion of the d and f orbitals. Finally, transition metal compounds require extensive treatment of electron correlation in order to obtain quantitatively useful information. Electron correlation is the energy that represents the instantaneous interactions between the electrons of the molecule (*3-5*).

The molecules in this study were described by relativistic effective core potentials (*21-23*) for the modeling calculations. This means that inner shell electrons (e.g., for metal atoms, 1-4s, 2-4p, 3-4d, and 4f) are replaced by a parameterized potential function. As a result, only outer shell electrons (e.g., for metal atoms, 5-6s, 5p, and 5d) are explicitly included in the calculation. Thus, this approximation not only accounts for relativistic effects but also significantly reduces the amount of computational work that is required per molecule. To describe the electron correlation of each system, Moller-Plesset second order perturbation (MP2) (*3-5*) single point calculations were performed at the Hartree-Fock optimized geometry.

All the calculations for this study were performed using GAMESS, General Atomic and Molecular Electronic Structure Systems, (*12*), an easily ported, parallel, quantum chemistry program that we have utilized successfully on other technical projects. Equilibrium geometries that correspond to energy minima and transition states were calculated at the Hartree-Fock level of theory within an effective core potential basis set (*21-23*). All possible oxidation states and spin states were described using single "high spin" configuration state functions, i.e. all unpaired electrons have alpha spin. However, as is generally accepted, higher spin states have a corresponding lower electronic energy and this trend is also observed for the molecules and the reactions of this study. Lower energy corresponds to molecules that are more likely to exist and reactions that are easier to accomplish. Thus, for the purposes of this discussion, we ignore the low spin cases since they are less likely to participate in the overall conversion of methane to methanol. Chemical intuition was used to develop initial guesses for the coordination geometry for each species but C_1 symmetry was employed in all the calculations.

This study comprised over seventy-five different compounds that interact via forty-five unique reactions. We did not allow for spin crossing in the reactions reported here, i.e. singlet state reactants produce singlet state products. The transition metals and oxidation states we examine here are Ta(+3 and +5), W(+4 and +6), and Re(+3, +5, and +7). The reaction energies, the differences in the sums of the electronic energies of the products and reactants (see equation 3), for reactions 12 and 13 are listed in Tables III and IV, respectively.

Table III. Electronic Reaction Energies for Reaction 12, kcal/mol

M-	Spin State	$\Delta E(HF)$	$\Delta E(MP2)$
Ta(OH)$_2$-	Triplet	15.1	5.9
Ta(OH)$_2$(O)-	Singlet	10.5	2.2
Ta(OH)$_4$-	Singlet	13.3	3.6
W(OH)$_3$-	Triplet	26.8	18.4
W(OH)$_3$(O)-	Singlet	15.5	7.2
W(OH)$_5$-	Singlet	12.5	0.1
Re(OH)$_2$-	Quintet	15.8	6.9
Re(OH)$_2$(O)-	Triplet	15.8	8.6
Re(OH)$_2$(O)$_2$-	Singlet	19.7	10.1
Re(OH)$_4$-	Triplet	19.3	8.3
Re(OH)$_4$(O)-	Singlet	20.4	13.7

Table IV. Electronic Reaction Energies for Reaction 13. kcal/mol

M-	Spin State	$\Delta E(HF)$	$\Delta E(MP2)$
Ta(OH)$_2$-	Triplet	24.1	33.7
Ta(OH)$_2$(O)-	Singlet	35.5	43.6
Ta(OH)$_4$-	Singlet	27.1	37.8
W(OH)$_3$-	Triplet	3.4	16.4
W(OH)$_3$(O)-	Singlet	15.1	28.4
W(OH)$_5$-	Singlet	19.2	35.9
Re(OH)$_2$-	Quintet	4.4	15.6
Re(OH)$_2$(O)-	Triplet	-0.3	13.9
Re(OH)$_2$(O)$_2$-	Singlet	-5.4	16.3
Re(OH)$_4$-	Triplet	10.2	25.4
Re(OH)$_4$(O)-	Singlet	-5.6	22.3

Grouping the Hartree-Fock electronic reaction energies as shown in Figures 2 and 3 makes similarities and trends more easily discernible. For example, Figure 2 shows that there is little difference in the energy of reaction for methane activation. All of the transition metal compounds are predicted to activate methane around 15 kcal/mol, although tantalum appears to be slightly better than tungsten or rhenium.

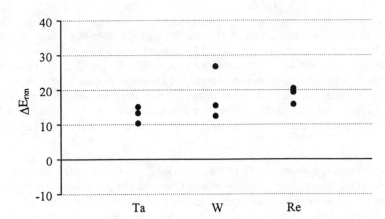

Figure 2. Hartree-Fock Electronic Reaction Energies for Reaction 12

However, scrutiny of Figure 3 clearly shows that the energy of reaction of alcohol liberation is a function of the transition metal compound. In fact, the graph shows that cleavage of the metal-oxygen bond is significantly less favorable for tantalum compounds than for compounds of tungsten or rhenium.

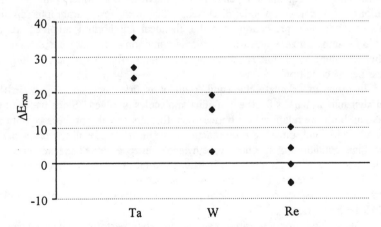

Figure 3. Hartree-Fock Electronic Reaction Energies for Reaction 13

From this information, it is clear that the activation of a carbon hydrogen bond in methane is not as important a differentiating characteristic within this third row transition metal series as the hydrogenolysis of the bound methoxy. The results of this investigation predict that cleavage of the transition metal oxygen bond by molecular hydrogen (reaction 13), appears to differentiate between the transition metals more so than the methane activation step. Significantly, tantalum is predicted to be very unfavorable for alcohol liberation in comparison to either tungsten or rhenium. In fact, rhenium compounds are predicted to be the best candidates for use in a catalytic cycle for methane conversion. This conclusion is supported by recent experimental work that was pursued in tandem with our study (*20*).

Process Thermodynamics. It is the responsibility of chemistry modelers to discern, at the microscopic level, the chemistry involved in processes. Putting that information to work in engineering models, such as Aspen, to simulate molecular ensembles, or even the bulk state, is still a challenge. Even so, this multiscale approach to process design is being actively pursued by farsighted companies.

Consider for example the alkylation (and subsequent transalkylation) reactions for ethylbenzene production.

$$C_2H_4 + C_6H_6 \cdots\!\!> C_8H_{10} \qquad\qquad (14)$$

These reactions are of interest to Amoco because ethylbenzene is the feedstock for styrene monomer. In addition, the technology was purchased from another company and the equilibrium relationships between conformers and process conditions were not completely disclosed. The next few paragraphs discuss chemistry modeling's predictions for the chemical equilibrium involved in the formation of ethylbenzene at various process conditions. In one sense, what we will be describing is chemistry modeling's technique for investigating alternate process design conditions.

In order to calculate the equilibrium, it is first necessary to obtain the thermodynamic properties for the individual molecules involved. Since we want to understand both the relationships between conformers and the process conditions of the alkylation and transalkylation reactions, we must include all the participants in the final equilibrium, specifically, ethylene, benzene, ethylbenzene, and the polyethylbenzenes through tetra-ethylbenzene, a total of twelve compounds. The thermodynamic properties for the individual molecules are computed as a function of temperature using the semiempiricial (AM1) quantum chemistry options within the GAMESS program (*7,12*).

The equilibrium composition calculations were performed using Chemkin, a Chemical Kinetics Package for the Analysis of Gas Phase Chemical Kinetics, obtained from Sandia National Laboratories (*24,25*). Chemkin is a FORTRAN tools suite that facilitates the formation, solution, and interpretation of problems involving elementary gas phase chemical reactions. As such, it is one of three basic components in the Sandia software set designed to facilitate simulations of elementary chemical reactions in flowing systems. (The other major elements are programs for transport properties and surface chemistry simulation.)

The data required for Chemkin consists entirely of thermodynamic properties, i.e. the absolute heat of formation, ΔH_f°, the absolute enthalpy, H°, absolute entropy, S°, and the heat capacity, C_p; for each species over some temperature range. A preprocessing program reads this raw thermodynamic data, converts it to a set of polynomial fits, and stores it in a database which is used in the Chemkin model. This database is identical in format to the NASA Chemical Equilibrium program (*26*). The other necessary input is the initial concentrations of each of the components in the system.

The thermodynamic fitting program uses the following forms for the thermodynamic functions (*24,25*):

$$C_p/R = a_1 + a_2T + a_3T^2 + a_4T^3 + a_5T^4 \qquad (15)$$

$$H^\circ/RT = a_1 + a_2T/2 + a_3T^2/3 + a_4T^3/4 + a_5T^4/5 + a_6/T \qquad (16)$$

$$S^\circ/R = a_1\ln T + a_2T + a_3T^2/2 + a_4T^3/3 + a_5T^4/4 + a_7 \qquad (17)$$

Furthermore, the fitting program fragments the temperature range into two pieces; thus, each molecule has fourteen fitting coefficients, i.e. a_1 - a_7 for each temperature interval. In our Chemkin model of reaction 14 the two temperature intervals are 300 - 1000K (low interval) and 1000K - 5000K (high interval).

In order to provide some comparisons of the thermodynamic data calculated from GAMESS(AM1) with that of other approaches, we have summarized in Table V the Gibb's free energy for reaction 14 as obtained from a thermodynamic database program maintained by the National Institute of Standards and Technology (*27*), labeled LIT for literature, as obtained by GAMESS(AM1), labeled AM1, and as obtained from Amoco's Styrene Technical Information Bulletin (*28*), labeled TIB, over the temperature range 300-1000K.

Table V. Gibb's Free Energy Reaction Energies for Reaction 14, kcal/mol

T(K)	ΔG(LIT)	ΔG(AM1)	ΔG(TIB)
300	-13.4	-14.6	-16.0
400	-10.4	-11.5	-13.0
500	-7.35	-8.53	-9.94
600	-4.37	-5.56	-6.94
700	-1.43	-2.61	-3.97
800	1.48	0.29	-1.03
900	4.34	3.16	1.88
1000	7.18	5.99	4.78

Overall, the values agree reasonably well, although both the literature and AM1 ΔG_{rxn} values become positive sooner than suggested by the TIB data. This result may have an impact on the individual final equilibrium concentrations for the components in the model. The effect on the overall equilibrium is not expected to be large. One way to improve these values would be to use a higher level of quantum chemical theory to determine the thermodynamics. Even so, the AM1 values fit reasonably well between the literature values and those from the TIB.

In order to ensure the validity of the calculated thermodynamic data, the Chemkin program was run using the plant conditions of the alkylator. For this Chemkin run, the temperature and pressure are held constant at the plant values, while the initial concentrations of ethylene and benzene are varied and the resulting equilibrium concentrations of each of the participating species is calculated.

The results of this computational experiment, as compared to the Styrene TIB data, are summarized in Figure 4. The Chemkin results are represented with dots; the TIB data with open diamonds. Each point on the curve corresponds to the product equilibrium composition for a particular alkylator feed molar ratio of ethylene and benzene. For example, the left most point (lower left hand corner) represents the product equilibrium composition starting with a molar ethylene to benzene ratio (E/B) of 0.1; the right most point (upper right hand corner) represents the product equilibrium composition starting with a molar E/B ratio of 1.0. The points are spaced at molar ratios of 0.1. As can be clearly seen, the agreement between the Styrene TIB and the Chemkin predictions is very good.

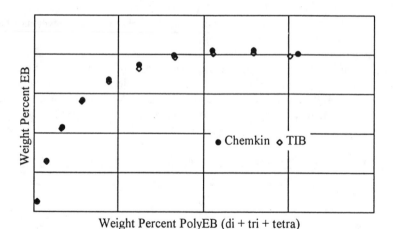

Weight Percent PolyEB (di + tri + tetra)

Figure 4. Resulting Equilibrium Weight Percent Ethylbenzene versus Resulting Equilibrium Weight Percent di-, tri-, and tetra-ethylbenzene as a Function of Ethylene and Benzene Molar Feed Ratio

To study the effects of temperature and pressure with this model, product equilibrium compositions resulting from the alkylation and transalkylation reactions in the process at various temperatures and pressures were calculated as a function of the alkylator feed ethylene to benzene molar ratio. Using these computationally derived equilibrium composition curves, a gas phase equilibrium reaction model for the alkylation and transalkylation reactions at various temperatures and pressures was successfully developed.

Conclusions

Within many modern industrial R&D programs, chemistry modeling groups apply computational chemistry methods as required by business group research and development projects (*29*). These calculations accurately predict molecular structures as well as a wide variety of molecular properties providing additional, alternative information upon which better technical business decisions can be based. Moreover, chemistry modeling simulations can resolve the intimate details of any general chemical reaction thereby, increasing the fundamental understanding of the reaction energetics, mechanisms, and kinetics. From this knowledge, chemical processes can be designed or optimized so that yields are improved and selectivity is enhanced. Ultimately, chemistry modeling will guide laboratory research programs by determining the most promising directions upon which to focus precious R&D monies amongst multiple initial choices.

Chemistry modeling projects, similar to those illustrated in this chapter, would not have been computationally feasible just a few years ago. At that time, quantum mechanical algorithms were not efficient enough and computing power was prohibitively expensive. But today, the impact that high performance computing is having on scientific research is profound and will continue to be substantial. RISC workstations and distributed computational methods and algorithms are effective, general tools capable of expanding the range of chemical systems for which useful calculations are possible. Consider that supercomputers and massively parallel computer systems are forecast to become at least two orders of magnitude faster, workstations even more magnitudes faster, within the next five years (*30*). At the same time, the cost of computing will decrease proportionally, so that what is only imaginable now will be not only within reach of leading edge computer technology but also priced at what would be considered a bargain today.

As computer power continues to increase and equivalent CPU costs continue to decrease, traditionally experimental areas of chemistry, such as solution kinetics and dynamics will become amenable to detailed computational study. Further, mixed phase reactions will be modeled so that direct comparison with experimental values can take place with confidence. Soon, calculations will provide the necessary data to questions that are either simply inaccessible or not easily addressed by experiment (*31*). For example, specific reactions which occur in the solid phase, or on top of solids, like those of a catalyst, are already being modeled (*32,33*). We have executed chemistry modeling projects which provide multicomponent equilibrium data that directly correspond to production plant (not pilot plant) conditions. This kind of interaction between industrial engineers and chemistry modelers promises to provide the long term effectiveness and efficiency of modern R&D for manufactured products.

These promising results are encouraging and suggest that a substantial acceleration of research and development efforts could be achieved through the proper integration of experimental, engineering, and chemistry modeling applications.   Consider that by the turn of the century, the software pieces necessary to take a new chemical product all the way from initial concept to three dimensional plant layout will be available.   Computational chemistry will have a tremendous impact in process assessment, process improvement, and process differentiation ventures of the future.   For example, thermodynamics and kinetic information can be used to assess the feasibility of a process (qualitatively answering the questions; 'Can it be made?' and 'Will the process work?') before any trial and error by experiment is done.   Using the explicitly precise approach inherent in chemistry modeling technology, mechanisms can be evaluated not only to obtain a much deeper and richer understanding of how a process works and can be made better but also to proscribe limits to the amount of improvement that might be achieved.   And finally, once a product is understood at the atomistic level, marketing questions concerning why it is better, what are the performance properties that differentiate it from the rest of the pack, can be used to a company's advantage.   We can confidently state that some of these issues are being addressed by chemistry modeling already.

This brief overview has reported on an important enabling technology, chemistry modeling, that is becoming an integral part of the modern industrial R&D process.   In cases in which it has been used at Amoco, it's led to solutions that we would probably not have otherwise found.   As computational science continues to be applied as an alternative, independent source of quantifiable, accurate technical information, ideas can be tested and verified through experimental or computational means, whichever is more direct or more efficient. This new methodology provides companies with a superior competitive advantage, allowing its personnel to make more time- and cost-effective business decisions; to "think harder and work smarter".   It is this team atmosphere that will lead to the synergy necessary to strongly accelerate a company's efforts to be "first to the future".   After all, the companies that best integrate the *generation* of new knowledge with the *use* of that knowledge will be the leaders in the next century.

Acknowledgments

Colleagues at Amoco Corporation, Pacific Northwest National Laboratory, Argonne National Laboratory, Ames Laboratory of Iowa State University, and Biosym Technologies, Inc. (now known as Molecular Simulations, Inc.) are recognized for their contributions to this report.

Literature Cited

1. See for example: ASPENPLUS and SPEEDUP, Aspen Technology; FLUENT, Fluent, Inc.; FLOW3D, AEA Technology, ST5, Heat Transfer Research, Inc.; PRO2, Simulation Science, Inc.
2. Parkinson, G.; Fouhy, K. *Chem. Eng.* **1996**, January, 30.
3. Hehre, W.; Radom, L.; Schleyer, P.; Pople, J. *Ab Initio Molecular Orbital Theory;* Wiley & Sons: New York, NY, 1986.
4. Szabo, A.; Ostlund, N. *Modern Quantum Chemistry;* MacMillan: New York, NY, 1982.
5. Levine, I. *Quantum Chemistry,* 4th ed.; Prentice-Hall: Englewood Cliffs, NJ, 1991.
6. *Molecular Mechanics;* Burkett, U.; Allinger, N., Eds.; ACS Monograph 177, Washington, D.C., 1982.
7. Dewar et al. *J. Am. Chem. Soc.* **1985**, 107, 3902.
8. Stewart, J. J. *Comp. Chem.* **1989**, 10, 209.
9. Parr, R. G.; Yang, W. *Density-Functional Theory of Atoms and Molecules;* Oxford: New York, NY, 1989.
10. Counts, R. *Quantum Chemistry Program Exchange (QCPE),* Indiana University, Department of Chemistry.
11. Schmidt, M.; Baldridge, K.; Boatz, J.; Jensen, J.; Koseki, S.; Gordon, M.; Nguyen, K.; Windus, T.; Elbert, S. *QCPE Bulletin,* **1990**, 10, 52.
12. Schmidt, M.; Baldridge, K.; Boatz, J.; Elbert, S.; Gordon, M.; Jensen, J.; Koseki, S.; Matsunaga, N.; Nguyen, K.; Su, S.; Windus, T.; Dupuis, M.; Montgomery, J. *J. Comput. Chem.* **1993**, 14, 1347.
13. Harding, L. In *Advances in Molecular Electronic Structure Theory;* Dunning, T., Ed.; JAI Press: Greenwich, CT, 1990, Vol. 1; 45-83.
14. Gardiner, Jr., W. C. *Rates and Mechanisms of Chemical Reactions;* Benjamin/Cummings: Menlo Park, CA, 1972.
15. Benson, S. W. *Thermochemical Kinetics;* Wiley & Sons: New York, NY, 1976; 2nd Ed.
16. Wilson, Jr., E. B.; Decius, J. C.; Cross, P. C. *Molecular Vibrations. The Theory of Infrared and Raman Vibrational Spectra;* Dover: New York, NY, 1980.
17. McQuarrie, D. *Statistical Thermodynamics;* Harper & Row: New York, NY, 1973.
18. Design Institute for Physical Property Research (DIPPR), Pennsylvania State University, July 1990.
19. CHETAH, The ASTM Chemical Thermodynamics & Energy Release Evaluation Program, American Society for Testing & Materials, January, 1990.
20. Basset, J.-M. and Niccolai, G. University of Lyon, France and Hagen, G. and Udovich, C., Amoco, private communications.
21. Stevens, W.; Basch, H.; Krauss, M. *J. Chem. Phys.* **1984**, 81, 6026.
22. Stevens, W.; Basch, H.; Krauss, M.; Jasien, P. *Can. J. Chem.* **1992**, 70, 612.
23. Cundari, T. R.; Stevens, W. J. *J. Chem. Phys.* **1993**, 98, 5555.

24. Kee, J.; Rupley, F. M.; Miller, J. A. Sandia Report SAND89-8009B **1994**.
25. Kee, J.; Rupley, F. M.; Miller, J. A. Sandia Report SAND87-8215B **1994**.
26. Gordon; McBride NASA SP-273 **1971**.
27. The National Institute of Standards and Technology (NIST) Standard Reference Database 25; a DOS based database and query program of thermodynamic quantities.
28. Amoco Corporation, private document.
29. Ruiz, J., Dow Chemical; Lauffer, D., Phillips 66; Koehler, M., Allied Signal; Dobbs, K., Dupont; private communications.
30. Zorpette, G. *IEEE Spectrum* **1992**, September, 28.
31. *Reviews in Computational Chemistry;* Lipkowitz, K.; Boyd, D., Eds.; VCH: New York, NY, 1990, Vol. 7 and other volumes in this excellent series.
32. Boudart, M.; Djega-Mariadassou, G. *Kinetics of Heterogeneous Catalytic Reactions;* Princeton University Press: Princeton, NJ, 1984.
33. *Theoretical and Computational Approaches to Interface Phenomena;* Sellers, H.; Golab, J., Eds.; Plenum: New York, NY, 1994.

Chapter 17

Implementation and Application of Computational Thermochemistry to Industrial Process Design at the Dow Chemical Company

David Frurip[1], Nelson G. Rondan[2], and Joey W. Storer[3,4]

[1]Analytical Sciences Laboratory, Dow Chemical Company, Building 1897F, Midland, MI 48667
[2]Computing, Modeling, and Information Sciences, Dow Chemical Company, Building 1707, Midland MI 48667
[3]Cray Research Inc., Eagan, MN 55121

A key factor in the safe and successful scaleup of a new chemical process is an understanding of the energy release from the intended (and sometimes from the unintended) chemistry. In our laboratory, we approach the problem of process heat determination in two ways. Usually we first attempt to determine the heat experimentally but this is not always practical, feasible or required. Another method, theoretical prediction is therefore necessary and in many ways advantageous. For many years this laboratory has extensively used and developed thermochemical prediction techniques. Since 1990, we have implemented computationally-intensive quantum mechanical techniques (G2 for example) to solve problems on a routine basis where the traditional techniques are lacking or simply fail. Pragmatism dictated that the tools to perform these computations be set up in such a way as to be transparent even to the occasional user. Combining appropriate computational resources, data manipulation software, and modeling expertise has allowed us to include these techniques among our standard predictive tools. In this paper we discuss the thermochemical prediction effort at Dow which led to the implementation of the quantum mechanical techniques. Several examples will be given which highlight the practical use of the quantum mechanical techniques in day-to-day industrial problem solving.

Introduction

Accurate thermochemical data are needed and used in a number of different ways at Dow. It is important to understand where thermochemical calculations and thermochemical data fit into the overall picture of a typical chemical process at a typical chemical company.

[4]Corresponding author

319

Critical Factors for a Safe and Successful Chemical Process. In general, a number of things must occur to ensure a safe and successful scale-up of an industrial chemical process. These include:

- The economics of the process must be favorable
- Personnel must be motivated
- The chemical synthesis needs to be well understood
- Engineering components must be designed properly
- Energy release hazards must be known and controlled.

The last factor is just as critical as the others. Without adequate understanding of the energy release potential of the process and a means to control that release, potential disaster awaits the first process upset. Numerous examples in the literature (and probably many more which are not) demonstrate the negative consequences of inadequate understanding and control of the energy release potential of chemical processes (*1*).

A major problem with the scaleup of a chemical process which liberates heat (usually dominated by the enthalpy term in the free energy relationship) is that although the heat removal capabilities scale as the surface area of the vessel, the total energy release potential scales as the volume of the material in the vessel. When the heat release rate exceeds the heat removal capabilities, a so-called "runaway" reaction results, sometimes very rapidly (due to the exponential nature of Arrhenius kinetics) taking the reaction temperature and pressure to levels which exceed the capability of containment. Burst relief devices, popped hoses, or even worse, burst vessels may result. This whole process is complicated by the fact that the desired chemistry may release enough heat to carry the reaction mass, under adiabatic conditions (it is not difficult to realize a *de facto* adiabatic state for a typical industrial size vessel) to a temperature where some other chemistry occurs, typically a decomposition reaction which may liberate significant quantities of gas.

Illustrating this latter point, Figure 1 shows a hypothetical differential scanning calorimetry (DSC) trace of a reaction mix known to react at a substantial rate near 100°C with a heat release of 50 cal/g. An estimate of the adiabatic temperature rise for this system (assuming a heat capacity of 0.5 cal/g/deg) is 100°C for a final temperature of 200°C. Note that the DSC trace shows another exothermic process (presumably a decomposition) detected near 180°C. Thus, if allowed to react under adiabatic conditions, the desired reaction may carry the reaction mass into the undesired state of the higher temperature chemistry.

Various testing and estimation strategies have been devised to help determine the so-called "Reactive Chemicals" potential of a given process, i.e. the potential for hazardous energy release. Where feasible, we have always approached this from both the experimental testing side (*2, 3*) and from the predictive or computational side. In the last few years we have also brought into play heat gain-heat loss modeling which is a combination of experimental determination of the heat release rate (usually obtained from calorimetry) and calculational tools to predict the critical conditions (size, temperature, geometry, etc.) where a process might lose control of the energy release (*4*). Several textbooks and other references are recommended to the interested reader as general guides to the experimental strategies used for Reactive Chemicals testing (*2, 5, 6*). Recently, we have published an article which discusses specific means by which we approach the problem of process heat release (*7*).

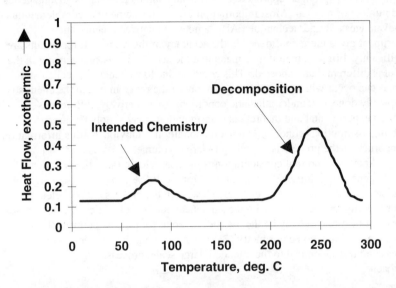

Figure 1. A hypothetical differential scanning calorimetry trace of a chemical mixture showing the intended chemistry (and its associated enthalpy release) and a higher temperature decomposition reaction (with its enthalpy release). Under adiabatic conditions in a large industrial reactor, the heat release from the intended reaction may carry the reaction mass to conditions where the decomposition reaction dominates.

The Two-Pronged Approach to the Determination of Process Heats. Process heats may be determined in one of two distinctly different, yet highly synergistic ways: (*1*) - experimentally via calorimetry or (*2*) - theoretically via calculation or prediction (in this paper we will use the terms prediction and calculation interchangeably). In our laboratories, we approach the problem of process heat determination from both directions. We usually determine the heat experimentally using any of a number of calorimetric techniques. These include accelerating rate calorimetry (ARC), simple mixing calorimetry, heat flow microcalorimetry, differential scanning calorimetry, VSP (vent sizing package), and reaction calorimetry (*7*). On occasion, specialized calorimetric techniques have been developed to address specific problems.

Experimental determination of reaction heats, hereafter referred to as $\Delta_r H$ is not always practical, feasible or required. Predictive techniques are therefore necessary. While some of these predictive techniques are based on empirical correlations, (Benson's method for example (*8*)), others are based on fundamental scientific principles and concepts. If no published or proprietary data exist for use with the desired predictive technique, estimations can be performed using Benson's group additivity method or (the focus of this paper) computationally-intensive quantum mechanical calculations.

This two-pronged approach (experimental and theoretical) is advantageous for a number of reasons. Most important, it lessens the somewhat complementary disadvantages of either technique. Also, agreement between theory and experiment gives more confidence in the accuracy of the results. If the results are significantly different, then it may mean that the actual chemistry occurring in the vessel is different than expected. This could be due to a change in mechanism, such as can occur when test conditions are out of the range in which the chemistry is typically done, e.g. under adiabatic conditions. Alternatively, differences between experimental and theoretical determinations may simply be due to differing assumptions concerning reaction conditions. Obviously, this information is important as the process is scaled up to larger volume.

Another important consideration is the heat release rate. If a chemical process generates significant total heat (i.e. for the 100% conversion scenario) but the kinetics of the process are relatively slow, then the process energy release is easier to mitigate than a case with the same heat but faster kinetics. In some cases, the rate of heat release may be the overriding factor for process control. Thus techniques like reaction calorimetry (7), which carry out the chemical process under conditions identical to the intended, large scale, process, may be the best approach.

Limitations. *In those cases where the **rate** of heat release is critical for the safe engineering design of the process, the **experimental approach is preferred due to the difficulty in computing kinetic barriers in the condensed phase for multi-component systems.*** The thermodynamic estimation schemes presented here do not address the kinetics of the reaction, and, as far as we know, no *generally* applicable kinetic estimation scheme exists for routine application to diverse chemistries in an industrial environment. Since the kinetics of a process may greatly depend on a number of controllable factors (stirring rate, reactant addition rate, etc.), a well designed experiment should be able to mimic these variables and thus measure an accurate and meaningful heat release rate.

There are two other limitations to the calculational approach. First, the estimated heats are determined for the chemistry which is *believed* to occur in the reaction vessel. There may be cases (and we have personal experience with several) where an important source of heat (salt formation for example) is neglected in the calculation. An experimental determination would probably detect this heat. Experimental techniques such as adiabatic calorimetry can yield invaluable information about heat and pressure releases for unknown chemical processes which take place at temperatures well above those intended in the process due to, for example, an undesired temperature excursion. Another limitation is the lack of accurate thermochemical data for many important species. Of course, this second factor is the subject of this paper and this monograph.

When predicting the heat of a chemical reaction, *it is always advisable to approach the problem from multiple directions using as wide a variety of techniques as feasible.* Thus, if the heat values predicted by the various techniques are reasonably consistent (within acceptable error limits), then a higher degree of confidence can be placed in the results.

An important consideration to both the experimental and calculational determination of reaction heats is the accuracy of the values determined. The uncertainty which an engineer is willing to accept is greatly dependent on how the numbers will be used. If data are required for sizing a heat exchanger, for example, a value of -2 ± 6 kcal/mol for a process heat presents more of a challenge than -30 ± 10 kcal/mol. The actual sign of a reaction enthalpy (i.e. exothermic or endothermic) is more easily obtainable experimentally. Unfortunately, for complex chemistries where thermodynamic data are missing from the literature, and where the standard estimation schemes fail, the calculational approach may lead to answers with unacceptable uncertainty limits. In such cases, consideration of a well designed experiment, by the techniques described above, may be the best solution.

When the Theoretical Approach is Favored. The above discussion notwithstanding, there are a number of factors favoring the use of the predictive approach to process heats. These are:

Safety Considerations: Many of the materials used in modern industrial chemistry can be quite toxic or reactive. If the process is at a first phase of scaleup, say at the early, exploration stage, then an estimation of the process heat may be adequate for the hazard evaluation. The exact size or scale of a process where one needs to decide on the approach (experimental versus calculational) is entirely dependent on the nature of the materials and process (e.g. explosives synthesis versus polymerization).

Experimental Challenges: Many times, actual reaction conditions of temperature and pressure make an experimental determination exceedingly difficult. For processes close to production scale, the time and effort in designing and building the calorimetric apparatus may be justified. For the first-step scale-up at the research stage, however, it may not be justified to do so. Also, the experimental approach always requires a chemical analysis to determine the extent of the reaction.

Side Reactions: A "side reaction" is defined as a chemical reaction which is not the intended or expected chemistry. Occasionally, these side reactions can dominate under some process upset condition such as elevated temperature, or too high a loading of catalyst, etc. Again, these conditions may be difficult to emulate in the calorimeter but are well suited for a theoretical treatment.

The focus of this paper is one tool (of many) we use for the prediction of reaction thermochemistry: quantum mechanics. The authors firmly believe that this tool represents a mini-revolution in thermochemistry, equal in impact to

Benson's group contribution method. Since we feel so strongly about this subject, we thought that an article on how we implemented these techniques at Dow might assist (or perhaps convince) others to do the same. We also felt it was important to give the technique developers (those typically insulated from industrial needs) some "real-world" experiences and suggestions which might help guide their future development efforts.

The remaining sections of this chapter are organized as follows. First we will briefly discuss the other tools for process heat estimation. This discussion will then serve as background for the implementation of the quantum mechanical techniques. To help illustrate, several real industrial examples will be given next. Finally, we will discuss the needs and future direction of this area.

Approaches to Thermochemical Data Estimation

Before discussing the quantum mechanical techniques we first discuss some fundamental thermochemical concepts and some traditional techniques.

For the hypothetical reaction:

$$aA + bB \longrightarrow cC + dD$$

the reaction enthalpy, $\Delta_r H°$ is determined from the standard enthalpies of formation, $\Delta_f H°$, of the species in the reaction according to:

$$\Delta_r H° = c\Delta_f H°(C) + d\Delta_f H°(D) - a\Delta_f H°(A) - b\Delta_f H°(B).$$

The standard enthalpy of formation, $\Delta_f H°$, is the enthalpy change upon formation of the material from the elements in their standard states at the temperature T. For example, the $\Delta_f H°$ for liquid CH_2Cl_2 at 25°C is the enthalpy change accompanying the reaction:

$$C(graphite) + H_2(g) + Cl_2(g) \longrightarrow CH_2Cl_2(liq)$$

Enthalpies of formation are determined experimentally a number of ways. Perhaps the most common, traditional way is through precise oxygen combustion measurements. The magnitude of formation heats can be highly dependent on the state of the material and many tabulations are available. Where data do not exist, estimation techniques are available (again this will be discussed below). *The key to estimation of reaction heats lies in the accurate knowledge of the standard enthalpies of formation, $\Delta_f H°$, of the reactants and products*.

The Typical Process for the Estimation of a Reaction Enthalpy. Given the types of complex chemistries occurring in most industries today, it is not surprising that accurate thermochemical data are most often not available for the exact

species participating in the subject reaction. For this reason, most people typically resort to an analog reaction for the purposes of a thermodynamic estimation. This technique allows us to write a reaction with simpler species, for which we might be able to obtain literature thermodynamic data. This technique has been discussed in more detail (*7, 9*).

The "analog reaction hypothesis" states that the heats of reaction for structurally similar reactions are identical. In effect the analog reaction hypothesis assumes that atoms, groups, substituents, *not* participating in the actual chemistry, and therefore, remaining intact structurally from the left-hand side to the right-hand side of the equation, do not affect the reaction thermodynamics significantly. For a typical example, consider the hypothetical reaction below:

In this case, a suitable analog could be:

Thermodynamic data for the species involved in the analog reaction are much easier to either find in the literature or to estimate using standard techniques discussed below. Note that this is not the only analog reaction one could choose.

As discussed earlier, it is always prudent to choose multiple analogs when feasible. If the heat values predicted by the various analog reactions are reasonably consistent (within acceptable error limits), then a higher degree of confidence can be placed in the results.

Unfortunately, many times, even with a simplified analog reaction, we are still unable to find literature data for the species involved. Thus, we must frequently resort to thermodynamic estimation methods.

Summary of Traditional Thermochemical Estimation Techniques. A number of traditional techniques are available for the estimation of thermochemical data.

Bond Energy Approach. The energy required to break a bond and separate the atoms to infinite distance in the gas phase at 0 K is a common definition of bond energies, but this is not a very useful definition for our purposes. Bond energy tables using this definition must NOT be used to predict heats of formation but may be used to predict heats of reaction (at zero Kelvin) by summing the bond energies for bonds broken and subtracting the sum of bond energies for bonds formed. Many times this method may be used to get an idea of the order of magnitude of the reaction heat.

$$\Delta_r H \approx \Sigma \text{ (energies of bonds broken)} - \Sigma \text{ (energies of bonds formed)}$$

A more useful approach is that of Benson (*10*) who deals with partial bond contributions to the gas-phase enthalpy of formation at 298.15 K. Basically, the bond contributions are groups which can be added together to predict either the enthalpy of formation of a molecule or the heat of reaction directly.

Benson's Method. Benson's second-order group contribution method was developed some 30 years ago as an improvement to bond energy (or bond contribution) methods (*10, 11*). This improvement is accomplished by accounting for:

- the effect of the bonded environment around a group
- ring effects (strain)
- isomerism (cis/trans, ortho/meta/para)
- non-bonded interactions (gauche effects)

The end result is a method that is more accurate than bond energy methods with typical uncertainties of 2-3 kcal/mol for CHNO containing materials. However, this method is more complex to use than bond energy methods and requires more data for parameterization (due to the larger number of groups) than bond energy methods. The user is referred to Benson's book (*8*) for a complete description of the method.

In Benson's method a group consists of a central multivalent atom (or atoms) and any monovalent ligands (H, halogens, CN, etc.) that comprise the main part of the group. Any remaining valences are then filled by the bonded environment. As a brief example, consider the Benson "CH_2" groups shown below where C represents a tetrahedral carbon, Cb represents an aromatic carbon. Note that the contribution of the CH_2 group to the enthalpy of formation is different in each bonded environment.

Benson group	Contribution to $\Delta_f H°$(kcal/mol, 25°C)
C - (2H, 2C)	-4.93
C - (2H, 2Cb)	-6.50
C - (2H, C, O)	-8.10
C - (2H, C, N)	-6.60

As a result of its accuracy and relative ease of use, Benson's method has been widely used throughout the literature and has been incorporated into many computer programs including CHETAH (*12*) and the NIST Structures and Properties Program (*13*).

Although the Benson second order group contribution method is very powerful, accurate, and extremely easy to use, a common problem is missing group values, especially in the chemical industry due to the frequency of structurally complex chemical species. More on this topic will be presented below.

Correcting Gas Phase Data to the Liquid State. Most thermochemical estimation schemes are for gas phase species although one exception is the NIST THERM/EST program, (*14*). Consequently, one is typically faced with the problem of having reasonably accurate gas phase estimates for a chemical process that actually takes place in the condensed phase. Many times the gas phase result is quite close to the condensed phase result due to a cancellation effect of the vaporization heats for products and reactants. This is generally the case for species which don't "change" their general bonding character from the reactant to the product side. For example, if a reaction takes place to form a carboxylic acid from non-acid reactants, the heat of vaporization correction would be significantly larger on the product side of the equation due to hydrogen bonding interactions and thus the gas phase heat of reaction estimation may differ substantially from the actual value.

A rule of thumb we have used over the years is the 100 cal/g rule. Many weakly interacting species have heats of vaporization close to 100 cal/g. For species which are strongly hydrogen bonded, their vaporization heats are larger than 100 cal/g. For example, alcohols, amines, and carboxylic acids are typically 150-200 cal/g. Halogenated species may have vaporization heats of 50-80 cal/g. The best approach is to look up the vaporization heat for a similar species and apply that result (on a cal/g basis) to the subject molecule.

Another rule of thumb is to ignore solution effects for organic systems in organic solvents. This assumes that the heat of solution of an organic material in an organic solvent is small (they are typically less than 2-4 kcal/mol). Thus, we may confidently substitute a enthalpy of formation estimate for a pure liquid for that of the same material in organic solution.

Heats of crystallization are more difficult to estimate but we have used Walden's rule in the past (*15*). This rule states that the heat of fusion (in cal/mol) is equal to 13 times the melting temperature (in Kelvin). This rule is based on an entropic argument, and is generally adequate for most "rigid" species but not for long chain-like structures.

The Quantum Mechanical Approach to Thermochemistry

Frequently, we are faced with a chemical reaction which makes use of the standard estimation schemes untenable. This usually results from the following sequence of events:

1. There are no literature thermodynamic data for one or more of the actual species in the reaction.

2. There are no literature thermodynamic data for one or more of the species in an analog reaction.

3. Experimental determination of the reaction heat is deemed unfeasible (e.g., the materials are too toxic, unstable, the reaction conditions too severe).

4. One or more group contributions (such as Benson groups) are unavailable and approximation or substitution of groups would lead to unacceptable uncertainty.

Until recently, when the above occurred, we were forced to rely on the experimental approach or to accept a large uncertainty in the prediction. In about 1990, we became aware of some of the newer quantum mechanical techniques which were focused on the accurate prediction of thermochemical data. In the jargon of the field, these techniques are claimed to meet so-called "chemical" accuracy.

Quantum mechanical calculations have, from the earliest days, sought to calculate an accurate, absolute energy of a particular state of a molecule. But, it has only been in the last few years that these calculations have been of sufficient accuracy and precision to be of use in lower accuracy engineering thermochemical applications for molecules of industrial interest. Today, researchers in the area of quantum thermochemistry are actively pursuing methods that balance computational efficiency against accuracy by studying the convergence of both one-electron basis set expansions and n-electron approximations to the correlation energy of the system. These researchers have recognized the current tremendous practical utility of these approaches. Currently, the available methods have been expanded to include more bonding and atom types. Most of the current methods are discussed in other chapters of this book.

The "Gaussian-2" or "G2" method developed by Pople, Curtiss, and Raghavachari appears to have received the most attention (*16a,b*). It is beyond the scope of this chapter to discuss this method in detail but a good description can be found in the original papers and in the chapter by Curtiss in this monograph.

The G2 method consists of the summation of the results from up to six separate calculations with standard basis sets and several levels of correlation. The series of six energy calculations are performed on a molecular conformation which has been "optimized", (i.e. a lowest energy structure identified). The vibrational frequencies are also calculated and are used to determine the zero-point energy. The calculated energies are then arithmetically manipulated to mimic the result of a hypothetical calculation performed at a very high level of theory and hence of accuracy.

At the present state-of-the-art, this type of calculation is limited, to five heavy (non-hydrogen) atoms from the first or second row in the periodic table. A five heavy atom molecule requires about six hours of CPU on a CRAY C-90; adding another heavy atom increases this time to ca. 20 hours. A slightly less accurate variant of the G-2 theory, the G2-MP2 technique, extends the practical size of the molecule by 1-2 heavy atoms (*16b*). We have performed selected calculations on larger systems (six and seven heavy atoms) but due to the computational costs, these larger calculations have been restricted. It is abundantly clear to us that in subsequent years, the cost for performing these calculations will continue to fall rapidly for two reasons. First, the hardware will continue to increase its performance (computational speed) for decreasing cost. Second, the technique developers will continue to devise better, more accurate, and more efficient ways to perform the thermochemical estimations.

Many calculations (for molecules up to four heavy atoms) can be performed on a standard workstation in a practical fashion. For these workstation calculations, the "wall clock" time to complete a job can compare favorably to the time waiting for an answer from a supercomputer due to the resource sharing aspects of the more powerful system.

The accuracy of the G2 technique is quite good. For 125 molecules (mainly 3-4 heavy atoms), the mean absolute deviations in calculated atomization energies (and hence in heats of formation) is 1.2 kcal/mol (maximum error 4 kcal/mol) (*16*). The beauty of the G2 technique is that the calculated enthalpy of formation is, for the most part, determined from first principles. However, there are two empirical parameters in the calculation. The calculated vibrational frequencies result from scaled force constants yielding a scaled ZPE and a so-called "higher level correction", dependent only on the total number of valence electrons in the system and includes one fitting parameter.

Although the size of molecule which can be attempted by these techniques is limited, we can accommodate the size restriction by using Benson's method for most of the molecule. Then missing Benson groups may be estimated easily and accurately without much computational cost. For example, suppose the Benson group enthalpy of formation value for CCl- (2C, N) was unavailable in the literature. One could estimate this group value by performing a G2 calculation on the species:

$$\begin{array}{c} CH_3 \\ | \\ Cl\!\!-\!\!\overset{\displaystyle |}{\underset{\displaystyle |}{C}}\!\!-\!\!CH_3 \\ NH_2 \end{array}$$

Using published values for the other Benson groups comprising this molecule, one can then extract the missing group needed. We refer to this technique as G2-Group Contribution (or "G2-GC"). This strategy was used by Seasholtz, Thompson, and Rondan to calculate the enthalpy of formation of eighteen small, prototype molecules containing imino (- C = N -) groups (17). From the results of the G2 calculations, Benson group values for the missing imino-related groups were derived. The results compare favorably to the limited experimental data, and were incorporated in the thermochemical program CHETAH 7.0 (12) to yield a self-consistent set of group values for this class of compounds.

The Typical Process of a G2 Calculation. The procedure for the typical G2 calculation in our laboratory is as follows:

1. Using a suitable interface to the Gaussian software (the quantum mechanical package which is widely used (18)), the molecule of interest is constructed with a reasonable geometry. We use and like the SPARTAN (19) software which allows a simple direct input of the molecular structure without the complexity of the so-called "Z-matrix" internal coordinate representation. This latter method can be quite frustrating to the neophyte and is the only practical method of input into the Gaussian package without an interface package such as SPARTAN.

2. At some lower level of theory (Hartree-Fock), the molecular energy is minimized with respect to the geometry. We perform this computation directly using SPARTAN on an SGI Workstation for convenience. If the final G2 calculational result is to represent the actual thermodynamic energy, it is important that care is exercised to ensure the structure used in the final set of more time consuming (and hence costly) computations is the lowest energy conformation and not a local minimum. Depending on the molecular type, we typically might "perturb" the conformation by rotating various bonds in the molecule and recomputing the new structure at the low level of theory to determine if it relaxes to the lowest energy arrangement. Difficulties do arise on occasion if there are two or more nearly equal energy conformers. In these cases, one might be forced to compute the energies of all structures at a higher level of theory (and hence more accurate) to determine the actual lowest energy structure.

3. Once the equilibrium geometry is determined, the next step is to run the series of individual computations (single point calculations, vibrational frequencies, and one optimization) required of the G2 or G2-MP2 methods. For most species, we transfer the optimized geometry from the workstation to the CRAY C-90 supercomputer. The Gaussian-94 package has automated this process by a simple specification of "G2" or "G2MP2" in the command line of the job. One potential problem can occur, however. If the geometry is not an equilibrium configuration, one or more vibrational frequencies may be imaginary. In the current release of Gaussian, the calculations continue regardless and the user may not become aware of this problem until after the (expensive) computations are

complete. Therefore it is a good practice to spend the time necessary to make absolutely sure that the geometry is truly the lowest energy from the outset.

4. After the calculation is complete, the energy results from the individual computations must be converted into standard thermochemical values. In our case, the most useful value is the standard enthalpy of formation. Although it would be straightforward for the Gaussian programmers to do this conversion automatically, they have not, and the user must do the final arithmetic manipulations manually. This may be due to the reluctance of the programmers to introduce experimental values for the atomic enthalpies of formation into the program. These values are known with relatively high accuracy and are widely available (NIST Tables (*13*) for example). It is the authors' opinion that the atomic enthalpies could easily be incorporated into the program. However, a spreadsheet program is probably the best current alternative and helps reduce errors except that much "cutting" and "pasting" is required.

Selected examples

Three examples are given below to illustrate the utility of the G2 and G2-MP2 methods at our company.

Example of Extraction of a Benson Group for a Heat of Reaction. The reaction heat of isopropylchloroformate (IPCF) with water was needed for some engineering design calculations.

Thermochemical data for IPCF were not found in our standard sources for data (reference 20 for example) and the Benson group was unavailable for the CO - (Cl, O) moiety. Thermochemical data are available for all the other species in the reaction. We decided that a calculation for the entire IPCF molecule was probably not practical (seven heavy atoms) nor was it necessary since only a single Benson group was missing. Therefore we decided to perform a G2 calculation on chloroformic acid:

This species contains two Benson groups, the one of interest and O - (H, CO), the latter is available in standard sources (*12*).

Following the protocol just described, we performed a G2 calculation on this species and obtained a standard enthalpy of formation of -101.6 kcal/mol. The literature value for the standard enthalpy of formation of the group O - (H, CO) is -58.1 kcal/mol (*12*). Simple arithmetic yields the value for the missing group $\Delta_f H°$ [CO - (Cl, O)] = -43.5 kcal/mol. This then allowed us to complete the heat of reaction calculation for the IPCF hydrolysis which agreed with a subsequent experimental determination to within 2 kcal/mol.

Potential Energy in a Molecule for Explosion Prediction. During distillation of gram quantities of a newly synthesized material, a small explosion occurred. As part of the subsequent investigation of the incident, it was decided that a so-called "Energy Release Evaluation" of the species should be conducted. This calculation seeks to determine the total potential chemical enthalpy available from the species by assuming products of the explosion that maximize the enthalpy of the decomposition reaction (*21*). Although not thermodynamically exact, this value has been shown to correlate with the ability of a material to propagate a rapid energy release (*12*). The species, isoxazolidine, is shown below:

Three thermochemical problems are immediately encountered in this species. First, the Benson groups N-(H, O, C) and O - (N, C) are not available. Also, the ring strain correction is unknown. Since this molecule contains only five heavy atoms, a G2 calculation is feasible. The G2 calculation for this species was accomplished overnight on a CRAY C-90 with the resulting $\Delta_f H°$ = -0.5 kcal/mol. In this example, only the enthalpy of formation was required to perform the Energy Release Evaluation using the CHETAH software. This calculation showed that the maximum heat of decomposition of this species is -1100 cal/g which is nearly 80% of the corresponding value for TNT (trinitrotoluene).

Extraction of a Group Contribution from a G2 Calculation. Although the Benson second order group contribution method is very powerful, accurate, and extremely easy to use, a profound limitation is the frequently encountered problem of missing group values. This is especially true in the modern chemical industry as the structural complexity of chemical species is increasing rapidly while the rate of experimental determination of thermochemical properties for these same moieties is constant or decreasing. As far as the authors are aware, the only research group in the U.S. which is actively deriving Benson group thermodynamic values is the Bartlesville group headed by Bill Steele under the AICHE DIPPR Project 874.

For the above reasons, we have recently begun a small project which determines some missing Benson group data using the G-2 or G2-MP2 methodologies. This project is similar to that begun at this laboratory earlier by Seasholtz et al. (*17*). Most recently, we have begun to use combinatorial means to determine all possible (but chemically feasible) combinations of atomic groupings in the Benson notation. Next we perform a search of our standard databases (CHETAH for example) to see if the group value exists in the literature. Also, we search for gas phase thermochemical data for species which contain the group and are structurally simple to allow an unambiguous extraction of the group value from the literature data.

Finding no (or limited) data in the literature, we then take the subject Benson group and construct a hypothetical molecule where the valencies of the pendant atoms all filled with either hydrogens or hydrocarbons. As an example, the Benson group Cd - (2N) is not available (Cd is a doubly bonded carbon) and no molecule was found in the literature containing this group and for which the appropriate thermochemical properties have been published. Thus we constructed the hypothetical molecule below:

This structure contains four Benson groups, for which three values are in the literature and the subject, unknown group. This is, of course, not the only species which can be constructed containing the unknown group but for the sake of computational efficiency, it will suffice. In Table I below, we show how we extracted the missing group value from the known values and the G2 result. The known group values were taken from the CHETAH program.

As a check on this result, we can try a simple extrapolation of known data. This is illustrated in Table II below. For the most part, replacing atoms in Benson groups in a systematic way results in a change in the enthalpy of formation of a nearly constant amount. Thus the difference (Δ below) may be added to the last member to extrapolate to the unknown. As far as the authors are aware, this technique has not been fully investigated for its accuracy. For this example, the result for the unknown, -2.6 kcal/mol, compares favorably with the G2-GC technique result of -3.8 kcal/mol and therefore adds confidence to the estimates. If time and budget allow it is certainly good practice to perform G2 calculations on other species containing the missing group.

Summary of Calculations for a Series of Molecules. As a matter of routine practice in the implementation of the G2 methodology in our laboratory, we attempted to perform calculations on species which were structurally similar to the subject (unknown) species but where experimentally measured thermodynamic data were available. If we could reproduce the experimental values within acceptable error limits for a closely related species, then we would have greater confidence in the results for the unknown. Thus, calculations for 25 such molecules were performed to check the validity and applicability of the G2 and G2-MP2 methodologies.

Table I. Extraction of thermochemical data for a Benson group using the results of a G2 calculation and known group increments. Literature data from CHETAH 7.0 (12).

Benson Group	$\Delta_f H°$, kcal/mol
N - (2H,C) x 2	+9.6
Cd - (2H)	+6.26
G2 result*	+12.09
Cd - (2N) unknown group	[-3.8] = 12.09 - (6.26 + 9.6)

* for $H_2C=(NH_2)_2$

Table II. Determining the enthalpy value for the unknown Cd - (2N) group by an empirical extrapolation procedure. The increment -6.47 kcal/mol is added to the Cd-(N-C) group value to obtain an estimate of the missing group. This result compares favorably with the G2 estimate in Table I. Literature data from CHETAH 7.0 (12).

Benson Group	$\Delta_f H°$, kcal/mol
Cd - (2C)	+10.34
$\quad\quad \Delta = -6.47$	
Cd - (N,C)	+3.87
Cd - (2N) unknown group	[-2.60] = +3.87 + Δ

Table III summarizes the compounds that we have calculated using the G2 and G2-MP2 theories. Note that in several instances, G2 results for some of the species listed (typically the smaller species) have already appeared in the literature. In these instances, we repeated the calculations primarily to confirm the validity of our procedural methodologies.

In Fig. 2 is a histogram plot of the differences between the calculated (by both G2 and G2-MP2 theories) heats of formation and the literature value (lithium alkyls excluded, see discussion below). Where more than one literature value was found, we used the average.

In general, the calculations have reproduced the literature data quite well and the differences are consistent with the reported errors from the original work of Curtiss et al. (16). Generally, the calculated values are within 4 kcal of the literature value for two thirds of the species studied. It must be noted, however, that except for the alkyl lithium species, the literature data for the species which the G2 and G2-MP2 techniques performed the poorest, may be of questionable validity and/or may be estimates themselves. For the case of the two lithium alkyls (butyl and ethyl lithium), one possible explanation for the observed discrepancy is the highly associated vapor state of these species which is not accounted for in the calculation of the isolated species. This association would tend to enhance the stability of the real gas and thus the observed errors are in the correct direction. Of course it also may be the case that either the G2 methodology may not be appropriate for these species or that the structures we used were metastable.

One important observation from these calculations is that the G2-MP2 methodology, which is significantly less computationally intensive than G2, is nearly as accurate as the G2 technique. This is the same conclusion as the developers of these methodologies came to on a set of relatively small molecules (*16*).

Conclusions and Discussion

Estimation techniques have been and will continue to be an important tool in the area of industrial thermochemistry. Although group contribution techniques are powerful, handy, easy to use, and generally quite accurate, the lack of data for a multitude of structural moieties limits their application. We have found that the quantum mechanical techniques fill an important role in industrial process design. As the "purely" computational techniques continue to develop and as the

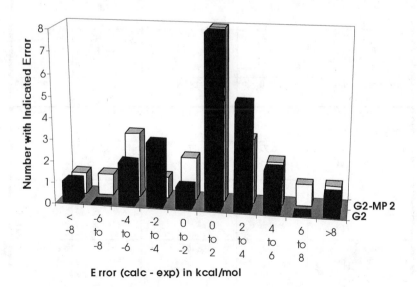

Fig. 2: Error Statistics for 23 Molecules (excludes Et-Li and Bu-Li)

TABLE III. Enthalpies of formation for a series of gaseous molecules according to the G2 and G2-MP2 theories. Calculations for these species were performed for various projects in our laboratory as a check on the validity of the methods or simply to confirm previously published calculational results.

Molecule	G2 Theory kcal/mol	G2-MP2 Theory kcal/mol	Literature kcal/mol	Literature Reference
ethylene $H_2C{=}CH_2$	12.9	13.4	12.5	20
propene $H_2C{=}CH{-}CH_3$	5.5	6.3	4.8	20
cyclopropane	13.8	14.5	12.7	20
cyclopropene	69.3	69.9	66.2	20
1,3-butadiene	28.4	29.2	26.5	20
cyclobutane	7.3	8.3	6.8	20
bicyclobutane	55.2	55.9	51.9	20
cyclobutene	40.6	41.5	37.5	20
ammonia NH_3	-10.8	-10.9	-11.0	22
phosgene $Cl_2C{=}O$	-54.9	-57.2	-51.9, -52.4	22 20
phosphoryl chloride $POCl_3$	-129.0	-133.8	-133.5	22
diazomethane $H_2C{=}N{=}N$	64.3	64.1	55.0	28
allene $H_2C{=}C{=}CH_2$	46.7	47.3	45.5	20
urea $H_2N{-}C(=O){-}NH_2$	-56.5	-57.0	-58.7	20
methyl fluoride FCH_3	-58.3	-58.5	-59	23

Table III. *Continued*

Molecule	G2 Theory kcal/mol	G2-MP2 Theory kcal/mol	Literature kcal/mol	Literature Reference
methylene difluoride	-110.8	-111.5	-106.8 -108.1	22 20
trifluoromethane	-170.6	-171.8	-166.0	24
carbon tetrafluoride	-228.6	-230.0	-223.1	20
ethyl lithium	27.0	27.9	13.9	22
n-butyl lithium	16.1	17.4	-6	9
aminotriazole	45.2	44.8	47.6	25
tetrahydropyrimidine	18.9	19.3	13.2	26
dimethyl chloroamine	8.1	7.4	19	27
methyl hydroxylamine	-12.7	-13.3	-12.0	10
acrylonitrile	46.2	46.7	43.2	20

computational hardware continues to become more efficient, greater use of the quantum mechanical methods will inevitably occur.

In this paper, we have presented a summary of the techniques we use routinely for thermochemical estimations. The recent implementation of quantum mechanical techniques has been relatively straightforward and we now use it routinely for the estimation of reaction heats. Finally, we present some thoughts on the needs in this area from our industrial perspective.

1. For a computational thermochemical estimation technique to be useful, it must be easy to use. The quantum mechanical techniques can, to a neophyte, be extremely awkward. The recent implementation of the G2 and G2-MP2 methodologies (along with other methodologies) in the GAUSSIAN 94 package went a long way to alleviate the problem. Also, the availability of graphical interfaces such as SPARTAN (*19*) has helped tremendously. In future releases, we recommend that the calculations perform the missing, final step of the calculation and produce a useful product for the thermochemist and engineer, that is the standard enthalpy of formation (making sure to list the experimental atomic enthalpies employed). Technique developers need to make sure their methodology becomes available in a form which can easily be implemented. Failure to do so may mean that a given method, even if it is shown to be very accurate and efficient, will be used rarely.

2. The technique developers need to strive toward continued improvement of accuracy and efficiency. Improvements in efficiency may happen automatically as computers become faster. For accuracy, most engineers would accept a reaction heat to within 4-5 kcal/mol, in some cases even $\pm$ 8-10 kcal/mol . The major problem, in the authors' opinion, with any estimation method is outliers. The users need to have clear guidelines as to when the techniques are likely to be successful and when they are likely to fail. For example, our application of these methods to the lithium alkyls points out a problem area.

3. To judge the accuracy of a given method in a fair and unbiased manner, a need exists to develop and make available a critically evaluated database of thermochemical data. This database can then be used by those developing the techniques as a standard test for comparing the accuracy of different models *in a fair and unbiased way.*

4. Practitioners and developers of the computational methodology need to be careful in their applications and criticism of the techniques to make sure that the best methods are applied to a given problem and that research efforts be directed in the most productive areas. As an example, recently Athanassios and Radom (*19*) described a "failing" of the G2 method for calculating combustion heats. The essence of their concern is that the errors in the G2 calculated enthalpies of formation of O_2, CO_2 and H_2O are compounded when a combustion enthalpy is determined since the errors are multiplied by the number of carbons and hydrogens in a molecule. Whereas their technical arguments are absolutely correct, it is our contention that except as an academic exercise, it would be folly for *anyone* to use the G2 methodology *alone* for a determination of a combustion heat *except for the determination of the enthalpy of formation of the reactant* since the enthalpies of formation of all the other appropriate species are quite well known.

Acknowledgments

The authors would like to thank the following for their continuing support and insight into the implementation of these computational methods at Dow: Dr. Tyler Thompson; Dr. Marybeth Seasholtz, Dr. Joe Downey, Dr. Marabeth LaBarge, Dr. James Ruiz, and Dr. Pat Andreozzi, all of the Dow Chemical Co.

Literature Cited

1. Bretherick, L., *"Bretherick's Handbook of Reactive Chemical Hazards"*, Fifth Ed., Butterworths, London, **1996**.
2. Grewer, T., *"Thermal Hazards of Chemical Reactions"*, Elsevier, New York, **1994**.
3. *"Guidelines for Chemical Reactivity Evaluation and Application to Process Design"*, CCPS, AIChE, New York, **1995**.
4. Hofelich, T. C.; Frurip, D. J.; and Powers, J. B., *"The Determination of Compatibility via Thermal Analysis and Mathematical Modeling"*, Proc. Safety Prog., **1994**, *Vol.13* (4), pp. 227-233.
5. Crowl, D.; Louvar, J. F., *"Chemical Process Safety: Fundamentals with Applications"*, Prentice Hall, **1990**; pp 390, Table 13-1.
6. Barton, J.; Rogers, R., *"Chemical Reaction Hazards-A Guide"*, Institution of Chemical Engineers, U.K., **1993**.
7. Frurip, D. J.; Chakrabarti, A.; Downey, J. R.; Ferguson, H. D.; Gupta, S. K.; Hofelich, T. C.; LaBarge, M. S.; Pasztor, A. J., Jr.; Peerey, L. M.; Eveland, S. E.; Suckow, R. A., *"Determination of Chemical Process Heats by Experiment and Prediction"*, International Symposium on Runaway Reactions and Pressure Relief Design, CCPS/AICHE, **1995**.
8. Benson, S. W., *"Thermochemical Kinetics"*, Second Edition, Wiley - Interscience, NY, **1976**.
9. Cox, J. D.; Pilcher, G., *"Thermochemistry of Organic and Organometallic Compounds"*, Academic Press, London, **1970**.
10. Benson, S. W., Cruickshank F. R., Golden, D. M., Haugen, G. R., O'Neal, H. E., Rodgers, A. S., Shaw, R, and Walsh, R., *Chem. Rev.*, **1969**, 69, 279.
11. Golden, D. M.; Benson, S. W., *J. Phys. Chem.*, **1973**, 77, 1687.
12. Downey, J. R.; Frurip, D. J.; LaBarge, M. S.; Syverud, A. N.; Grant, N. K.; Marks, M. D.; Harrison, B. K.; Seaton, W. H.; Treweek, D. N.; Selover, T. B., *"ASTM Data Series DS 51B"*, CHETAH 7.0, The ASTM Computer Program for Chemical Thermodynamic and Energy Release Evaluation (NIST Special Database 16), **1994**.
13. NIST Standard Reference Database 25, *"NIST Structures and Properties Database and Estimation Program, Version 2.0"*, **1992**.
14. NIST Standard Reference Database 18, *"NIST THERM/EST Program Version 5.0; Estimation of Chemical Thermodynamic Properties of Organic Compounds"*. See also Domalski, E. S. and Hearing, E. D., J. Phys. Chem. Ref. Data 22, **1993**, 805-1159.

15. Grant, D. J. W.; Higuchi, T., *"Solubility Behavior of Organic Compounds"*, (Techniques of Chemistry, Vol. 21), p.22, Wiley, **1990**.

16. (a): Curtiss, L. A.; Raghavachari, K.; Trucks, G. W.; Pople, J. A.; *J. Chem. Phys.*, **1991**, 94, 7221; (b): Curtiss, L. A.; Raghavachari, K.; Pople, J. A., *J. Chem. Phys.*, **1993**, 98(2), 1293.

17. Seasholtz, M. B.; Thompson, T. B.; Rondan, N. G.; *J. Phys. Chem.*, **1995**, 99 (51), 17838.

18. Gaussian 94 (Revision D.3), Frisch, M. J.; Trucks, G. W.; Schlegel, H. B.; Gill, P. M. W.; Johnson, B. G.; Robb, M. A.; Cheeseman, J. R.; Keith, T. A.; Petersson, G. A.; Montgomery, J. A.; Raghavachari, K.; Al-Laham, M. A.; Zakrzewski, V. G.; Ortiz, J. V.; Foresman, J. B.; Peng, C. Y.; Ayala, P. A.; Wong, M. W.; Andres, J. L.; Replogle, E. S.; Gomperts, R.; Martin, R. L.; Fox, D. J.; Binkley, J. S.; Defrees, D. J.; Baker, J.; Stewart, J. P.; Head-Gordon, M.; Gonzalez, C.; Pople, J. A., Gaussian, Inc., Pittsburgh PA, **1995**.

19. Spartan 4.1, Wavefunction Inc. 18401 Von Karman Ave., Suite 370, Irvine, CA 92612 USA

20. Pedley, J. B.; Naylor R. D.; Kirby, S. P., *"Thermochemical Data of Organic Compounds,"* Second Edition, Chapman and Hall, London, **1986**.

21. Frurip, D. J.; *Plant/Oper. Prog.*, **1992**, 11(4), 224.

22. Wagman, D. D.; Evans, W. H.; Parker, V. B., et al., *"The NBS Tables of Chemical Thermodynamic Properties,"* J. Phys. Chem. Ref. Data, **1982**, 11, Suppl. 2; Errata, *J. Phys. Chem. Ref. Data*, **1990**, 19, 1042.

23. Karpus, S. G.; Liebman, J. F., *J. Am. Chem. Soc.*, **1985**, 107, 6089.

24. Kudchadker, S. A.; Kudchadker, A. P.; Wilhoit, R. C.; Zwolinski, B. J., *J. Phys. Chem., Ref. Data,* **1978**, 7, 417.

25. Williams, M. M.; McEwan, M. S.; Henry, R. A., *J. Phys. Chem.*, **1957**, 61, 261.

26. Steele, W.; BDM Oklahoma, AIChE / DIPPR Project 871 Result, **1994**.

27. Liebman, J. F., *J. Phys. Chem. Ref. Data*, **1988**, 17 (Suppl. 1).

28. Vogt, J.; Williamson, A. D.; Beauchamp, J.L., *J. Am. Chem. Soc.*, **1978**, 100, 3478.

29. Athanassios, N.; Radom, L., *J. Phys. Chem.*, **1994**, 98, 3092.

Chapter 18

Ab Initio Calculations for Kinetic Modeling of Halocarbons

R. J. Berry[1], M. Schwartz[1,2], and Paul Marshall[1,2]

[1]Center for Computational Modeling of Nonstructural Materials, Wright Laboratory, Materials Directorate, Wright-Patterson Air Force Base, OH 45433
[2]Department of Chemistry, University of North Texas, Denton, TX 76203

The thermochemistry and reaction kinetics of halogenated hydrocarbons have been investigated by *ab initio* methods in order to improve our understanding of their flame chemistry and likely roles in flame suppression. Bond additivity corrections at the G2, G2(MP2), CBS-4 and CBS-Q levels of theory were developed for fluorinated and chlorinated C_1 and C_2 species, including saturated and unsaturated compounds. The resulting enthalpies of formation are in excellent agreement with experimental values. Transition states for the reactions of H atoms with hydrofluoromethanes were characterized at up to the G2 level of theory, and application of transition state theory yielded rate constants in good accord with experimental results. A similar analysis for H and OH reactions with CH_3I also agrees with the known thermochemistry and kinetics. These investigations provide insight into the major product channels and the temperature dependence of the rate constants. The implications for flame suppression by haloalkanes are discussed.

During the past two decades, it has become apparent that the release of volatile chlorofluorocarbons (CFCs) and halon fire suppressants (e.g. CF_3Br, CF_2ClBr, CF_2Br-CF_2Br) is a major cause of depletion of the stratosphere's ozone layer (*1-3*). Hence, there have been numerous restrictions placed upon the industrial use of CFCs and halons, which has led to concerted efforts to find new, "ozone-friendly" replacements (*4*). Potential replacement agents include hydrofluorocarbons and iodocarbons, which are degraded in the troposphere and, hence, pose no significant risk to the ozone layer. They also possess low global warming potentials.

The effectiveness of a proposed flame suppressant can be reliably predicted via kinetic modeling provided accurate thermochemical and kinetic data are available for the hundreds of reactions associated with the suppressant and its interactions with flame species. Such data are often difficult to obtain from experiments, especially at combustion temperatures. Experimental determinations are further hampered by the transient nature of many of the intermediate species involved, which makes it difficult to isolate individual reactions and species for measurements. An *ab initio* computational alternative is desirable because it would enhance our understanding of these complex processes at a molecular level and provide a cost-effective way of screening/predicting suitable replacement agents for fire suppression.

Thermochemistry of Haloalkanes

Enthalpies of Formation from *ab Initio* Energies. *Ab initio* quantum mechanics has proven to be a most valuable tool for estimating thermochemical quantities (dissociation energies, enthalpies of formation, heat capacities, entropies, etc.) of gas phase molecules, radicals and ions in systems where experimental data are either unavailable or unreliable. Briefly outlined below is the procedure by which one may utilize quantum mechanical energies to furnish a direct estimate of the enthalpy of formation of a fluoroalkane.

The molecular atomization energy, ΣD_0, is obtained from the calculated energies of the molecule and constituent atoms via the relation:

$$\Sigma D_0(C_xH_yF_z) = x\ E(calc, C) + y\ E(calc, H) + z\ E(calc, F) - E_0(calc, C_xH_yF_z) \qquad (1)$$

E_0 for the molecule contains the zero-point energy (ZPE), obtained from calculated vibrational frequencies after adjustment by the appropriate scale factor (0.8929 for HF/6-31G(d) frequencies). The enthalpy of formation at 0 K can then be computed from the atomization energy and the experimental enthalpies of formation of the constituent atoms via:

$$\Delta_fH^o(C_xH_yF_z, 0\ K) = x\ \Delta_fH^o(C_{gas}, 0\ K) + y\ \Delta_fH^o(H_{gas}, 0\ K) + z\ \Delta_fH^o(F_{gas}, 0\ K)$$
$$- \Sigma D_0(C_xH_yF_z) \qquad (2)$$

Finally, the room temperature enthalpy is obtained from:

$$\Delta_fH^o(298.15\ K) = \Delta_fH^o(0\ K) + \delta H^o(C_xH_yF_z) -x\ \delta H^o(C) - 0.5y\ \delta H^o(H_2)$$
$$- 0.5z\ \delta H^o(F_2) \qquad (3)$$

$\delta H^o = H^o(298.15\ K) - H^o(0\ K)$ represents the thermal contribution to a species' enthalpy. It is obtained from experimental data for the elements. For the compound, one computes this quantity from the calculated rotational constants and scaled vibrational frequencies using standard statistical mechanical formulae.

Current state of the art *ab initio* methods such as Pople's G2 [Gaussian-2] (*5,* see chapter by Curtiss and Raghavachari) and simpler G2(MP2) (*6*) methods, and Petersson's CBS-4 and CBS-Q methods (*7,* see chapter by Petersson), have been proposed for the accurate determination of enthalpies of formation using the procedure outlined above. The estimated accuracies of these methods, as ascertained by comparison with the "G2 test set" (*5*) of 55 molecules with accurately known atomization energies, are very good. For the test set, RMS deviations of the computed energies from experiment for the four protocols are 6.3, 6.7, 9.2, and 4.2 kJ mol⁻¹, respectively (*7*). The agreement is not surprising since the four methods were optimized to minimize disagreement with experimental data including the "G2 test set." However, these data include only one molecule with a carbon-halogen bond, CH_3Cl. Therefore, a systematic investigation of the ability of these methods to predict accurate enthalpies of formation of chlorofluorocarbons was conducted.

Chlorofluoromethanes. A previous study by Ignacio and Schlegel (*8*) utilized HF/6-31G(d) optimized geometries and vibrational frequencies, MP4/6-31G(d,p) energies and isodesmic reactions (using the experimental enthalpies of CH_4, CF_4 and CCl_4) to compute the enthalpies of formation of the remaining chlorofluoromethanes. Based on the results they estimated the accuracy of this method to be ±13 kJ mol⁻¹.

In order to test the capabilities of the currently utilized high level "compound methods" to calculate the enthalpies of halocarbons with improved accuracy, the G2 (*5*), G2(MP2) (*6*), CBS-Q and CBS-4 (*7*) methods were applied (*9*) to the chlorofluoromethanes which included the four fluoromethanes, the four chloromethanes and the six compounds containing both chlorine and fluorine. In I the calculated G2 and CBS-Q enthalpies are compared with experiment. Experimental enthalpies of the fluoromethanes were taken from Kolesov's compilation (*10*). For the remaining molecules the recommended JANAF (*11*) enthalpies were chosen for comparison. Deviations of the calculated enthalpies from experiment, [$\Delta_f H^o$(calc) - $\Delta_f H^o$(expt)], are listed in parentheses. An alternative, although less complete, compilation of enthalpies of formation of the CFCs has been published by Pedley *et al.* (*12*). The literature values from this reference are within 2 kJ mol⁻¹ of those in the JANAF tables (*11*) for most species. However, for CF_2Cl_2 and $CFCl_3$, the enthalpies in the former reference are lower by 14 and 20 kJ mol⁻¹, respectively.

The computed enthalpies in I exhibit substantial deviations from experiment, with RMS errors of 15 and 19 kJ mol⁻¹ for G2 and CBS-Q, respectively. These deviations are systematic, with almost all calculated enthalpies lying lower than reported experimental values, as evidenced by the large negative mean deviations. These results are in sharp contrast to many earlier investigations of non-halogenated organic species; e.g. RMS deviations from experimental enthalpies for the "G2 test set" were typically 4-9 kJ mol⁻¹ for these methods (*vide supra*).

To explore the distribution of errors in these series in greater detail, it is instructive to plot the deviation from experiment, [$\Delta_f H^o$(calc) - $\Delta_f H^o$(expt)], as a function of one type of carbon-halogen bond while holding the number of the other C-X bond types constant, e.g. a plot of error versus n_{CF} (number of CF bonds) in the series CH_3Cl, CH_2FCl, CHF_2Cl, CF_3Cl. This plot is displayed for the CBS-Q method in 1 (one finds similar trends using the other methods). One observes quite clearly that the negative error increases monotonically with increasing number of either C-F or C-Cl bonds; the trend is "roughly" linear with C-X bond.

One approach to correct systematic errors in *ab initio* estimates of enthalpies of formation is to employ the concept of Bond Additivity Corrections (BACs), developed by Melius and coworkers (*13,14*, see chapter by Zachariah and Melius) for MP4/6-31G(d,p) enthalpies. In this method, it is assumed that the deviation of calculated enthalpies from experiment is a linear function of the number of each type of bond in the molecule, as indicated in the following expression:

$$\Delta_f H^o(BAC) = \Delta_f H^o(calc) - \Sigma\, n_i\, \Delta_i$$
$$= \Delta_f H^o(calc) - [n_{CH}\, \Delta_{CH} + n_{CF}\, \Delta_{CF} + n_{CCl}\, \Delta_{CCl}] \tag{4}$$

Table I. Calculated and Experimental Enthalpies of Formation in Chlorofluoromethanes[a]

Species	Expt.	G2[b]	CBS-Q[b]	G2[b] [BAC]	CBS-Q[b] [BAC]
CH_4	-74.9±0.4	-77.7(-2.8)	-74.0(0.9)	-77.7(-2.8)	-74.0(0.9)
CH_3F	-232.6±8.4	-244.1(-11.5)	-238.7(-6.1)	-237.6(-5.0)	-235.2(-2.6)
CH_2F_2	-452.2±1.8	-463.7(-11.5)	-457.6(-5.4)	-450.7(1.5)	-450.6(1.6)
CHF_3	-697.6±2.7	-714.0(-16.4)	-706.7(-9.1)	-694.5(3.1)	-696.2(1.4)
CF_4	-933.0±1.7	-956.5(-23.5)	-947.7(-14.7)	-930.5(2.5)	-933.7(-0.7)
CH_3Cl	-83.7±2.1	-85.5(-1.8)	-86.3(-2.6)	-82.7(1.0)	-77.8(5.9)
CH_2Cl_2	-95.5±1.3	-98.1(-2.6)	-105.6(-10.1)	-92.4(3.1)	-88.6(6.9)
$CHCl_3$	-103.2±1.3	-107.6(-4.4)	-125.3(-22.1)	-99.2(4.0)	-99.8(3.4)
CCl_4	-96.0±2.1	-107.7(-11.7)	-137.3(-41.3)	-96.5(-0.5)	-103.3(-7.3)
CH_2FCl	-261.9±13.0	-273.3(-11.4)	-272.3(-10.4)	-264.0(-2.1)	-260.3(1.6)
CHF_2Cl	-481.6±13.0	-498.1(-16.5)	-495.9(-14.3)	-482.3(-0.7)	-480.4(1.2)
CF_3Cl	-707.9±3.3	-731.8(-23.9)	-728.2(-20.3)	-709.4(-1.5)	-709.2(-1.3)
$CHFCl_2$	-283.3±13.0	-295.8(-12.5)	-301.4(-18.1)	-283.7(-0.4)	-280.9(2.4)
CF_2Cl_2	-491.6±8.0	-513.9(-22.3)	-517.4(-25.8)	-495.3(-3.7)	-493.4(-1.8)
$CFCl_3$	-288.7±6.3	-305.2(-16.5)	-318.8(-30.1)	-290.3(-1.6)	-289.8(-1.1)
RMS	±6.9	14.5	18.8	2.6	3.4
AVG	—	-12.6	-15.3	-0.2	0.7

[a] $\Delta_f H^0$ at 298.15 K in kJ mol^{-1} units
[b] Values in parentheses represent deviations from experiment
SOURCE: Reproduced from reference 9. Copyright 1996 American Chemical Society.

Linear regression was used to fit equation 4 to the experimental data on the series of 15 CFCs. This provided values for the BAC corrected enthalpies of formation in I and the three BAC parameters, Δ_{CH}, Δ_{CF} and Δ_{CCl} (II (A)) for all four *ab initio* methods. The values of Δ_{CH} were quite small, and were therefore set equal to zero. One sees from I that the BAC corrected enthalpies of formation are in extremely good agreement with experiment. The residual RMS deviations of 2.6 and 3.4 kJ mol^{-1}, are almost an order of magnitude lower than errors in the uncorrected enthalpies and, indeed, lie significantly below the RMS experimental uncertainty of 6.9 kJ mol^{-1}. The BACs have also removed the systematic underprediction of the enthalpies of

Table II. Fluoroalkane Bond Additivity Parameters and Resulting Errors[a]

Method	Δ_{CF}	f_C	RMS1[b]	AVE1[b]	RMS2[c]	AVE2[c]
(A) Chlorofluoromethanes ($\Delta_{CF} + \Delta_{CCl}$)[d]						
G2	-6.51±0.41	0.0	14.5	-12.6	2.6	-0.2
G2(MP2)	-7.98±0.38	0.0	21.5	-19.3	2.4	0.0
CBS-4	-1.28±0.74	0.0	20.7	-15.3	4.7	0.6
CBS-Q	-3.51±0.55	0.0	18.8	-15.3	3.4	0.7
(B) Fluoroethanes (Δ_{CF})						
G2	-6.51±0.41	0.0	28.4	-25.8	7.6	-6.3
G2(MP2)	-7.98±0.38	0.0	33.4	-29.7	7.3	-5.8
CBS-4	-1.28±0.74	0.0	10.9	-10.0	8.6	-6.5
CBS-Q	-3.51±0.55	0.0	20.9	-18.4	9.7	-7.7
(C) Fluoroethanes ($\Delta_{CF} + f_C$)						
G2	-6.51±0.41	1.24±0.08	28.4	-25.8	5.3	-1.6
G2(MP2)	-7.98±0.38	1.20±0.06	33.4	-29.7	4.9	-1.0
CBS-4	-1.28±0.74	1.73±1.50	10.9	-10.0	7.2	-2.6
CBS-Q	-3.51±0.55	1.78±0.29	20.9	-18.4	5.3	-0.8
(D) Fluoroethylenes						
G2	-6.51±0.41	1.24±0.08	21.4	-18.2	4.7	-2.1
G2(MP2)	-7.98±0.38	1.20±0.06	25.2	-21.0	4.9	-1.9
CBS-4	-1.28±0.74	1.73±1.50	10.5	-8.7	5.8	-3.8
CBS-Q	-3.51±0.55	1.78±0.29	17.3	-13.7	5.2	-2.0

[a] In units of kJ mol^{-1}

[b] Errors in *ab initio* enthalpies

[c] Errors in BAC corrected enthalpies

[d] Using a Δ_{CCl} of -2.80±0.41, -6.54±0.38, -10.62±0.74 and -8.50±0.55, respectively, for the G2, G2(MP2), CBS-4 and CBS-Q methods

SOURCE: Reproduced with permission from reference 19. Copyright 1997 Elsevier.

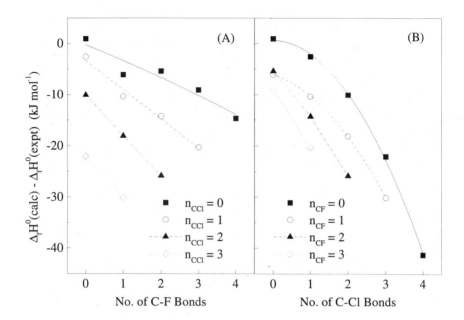

Figure 1. Deviations of CBS-Q enthalpies of formation of chlorofluoromethanes from experiment plotted as a function of the number of: (A) C-F bonds and (B) C-Cl bonds (Reproduced from reference 9. Copyright 1996 American Chemical Society.)

formation, as revealed by the extremely small average errors in the corrected results. Hence, BAC corrections for the G2 and CBS-Q methods were successfully employed to predict highly accurate enthalpies of formation in the chlorofluoromethanes. As shown by the statistical data presented in II (A), similar improvements were also obtained with the G2(MP2) and CBS-4 methods (9). The curvature exhibited in 1(B) and the increasingly negative slopes of the lines in 1(A) are manifestations of "heavy atom" interactions between the C-F and C-Cl bonds (9).

The application of spin-orbit coupling corrections to atomic energies can improve the agreement of G2 enthalpies of formation with experiment. These corrections (9) can remove some, but not all of the systematic errors in chloromethanes, and are far too small to account for the observed deviations in fluoromethanes and the mixed species. Therefore, spin-orbit corrections were not performed explicitly, but are absorbed by the empirical BAC parameters.

Fluoroethanes and Ethylenes. Next, the G2, G2(MP2), CBS-Q and CBS-4 quantum mechanical protocols were utilized to compute the enthalpies of formation of C_2 fluorocarbons. Contained in III (A) is a comparison of the computed G2 enthalpies for fluoroethanes with experiment (15-18). The deviations in the *ab initio* enthalpies are also presented in the Table and plotted (for the G2 and CBS-Q methods only) in 2 (closed squares and solid lines).

One observes from the RMS and average errors in the table that the G2 enthalpies exhibit large negative deviations from experiment. Furthermore, from 2, one sees that these negative errors are systematic with an approximately linear dependence upon the number of C-F bonds in the molecule.

Table III. C_2 Fluorocarbons: Calculated vs. Experimental Enthalpies[a]

Species	Expt.	G2[b]	G2[BAC][b]
(A) Fluoroethanes			
CH_3-CH_3	-84.1±0.4	-86.0 (-1.9)	-86.0 (-1.9)
CH_3-CH_2F	-263.2±1.6	-279.7 (-16.5)	-271.6 (-8.4)
CH_2F-CH_2F	-433.9±11.8	-459.8 (-25.9)	-443.7 (-9.8)
CH_3-CHF_2	-500.8±6.3	-516.4 (-15.6)	-500.3 (0.5)
CH_2F-CHF_2	-664.8±4.2	-687.0 (-22.2)	-662.8 (2.0)
CH_3-CF_3	-745.6±1.6	-772.1 (-26.5)	-747.9 (-2.3)
CHF_2-CHF_2	-877.8±17.6	-906.6 (-28.8)	-874.3 (3.5)
CH_2F-CF_3	-895.8±4.2	-934.2 (-38.4)	-901.9 (-6.1)
CHF_2-CF_3	-1104.6±4.6	-1145.9 (-41.3)	-1105.5 (-0.9)
CF_3-CF_3	-1342.7±6.3	-1383.7 (-41.0)	-1335.3 (7.4)
RMS	±7.7	28.4	5.3
AVG	—	-25.8	-1.6
(B) Fluoroethylenes			
CH_2=CH_2	52.4±0.8	53.3 (0.9)	53.3 (0.9)
CH_2=CHF	-140.1±2.5	-146.2 (-6.1)	-138.1 (2.0)
CHF=CHF[Z]	-297.1±10.0	-315.1 (-18.0)	-299.0 (-1.9)
CHF=CHF[E]	-292.9±10.0	-318.4 (-25.5)	-302.3 (-9.4)
CH_2=CF_2	-336.4±4.0	-359.2 (-22.8)	-343.1 (-6.7)
CHF=CF_2	-491.0±9.0	-512.5 (-21.5)	-488.3 (2.7)
CF_2=CF_2	-658.5±2.9	-693.2 (-34.7)	-660.9 (-2.4)
RMS	±6.7	21.4	4.7
AVG	—	-18.2	-2.1

[a] $\Delta_f H^0$ in units of kJ mol^{-1}
[b] Values in parentheses represent deviations from experiment
SOURCE: Reproduced with permission from reference 19. Copyright 1997 Elsevier.

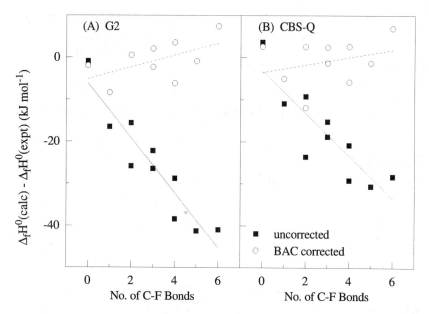

Figure 2. Deviations of the computed enthalpies of formation of fluoroethanes from experiment as a function of the number of C-F bonds. (A) G2 and (B) CBS-Q (Reproduced with permission from reference 19. Copyright 1997 Elsevier.)

Since the deviations are linearly dependent upon n_{CF}, one may, in principle, again apply the BAC correction to obtain corrected results. Furthermore, because of the close similarity of the fluoroethanes (C_2s) to the fluoromethanes (C_1s) in the earlier work, it would be reasonable to expect that the same C-F BAC parameter should also correct the systematic errors found in this study. The results of this procedure are displayed in category B of II. The second column contains the values of Δ_{CF} (9) with their estimated standard deviations. The last two pairs of columns represent the RMS and average errors without and with the application of the C-F BAC parameter. As noted above, the values of RMS2 and AVE2 in II (A) demonstrate that the application of C-F and C-Cl BACs removes virtually all systematic errors in the chlorofluoromethanes. From RMS2 in II (B), one sees, of course, that transferring the C-F BAC to fluoroethanes yields a substantial decrease in RMS deviations in the C_2s. However, in contrast to the C_1s, the comparatively large negative values of AVE2 reveal that not all of the negative systematic error has been removed by the bond additivity correction. While the problem could be remedied by refitting to minimize the RMS residuals in the fluoroethanes, this yields a larger value for Δ_{CF}, which would overcorrect enthalpies in the C_1s. Further, it is physically unrealistic that the error due to a C-F bond should differ in the two series.

In order to explore the source of the remaining systematic errors in the fluoroethanes, results of the earlier investigation of the chlorofluoromethanes (9) were re-examined. In that work, it was discussed at some length how trends and curvature in plots of $\Delta_f H^o$(calc) - $\Delta_f H^o$(expt) vs. n_{C-F} (at fixed n_{C-Cl}) and vs. n_{C-Cl} (at fixed n_{C-F}) provided definitive evidence of "heavy atom" interactions. In the earlier work, it was decided not to include heavy atom interactions since the RMS residuals using linearly independent BACs were already below the experimental uncertainties in the chlorofluoromethanes. In contrast, for the fluoroethanes, the comparatively large residual errors (RMS2 in II (B)) and negative average deviations (AVE2) indicate that the introduction of a heavy atom interaction parameter is necessary in this series to account for increased errors due to the presence of a second carbon atom attached to the carbon containing the C-C bond. Thus an interaction parameter, f_c (*19*), was incorporated into the BAC equation which, assuming no C-H bond error, becomes:

$$\Delta_f H^o(BAC) = \Delta_f H^o(calc) - n_{CF} \Delta_{CF} f_C \qquad (5)$$

The value of the parameter is optimized (holding Δ_{CF} constant at the value determined for the CFCs) to minimize the RMS deviation from experiment in the fluoroethanes.

The resultant values of the interaction parameter with its error estimate are shown in the third column of II (C). RMS2 of category C shows that the residuals using a C-F BAC with an interaction parameter are reduced significantly (relative to category B) for all but the CBS-4 method. Further, the much smaller negative values of AVE2, relative to category B, indicate that almost all of the remaining systematic error has been removed from the BAC corrected enthalpies of formation. Values of RMS2 are well within the RMS experimental uncertainty (III (A)). The removal of systematic error is also demonstrated in 2, in which it is seen that errors in the BAC corrected enthalpies of formation (open circles and dashed line) are clustered about $\Delta_f H^o$(calc) - $\Delta_f H^o$(expt) = 0.

These studies have been extended to the fluoroethylenes ($C_2H_xF_{4-x}$, x=0-4) and fluoroacetylenes ($C_2H_xF_{2-x}$, x=0-2) using the G2, G2(MP2), CBS-Q and CBS-4 methods (*20*). It was decided to use the same parameter values for Δ_{CF} and f_C as were used for the fluoroethanes. As shown in III (B) for the G2 method, systematic errors in the calculated enthalpies (third column) are almost completely removed by application of the BAC and interaction parameter (fourth column). It is very satisfying to find that, for both methods, the RMS deviations in BAC corrected enthalpies of the fluoroethylenes are lower than RMS uncertainties in the experimental data (II(D)).

Extension of *ab Initio* Methods to Heavier Halogens. The G2 method was originally defined for the elements H through Cl. However, Radom and coworkers have recently developed basis sets for Br and I that permit G2 analysis of species containing these halogens (*21*). These authors described both effective core potential and all-electron basis sets that lead to G2[ECP] and G2[AE] energies, respectively. The G2[ECP] method is much faster computationally, as only the valence electrons are treated explicitly, and there is claimed to be little loss of accuracy as compared to G2[AE] results (*21*). These basis sets have been employed to analyze the thermochemistry of bromine and iodine monoxides (22) and hydroxides (23,24) via G2 and related approaches. Hassanzadeh and Irikura have analyzed similar systems using CCSD(T) calculations with large basis sets (25), and Lee has investigated a series of triatomic bromine compounds at a similar level of theory (26). An early application of high-level methods to Br and I-containing compounds was by Kellö and Sadlej, who studied cyanogen halides (27). Typically, a variety of isodesmic or homodesmic reactions have been employed to relate the unknown heats of formation to experimental quantities.

A complication is that radicals with significant unpaired electron spin density close to Br or I nuclei exhibit significant spin-orbit coupling, a relativistic effect which is excluded from standard *ab initio* calculations, but which can lead to large energy corrections. This effect leads to splitting of electron levels. For example, standard calculations on ground-state halogen atoms yield the weighted mean energy of the two ^2P_J levels (J = 1/2 and 3/2), weighted by the 2J+1 degeneracy. However, at 0 K only the lower energy $^2P_{3/2}$ level is populated. This correction is usually neglected for F and Cl atoms, because the ^2P_J levels lie close together and the errors introduced are only 1.6 and 3.5 kJ mol^{-1}, respectively. These corrections are implicit in the C-F and C-Cl BACs discussed here, and have recently been shown to improve agreement with experimental energies of Cl-containing compounds (28). By contrast, for I atoms at 0 K the required correction is 30.3 kJ mol^{-1} and must be included if chemical accuracy is to be achieved. The magnitude of the spin-orbit splitting between ground-state sub-levels for radicals may be computed (21) or, if available, empirical data can be employed. On the assumption that the spin-orbit states have similar geometries and vibrational frequencies, the influence of low-lying electronic states on thermodynamic functions such as C_p, S and $H_T - H_0$ is straightforward to evaluate (11), once the energy splittings and degeneracies are established.

Haloalkane Kinetics

Product Channels for H + Fluoromethanes. The decomposition of fluoromethanes in hydrocarbon flames has recently been investigated (29) using numerical simulations in conjunction with flame speed measurements. In stoichiometric methane/air flames at high (8%) inhibitor concentrations, the relative amounts of fluoromethane destroyed via reaction with H atoms, unimolecular decomposition and reaction with OH radicals were found to be 4:3:3 and 6:3:1, respectively, for CH_2F_2 and CHF_3.

Destruction by H attack, which is the leading fluoromethane consumption pathway, has been the subject of several experimental and theoretical investigations.

$$H + CH_{4-x}F_x \rightarrow products \qquad x=1,2,3,4 \qquad (6\text{-}9)$$

There are three possible reaction channels; abstraction of F, abstraction of H, and substitution. For CH_3F all three channels are thermochemically reasonable (11):

$$H + CH_3F \quad \rightarrow CH_3 + HF \qquad \Delta_r H^\circ_{298} = -111 \text{ kJ mol}^{-1} \qquad (6a)$$

$$\rightarrow CH_2F + H_2 \qquad \Delta_r H^\circ_{298} = -16 \text{ kJ mol}^{-1} \qquad (6b)$$

$$\rightarrow CH_4 + F \qquad \Delta_r H^\circ_{298} = -12 \text{ kJ mol}^{-1} \qquad (6c)$$

Thus, the net decomposition rate, $k_6 = k_{6a} + k_{6b} + k_{6c}$. In their review Baulch *et al.* (*30*) suggested F-abstraction is the dominant channel. Their rate constant recommendation is largely based on the work of Westenberg and deHaas (*31*) who monitored the disappearance of CH_3F in the presence of a large excess of H, so that knowledge of the products was not necessary for the determination of the total rate constant. Similarly, Parsamyan and Nalbandyan (*32*) assumed F-abstraction to be the main pathway while analyzing their $H + CH_2F_2$ data. On the other hand BAC-MP4 calculations of Westmoreland *et al.* (*33*) predict H-abstraction to be the most important pathway for both reactions 6 and 7.

Recently, high level *ab initio* methods such as G2(MP2) and G2(ZPE=MP2) have been employed to compute the reaction barriers for these pathways; full details may be found elsewhere (*34*). H-abstraction is computed to have the lowest energy barrier, $E_0^\ddagger$, for reactions 6-8. For reaction 9 the barrier to F-substitution is very high which leaves F-abstraction as the only accessible channel.

Controversy also exists over the value of the rate constants. For example, measurements of k_6 differ by over two orders of magnitude at 600 K, where different measurement techniques overlap (*31,35*). To resolve these discrepancies the G2(MP2) results (*34*) were used in transition state theory (TST) calculations to predict the reaction rate constants, k_{TST}, as a function of temperature. The resulting Arrhenius plots of the rate constant are compared with experiment in 3.

$$k_{TST} = \Gamma \frac{k_B T}{h} \frac{Q_{TS}^\ddagger}{Q_H Q_{CH_{4-x}F_x}} \exp\left(-\frac{E_0^\ddagger}{RT}\right) \qquad (10)$$

where Γ is the quantum mechanical tunneling correction derived at the zero-curvature level (*36*) and the partition functions, Q, include rotational symmetry numbers. This is the simplest form of TST. Q was derived on the usual basis of the rigid rotor - harmonic oscillator approximation. Any anharmonicity would lead to slightly higher values of Q, with a likely larger effect on the TS because of its looser modes. Thus, k_{TST} may be underestimated. However, an opposing factor is that neglect of variational effects implies k_{TST} will tend to be too high. The zero-curvature tunneling model neglects two opposing effects. One is that this one-dimensional model does not allow for multi-dimensional corner-cutting tunneling paths, which would increase Γ. On the other hand, the MP2 imaginary frequencies employed here are probably too high, which will lead to overestimated Γ values. A new method for improved calculations of the energy barrier is outlined in Petersson's chapter in this book. The effects of these neglected factors appear either to be small or to cancel because, as shown below, the canonical TST model generally agrees well with experiment.

For H + CH_3F, the minor F-abstraction and substitution channels are predicted to have small rate constants, hence the total rate constant k_6 is essentially equal to k_{6b}. This agrees with the recommended rate constant in the 600-1000 K range within the factor of 5 experimental error limits (30). If the entire difference between theory and experiment is assigned to errors in the *ab initio* $E_0^{\ddagger}$ for this reaction, then this corresponds to a computational error of only 4 kJ mol^{-1} (at 800 K). Employing the critical evaluation by Baulch *et al.* (30), some of the earlier measurements are seen to be in error (37,38).

The computed rate constant for reaction 7 was within a factor of 2 of the value reported by Parsamyan and Nalbandyan (32). For reaction 8 the computed rate constant compares well with the reported experimental determinations (39-41). For reaction 9 (H + CF_4), the calculated rate constant is of the same order of magnitude as the measurements by Kochubei and Moin (42). However, the measurements imply that the computed G2(MP2) barrier has been overestimated by about 22 kJ mol^{-1}.

Thus the *ab initio* potential energy surfaces suggest that, contrary to some earlier assumptions, H atoms react predominantly with the C-H bonds in fluoromethanes and that the dominant product is H_2. F atom abstraction is unfavorable

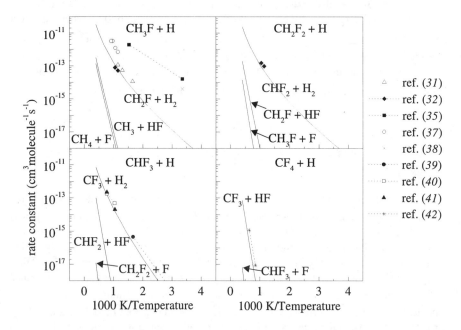

Figure 3. H + fluoromethanes: G2(MP2) Arrhenius plots (solid lines) vs. experiment (Reproduced with permission from reference 34. Copyright 1997 Elsevier.)

kinetically, even though HF formation is the most exothermic pathway. G2(MP2) based TST results are in good accord with experimentally determined rate constants for H attack on CH_3F, CH_2F_2 and CHF_3. CF_4 is several orders of magnitude less reactive than the other fluoromethanes towards atomic hydrogen, because it does not contain labile C-H bonds. The high C-F bond strength also makes unimolecular dissociation unfavorable under combustion conditions. Slow F-atom abstraction is the only plausible pathway for H-atom attack on CF_4. Thus CF_4 is essentially inert in a flame and the flame suppressant activity of CF_4 is largely physical, i.e. through cooling and dilution. By contrast, the other fluoromethanes react quickly with H atoms to yield F-containing radicals that undergo further chemistry. This allows for the possibility of chemical flame suppression radical scavenging by CH_3F, CH_2F_2 and CHF_3.

The Reactions of H Atoms and OH Radicals with Iodomethane. As a demonstration of the application of G2 analysis to iodine chemistry, results obtained via the G2[AE] method of Radom and coworkers (*21*) are presented for the reactions of H and OH with CH_3I. These data were used to assess the combustion chemistry of CH_3I, an important intermediate in the chemistry of iodine-mediated flame suppression by agents such as CF_3I (*43*). First the thermochemistry was examined, as a check of the accuracy of the G2 method, and then the reaction kinetics were analyzed and compared with experiment. One aim was to distinguish between various possible sets of products, which for some systems has been a difficult experimental problem. The present work describes an *ab initio* analysis of the kinetics of the reactions

$$CH_3I + H \quad \rightarrow CH_2I + H_2 \quad \quad \quad (11a)$$

$$\rightarrow CH_3 + HI \quad \quad \quad (11b)$$

$$CH_3I + OH \quad \rightarrow CH_2I + H_2O \quad \quad \quad (12a)$$

$$\rightarrow CH_3 + HOI \quad \quad \quad (12b)$$

The results are employed to derive high-temperature rate constants via TST and to assess branching ratios for H vs I abstraction. Contributions from the I-displacement channels were assumed to be negligible based on the G2(AE) results reported for CH_3I + H $\rightarrow CH_4$ + I (*44*).

The transition state for reaction 11a was first investigated by Schiesser *et al.* using moderately sized ECP basis sets (*45*), while details of the G2 analysis of channels 11a-12b may be found elsewhere (*46*). As a check on the accuracy of the G2 methodology, $\Delta_r H^o_{298}$ was computed for the C-H and C-I bond breaking reactions 11a and 11b. The results are -9.0 and -61.7 kJ mol^{-1}, in excellent accord with the experimental values of -5.3 ± 6.7 and -62.2 ± 0.9 kJ mol^{-1} (*47*), and thus the thermochemistry of the C-H and C-I bonds is seen to be well-described at the G2 level.

As in the H + fluoromethane example discussed above, the rate constants k were computed via canonical TST (see equation 10). The scaled HF/6-31G(d) and MP2=full/6-31G(d) frequencies of the TSs are similar, except for the low frequency C-I-H bending mode. This apparently reflects the neglect or overestimation of the influence of electron correlation on this mode, which has a significant influence on the TST rate constant. As a compromise, the geometric mean of TST calculations based on both sets of frequencies was derived. The rate constants are plotted in 4. The results for H + CH_3I are consistent with the known bond strengths D_{298} (47), where the weaker C-I bond (D_{298} = 239 kJ mol^{-1}) is much more reactive than the C-H bond (D_{298} = 431 kJ mol^{-1}). This is in accord with previous experimental studies which concluded or assumed that HI was the dominant product (48-51), and yields a k_{11} which agrees well with the room temperature experimental data.

TS structures for attack by the OH radical at the H and I atoms of CH_3I were also investigated. In the latter case no barrier beyond the endothermicity was discernible on the HF/6-31G(d) potential energy surface. Consideration of the QCISD(T)/6-311G(d,p) energies calculated at selected points along the MP2/6-31G(d) intrinsic reaction coordinate showed that the energy is below that of HOI + CH_3 at all points. A distinct TS therefore cannot be localized and canonical TST cannot be applied to this channel. k_{12b} was roughly estimated via the relation $k = A \exp(-E_a/RT)$

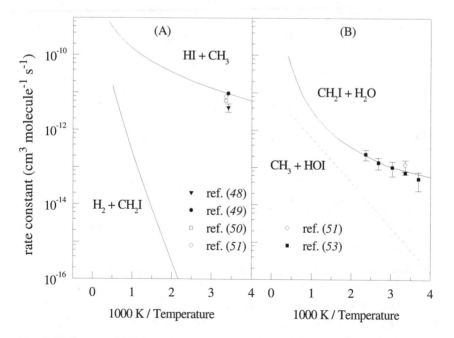

Figure 4. Computed TST results (solid lines) vs. experiment for: (A) H + CH_3I. (B) OH + CH_3I. The dashed line for the CH_3 + HOI channel shows an empirical estimate. (Reproduced with permission from reference 46. Copyright 1997 Elsevier.)

by assuming a pre-exponential factor $A = 1 \times 10^{-11}$ cm^3 molecule^{-1} s^{-1}, equal to that measured for $CF_3I + OH$ (*52*), and an activation energy $E_a = 23$ kJ mol^{-1}. This is equal to the endothermicity at 298 K based on D_{298}(HO-I) = 216 kJ mol^{-1} (*52*).

4(B) shows that, by contrast to H atom attack, the bond that is more reactive towards OH is the stronger C-H bond and that $CH_2I + H_2O$ formation, the more exothermic channel, is expected to dominate at all temperatures. A check on the reliability of the kinetic calculations is comparison with the measurements of the total removal rate constant $k_{12} = k_{12a} + k_{12b}$ made by Brown *et al.* (*53*). They are seen to be in excellent agreement, as is the room temperature k_{12} value obtained by Gilles *et al.* (*51*). Again, the geometric mean of the HF and MP2-based TST results gives good accord with experiment and provides a TST extrapolation of these measurements to combustion conditions. The k_{12} value from the extrapolation to 2000 K is about 25 times greater than the value obtained from a simple linear Arrhenius extrapolation of the data of Brown *et al.* (*53*), which reflects the significant curvature of the theoretical Arrhenius plot. This predicted curvature is similar to that measured for the analogous reaction of OH with CH_4 (*54*). The TST analysis also indicates that HOI formation by OH attack on CH_3I is of only minor importance in both flames and the atmosphere.

The TST rate constants for reactions 11 and 12 and experimental data (*51,55*) for

$$O + CH_3I \quad \rightarrow IO \text{ and other products} \tag{13}$$

$$CH_3I + Ar \quad \rightarrow CH_3 + I + Ar \tag{14}$$

were employed in a numerical simulation of an adiabatic premixed stoichiometric CH_4/air flame at atmospheric pressure, modeled with CHEMKIN (*56*) using the GRI-Mech mechanism (*57*). The results apply to trace quantities of CH_3I (i.e. the C/H/O chemistry was assumed to be unaffected). As may be seen from 5, the dominant removal pathway is via H-atom attack. The next most important removal pathway, about an order of magnitude less effective than reaction 11, is attack by OH. Thus CH_2I will be a minor product of CH_3I decay, with HI as the dominant product.

Summary

Ab initio methods have been applied to investigate the thermochemistry of halogenated alkanes and the kinetics of major pathways for their reactions in flames. Where the computed thermochemistry and rate expressions can be compared with experimental values, the results are in good accord and provide a means to extrapolate limited temperature information over wider temperature ranges. The *ab initio* calculations of enthalpies of formation and rate constants are sufficiently accurate to survey a wide range of possible steps in combustion mechanisms, in order to help limit the number of processes to be included in combustion models and to aid in the identification of key reactions that merit more detailed laboratory study.

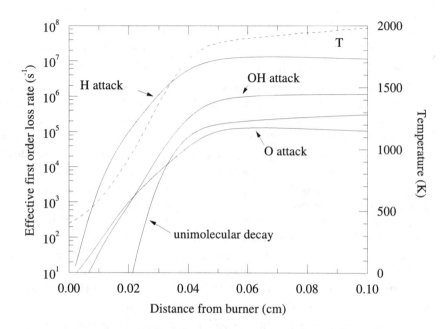

Figure 5. Reciprocal lifetimes for traces of CH_3I in an adiabatic premixed stoichiometric CH_4/air flame at atmospheric pressure with respect to unimolecular decomposition and attack by H, O and OH (solid lines, left axis), and the temperature profile (dashed line, right axis).
(Reproduced with permission from reference 46. Copyright 1997 Elsevier.)

Acknowledgments

The authors thank the Air Force Office of Scientific Research and the Wright Laboratory, Materials Directorate at Wright-Patterson Air Force Base. M.S. and P.M. acknowledge the Robert A. Welch Foundation (Grant Nos. B-657 and B-1174) and the UNT Faculty Research Fund for financial support.

Literature Cited

1. Rowland, F. S. *Environ. Sci. Technol.* **1991**, *25*, 622.
2. McFarland, M.; Kaye, J. *Photochem. Photobiol.* **1992**, *55*, 911.
3. Banks, R. E. *J. Fluorine Chem.* **1994**, *67*, 193.
4. (a) Todd, C. S. *The Use of Halons in the United Kingdom and the Scope for Substitution*; HMSO: London, 1991, and references therein. (b) Grosshandler, W. L.; Gann, R. G.; Pitts, W. M. *Evaluation of Alternative In-Flight Fire Suppressants for Full-Scale Testing in Simulated Aircraft Engine Nacelles and Dry Bays*; NIST Spec. Publ. 861; NIST: Washington DC, 1994.
5. Curtiss, L. A.; Raghavachari, K.; Trucks, G. W.; Pople, J. A. *J. Chem. Phys.* **1991**, *94*, 7221.

6. Curtiss, L. A.; Raghavachari, K.; Pople, J. A. *J. Chem. Phys.* **1993**, *98*, 1293.
7. Ochterski, J. W.; Petersson, G. A.; Montgomery, J. A., Jr. *J. Chem. Phys.* **1996**, *104*, 2598.
8. Ignacio, E. W.; Schlegel, H. B. *J. Phys. Chem.* **1992**, *96*, 5830.
9. Berry, R. J.; Burgess, D. R., Jr.; Nyden, M. R.; Zachariah, M. R.; Melius, C. F.; Schwartz, M. *J. Phys. Chem.* **1996**, *100*, 7405.
10. Kolesov, V. P.; *Russ. Chem. Rev.* **1978**, *47*, 1145.
11. Chase, M. W., Jr.; Davies, C. A.; Downey, J. R., Jr.; Frurip, D. J.; McDonald, R. A.; Syverud, A. N. "JANAF Thermochemical Tables, Third Edition". *J. Phys. Chem. Ref. Data*, **1985**, *14*, Suppl. 1.
12. Pedley, J. B.; Naylor, R. D.; Kirby, S. P. *Thermochemical Data of Organic Compounds*; Chapman & Hall: London, UK, 1986.
13. (a) Melius, C. F. *Springer-Verlag DFVLR Lecture Notes*; Springer-Verlag: Berlin, 1990. (b) Melius, C. F. *Chemistry and Physics of Energetic Materials*; S. N. Kluwer Academic: New York, 1992.
14. (a) Ho, P.; Melius, C. F. *J. Phys. Chem.* **1990**, *94*, 5120. (b) Allendorf, M. D.; Melius, C. F. *J. Phys. Chem.* **1993**, *97*, 72.
15. Tsang, W.; Hampson, R. F. *J. Phys. Chem. Ref. Data* **1986**, *15*, 1087.
16. Chen, S. S.; Rodgers, A. S.; Chao, J.; Wilhoit, R. C.; Zwolinski, B. J. *J. Phys. Chem. Ref. Data* **1975**, *4*, 441.
17. Burgess, D. R., Jr.; Zachariah, M. R.; Tsang, W.; Westmoreland, P. R. *Prog. Energy Combust. Sci.* **1996**, *21*, 1034.
18. Lacher, J. A.; Skinner, H. A. *J. Chem. Soc. (A)* **1968**, 1034.
19. Berry, R. J.; Ehlers, C. J.; Burgess, D. R., Jr.; Zachariah, M. R.; Nyden, M. R.; Schwartz, M. *J. Mol. Struct. (Theochem)*, in press.
20. Berry, R. J.; Schwartz, M. *Struct. Chem.*, submitted.
21. Glukhovtsev, M. N.; Pross, A.; McGrath, M. P.; Radom, L. *J. Chem. Phys.* **1995**, *103*, 1878.
22. McGrath, M. P.; Rowland, F S. *J. Phys. Chem.* **1996**, *100*, 4815.
23. McGrath, M. P.; Rowland, F. S. *J. Phys. Chem.* **1996**, *98*, 4774.
24. Glukhovtsev, M. N.; Pross, A.; Radom, L. *J. Phys. Chem.* **1996**, *100*, 3498.
25. Hassanzadeh, P.; Irikura, K. K. *J. Phys. Chem.* **1997**, *101*, 1580.
26. Lee, T. J. *J. Phys. Chem.* **1995**, *99*, 15074.
27. Kellö, V.; Sadlej, A. *Mol. Phys.* **1992**, *75*, 209.
28. Curtiss, L. A.; Raghavachari, K.; Redfern, P. C.; Pople, J. A. *J. Chem. Phys.* **1997**, *106*, 1063.
29. Linteris, G. T. In *Halon Replacements: Technology and Science*; Miziolek, A. W.; Tsang, W., Eds.; ACS Symp. Ser. 611; ACS: Washington DC, 1995; p 260.
30. Baulch, D. L.; Duxbury, J.; Grant, S. J.; Montague, D. C. *J. Phys. Chem. Ref. Data* **1981**, *10*, suppl. 1.
31. Westenberg, A. A.; deHaas, N. *J. Chem. Phys.* **1975**, *62*, 3321.
32. Parsamyan, N. I.; Nalbandyan, A. B. *Arm. Khim. Zhur.* **1968**, *21*, 1003.
33. Westmoreland, P. R.; Burgess, D. F., Jr.; Tsang, W.; Zachariah, M. R. *25th Symp. (Int.) Combust.;* The Combustion Institute: Pittsburg, PA, 1994; p 1505.
34. Berry, R. J.; Ehlers, C. J.; Burgess, D. F., Jr.; Zachariah, M. R.; Marshall, P. *Chem. Phys. Lett.*, **1997**, *269*, 107.
35. Aders, W. -K.; Pangritz, D.; Wagner, H. Gg. *Ber. Bunsenges. Phys. Chem.* **1975**, *79*, 90.

36. Steckler, R.; Hu, W.-P.; Liu, Y.-P.; Lynch, G. C.; Garrett, B. C.; Isaacson, A. D.; Lu, D.-H.; Melissas, V. S.; Truong, T. N.; Rai, S. N.; Hancock, G. C.; Lauderdale, J. G.; Joseph, T.; Truhlar, D. G. *POLYRATE* version 6.5 (University of Minnesota, Minneapolis, 1995).

37. Hart, L. W.; Grunfelder, C.; Fristrom, R. M. *Combust. Flame* **1974**, *23*, 109.

38. Jones, W. E.; Ma, J. L. *Can. J. Chem.* **1986**, *64*, 2192.

39. Amphlett, J. C.; Whittle, E. *Trans. Faraday Soc.* **1967**, *63*, 2695.

40. Skinner, G. B.; Ringrose, G. H. *J. Chem. Phys.* **1965**, *43*, 4129.

41. Richter, H.; Vandooren, J.; Van Tiggelen, P. J. *25th Symp. (Int.) Combust.*; The Combustion Institute: Pittsburg, PA, 1994; p 825.

42. Kochubei, V. F.; Moin, F. B. *Kinet. Catal.* **1971**, *11*, 86. English translation p. 712.

43. Noto, T.; Babushok, V.; Burgess, D. R., Jr.; Hamins, A.; Tsang, W.; Miziolek, A. *26th Symp. (Int.) Combust.;* The Combustion Institute: Pittsburg, PA, in press.

44. Yuan, J.; Wells, L; Marshall, P. *J. Phys. Chem. A,* in press.

45. Schiesser, C. H.; Smart, B. A.; Tran, T.-A. *Tetrahedron* **1995**, *51*, 3327.

46. Marshall, P.; Misra, A.; Berry, R. J. *Chem. Phys. Lett.* **1997**, *265*, 48.

47. McMillen, D. F.; Golden, D. M. *Ann. Rev. Phys. Chem.* **1982**, *33*, 493. Data extrapolated to 0 K using *ab initio* H_{298}-H_0 values.

48. Leipunskii, I. O.; Morozov, I. I.; Tal'roze, V. L. *Dokl. Phys. Chem.* **1971**, *198*, 547. Russ. orig. p. 136.

49. Levy, M. R.; Simons, J. P. *J. Chem. Soc. Faraday Trans. 2* **1975**, *71*, 561.

50. Sillesen, A.; Ratajczak, E.; Pagsberg, P. *Chem. Phys. Lett.* **1993**, *201*, 171.

51. Gilles, M. K.; Turnipseed, A. A.; Talukdar, R. K.; Rudich, Y.; Villalta, P. W.; Huey, L. G.; Burkholder, J. B.; Ravishankara, A. R. *J. Phys. Chem.* **1996**, *100*, 14005.

52. Berry, R. J.; Yuan, W.-J.; Misra, A.; Marshall, P. unpublished work.

53. Brown, A. C.; Canosa-Mas, C. E.; Wayne, R. P. *Atmos. Environ.* **1990**, *24*, 361.

54. Madronich, S.; Felder, W. *20th Symp. (Int.) Combust.;* The Combustion Institute: Pittsburg, PA, 1984; p 703.

55. Saito, K.; Tahara, H.; Kondo, O.; Yokubo, T.; Higashihara, T.; Murakami, I. *Bull. Chem. Soc. Jpn.* **1980**, *53*, 1335.

56. Kee, R. J.; Rupley, F. M.; Miller, J. A. *Chemkin-II: A Fortran Chemical Kinetics Package for the Analysis of Gas-Phase Chemical Kinetics* (Sandia National Laboratories Report No. SAND89-8009B, 1991).

57. Frenklach, M.; Wang, H.; Yu, C.-L.; Goldenberg, M.; Bowman, C. T.; Hanson, R. K.; Davidson, D. F.; Chang, E. J.; Smith, G. P.; Golden, D. M.; Gardiner, W. C.; Lissianski, V. *GRI-Mech 1.2*, http://www.gri.org.

Chapter 19

Use of Density Functional Methods To Compute Heats of Reactions

Peter Politzer and Jorge M. Seminario

Department of Chemistry, University of New Orleans, New Orleans, LA 70148

The use of density functional procedures for computing bond
dissociation energies, heats of chemical reactions, and heats of
formation is examined and discussed. The DF/B3P86/6-31+G**
combination of functionals and basis set has been found to be quite
effective for the first two purposes, and is being extended to heats of
formation, although these are presently being obtained with good
accuracy at the DF/BP86/6-31G** level. The decomposition of the
energetic compound ammonium dinitramide is considered in some
detail. It is suggested that the initial step may be sublimation to
$NH_3 + HN(NO_2)_2$, followed by loss of NO_2 from either $HN(NO_2)_2$ or
its tautomer $O_2N-NN(O)OH$. Computed $\Delta H(298\ K)$ are listed for 32
possible intermediate steps.

In designing and evaluating new propellants and fuels, it is important to have as much
information as possible concerning the mechanisms and energetics of their
combustion and decomposition processes. This requires a knowledge of relevant
heats of formation, activation barriers, bond dissociation energies, etc. If the
compound in question is one that has been proposed but not yet synthesized, or if the
amount available is too little for adequate laboratory characterization, then the only
means for acquiring the needed data is through computations.

This can pose a significant challenge, since the quantities desired are generally
obtained by taking the difference of two much larger numbers; any errors in the latter
may therefore be greatly magnified in the former. Molecular energies are normally
two to three orders of magnitude larger than are the energy changes in chemical
reactions, so the situation is akin to determining the weight of a truck driver by
weighing the truck with and without the driver in it!

Another complicating factor is that energetic molecules, such as would be used in
propellant formulations, typically contain large proportions of the most "electron-
rich" atoms (nitrogen, oxygen and fluorine), which have higher outer-shell electron
densities (1). It is particularly important in the case of such molecules that the
computational treatment take account of electronic correlation effects (2-6). However
correlated *ab initio* procedures, such as Møller-Plesset perturbation methods and

configuration interaction, are not practical for molecules as large as many of those that are of interest. In contrast, density functional techniques, which also include electronic correlation, are much less demanding of computer resources and can accordingly be used to treat relatively large systems (7-11).

In principle, therefore, density functional procedures offer a means for quantitatively analyzing reaction energetics, provided that sufficient accuracy can be achieved. It is the question of accuracy that shall be addressed in this chapter.

Density Functional Methods

Local Approximation. The term "density functional" (DF) encompasses a variety of computational procedures, which differ in the representation of electronic exchange and correlation contributions. The local density approximation, which treats these in terms of the uniform electron gas model (7,8), generally overestimates dissociation energies by a significant amount (12-16). On the other hand, the local density approach can be quite acceptable if a molecule is simply undergoing rearrangement rather than fragmentation; thus it has been applied successfully to computing energy changes in isomerizations, activation barriers in unimolecular reactions, and rotational barriers (14,17-21). If a reaction of interest does involve bond-breaking, a useful option can be to carry out the time-consuming geometry optimizations at the local density functional level, which is generally satisfactory for this purpose (21-23), followed by a single-point energy calculation with a higher-level but computationally more expensive density functional procedure.

Non-local Methods. The more accurate (non-local) density functional techniques are those that explicitly recognize the non-uniformity of the electronic density $\rho(\mathbf{r})$. Their exchange and correlation functionals generally include terms involving the gradient of the density, $\nabla\rho(\mathbf{r})$; for recent discussions, see references 9 - 11. Among these functionals are the Becke (B) (24) and the Becke-3 (B3) (25) for exchange and the Perdew 86 (P86) (26), the Lee-Yang-Parr (LYP) (27) and the Perdew-Wang 91 (PW91) (28) for correlation. All of these are available as options in the Gaussian 94 package of programs (29). Commonly used basis sets are the 6-31G* and 6-31G**, to which diffuse functions may be added, e.g. 6-31+G* and 6-31+G**.

In Table I, dissociation energies computed by a variety of methods for the indicated bonds in H_3C-NO_2, H_3C-OH and H_3C-CN are compared to the experimental values. In such calculations, the first step is to optimize the geometries of the molecules and their fragments. This gives their energy minima at 0 K. These are converted to enthalpies at 298 K by adding the translational, rotational and vibrational contributions, using the ideal gas model, and then applying the definition $H = E + PV = E + RT$ (30). As is customary, the dissociation energy of the bond A–B is defined as ΔH for the bond-breaking process A–B $\rightarrow$ A + B, at 298 K (31,32). Hartree-Fock results are not included because they are generally poor, typically too low by 25 - 40 kcal/mol (30); this is because the effect of correlation is significantly greater in A–B than in the separated fragments A + B. (For reactions other than dissociations, Hartree-Fock ΔH values can sometimes be quite accurate, if the changes in correlation energies are small or cancel (30).)

Table I is included in order to demonstrate the sorts of results that are obtained by correlated computational methods. One point to note is that higher-level procedures, while giving more accurate energies for individual molecules, do not necessarily give better ΔE and ΔH for reactions involving these molecules; for example, MP2 values are often closer to the experimental than are MP3, MP4 and coupled cluster, CCSD(T) (33). Indeed MP2 can be quite effective for calculating dissociation energies (30,33), although there are also instances such as H_3C-CN, for which it gives a very poor result (Table I). In any case, as mentioned earlier, the

Table I. Calculated H_3C-NO_2, H_3C-OH and H_3C-CN dissociation energies at 298 K [a]

Method	$\Delta H(298\ K)$, kcal/mol		
	H_3C-NO_2	H_3C-OH	H_3C-CN
MP2/CEP-TZDP++//MP2/CEP-TZDP	59.6	92.1	141.0
MP3/CEP-TZDP++//MP2/CEP-TZDP	59.6	83.6	131.8
MP4/CEP-TZDP++//MP2/CEP-TZDP	56.0	87.8	134.8
CCSD(T)/CEP-TZDP++//MP2/CEP-TZDP	57.3	85.9	121.4
DF/BLYP/6-31G**//BLYP/6-31G**	53.5	91.5	---
DF/B3LYP/6-31G**//B3LYP/6-31G**	56.2	89.4	123.3
DF/B3LYP/6-31+G*//B3LYP/6-31+G*	56.2	87.2	121.7
DF/B3LYP/6-31+G**//B3LYP/6-31+G**	55.7	86.9	---
DF/B3PW91/6-31G**//B3PW91/6-31G**	57.0	90.0	124.6
DF/B3PW91/6-31+G*//B3PW91/6-31+G*	57.2	88.4	123.3
DF/B3P86/6-31G**//B3P86/6-31G**	60.0	93.2	---
DF/B3P86/6-31+G*//B3P86/6-31+G*	60.2	91.7	126.2
DF/B3P86/6-31+G**//B3P86/6-31+G**	59.8	91.4	125.9
experimental	60.6[b]	92.3[b]	121[b]

[a]The MP and CCSD(T) results are taken from reference 33, in which the basis set is described; the density functional (DF) results are from reference 34.
[b]Reference 36.

correlated *ab initio* techniques are not practical for most molecules of interest in the present context.

The density functional results in Table I tend to be reasonably good, most of them being within 4 kcal/mol of the experimental and nearly half being within 2 kcal/mol. However no consistent trends are evident among the various procedures. For instance, the BLYP/6-31G** is the worst for H_3C-NO_2 but the best for H_3C-OH. The B3P86 methods are overall the most effective ones for H_3C-NO_2 and H_3C-OH, but the poorest ones for H_3C-CN. Increasing the size of the basis set is no guarantee of improved results, as can be seen from the B3LYP/6-31G** and B3LYP/6-31+G** values for H_3C-NO_2 and H_3C-OH.

Under these circumstances, it seemed desirable to undertake a comparative evaluation of various combinations of exchange and correlation functionals and basis sets, to permit at least an initial decision as to which one(s) are, overall, the best suited for calculating reaction energetics. Accordingly we used the nine DF combinations listed in Table I to optimize geometries and compute bond dissociation energies for 13 molecules (*34*): $H-CH_3$, $H-CN$, $H-NH_2$, $H-NF_2$, $H-SH$, H_3C-CH_3, H_3C-NH_2, H_3C-NO_2, H_3C-OH, H_3C-F, H_3C-SH, F_2N-NF_2 and $Cl-CN$. On the basis of these results, when compared to experimental data, we selected six functional and basis set combinations (the ones used for H_3C-CN in Table I) and applied them to 15 bonds in an additional 13 molecules: $H-ONO$, $H-ONO_2$, $H-SiH_3$, H_3C-CN, H_3C-ONO_2, H_3C-SiH_3, H_3CO-NO_2, $(H_3C)_2N-NO_2$, H_2N-NH_2, $HO-OH$, $HO-NO$, O_2N-NO_2, $F-NF_2$, $O=CO$ and $HC\equiv CH$.

Overall, the highest level of accuracy was achieved with the B3P86/6-31+G** procedure, followed by the B3P86/6-31+G* and the B3PW91/6-31G** (*34*). All three of these had average errors, for 28 bonds, of less than 2.5 kcal/mol. They all have problems with O=CO, for which both of the B3P86 calculations overestimate the dissociation energy by 6.2 kcal/mol, and the B3PW91 by 5.1 kcal/mol. The latter is even worse for $H-NF_2$, being too low by 6.3 kcal/mol.

As a more stringent test of our most succesful density functional procedure, the B3P86/6-31+G**, we computed $\Delta H(298\ K)$ for 13 different chemical reactions, involving as many as eight molecules in the balanced equation (35). Since these reactions involve rearrangements of atoms, there is less chance for cancellation of errors within groups that are maintaining their identities; indeed the numbers of molecules involved provides opportunities for accumulation of errors. On the average, our calculated $\Delta H(298 \text{ K})$ were within 6 kcal/mol of the experimental (35); our worst result was for the reaction,

$$2NH_3 + 2O_2 \rightarrow N_2O + 3H_2O$$

for which we find $\Delta H(298 \text{ K}) = -117.4$ kcal/mol instead of the experimental -131.8 kcal/mol (36).

It is very gratifying to observe the degrees of accuracy obtained at the B3P86/6-31+G** level for bond dissociations and for general chemical reactions, given that the computed ΔH are invariably very small differences between extremely large numbers. Furthermore, it should be kept in mind that the measured heats of formation upon which the experimental ΔH are based are also subject to errors (36), which can similarly accumulate and/or partially cancel. An additional test (and application) of density functional B3P86/6-31+G** calculations, pertaining to the decomposition of ammonium dinitramide, shall be reported in the next section.

Ammonium Dinitramide Decomposition

Ammonium dinitramide, $NH_4N(NO_2)_2$, is an ionic compound which has attracted considerable interest in the last few years as a potential oxidizing component of propellant formulations (37-46); in particular, its thermal decomposition has been extensively investigated, and a number of reactions have been suggested as being intermediate steps (37-41,44,46). Some of these are listed in Table II, together with our heats of reaction computed at the DF/B3P86/6-31+G** level. We have included experimental $\Delta H(298 \text{ K})$ values in as many instances as possible. The average error is less than 3 kcal/mol.

It is generally believed that an early step in the decomposition process is N–NO$_2$ bond scission, producing NO$_2$ (37-41,44,46). This could involve the original dinitramide ion, reaction (29) in Table II,

$$N(NO_2)_2^- \longrightarrow NNO_2^- + NO_2$$

for which our calculated $\Delta H(298 \text{ K})$ is 49.7 kcal/mol. However this process, reaction (29), is for $N(NO_2)_2^-$ in the gas phase, which means that it must be preceded by sublimation of ammonium dinitramide to the gas phase ions, reaction (1); this requires an input of 144 kcal/mol (Table II).

A less demanding alternative is sublimation to NH_3 and $HN(NO_2)_2$, reaction (2). If this can proceed essentially directly, without the ions being a significant intermediate stage, then the energetic requirement is only 44 kcal/mol (Table II). Subsequent loss of NO$_2$ could be either from $HN(NO_2)_2$, reaction (11), or from its tautomer, $O_2N–NN(O)OH$, reaction (5). The two are related through the equilibrium,

Table II. Calculated (DF/B3P86/6-31+G**) heats of reaction at 298 K [a]

Reaction	ΔH (298 K), kcal/mol	
	calc.	exp.[b]
(1) $NH_4N(NO_2)_2$ (solid) $\longrightarrow$ NH_4^+ + $N(NO_2)_2^-$	(144)[c]	
(2) $NH_4N(NO_2)_2$ (solid) $\longrightarrow$ NH_3 + $HN(NO_2)_2$	(44)[c]	
(3) NH_4^+ + $N(NO_2)_2^-$ $\longrightarrow$ NH_3 + $HN(NO_2)_2$	−99.9	
(4) $HN(NO_2)_2$ $\longrightarrow$ $O_2N-NN(O)OH$	1.8	
(5) $O_2N-NN(O)OH$ $\longrightarrow$ $NN(O)OH$ + NO_2	36.3	
(6) $O_2N-NN(O)OH$ $\longrightarrow$ N_2O + $HONO_2$	−33.8	
(7) $NN(O)OH$ $\longrightarrow$ N_2O + OH	−19.5	
(8) $HONO_2$ $\longrightarrow$ NO_2 + OH	50.6	49
(9) NH_3 + NO_2 $\longrightarrow$ NH_2 + $HONO$	31.3	29
(10) $HONO$ $\longrightarrow$ NO + OH	52.8	50
(11) $HN(NO_2)_2$ $\longrightarrow$ $HNNO_2$ + NO_2	40.7	
(12) $HNNO_2$ $\longrightarrow$ NH + NO_2	36.8	
(13) $HNNO_2$ $\longrightarrow$ N_2O + OH	−22.2	
(14) $HNNO_2$ + NO $\longrightarrow$ NO_2 + $HNNO$		
(15) $HNNO_2$ + OH $\longrightarrow$ $2NO$ + H_2O		
(16) $HNNO$ + OH $\longrightarrow$ N_2O + H_2O		
(17) H + $HONO$ $\longrightarrow$ HNO + OH	2.2	2
(18) H + $HONO$ $\longrightarrow$ H_2 + NO_2	−30.2	−25
(19) OH + NH_3 $\longrightarrow$ NH_2 + H_2O	−10.1	−11
(20) $2NO_2$ + H_2O $\longrightarrow$ $HONO$ + $HONO_2$	−9.2	−9
(21) NH + NO_2 $\longrightarrow$ HNO + NO	−37.6	
(22) NH_3 $\longrightarrow$ NH_2 + H	109.0	108.2

(23) $H + N_2O \longrightarrow N_2 + OH$ −56.1   −62.4

(24) $NH_2 + NO_2 \longrightarrow H_2N-NO_2$ −54.3

(25) $H_2N-NO_2 \longrightarrow H_2O + N_2O$ −27.9

(26) $NH_2 + NO \longrightarrow N_2 + H_2O$ −113.5   −124.7

(27) $2NO_2 \longrightarrow N_2O_4$ −14.3   −14

(28) $N_2O_4 \longrightarrow NO^+ + NO_3^-$ 175.8

(29) $N(NO_2)_2^- \longrightarrow NNO_2^- + NO_2$ 49.7

(30) $NNO_2^- \longrightarrow NO^- + NO$ 68.2

(31) $NH_4^+ + NNO_2^- \longrightarrow HNNO_2 + NH_3$ −108.9

(32) $NH_4^+ + NNO_2^- \longrightarrow N_2O + NH_3 + OH$ −131.1

(33) $NH_4^+ + NNO_2^- \longrightarrow N_2O + NH_2 + H_2O$ −141.2

(34) $NH_4^+ + HONO \longrightarrow N_2 + 2H_2O + H^+$ 134.2

---

[a]All reactants and products appearing in this table are in the gaseous phase except
$NH_4N(NO_2)_2$, which is in the solid phase.
[b]Reference 36.
[c]Reference 47.

with $HN(NO_2)_2$ being slightly more stable, reaction (4). The $N-NO_2$ dissociation
energies in $HN(NO_2)_2$ and $O_2N-NN(O)OH$ are 40.7 and 36.3 kcal/mol, reactions (11)
and (5), respectively. (These are significantly less than the analogous dissociation
energy for $N(NO_2)_2^-$.) Thus it may be that sublimation to $NH_3 + HN(NO_2)_2$ is the
initial step in the decomposition of ammonium dinitramide. It is relevant to note that
an analogous process is believed to begin the decomposition of ammonium
perchlorate (48-51):

$$NH_4ClO_4 \text{ (solid)} \longrightarrow NH_3 + HClO_4$$

Heats of Formation

The heat of formation is one of the key factors to consider in evaluating a new
energetic compound, whether existing or proposed. It permits the determination of
the energy released upon decomposition or combustion, and enters into the
calculation of key explosive and propellant properties, such as detonation velocity
and pressure (50,52) and specific impulse (53).

We have developed a procedure for computing heats of formation for both gas phase and solid phase compounds (54,55). The initial step is the calculation of ΔE (0 K) for the gas phase formation of the compound from its elements, using optimized geometries. Since the most stable elemental form of carbon is graphite, we use its experimental heat of sublimation to account for its conversion to the gas phase. ΔE(0 K) is computed at the density functional BP86/6-31G** level, and then converted to ΔH(298 K) in the manner described earlier. Finally, we add correction terms for the carbons, nitrogens and oxygens (54); these were determined by requiring that our ΔH(298 K) values optimally reproduce the experimental standard heats of formation at 298 K for a data base of 54 gas phase compounds of various types, primarily organic.

The vibrational contribution to ΔH(298 K) can of course be determined from the normal mode frequencies; however the calculation of these for large molecules can be quite demanding of computer resources. We have found that the vibrational energy can also be obtained very easily and with good accuracy from the molecular stoichiometry (56). Accordingly our heat of formation procedure no longer involves computing the frequencies, unless it is desired to confirm that the structure does correspond to a local minimum in the energy (indicated by the absence of imaginary frequencies (30)).

The procedure that has been described produces gas phase heats of formation. These can be used to determine the liquid and solid phase values, which are often of greater practical importance, if the heats of vaporization and sublimation are known:

$$\Delta H_f(298\ K,\ liquid) \ = \ \Delta H_f(298\ K,\ gas) \ - \ \Delta H_{vap}(298\ K) \qquad [1]$$

$$\Delta H_f(298\ K,\ solid) \ = \ \Delta H_f(298\ K,\ gas) \ - \ \Delta H_{sub}(298\ K) \qquad [2]$$

We have shown that both $\Delta H_{vap}(298\ K)$ and $\Delta H_{sub}(298\ K)$ can be expressed quantitatively in terms of electrostatic potentials on molecular surfaces (57,58); the approach is one that we have used successfully to develop relationships that permit the prediction of a variety of liquid, solid and solution properties (57,59,60). We can accordingly evaluate $\Delta H_{vap}(298\ K)$ and $\Delta H_{sub}(298\ K)$ from the calculated surface potentials, and then convert our $\Delta H_f(298\ K,\ gas)$ to $\Delta H_f(298\ K,\ liquid)$ and/or $\Delta H_f(298\ K,\ solid)$ by means of equations [1] and [2].

The accuracies of our computed heats of formation, both gas and solid phase, have been tested by comparison to experimental data for more than a dozen compounds; the average error was less than 4 kcal/mol. We have now used this procedure to calculate heats of formation for nearly 60 proposed or newly-synthesized energetic compounds; many of these results are presented and discussed elsewhere (55,61-64).

As discussed earlier in this chapter, our very recent work supports the use of the DF/B3P86/6-31+G** method for the calculation of reaction energetics. Accordingly we are in the process of recomputing our heat of formation data base with this density functional procedure in order to obtain a new set of correction terms. The data base is also being expanded, to encompass more types of compounds (65). Future heat of formation calculations will use the B3P86/6-31+G** combination of functionals and basis set.

Summary

The use of density functional methods for computing reaction energetics, i.e. ΔH(298 K), has been examined, and evidence has been presented which indicates the effectiveness, for this purpose, of the DF/B3P86/6-31+G** procedure. The decomposition of the energetic compound ammonium dinitramide was considered in some detail, and calculated ΔH(298 K) values are given for 32 possible intermediate reactions. It was suggested that the initial step in the decomposition of ammonium dinitramide might be its sublimation to $NH_3 + HN(NO_2)_2$, which could be followed by loss of NO_2 from either $HN(NO_2)_2$ or its tautomer $O_2N-NN(O)OH$. Finally, a density functional procedure for determining gas phase heats of formation was reviewed, as well as techniques for converting these to liquid and solid phase values. Overall, the results that have been discussed testify to the capabilities of density functional computational methods to provide meaningful reaction energies and enthalpies for processes of practical significance.

Acknowledgments

We greatly appreciate the assistance of Dr. Jane S. Murray, and the financial support provided by the Ballistic Missile Defense Organization and the Office of Naval Research through contract N00014-95-1-1339, Program Officers Dr. Leonard H. Caveny (BMDO) and Dr. Richard S. Miller (ONR).

Literature Cited

1. Politzer, P.; Murray, J. S.; Grice, M. E. In *Chemical Hardness*; Sen, K. D., Ed.; Structure and Bonding No. 80; Springer-Verlag: Berlin, 1993, p. 101.
2. DeFrees, D. J.; Levi, B. A.; Pollack, S. K.; Hehre, W. J.; Binkley, J. S.; Pople, J. A. *J. Am. Chem. Soc.* **1979**, *101*, 4085.
3. Clabo, D. A.; Schaefer III, H. F. *Int. J. Quant. Chem.* **1987**, *31*, 429.
4. Coffin, J. M.; Pulay, P. *J. Phys. Chem.* **1991**, *95*, 118.
5. Seminario, J. M.; Concha, M. C.; Politzer, P. *J. Comp. Chem.* **1992**, *13*, 177.
6. Phillips, D. H.; Quelch, G. E. *J. Phys. Chem.* **1996**, *100*, 11270.
7. Parr, R. G.; Yang, W. *Density-Functional Theory of Atoms and Molecules*; Oxford University Press: New York, 1989.
8. *Density Functional Methods in Chemistry*; Labanowski, J. K.; Andzelm, J. W., Eds. ; Springer-Verlag: New York, 1991.
9. *Modern Density Functional Theory*; Seminario, J. M.; Politzer, P., Eds.; Elsevier: Amsterdam, 1995.
10. *Recent Developments and Applications of Modern Density Functional Theory*; Seminario, J. M., Ed.; Elsevier: Amsterdam, 1996.
11. Parr, R. G.; Yang, W. *Ann. Rev. Phys. Chem.* **1995**, *46*, 701.
12. Gunnarsson, O.; Jones, R. O. *Phys. Rev. B* **1985**, *31*, 7588.
13. Becke, A. D. *J. Chem. Phys.* **1986**, *84*, 4524.
14. Grodzicki, M.; Seminario, J. M.; Politzer, P. *J. Chem. Phys.* **1991**, *94*, 1668.
15. Wimmer, E. In *Density Functional Methods in Chemistry*; Labanoswki, J. K.; Andzelm, J. W., Eds.; Springer-Verlag: New York, 1991, p. 7.
16. Ziegler, T. *Chem. Rev.* **1991**, *91*, 651.
17. Seminario, J. M.; Grodzicki, M.; Politzer, P. In *Density Functional Methods in Chemistry*; Labanoswki, J. K.; Andzelm, J. W., Eds.; Springer-Verlag: New York, 1991, p. 419.
18. Habibollahzadeh, D.; Grodzicki, M.; Seminario, J. M.; Politzer, P. *J. Phys. Chem.* **1991**, *95*, 7699.
19. Habibollahzadeh, D.; Murray, J. S.; Grice, M. E.; Politzer, P. *Int. J. Quant. Chem.* **1993**, *45*, 15.
20. Jursic, B. S. *Chem. Phys. Lett.* **1996**, *256*, 213.
21. Politzer, P.; Seminario, J. M. *Trends in Phys.Chem.* **1992**, *3*, 175.
22. Seminario, J. M.; Concha, M. C.; Politzer, P. *Int. J. Quant. Chem., Quant. Chem. Symp.* **1991**, *25*, 249.

23. Redington, P. K.; Andzelm, J. W. In *Density Functional Methods in Chemistry*; Labanoswki, J. K.; Andzelm, J. W., Eds.; Springer-Verlag: New York, 1991, p. 411.
24. Becke, A. D. *Phys. Rev. A* **1988**, *38*, 3098.
25. Becke, A. D. *J. Chem. Phys.* **1993**, *98*, 5648.
26. Perdew, J. P. *Phys. Rev. B* **1986**, *33*, 8822.
27. Lee, C.; Yang, W.; Parr, R. G. *Phys. Rev. B* **1988**, *37*, 785.
28. Perdew, J. P.; Wang, Y. *Phys. Rev. B* **1992**, *45*, 13244.
29. Frisch, M. J.; Trucks, G. W.; Schlegel, H. B.; Gill, P. M. W.; Johnson, B. G.; Robb, M. A.; Cheeseman, J. R.; Keith, T. A.; Petersson, G. A.; Montgomery, J. A.; Raghavachari, K.; Al-Laham, M. A.; Zakrezewski, V. G.; Ortiz, J. V.; Foresman, J. B.; Cioslowski, J.; Stefanov, B. B.; Nanayakkara, A.; Challacombe, M.; Peng, C. Y.; Ayala, P. Y.; Chen, W.; Wong, M. W.; Andres, J. L.; Replogle, E. S.; Gomperts, R.; Martin, R. L.; Fox, D. J.; Binkley, J. S.; Defrees, D. J.; Baker, J.; Stewart, J. P.; Head-Gordon, M.; Gonzalez, C.; Pople, J. A. *Gaussian 94 (Revision B.3)*; Gaussian, Inc.: Pittsburgh, PA, 1995.
30. Hehre, W. J.; Radom, L.; Schleyer, P. v. R.; Pople, J. A. *Ab Initio Molecular Orbital Theory*; Wiley-Interscience: New York, 1986.
31. Benson, S. W. *J. Chem. Educ.* **1965**, *42*, 502.
32. McMillen, D. F.; Golden, D. M. *Ann. Rev. Phys. Chem.* **1982**, *33*, 493.
33. Basch, H. *Inorg. Chim. Acta* **1996**, *252*, 265.
34. Wiener, J. J. M.; Politzer, P. *J. Mol. Struct. (Theochem)*, in press.
35. Politzer, P.; Wiener, J. J. M.; Seminario, J. S. In *Recent Developments and Applications of Modern Density Functional Theory*; Seminario, J. M., Ed.; Elsevier: Amsterdam, 1996; Vol. 4, Chap. 22.
36. Lias, S. G.; Bartmess, J. E.; Liebman, J. F.; Holmes, J. L.; Levin, R. D.; Mallard, W. G. *J. Phys. Chem. Ref. Data* **1988**, *17*, suppl. 1.
37. Schmitt, R. J.; Krempp, M.; Bierbaum, V. M. *Int. J. Mass. Spectr. Ion Proc.* **1992**, *117*, 621.
38. Rossi, M. J.; Bottaro, J. C.; McMillen, D. F. *Int. J. Chem. Kinet.* **1993**, *25*, 549.
39. Doyle, R. J., Jr. *Org. Mass. Spectr.* **1993**, *28*, 83.
40. Michels, H. H.; Montgomery, J. A., Jr. *J. Phys. Chem.* **1993**, *97*, 6602.
41. Brill, T. B.; Brush, P. J.; Patil, D. G. *Combust. Flame* **1993**, *92*, 178.
42. Politzer, P.; Seminario, J. M.; Concha, M. C.; Redfern, P. C. *J. Mol. Struct. (Theochem)* **1993**, *287*, 235.
43. Politzer, P.; Seminario, J. M. *Chem. Phys. Lett.* **1993**, *216*, 348.
44. Mebel, A. M.; Lin, M. C.; Morokuma, K.; Melius, C. F. *J. Phys. Chem.* **1995**, *99*, 6842.
45. Russell, T. P.; Piermarini, G. J.; Block, S.; Miller, P. J. *J. Phys. Chem.* **1996**, *100*, 3248.
46. Wight, C.A.; Vyazovkin, S.; *Proc JANNAF Combustion Subcommittee Mtg.*, Monterey, CA, Nov. 1996.
47. Politzer, P.; Seminario, J. M.; Concha, M. C., to be published.
48. Jacobs, P. W. M.; Whitehead, H. M. *Chem. Rev.* **1969**, *69*, 551.
49. Hackman, E. E., III; Hesser, H. H.; Beachell, H. C. *J. Phys. Chem.* **1972**, *76*, 3545.
50. Urbanski, T. *Chemistry and Technology of Explosives*; Pergamon Press: New York, 1984; Vol. 4.
51. Brill, T. B.; Brush, P. J.; Patil, D. G. *Combust. Flame* **1993**, *94*, 70.
52. Kamlet, M. J.; Jacobs, S. J. *J. Chem. Phys.* **1968**, *48*, 23.
53. Politzer, P.; Murray, J. S.; Grice, M. E.; Sjoberg, P. In *Chemistry of Energetic Materials*; Olah, G. A.; Squire, D. R., Eds.; Academic Press: New York, 1991, Chap. 4.

54.   Habibollahzadeh, D.; Grice, M. E.; Concha, M. C.; Murray, J. S.; Politzer, P. *J. Comp. Chem.* **1995**, *16*, 654.

55.   Politzer, P.; Murray, J. S.; Grice, M. E. In *Decomposition, Combustion, and Detonation Chemistry of Energetic Materials*; MRS Symp. Ser. 418; Brill, T. B.; Russell, T. P.; Tao, W. C.; Wardle, R. B., Eds.; Materials Research Society: Pittsburgh, PA, 1996.

56.   Grice, M. E.; Politzer, P. *Chem. Phys. Lett.* **1995**, *244*, 295.

57.   Murray, J. S.; Politzer, P. In *Quantitative Treatments of Solute/Solvent Interactions*; Murray, J. S. ; Politzer, P., Eds.; Elsevier: Amsterdam, 1994, Chap. 8.

58.   Politzer, P.; Murray. J.S.; Grice, M. E.; DeSalvo, M., Miller, E. *Mol. Phys.*, submitted.

59.   Murray, J. S.; Brinck, T.; Lane, P.; Paulsen, K.; Politzer, P. *J. Mol. Struct. (Theochem)* **1994**, *307*, 55.

60.   Politzer, P.; Murray, J. S.; Brinck, T.; Lane, P. In *Immunoanalysis of Agrochemicals; Emerging Technologies*; ACS Symposium Series 586; Nelson, J. O.; Karu, A. E.; Wong, R. B., Eds.; ACS: Washington, DC, 1995, Chap. 8.

61.   Grice, M. E.; Habibollahzadeh, D.; Politzer, P. *J. Chem. Phys.* **1994**, *100*, 4706.

62.   Politzer, P.; Lane, P.; Grice, M. E.; Concha, M. C.; Redfern, P. C. *J. Mol. Struct. (Theochem)* **1995**, *338*, 249.

63.   Politzer, P.; Lane, P.; Sjoberg, P.; Grice, M. E.; Shechter, H. *Struct. Chem.* **1995**, *6*, 217.

64.   Grice, M. E.; Politzer, P. *J. Mol. Struct.* **1995**, *358*, 83.

65.   Politzer, P; Grice, M. E., to be published.

Chapter 20

Periodic Trends in Bond Energies: A Density Functional Study

Tom Ziegler

Department of Chemistry, University of Calgary, Calgary,
Alberta T2N 1N4, Canada

Density functional theory (DFT) makes it possible to obtain reasonable
(±25 kJ/mol) estimates of bond energies for compounds involving all
elements in the periodic table. This account assesses the accuracy of
thermochemical data obtained by DFT and rationalises some of the
periodic trends obtained from systematic studies involving
homologous series of compounds.

It is now possible to calculate bond energies with reasonable accuracy by the aid
of quantum mechanical methods. This account reviews some of the results obtained
by methods based on density functional theory (DFT) (1). The data available for A-B
bond energies exhibit interesting periodic trends as A (or B) changes position in the
periodic table. The origin of these periodic trends will be discussed in terms of
fundamental concepts based on the Pauli exclusion principle and Einstein's special
theory of relativity. The exposition will progress from single- to multiple- bonded
systems with special emphasis on transition metal compounds. This account is not
exhaustive and references should be made to previous DFT reviews (2) as well as the
excellent study by Frenking et al. (2c) based on ab initio methods and other DFT
investigations in this volume .

The Single Bond

Main Group Compounds. Table I displays calculated and experimental $X-CH_3$
bond energies (3a) for the halogen series X= F,Cl,Br, I. We note that the DFT based
local density approximation (1b) (LDA) overestimates bond energies, a fact that is
well established by now (1a). The gradient corrected BP86 scheme (4) affords on the
other hand estimates that are within 5 kcal/mol of experiment. Both DFT based
methods include to some degree electron correlation. Among the *ab initio* methods
(5), Hartree Fock (HF) - without correlation- and the second order Møller-Plesset
perturbation method (MP2) -with some correlation - underestimate the bond energies.
The agreement with experiment is not improved by going to higher order of
perturbation in the MP4 scheme. For this series the gradient corrected DFT theory
(BP86) affords the best fit to experiment. In most cases BP86 supplies better bond
energies than the HF,MP2 and MP4 methods.

Table I. Calculated[a] and Experimental[d] X-CH$_3$ Bond Enthalpies ΔH_0^o (kJ/mol), (X=F, Cl, Br, I).

Halogen (X)	LDA	BP86	HF	MP2	MP4	Exp	Overlap[b]	Energy Gap[c]
F	578.5	469.0	250.5	437.3	420.5	465.2	0.26	7.5
Cl	424.3	338.1	195.6	327.3	316.8	342.3	0.34	3.8
Br	357.4	278.4	142.1	270.9	261.2	286.7	0.35	3.0
I	297.6	222.3	89.0	212.8	205.2	232.8	0.36	2.1

[a]Ref. 3a. [b]Overlap between np$_\sigma$ and M$_\sigma$, see **1**. [c]Energy gap (eV) between np$_\sigma$ and M$_\sigma$, see **1**. [d]Ref. 3b

It follows from Table I that the X-CH$_3$ bond decreases in strength from fluorine towards iodine. This trend can be understood by observing that the X-CH$_3$ linkage is established by the interaction between the singly occupied np$_\sigma$ and M$_\sigma$ orbitals, **1**. In this interaction density is transferred from M$_\sigma$ of the CH$_3$ fragment - with the higher energy - to np$_\sigma$ of the halogen atom with the lower energy, **1**. The transfer is most favourable for fluorine with the np$_\sigma$ orbital of lowest energy, Table I. In a more covalent bond the <np$_\sigma$|M$_\sigma$> overlap might have been trend setting. However, <np$_\sigma$|M$_\sigma$> is seen to run counter to what one would expect if it was the determining factor for the X-CH$_3$ bond strength, Table I.

1   **2**

The increase in the energy of the halogen based np$_\sigma$ orbital towards higher n is a consequence of the ***intra-atomic Pauli core repulsion***. (6). As more and more p-type core orbitals are added to the halogen (2p,3p, and 4p in the case of iodine), the np$_\sigma$ valence orbital develops a nodal structure in the core region and expands in the valence region - in both cases in order to obey the Pauli exclusion principle which requires core and valence electrons of like spin not to be found at the same position. The expansion in the valence region, and the development of a nodal structure in the core region, raises the energy by respectively reducing the nuclear attraction and increasing the kinetic energy (6). The expansion of np$_\sigma$ due to the intra-atomic Pauli core repulsion is also responsible for the increase in the <np$_\sigma$|M$_\sigma$> overlap towards heavier halogens.

There is one more factor of importance for the trend in the X-CH$_3$ bond energies. As more and more core orbitals are added to the halogen, these occupied orbitals will interact repulsively with M$_\sigma$, **2**, and destabilize the X-CH$_3$ bond progressively from fluorine towards iodine. The destabilization is again a consequence of the Pauli exclusion principle. Essentially (2a), M$_\sigma$ will develop a nodal structure in the halogen core region similar to that of np$_\sigma$ in order to exclude core and valence electrons of like spin from the same region in X-CH$_3$. The nodal structure will in turn increase the kinetic energy. The destabilizing interaction in **2** is termed ***inter-atomic Pauli core repulsion***. (6). The importance of intra- and inter- atomic Pauli repulsion for periodic trends in thermochemistry has been discussed extensively by Kutzelnigg (6).

Table II. Cl-MH$_3$ Bond Dissociation Enthalpies[a] ΔH_0^o (kJ/mol), (M=C,Si,Ge,Sn).

Bond	BP86	HF	MP2	MP4	G2	Exp.[d]	Overlap[b]	Energy Gap[c]
C-Cl	338.1	195.6	327.3	316.8	346.9	348.6	0.34	3.8
Si-Cl	441.4			437.7	455.6	472.3	0.34	4.6
Ge-Cl	402.1						0.33	4.7
Sn-Cl	391.2						0.33	4.9

[a]Ref. 7a. [b]Overlap between np$_\sigma$ and M$_\sigma$, see **1**. [c]Energy gap (eV) between np$_\sigma$ and M$_\sigma$, see **1**. [d]Ref.7b

Table II deals with the Cl-MH_3 bond where we now fix the halogen part at chlorine and vary the group-14 component from carbon to tin. The limited experimental data set indicates again that BP86 provides a better fit to experiment than HF,MP2 and MP4. On the other hand, the highly accurate G2 *ab initio* scheme (8) fares somwhat better than BP86. It is interesting to note that carbon forms a weaker bond to chlorine than the higher group-14 homologues. This is in line with the fact that the energy gap between np_σ and M_σ is smallest for carbon. For the other elements, both the energy gap and the $<np_\sigma|M_\sigma>$ overlap are quite similar. The decrease in the Cl-MH_3 bond strength from M=Si to M=Sn is instead set by the inter-atomic Pauli core repulsion (7a). That is, the $3p_\sigma$ valence orbital of chlorine interacts repulsively with the M-core orbitals of the MH_3 fragment.

The final main group example deals with the homopolar H_3M-MH_3 (M=C,Si, Ge,Sn,Pb) molecules, Table III, where two MH_3 fragments are joined together. In this case both BP86 theory (9) and experiment exhibit a decrease in the bond strength from carbon to lead. The trend setting factor here is not the $<M_\sigma|M_\sigma>$ overlap (9) but rather the expected increase in the inter-atomic Pauli core repulsion as more and more core orbitals are added to the group-14 element. Thus, in this case M_σ of one MH_3 fragment interacts repulsively with core orbitals on M of the opposite fragment.

Table III. H_3M-MH_3 Bond Dissociation Enthalpies[a] ΔH_0^o (kJ/mol), (M=C,Si,Ge,Sn)

Bond Method	C-C	Si-Si	Ge-Ge	Sn-Sn	Pb-Pb
BP86	363	293	257	220	185
Exp[a].	367	321	276		

[a]Ref. 9a.

Coinage Metal Compounds.   Dimers and hydrides of the coinage metals have been the subject of numerous studies (10) as they are among the simplest metal-metal and metal-ligand bonded systems. Table IV gives calculated M-H and M-M (M=Cu,Ag,Au) bond energies based on the non-relativistic BP86 scheme (BP86-NR) presented in the previous section as well as a quasi-relativistic extension (BP86-QR) due to Snijders et al. (11b,c). We note that the M-M and M-H bond energies decrease from copper to gold in the non-relativistic case, in disagreement with experiment. The decrease can largely be attributed to inter-atomic Pauli core repulsion between $1s_H$ (or ns_M) and the growing number of core orbitals on the opposite metal centre(10b,c).

Table IV. Electronic M-H and M_2 Bond Dissociation Enthalpies[a] (kJ/mol)

Compound Method	CuH	AgH	AuH	Cu_2	Ag_2	Au_2
BP86-NR	279.2	222.0	218.6	199.0	145.0	137.5
BP86-QR	287.1	247.8	322.7	210.7	163.0	222.0
Exp[b].	276 ±10	230 ±10	309±15	188±10	155±10	217±10

[a]Ref. 11a. [b] Ref. 11d

Adding relativistic effects (BP86-QR) substantially strengthens the Au-H and Au-Au bonds to the extent where they become the strongest in each of the homologous series. Instead silver is seen to form the weakest bond, in agreement with experiment, Table IV. The origin of the relativistic bond stabilisation - and bond contraction- has been discussed previously (10). Essentially, valence electrons in s-orbitals of heavy elements can obtain high instantaneous velocities near the nuclei. The high velocities will increase the mass of the electron. The electron mass increase will in turn diminish the inter-atomic core Pauli repulsion by reducing the kinetic energy (10b,c). The effect is most pronounced for gold since its larger nuclear charge allows for the highest instantaneous velocities close to the nucleus.

Transition Metal Compounds.   The coinage metal compounds discussed above
are unique in that ns is the dominant valence bonding orbital on the metal. This
results in spectacular relativistic effects since valence ns electrons can come close to
the nucleus and thus acquire high instantaneous (relativistic) velocities, at least in the
case of gold. Most other 'real' transition metals (12) primarily make use of d-orbitals
in bond formation, **3**. Figure 1 displays calculated (13a) MCl_3-L bond energies,
where M represents the group-4 triad M=Ti,Zr, and Hf of early transition metals and
L runs trough the series L= PH_2,SiH_3,H,CH_3,SH,NH_2,CN,OCH_3,OH of 'one'electron
ligands. The Cl_3M-L linkage is established by the interaction between the singly
occupied L_σ and M_σ orbitals, **3**, in a way similar to **1**. In this interaction, density is
transferred from M_σ of higher energy to the L_σ ligand orbital of lower energy. For a
given metal the Cl_3M-L bond energy increases with the energy gap between M_σ and
L_σ. Thus the more electronegative ligands NH_2,OCH_3 and OH form the stronger
bonds. The Cl_3M-L bonds are stabilized to some degree by a secondary π−interaction,
4, involving occupied π−type ligand lonepairs and empty M_π metal orbitals. The latter
are unoccupied for the early electron poor transition metals.

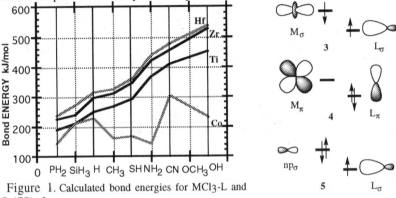

Figure 1. Calculated bond energies for MCl_3-L and
$Co(CO)_4$-L

The Cl_3M-L bond energies are seen to increase in going from the 3d-member
titanium to the 5d-member hafnium, with the largest jump between titanium and
zirconium. The most important factor here is the inter-atomic Pauli core repulsion
involving L_σ and the core-like ns,np metal orbitals with n=3,4,5 for Ti,Zr, and Hf
respectively, **5**. In the case of titanium, the 3s and 3p orbitals are of the same radial
extent as the 3d orbital M_σ. Hence a sizable bonding interaction, **3**, for titanium will
also result in strong inter-atomic Pauli core repulsions, **5**. For zirconium and hafnium,
the d-type M_σ orbital is more diffuse than the ns and np core-like orbitals, resulting in
less destabilization, **5**. The d-orbital of titanium is more compact relative to ns and np
than the corresponding nd orbitals of zirconium and hafnium, as it is free from intra-
atomic Pauli core repulsion due to d-orbitals with a lower n quantum number. Also
shown in Figure 1 are the corresponding L-$Co(CO)_4$ bond energies for the late
transition metal cobalt. We note for hydrogen that the Co-H energy is somewhat
smaller than that of the group-4 metals. This is so since the M_σ orbital is of lower
energy at the end of the transition series where the effective charge is largest. Thus
the electron transfer in **3** is less favorable. For many of the other ligands we note that
the Co-L bond energies are substantially lower than what one would expect from
considering the difference in strength between the Co-H and the Ti-H bonds. In these
cases the secondary π−interaction, **4**, has a large destabilizing influence on the Co-L
bond as the M_π d-orbital now is occupied and involved in a four-electron two orbital
destabilizing interaction with L_π. The only ligand somewhat out-of-line is CN since it
has an additional L_π^* orbital capable of accepting density from M_π .

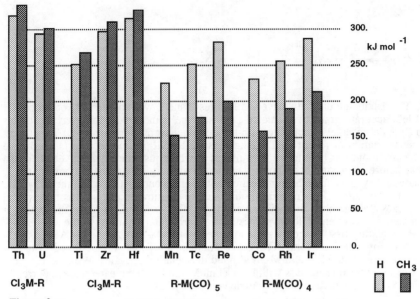

Figure 2. A comparison of M-H and M-CH3 bond energies for different metals

It stands to reason that metal-hydrogen and metal-alkyl bond energies are among the most important thermochemical parameters in organometallic chemistry. Figure 2 displays trends in calculated (13b) M-H and M-CH$_3$ energies for early as well as late transition metals. We note first of all that both M-H and M-CH$_3$ bond energies increase down a triad of metals on account of the decrease in the inter-atomic Pauli core repulsions, **5**, already discussed in connection with Figure 1. Also M-H bonds are seen to be stronger than M-CH$_3$ bonds for electron rich late to middle transition metals as the σ_{CH3} bonding orbitals interact with occupied d_π orbitals on the metal to destabilize the M-CH$_3$ bond compared to the M-H linkage where such interactions are impossible. On the other hand,. for early transition metals M-H and M-CH$_3$ bonds are of similar strength since the d_π orbitals now are empty and the interaction between σ_{CH3} and d_π stabilising (13b). In fact, the stabilizing interaction makes in some cases the M-CH$_3$ bond stronger, Figure 2.

$$M+ \underset{H}{\overset{H}{|}} \longrightarrow M \overset{H}{\underset{H}{\diagup\diagdown}}$$
6a

$$M+ \underset{H}{\overset{CH_3}{|}} \overset{?}{\longrightarrow} M \overset{CH_3}{\underset{H}{\diagup\diagdown}}$$
6b

$$M+ \underset{CH_3}{\overset{CH_3}{|}} \overset{}{\not\longrightarrow} M \overset{CH_3}{\underset{CH_3}{\diagup\diagdown}}$$
6c

The fact that M-H bonds are stronger than M-alkyl bonds for late to middle transition metals, Figure 2, has profound implications for elementary reaction steps in organometallic chemistry (13b,14). Thus, H$_2$ adds readily to a metal centre, **6a**, as the two M-H bonds gained are stronger than a single H-H bond (400 kJ/mol). On the other hand a C-C alkyl bond never adds to a metal centre, **6c**, as the C-C bond broken (400 kJ/mol) is stronger than the two M-C bonds formed. The addition of a C-H bond , **6b**, is an intermediate case observed in a few cases.

The migration of an alkyl group from a metal centre to the carbon of a cis-carbonyl is quite a facile process, **7a**, as the M-alkyl bond is relatively weak. On the other hand, the corresponding migration of a hydrogen, **7b**, is hardly ever observed due to the strength of the M-H bond.

7a   7b

The calculated data displayed in Figures 1 and 2 were based (13) on an earlier and less accurate gradient corrected DFT scheme than BP86. However, the trends depicted are not likely to change with the introduction of more recent DFT methods. Direct validation of recent DFT methods as well as *ab initio* schemes (12a-d) have been carried out on "bare metal" complexes in which just one ligand is attached to a metal centre in a single σ-bond. Dissociation energies for these coordinatively unsaturated systems have been measured with high accuracy (12f-g) in the gas phase for many systems. For 'normal' coordinatively saturated ML_n transition metal complexes average bond dissociation energies, $ML_n \rightarrow M+nL - n\Delta H_{av}$, are available for a number of systems. However, accurate experimental estimates of the chemically more interesting first ligand dissociation energy, $ML_n \rightarrow ML_{n-1} + L - \Delta H_1$, is only available for a few σ-bonds (12f). We compare in Table V calculated (14) and experimental first M-H and M-alkyl bond dissociation energies. We note that the agreement with experiment is within 25 kJ/mol, which is the mean error we currently attribute to the calculation of M-L single bonds by BP86.

Table V. Calculated[a] and Experimental[c] M-H and M-alkyl Bond Energies[b] for Transition Metal Systems.

	D(M-R)	
Molecule	BP86	Exp.
$(CO)_5Mn$-H	288.13	284.2±4
$(CO)_5Mn$-CH_3	207.93	192±11
$(CO)_5Mn$-CF_3	223.96	203±6
$(CO)_5Mn$-$C(O)CH_3$	188.75	185±8
$(CO)_4Co$-H	283.11	280.1±4
$(CO)_4Co$-CH_3	197.63	

[a]Ref 14. [b] Energies in kJ/mol. [c]Ref. 12f

Table VI. Calculated[a] and Experimental[b] M-H and M-CH_3 Bond Energies[c] for f-Block Elements

Bond	D(M-X)-NR	D(M-X)-REL	Exp.
Cl_3Th-H	125.8	317.7	~ 335
Cl_3Th-CH_3	149.6	333.6	~ 335
Cl_3U-H	43.9	293.0	319.4
Cl_3U-CH_3	70.2	301.8	302.6
Cl_3Hf-H	310.6	318.1	
Cl_3Hf-CH_3	323.5	331.9	

[a] Ref. 11c,e. D(M-X)-NR are non relativistic bond energies whereas D(M-X)-REL includes relativistic effects according to the method of Ref. 11c. [b] Ref. 12e. [c]KJ/mol

Compounds of f-block Elements. The final type of σ-bond discussed here involves hydrogen or methyl bound to the f-block elements thorium and uranium (11c,e). It follows from Table VI that relativistic effects considerably strengthen the bonds to thorium and uranium. In the non-relativistic case the 5f-orbitals of lower n-quantum number are of lower energy than the 6d shell, and primarily involved in the M-H and M-CH_3 (M=Th,U) bonds. Since the 5f orbitals are rather contracted the 5f overlaps with the $1s_H$ or σ_{CH_3} orbitals are small and the M-H and M-CH_3 bonds (M =Th,U) weak. The inclusion of relativity will again reduce the kinetic energy of the s-type orbitals (including 7s) and contract their radial extent. This will reduce the effective nuclear charge seen by 5f and 6d and increase their energy (15) This indirect relativistic destabilization is largest for 5f and 6d now becomes the predominant valence orbital. The 6d has better overlaps with $1s_H$ or σ_{CH_3}. Thus, the calculated M-H and M-CH_3 (M= Th,U) bonds become stronger in the relativistic case(11e) and quite similar to the experimental estimates. We see further by comparison that the analogous Hf-H and Hf-CH_3 bonds are only influenced little by relativity. Hafnium binds primarily through 5d with and without relativity. The M-CH_3 and M-H bonds are quite comparable in strength for the f-block elements and the 5d element hafnium, with the f-elements forming bonds that are ~ 25 kJ/mol stronger.

The Double Bond

We shall now turn to a discussion of double bonds. To this end we shall make reference to the extended transition state method (2a,16) (ETS) by which it is possible to break up the total bond energy ΔE_{AB} into stabilising contributions from $\sigma,\pi,\delta-$ bonds, **8b-d**, as well as repulsive 3- and 4-electron orbital destabilizing interactions of the type we have encountered in the inter-atomic Pauli core repulsion, **2,5**.

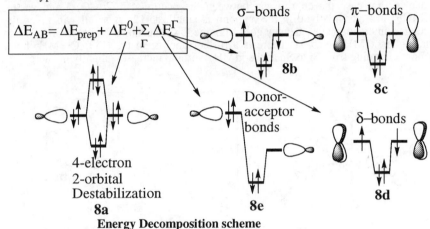

$$\Delta E_{AB}= \Delta E_{prep}+ \Delta E^0+\sum_{\Gamma} \Delta E^{\Gamma}$$

σ−bonds   π−bonds

8b

8c

4-electron
2-orbital
Destabilization
8a

Donor-
acceptor
bonds

δ−bonds

8e   **8d**

Energy Decomposition scheme

Double Bonds Between Main Group Elements. We shall begin our discussion of periodic trends in double bonds by considering calculated and experimental M=M bond energies in the homologous series of ethylene type group-14 compounds M_2H_4, **9**, (M=C,Si,Ge,Sn, and Pb), Table VII. We note first of all that BP86 is in as good agreement with experiment as the high level G1 and G2 *ab initio* methods, Table VII. In general, the M=M bond is seen to decrease in strength as we decend the group towards lead. It has previously been suggested that this trend is due to a weakening of the π-bond. In fact, our ETS analysis as well as other studies (16) have shown that the relative importance of the π-bond increases compared to that of the σ-component towards the heavier congeners. Instead, it is the inter-atomic Pauli core repulsions, **10**, that is responsible for the weak bond in the higher homologues. That is, the two group-14 elements can not come close enough together for the heavier group-14 elements to form strong σ-bonds without encountering an extensive inter-atomic Pauli core repulsions. As a result, a much longer equilibrium distance is adopted than what would give the optimal σ-interaction (16).

Table VII. Calculated and Experimental Bond Energies[a,b] in M_2H_4 (M=C,Si,Ge,Sn, and Pb)

Bond	BP86	ab initio	$\Delta E_\sigma/\Delta E_\pi$	Exp
C=C	739	735 (G1)	2.78	719
Si=Si	250	246 (G2)	2.36	265
Ge=Ge	180	154 (HF)	2.09	
Sn=Sn	121	119 (HF)	2.10	54
Pb=Pb	42		1.41	

[a]Energies in kJ/Mol. [b]Ref. 17

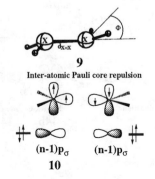

9

Inter-atomic Pauli core repulsion

$(n-1)p_\sigma$ $(n-1)p_\sigma$

10

Ligand to Metal Double Bonds. The $Cr(CO)_5$ fragment is isolobal with MH_2 (M=C,Si,Ge,Sn,Pb). It can as a result replace one of the MH_2 units in the M_2H_4 molecules, **9**, to form the carbene type series, **11**, $Cr(CO)_5MH_2$ (M=C,Si,Ge,Sn,Pb) . Table VIII displays the calculated $(CO)_5Cr=MH_2$ bond energies decomposed into contributions from the steric (**8a**) interaction, ΔE^o, and the σ- and π-bonding interactions, ΔE_σ and ΔE_π, respectively. It is clear that the π-bonding contribution, ΔE_π, is trend setting. It is largest for $(CO)_5Cr=CH_2$ with the strongest $Cr=MH_2$ bond since in that case the energy gap between the occupied $\pi(Cr(CO)_5)$ and vacant $\pi(MH_2)$ orbitals involved in the π-bond is smallest, **12**. A small gap gives the strongest interaction since density goes from the lower occupied orbital $\pi(Cr(CO)_5)$ to the higher $\pi(MH_2)$ orbital. There is a further decline in the $(CO)_5Cr=MH_2$ bond strength in going from Si to Sn as the $\pi(MH_2)$ orbital increases slightly in energy.

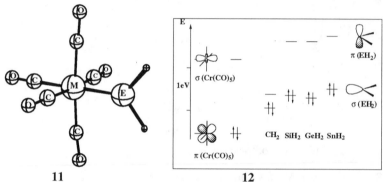

11	**12**

We have also studied the influence of the metal center on the carbene like double bond in the $(CO)_5M=CH_2$ series (M=Cr,Mo,W), Table IX.

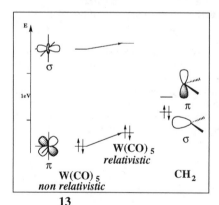

13

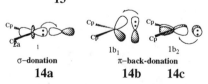

σ–donation	π–back-donation	
14a	**14b**	**14c**

Table VIII. Calculated Electronic Bond Enthalpies[a,b,c] for $(CO)_5Cr= MH_2$ (M=C,Si,Ge,Sn).

Bond	ΔE^o	ΔE_σ	ΔE_π	ΔE
$Cr=CH_2$	-188	267	202	281
$Cr=SiH_2$	-173	332	82	241
$Cr=GeH_2$	-153	268	72	186
$Cr=SnH_2$	-171	282	51	163

[a] $\Delta E=\Delta E^o+\Delta E_\sigma+\Delta E_\pi$. [b] kJ/mol. [c] Ref.18

Table IX. Calculated Electronic Bond Enthalpies [a,b,c,d] for $(CO)_5M= CH_2$ (M=Cr,Mo,W).

Bond	ΔE^o	ΔE_σ	ΔE_π	ΔE
$Cr=CH_2$	-188	267	202	281
$Mo=CH_2$	-186	236	204	254
$W=CH_2$ (NR)	-175	231	198	254
$W=CH_2$ (Rel)	-149	233	221	305

[a] $\Delta E=\Delta E^o+\Delta E_\sigma+\Delta E_\pi$. [b] kJ/mol. [c] Ref. 19. [d] For Mo and Cr relativistic effects are included. They are small.

We note that with relativistic effects included, molybdenum forms the weakest bond, whereas the M=CH$_2$ linkage is strongest for tungsten. Relativity adds 50 kJ/mol to the W=CH$_2$ bond by increasing the ΔE_π component due to the back-donation from $\pi(W(CO)_5)$ to $\pi(CH_2)$, **13**. This donation is enhanced as relativity destabilizes the $\pi(W(CO)_5)$ d-based tungsten orbital, **13**. The destabilization is due to the contraction of the s-type orbitals which will reduce the effective nuclear charge experienced by the 5d-levels. We shall shortly see that the indirect relativistic destabilization of the 5d-levels is very important in stabilizing bonds that involves back-donation from these levels to vacant ligand orbitals.

We have also investigated (20)the influence of both ligand and metal on the strength of the double bond in the series Cp$_2$M=E (E=O,S,Se,Te; M=Ti,Zr), Table X. Here the σ−bond is established as a donation, **14a**, of density from the occupied d$_\sigma$ orbital on Cp$_2$M to the empty p$_\sigma$ orbital on E. On the other hand, the π−bond is due to back-donation of charge from one occupied p$_\pi$ orbital on E to the empty 1b$_1$ dπ orbital on Cp$_2$M, **14b**. It is clear from Table X that the M=E bond decreases in strength with decreasing electronegativity of E and that this trend is determined by the σ−donation **14a**. Simply, the transfer of charge from the metal is most favourable for oxygen with the most stable p$_\sigma$ orbital. Also shown in Table X are Zr=E energies for the two limiting group-16 elements, oxygen and tellurium. We note that the Zr=E bond seems to be stronger than the corresponding Ti=E bond. This is primarily due to the steric interaction ΔE^0. It can again be traced back to the strong inter atomic Pauli core repulsion from the 3s and 3p orbitals of the 3d-element (titanium) ,**5**.

Table X. Calculated Electronic Bond Enthalpies[a,b,c] for Cp$_2$M=E (E=O,S,Se,Te; M=Ti,Zr).

Bond	ΔE^0	ΔE_σ	ΔE_π	ΔE
Ti=O	-1365	1802	165	602
Ti=S	-870	1121	140	391
Ti=Se	-754	972	119	337
Ti=Te	-674	806	126	258
Zr=O	-919	1452	95	628
Zr=Te	-411	693	86	368

[a]$\Delta E=\Delta E^0+\Delta E_\sigma+\Delta E_\pi$. [b]kJ/mol.[c]Ref.20

Table XI. Calculated Electronic Bond Enthalpies [a,b,c]for (CO)$_4$M=M(CO)$_4$ (M=Fe,Ru,Os).

Bond	ΔE^0	ΔE_σ	ΔE_π	ΔE
Fe=Fe	-85	245	46	206
Ru=Ru	-78	239	88	249
Os=Os	-67	260	108	301

[a]$\Delta E=\Delta E^0+\Delta E_\sigma+\Delta E_\pi$. [b]kJ/mol.[c]Ref. 21

Orbital Energies of Cp$_2$M (left) and E (right).

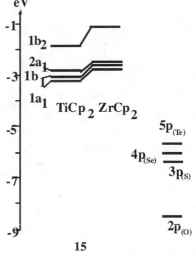

15

Metal-Metal Double Bonds. Each of the MH$_2$ (M=C,Si,Sn,Pb) fragments in H$_2$M=MH$_2$, **9**, are isolobal with M(CO)$_4$ (M=Fe,Ru,Os). We give in Table XI the calculated energies for the M=M double bonds in (CO)$_4$M=M(CO)$_4$. We note that the 3d-element iron forms the weakest double bonds. This is again due to the fact that the first transition series elements have core type 3s and 3p orbitals of the same radial extent as the 3d valence orbital leading to inter atomic Pauli core repulsion, **5**. The repulsion will prevent the two iron atoms from coming close enough to form strong enough π-bonds. It is a general trend that metal-metal bonds of the first transition metal series are weaker than those of the higher homologous (22) as a result of the inter atomic Pauli core repulsion involving the 3s and 3p metal orbitals.

Multiple Bonds

We deal in the last section with multiple bonds of order higher than two, including synergistic metal-ligand bonds in which the metal fragment and the ligand both serve as electron donors as well as acceptors.

Bonds between main group elements. Table XII provides calculated (23) $(CH_3)_3P$-E bond energies for the complete series of chalcogenides, E=O,S,Se, and Te. The $(CH_3)_3$P-E bond is established by electron donation from a PR_3 σ-orbital to an empty p_σ orbital on the chalcogenide, **16a**, as well as π-back donations from p_π of E, **16b**, to a degenerate pair of low-lying PC σ^* orbitals on PMe_3. The P-E bond strength is seen to decrease from oxygen to tellurium. This is primarily due to a corresponding decrease in the contribution, ΔE_σ, from the σ-bond interaction, **16a**, as the energy gap in **16a** diminishes. The π-back donation ΔE_π, **16b**, is not negligible for any of the systems. It attains further a proportionally larger importance towards the end of the family as it decreases much more slowly than ΔE_σ. It might be surprising that ΔE_π is largest for the more electronegative element oxygen. This is a synergistic effect. As charge is built up on oxygen in the σ-donation there is a strong need to relieve the excess density by π-back donation. We have extended the study by keeping the chalcogenide constant while changing the pnicogen to nitrogen and arsenic. The N-O bond is seen to be weak with a negligible contribution from ΔE_π. Essentially, the energy difference between oxygen and nitrogen is not large enough to warrant a substantial σ-donation (and induce a π-back donation).

Table XII. $(CH_3)_3$X-Y Bond Energies[a,b,c]

X-Y Bond	ΔE^o	ΔE_σ	ΔE_π	ΔE
P-O	-1068	1453	159	544
P-S	-360	591	106	337
P-Se	-283	459	90	266
P-Te	-198	301	81	184
N-O	-664	859	2	197
As-O	-857	1094	106	343

[a]Ref. 23.. [b]$\Delta E = \Delta E^o + \Delta E_\sigma + \Delta E_\pi$. [c] kJ/mol

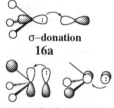

σ-donation
16a

π-back-donation
16a

The metal-carbonyl bond. The nature and strength of metal-carbonyl bonds has been studied extensively. In a few cases accurate experimental estimates are available for the first M-CO dissociation energy in binary metal-carbonyls, Table XIII. The first generation of DFT methods (HFS and LDA) typically overestimated M-CO bonds to the point where they were of little use in organometallic thermochemistry, Table XIII. However, the inclusion of gradient corrections as suggested by Becke (4b) and Perdew (4a) greatly increased the agreement with experiment, see BP86 of Table XIII.

Table XIII. First M-CO Bond Dissociation Energies[a,b,c,d,e].

Molecule	HFS	LDA	BP86	Exp
$Cr(CO)_6$	278	276	193	192/155
$Mo(CO)_6$	226	226	166	170
$W(CO)_6$ (NR)	201	202	162	
$W(CO)_6$ (QR)	257	249	183	192
$Fe(CO)_5$	267	263	192	176
$Ni(CO)_4$	194	192	121	104

[a]Ref.24·a [b]$\Delta E = \Delta E^o + \Delta E_\sigma + \Delta E_\pi$. [c] kJ/mol. [d]Except for $W(CO)_6$, all bond energies include relativistic effects. However, they are only important for tungsten. [e]Experimental data from Ref. 24b, except $Ni(CO)_4$ taken from Ref. 24c

It can be seen from Table XIII that the M-CO bond energy within the homologous series of hexacarbonyls has its minimum for the 4d-member of the triad, molybdenum. Further on, this trend is caused by a relativistic stabilisation of the W-CO bond to the point were it becomes stronger than the corresponding Mo-CO linkage. Figure 3 underlines further the importance of relativity for periodic trends in M-CO bond energies. For the $M(CO)_4$ (M=Ni,Pd,Pt), $M(CO)_5$ (M=Fe,Ru,Os), and $M(CO)_6$ (M=Cr,Mo,W) series we find in all cases that the M-CO bond is weakest for the 4d member. As indicated to the right of Figure 3, this is largely due to a relativistic stabilisation of the M-CO bond for the 5d homologues. Figure 4 underlines further for the same series that the weak M-CO bond for M=Mo,Ru, and Pd are associated with long M-CO distances. For the heavier congeners (M=W,Os,Pt) the relativistic M-CO bond stabilisation is associated with a M-CO bond contractions, right side of Figure 4.

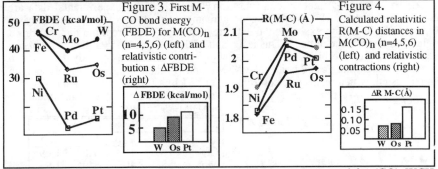

Figure 3. First M-CO bond energy (FBDE) for $M(CO)_n$ (n=4,5,6) (left) and relativistic contributions ΔFBDE (right)

Figure 4. Calculated relativitic R(M-C) distances in $M(CO)_n$ (n=4,5,6) (left) and relativistic contractions (right)

The origin of the relativistic effect is the same as that encountered for $(CO)_5WCH_2$ previously, **13**. That is, the 5d-orbitals are destabilized and better able to donate density-in the present case to the vacant π^*_{CO} ligand orbitals (24).

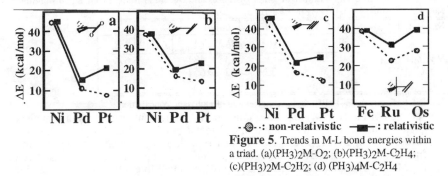

- ⊙ - : non-relativistic   ▬ : relativistic

Figure 5. Trends in M-L bond energies within a triad. (a)$(PH_3)_2M-O_2$; (b)$(PH_3)_2M-C_2H_4$; (c)$(PH_3)_2M-C_2H_2$; (d) $(PH_3)_4M-C_2H_4$

Figure 5 provides further examples (25) of how the 4d-member within the triad forms the weakest and most labile bonds to ligands after relativistic effects have been included (solid lines). Without relativity the 5d element forms the weakest metal-ligand bonds (dotted lines).

17a   **17b**   **18a**   **18b**

As a last example underlining the importance of relativistic effects in the metal-ligand bond, we mention here complexes with H_2 as a ligand (26) In **17a** H_2 is complexed to the $M(PR_3)_3(H)_2$ fragment. The bond involves again ligand to metal donation, **18a**, and metal to ligand back-donation, **18b**. For M=Cr and Mo the dihydrogen complex **17a** is stable. However for the heavier congener tungsten, relativity enhances the back-donation into $\sigma^*_{H_2}$ to the point where the H-H bond breaks and the hydride complex **17b** become the more stable conformer.

The metal-metal bond. The description of the metal-metal bond in terms of bonding ,**19**, and anti-bonding, **20**, orbitals with local σ,π, and δ symmetries has been immensely useful in codifying the increasing number of synthesized binuclear complexes with multiple metal-metal bonds.

The metal dimers Cr_2 and Mo_2 are among the simplest possible systems with multiple metal-metal bonds, and among the few species for which accuracte experimental M-M bond energies are known. The dimers have a hextuple bond with a $1\sigma^2 2\sigma^2 \pi^4 \delta^4$, **19**, electron configuration. Calculated and experimental data are compared in Table XIV . We find good agreement between experiment in the cases of Cr_2 and Mo_2 where experimental data are available. The chromium dimer forms a rather weak metal-metal bond. This is a result of the inter-atomic Pauli core repulsion between 3s,3p core orbitals on one metal with 3d on the other as already discussed in connection with the Fe=Fe bond. The M-M bond is much stronger for molybdenum and tungsten, with a relativistic contribution of 85 kJ/mol for the W-W bond.

Table XIV. Bond Energies [a,b] and Distances for M_2(M=Cr,Mo,W)

M_2	D(M-M)		R(M-M)	
	Calc.	Exp.	Calc.	Exp.
Cr_2	169	150	1.65	1.69
Mo_2	388	403	1.95	1.93
W_2	425	-	2.03	-

[a]kJ/mol. [b]Ref.22d.

Table XV. Optimized Metal-Metal Bond Distances and Decomposition of Calculated Bonding Energies[a,b,c] for $M_2Cl_4(PH_3)_4$ (M=V,Nb,Ta)

$M_2Cl_4(PH_3)_4$	R(M-M)	D(M-M) calculated				
	Calc.	ΔE^o	ΔE_σ	ΔE_π	ΔE_R[d]	ΔE
$V_2Cl_4(PH_3)_4$	2.04	-667	253	481		67
$Nb_2Cl_4(PH_3)_4$	2.30	-721	562	651		259
$Ta_2Cl_4(PH_3)_4$	2.52	-577	404	562	8	297

[a]kJ/mol. [b]$\Delta E=\Delta E^o+\Delta E_\sigma+\Delta E_\pi+\Delta E_\delta+\Delta E_R$. [c]Ref 22d [d] Relativistic contribution

Compounds of the type $M_2Cl_4(PR_3)_4$ contain the quintessential features of binuclear complexes with unsupported multiple metal-metal bonds. Table XV gives D(M-M) bond energies for $M_2Cl_4(PR_3)_4$ (M=V,Nd,Ta) with a triple bond and the electron configuration $1\sigma^2 \pi^4$. Bond energies for the quadruply bonded $M_2Cl_4(PR_3)_4$ (M=Cr,Mo,W) systems with the $1\sigma^2 \pi^4 \cdot \delta^2$ are given in Table XVI. Also discussed in Table XVII are the triple bonded $M_2Cl_4(PR_3)_4$ (M=Mn,Tc,Re) complexes with a $1\sigma^2 \pi^4 \cdot \delta^2 \delta^{*2}$ configuration (27).

Table XVI. Optimized Metal-Metal Bond Distances and Decomposition of Calculated Bonding Energies[a,b,c] for $M_2Cl_4(PH_3)_4$ (M=Cr,Mo,W)

$M_2Cl_4(PH_3)_4$	R(M-M)	D(M-M) calculated					
	Calc.	ΔE^o	ΔE_σ	ΔE_π	ΔE_δ	ΔE_R[d]	ΔE
$Cr_2Cl_4(PH_3)_4$	1.89	-655	274	473	10		102
$Mo_2Cl_4(PH_3)_4$	2.16	-643	345	620	49		371
$W_2Cl_4(PH_3)_4$	2.29	-419	316	525	24	14	460

[a]kJ/mol. [b]$\Delta E=\Delta E^o+\Delta E_\sigma+\Delta E_\pi+\Delta E_\delta+\Delta E_R$. [c]Ref 22d [d] Relativistic contribution

It follows that metal-metal bonds of 3d-elements are weak (60-100 kJ/mol), and the reason is again the destabilizing influence of the 3s,3p core-like orbitals on the metals. We find for a pair of homologous binuclear complexes involving 4d and 5d elements, that the 5d element invariably has the stronger metal-metal bond. The 5d elements have weaker σ- and π–interactions, ΔE_σ and ΔE_π, but also a less repulsive steric interaction, ΔE^0, all because of a longer M-M bond compared to the 4d-homologues. The reduction in the steric interaction, ΔE^0, is the prevailing factor in making the M-M bonds of the heavier homologues the strongest. We note that the σ-interaction ΔE_δ is weak. It adds very little to the M-M bond strength

Table XVII. Optimized Metal-Metal Bond Distances and Decomposition of Calculated Bonding Energies[a,b,c] for $M_2Cl_4(PH_3)_4$ (M=Mn,Tc,Re)

$M_2Cl_4(PH_3)_4$	R(M-M)	D(M-M) calculated					
	Calc.	ΔE^0	ΔE_σ	ΔE_π	ΔE_δ	$\Delta E_R{}^d$	ΔE
$Mn_2Cl_4(PH_3)_4$	1.92	-650	291	451	7		99
$Tc_2Cl_4(PH_3)_4$	2.17	-614	355	591	5		337
$Re_2Cl_4(PH_3)_4$	2.29	-427	328	520	9	11	441

[a]kJ/mol. [b]$\Delta E=\Delta E^0+\Delta E_\sigma+\Delta E_\pi+\Delta E_\delta+\Delta E_R$. [c]Ref 22d [d] Relativistic contribution

Concluding Remarks.
   We have analysed some of the factors responsible for periodic trends in bond energies involving main group elements and transition metals. We have shown that the Pauli exclusion principle is responsible for many of the observed trends through intra- and inter- atomic core repulsion. Relativistic effects are also important as valence electrons near the nucleus can obtain high instantaneous velocities. Relativistic effects will reduce intra- and inter- atomic core repulsion and thus strengthen the chemical bond.

Acknowledgment. The work presented here was supported by Natural Sciencess and Engineering Research Council of Canada (NSERC) as well as the donors of the Petroleum Research Fund, administered by the American Chemical Society. Thanks also to Drs. Elzbieta Folga, Jian Li, Heiko Jacobsen, Liqun Deng, Nicole Sandblom and Matthias Bickelhaupt for their contributions.

Literature Cited.

1.   (a) Ziegler, T. Chem. Rev. **1991**, 91, 651.
     (b) Parr, R. G.; Yang, W. *Energy Density Functional Theory of Atoms and Molecules* ,Oxford University, New York, **1989**.
     (c)Baerends,E.J. J.Phys.Chem. **1997**, in press.
2.   (a)Ziegler,T. "A General Energy Decomposition Scheme for the Study of Metal-Ligand Interactions in Complexes, Clusters and Solids", NATO-ASI Series ; D. Salahub (editor), 1992, C 378, 367.
     (b)Ziegler,T.; Tschinke,V. "Density Functional Calculations on Transition Metal Complexes", T.J.Marks (Ed.), ACS Symposium Series 428, 1990, 279
     (c)Frenking, G.; Antes,I.;Böhme,M.;Dapprich,S.;Ehlers,A.W.;
     Jonas,V.;Neuhaus,A.; Stegmann,M.O.R.; Veldkamp,A.; Vyboishchikov, S.F., in Reviews in Computational Chemistry, Vol. 8, Lipkowitz, K.B. and Boyd, D.B. (Eds), VCH, New York, (1996)
3.   (a) L.Deng, V.Branchadell; T.Ziegler , J.Am.Chem.Soc. **1994**,116, 10645
     (b)*CRC Hand book of Chemistry and Physics* , edited by Lide, D.R. 74th Edition, **1993-1994**.
4.   (a) Perdew, J. P. Phys. Rev. **1986**, B13, 8822.
     (b) Becke, A. D. Phys. Rev. **1988**, A38, 3098.
5.   Hehre, W.J.; Radom, L.; Schleyer, P. v. R.; Pople, J. A. In *Ab Initio Molecular Orbital Theory* ; Wiley-Interscience: New York, **1986**.

6. W. Kutzelnigg, Angew. Chem., Int. Ed. Engl, **1984**,23, 272
7. (a)Bickelhaupt,F.M.; Ziegler,T.; Schleyer,P.v.R. Organometallics **1996**,15,147
(b)Lias, S.G.; Bartmess, J.E.; Liebman, J.F.; Holmes, J.L.; Levin R.D.;
Mallard,W.G. J. Phys. Chem. Ref. Data **1988**, 17, Suppl. No. 1.
8. Su, M.-D.; Schlegel,H.B. J.Phys.Chem. **1993**,97,8732
9. Jacobsen,H;Ziegler,T. J.Am.Chem.Soc. **1994**, 116,3667
10. (a)Pyykkö,P. Chem.Rev. **1988**,88,563
(b)Ziegler,T.;Snijders,J.G.;Baerends,E.J. J.Chem.Phys. **1981**,74,1271
(c)Ziegler,T.;Snijders,J.G.;Baerends,E.J. Chem.Phys.lett. **1980**,75,1
(d)Ziegler,T.; Snijders,J.G.; Baerends,E.J. "The Challenge of d and f Electrons",
Salahub,D.R.;Zerner,M.C. (Ed.), ACS Symposium Series 394, 1989, 322.
11. (a)Ziegler, T. Unpublished results
(b)Snijders,J.G.; Baerends,E.J.;Ros.P. Molec. Phys. **1979**,38,1909
(c) Ziegler,T.;Tschinke,V.;Baerends,E.J.;Snijders,J.G.;Ravenek,W. J.Phys.Chem.
1989,93,3050
(d)Huber,K.P.; Herzberg,G. ' Molecular Spectra and Molecular structure' , vol 4.
Van Nostrand Reinhold, New York, **1979**.
(e)Wezenbeel,E.M.; Baerends,E.J.; Ziegler,T. Inorg.Chem. **1995**,34,238
12 (a)Holthausen,M.C; Koch,W. J.Am.Chem.Soc. **1996**,118,9932
(b)Ricca,A.;Bauschlicher,C.W.,Jt. J.Phys.Chem. **1995**,99,5922
(c)Ziegler,T.;Li,J. Can.J.Chem. **1994**,72,783
(d)Blomberg,M.R.A.;Siegbahn,P.E.M;Lee,T.L.;Rendell,A.P.;Rice,J.E.
J.Chem.Phys. **1991**,5898
(e)Bruno, J.W. ; Stecher, H.A. ; Morss, L.R.; Sonnenberger, D.C. ; J.Marks, T.
J.Am.Chem.Soc. **1986**,**108**,7275 .
(f)Simoes,J.A.M.; Beauchamp, J.L. Chem.Rev. **1990**,90,629
(g)B.S. Freiser (ed.) in Organometallic Ion Chemistry, Kluwer Academic
Publisherss, Dordrecht **1995**
13 (a) Ziegler,T.;Tschinke,V.;Versluis,L.;Baerends,E.J.;Ravenek,W. Polyhedron,
1988,7,1625
(b)Ziegler,T.;Tschinke,V.;Becke,A. J.Am.Chem.Soc. **1987**,109,1351
14 E.Folga; T.Ziegler J.Am.Chem.Soc. **1993**,115,5169.
15 Schwarz,W.H.E.; van Wezenbeek,E.M.; Baerends,E.J.;Snijders,J.G.J.Phys
1989,B22 1515
16 Ziegler, T.;Rauk,A. Theoret. Chim. Acta (Berl.), **1977**,46, 1.
17 Jacobsen, H.; .Ziegler, T. J.Am. Chem. Soc., **1994**,116,3667
18 Jacobsen, H.; Ziegler, T. Inorg.Chem. **1996**, 35,773
19 Jacobsen,H.;Schreckenbach,G.; Ziegler,T. J.Phys.Chem, **1994**, 98, 11406
20 Fischer,J.M.;Piers,W.E.; Ziegler,T.; MacGillivray,L.R.; Zaworotko, M.J.
Chem.Eur.J. **1996**,2,1220
21 Jacobsen,H; Ziegler.,T. J.Am.Chem.Soc. 1996, 118,4631
22 (a) Ziegler,T. J. Am. Chem. Soc.**1983**, 105, 7543
(b)Ziegler,T. J. Am. Chem. Soc. **1984**, 106, 5901
(c)Ziegler,T. J. Am. Chem. Soc., **1985**, 107, .
(d)Ziegler, T.; Tschinke,V.; Becke,A. Polyhedron, **1987**, 6, 685 .
23 Sandblom,N; Chivers,T.;Ziegler,T. Can.J.Chem, **1996**, 74, 2363
24 (a)J.Li, G.Schreckenbach and T.Ziegler J.Am.Chem.Soc., **1995**,117,486
(b)Lewis, K.E.; Golden, D.M.; Smith, G.P. *J. Am. Chem. Soc.* **1984**, *106*, 3905.
(c)Stevens, A.E.; Feigerle,C.S.; Lineberger,W.C. *J. Am. Chem. Soc.* **1982**, *104*,
5026.
25 Li,J.; Schreckenbach,G.;Ziegler,T. Inorg.Chem. **1995**,34,3245
26 (a)Li,J.; Ziegler,T. Organometallics **1996**, 15, 3844
(b) Li,J.; Dickson,R.M., Ziegler,T. J.Am.Chem.Soc. **1995**,117,11482
27 The two δ components are not degenerate and $1\sigma^2 \pi^{4\cdot}\delta^2 \delta^{*2}$ is of lower energy
than $\pi^{4\cdot}\delta^4$, see Ref. 22d.

APPENDICES

Appendix A

Software and Databases for Thermochemistry

David J. Frurip[1] and Joey W. Storer[2]

[1]Analytical Sciences, Reactive Chemical/Thermal Analysis/Physical Properties
Discipline, The Dow Chemical Company, Building 1897F, Midland, MI 48667
[2]Computing Modeling and Information Sciences, The Dow Chemical Company,
Building 1707, Midland, MI 48667

The following is a list of publicly available software tools, electronic data bases, web sites and published literature for use in thermochemical applications. Whereas we attempted to be comprehensive in our listing, we were understandably biased by our historical uses and needs at the Dow Chemical Company which has traditionally been focused on organic chemistry. We apologize in advance for not including below any source of data, web site, or software tool which might be publicly available but either unknown or not used extensively by us. It is our hope that this listing might assist the novice user in acquiring basic resources for thermochemical data and computational predictions. The brief annotations following each entry are, in many cases, taken largely from the promotional material of the product. In other cases they are the subjective opinions of the authors. When approximate prices are indicated below, they do not reflect the steep discounts often available to educational institutions.

By far, the most comprehensive listings of general molecular modeling software are the appendices to the periodically published "Reviews in Computational Chemistry" edited by Lipkowitz and Boyd (1). The reader is strongly encouraged to consult the latest listing for more information.

The organization of this appendix is as follows:

Software Tools for Empirical Estimation of Thermochemistry
Software Tools for Quantum Mechanical Prediction of Thermochemistry
Published Literature Sources for Thermodynamic Data
Electronic Databases for Thermochemistry
Web Sites with Thermochemical Themes

Software Tools for Empirical Estimation of Thermochemistry

The software listed below can provide the user with an efficient means to use the empirical group additivity thermodynamic estimation methods. The Structures and Properties program is perhaps the best implementation of Benson's method (see the chapter in this book by Benson and Cohen) in that a graphical interface allows the user to draw the molecule with the mouse and the program deconvolutes the structure into the component groups. The current version of CHETAH does not have this graphical capability, but this weakness is compensated by inclusion of the largest publicly available database of Benson groups. Also, unlike the Structures and Properties software, CHETAH does allow the user to substitute a similar group if one happens to be missing. The THERM/EST program is noteworthy because it calculates thermodynamic data for species in the solid, liquid, and gaseous states. The write-ups below are, for the most part, taken directly from the corresponding web pages for the subject programs. All three programs run on desktop personal computers (PCs) and are available from NIST. More information may be obtained from:

Standard Reference Data Program
National Institute of Standards and Technology
Bldg. 820, Rm. 113
Gaithersburg, MD 20899-0001
(301) 975-2208 (VOICE) / (301) 926-0416 (FAX) / srdata@nist.gov (E-MAIL)

CHETAH (http://www.nist.gov/srd/spec16.htm) Price: $350.00. NIST SPECIAL DATABASE 16: CHETAH - Chemical Thermodynamic and Energy Release Program (Version 7.0). From the National Institute of Standards and Technology (NIST) and the American Society for Testing and Materials (ASTM).

CHETAH 7.0 is a tool for predicting both thermochemical properties and certain reactive chemical hazards associated with chemicals and their reactions. For thermochemical estimations, CHETAH is designed to conveniently and accurately calculate gas phase properties such as heat capacity, enthalpy, enthalpy of formation, entropy, and Gibbs energy of reaction as functions of temperature. The output of the "Energy Release Evaluation" option provides information about the potential ability of a material to decompose with violence if subjected to a severe impact.

NIST Structures and Properties (http://www.nist.gov/srd/ion.htm) Price: $200. This database features a complete implementation of Benson's Group Additivity estimation method for gas phase heats of formation, entropies, and heat capacities. The program also allows estimation of vapor pressures and boiling points using a group method. Properties are estimated solely from structures that are drawn on the screen using an integrated easy-to-use structure-drawing module. No knowledge of estimation methods is required. Either a mouse or a keyboard can be used for drawing.

Estimations include: automatic perception of rings and long-range interactions; determination of symmetry number correction, even for complex ring systems; inclusion of 50 new group values and 80 new ring corrections derived from recent literature data.

The accompanying database of chemical and thermochemical properties includes enthalpies and entropies of formation for approximately 5,000 compounds, taken from three Standard Reference Databases - NIST Positive Ion Energetics Database, NIST Chemical Kinetics Database, and NIST JANAF Thermochemical Tables. Because the database has been combined with the Positive Ion Energetics Database, molecular structures and names, as well as ionization energies of 11,000 molecules are included.

The program allows the user to set up a balanced chemical reaction and calculate the energetics of the reaction, or equilibrium constant, using either estimated or experimental data. You may also replace thermochemical estimates with your own data in such calculations.

This program is currently sold as "19A and 19B, NIST Positive and Negative Ion Energetics Database with Structures and Properties Software and Data."

THERM/EST (http://www.nist.gov/srd/thermest.htm) Price: $100.00. NIST Estimation of the Thermodynamic Properties for Organic Compounds at 298.15 K - Compounds Containing the Elements C,H,N,O,S, and Halogens. This database carries out 2nd order additivity-type estimations for thermodynamic properties of organic molecules at 298.15 K in the gas, liquid, and solid phases. A database with experimental data for approximately 1512 compounds is included.

The thermodynamic properties calculated are: enthalpy of formation, heat capacity, entropy, entropy of formation, Gibbs energy of formation, equilibrium constant for the formation reaction. The thermodynamic properties have been developed for the gas, liquid, and solid phases. Data are provided for various organic families such as alkanes, alkenes, alcohols, ethers, amines, thiols, sulfides, etc. Other features include the capacity to search the database by compound name or formula and a bibliography containing nearly 1000 references.

Software Tools for Quantum Mechanical Prediction of Thermochemistry

Computational thermochemistry using methods of quantum mechanical molecular orbital theory is a straight forward task where energy results, often from several calculations, are combined to give estimates of thermochemical values. Statistical mechanics is also applied to correct for temperature, see Appendix B. Examples of how to carry out these calculations can be found in Appendix C. Here we simply list for the novice some of the software tools and computer hardware requirements needed to begin making thermochemical estimates. The commercialized programs are generally easier to use than those under development in academia.

The atomistic methods available for computing thermochemical estimates fall into three categories: molecular mechanics, semiempirical quantum mechanics, and ab initio quantum mechanics. The ab initio approaches may be subdivided into those based entirely on molecular orbitals and those based on computing the electron density, i.e. density function theory (DFT). Thermochemical estimates from force fields in molecular mechanics applications are particularly good for hydrocarbons. These methods are very fast computationally, can run on a personal computer, and are priced near $1000. Two applications in this area that are listed below are MM3 and PCMODEL. Semiempirical methods are parametric in the description of electron-electron interactions. The parameters are optimized to give good geometries and heats of formation for molecules containing first row elements. These methods are very fast computationally, can run on a personal computer, and are priced in the range of $1000 to $5000. The various molecular orbital approaches to computational thermochemistry differ generally in the choice of basis set to describe the molecular orbitals and then the treatment of electron correlation. Often, the better the basis set and the better the treatment of electron correlation the better will be the resulting thermochemical estimate. While these approaches are systematically improvable they can become quite expensive. Nevertheless, with molecular orbital methods you tend to get what you pay for. Density functional theories using the generalized gradient approach have implicitly selected the form of the correction for electron correlation. Thus, with DFT, you *might* get a very good description of the electron correlation at a low cost compared to the molecular orbital methods. Commercial quantum mechanics packages range widely in price but cost on the order of $20,000. They traditionally run on UNIX workstations or supercomputers but are now becoming available in versions for personal computers (see below).

There are several important considerations before making an effort in computational thermochemistry. Here we address the quantum mechanical methods in particular because they provide the best overall accuracy. This section will address a few of the practicalities of performing these computations on a routine basis from both a hardware and software perspective.

Resource demand for computational thermochemistry with quantum mechanics. The G2 and G2/MP2 methods of Curtiss, Raghavachari, Trucks, and Pople are among the preferred methods for computational thermochemical estimates at the Dow Chemical Company(2). These methods are also among the more demanding of computer resources. Problems in quantum mechanics in general and computational thermochemistry in particular are subject to exponential scaling laws that can quickly limit the size of molecule that can be studied on a particular computer in a reasonable span of time. For example, one stage of the G2 series of calculations has a seventh order scaling dependence. The result is that a small increase in the size of the molecule requires a large increase in computational resources. For a more complete list of resource requirements as a function of method see Foresman and Frisch (3). A research program in computational thermochemistry should consider the availability of computer resources as a function of problem size. Without an evaluation of resources and problem size such a research program can become abruptly stalled.

Computer hardware for computational thermochemistry. Given the scaling issue described above, some attention should be devoted to the selection of the computer hardware necessary to complete the desired computations. Currently, there are several vendors with systems that can be expanded to become parallel compute servers with increasing memory, cpu, i/o performance (i.e., i/o channel and disk speed), and disk space. The advantage of these commodity based servers is that one can start small and grow over time while taking advantage of parallelism inherent to many of the calculations. Configuring a computer system for computational thermochemistry will generally result in a robust system able to handle many other tasks. As mentioned above, even a personal computer could successfully perform a G2 calculation (see Appendix C). As with any hardware decision one must weigh the risk of a failed computation and wall clock performance against the time criticality of the result. Notably however, even on the most expensive supercomputer, the cost of a thermochemical estimate for a medium sized organic molecule is comparable to or less than the cost of the actual experiment and yields similar accuracy.

Quantum chemistry software for computational thermochemistry. There is currently no single application that accomplishes everything needed to derive the heat of formation of a molecule from *ab initio* quantum mechanics. This issue is discussed in the chapter by Frurip, Rondan, and Storer. Consequently, we have recommendations for combinations of software applications which when put together provide all the necessary capabilities. These recommendations are broken up into two basic needs: the need to generate the proper coordinates describing the make-up and structure of the molecule; and the need to refine the molecular structure quantum mechanically and compute the component energetics used to derive the thermochemical estimate. As with computer hardware, quantum chemistry software is changing rapidly toward improved ease-of-use and performance. While the software packages in use at the Dow Chemical Company have proven robust over at least 4 years (a long time for software), new applications are always being developed and should be reviewed on a regular basis. A fast method to input and visualize the structure of a molecule resulting in a set of coordinates can be achieved with numerous programs available on multiple platforms including personal computers, UNIX workstations, and supercomputers. The coordinates are usually generated with a program that provides a graphical user interface to the researcher that includes software modules specifically designed for the rapid build-up of molecular structures. Ideally, the resulting coordinates are saved to disk in a format compatible with the quantum chemistry application. At Dow Chemical the program Spartan from Wavefunction Inc. is the primary interface used for generating molecular structures subsequently submitted for quantum thermochemical estimation. While we recommend Spartan for its ease-of-use and visualization capabilities, one might just as easily take advantage of the programs listed below. Some of these applications have a much broader applicability and in these cases we simply identify their multiple functionality.

Cerius²/Molecular Simulations Inc. (http://www.msi.com/) Molecular
Simulations Inc., 9685 Scranton Road, San Diego, CA 92121, 619-485-9990.
Cerius2 provides an easy-to-use simulation and modeling environment, offering a
broad range of scientific application modules. The interface to the quantum
mechanics workbench includes a module for Gaussian as well as other programs.

Chem 3D Pro. (http://www.camsci.com/) Cambridge Scientific Computing Inc.
75 Massachusetts Avenue, Suite 61, Cambridge, MA 02139, U.S.A. Tel. 800-450-
7606 (ext. 399), 617-491-6862, fax 617-491-8208, e-mail info@camsci,com,
support@camsci.com. This personal computer program for molecular mechanics
(MM2) now offers an interface to Gaussian94*W* which is the version of the
Gaussian program for Windows95.

Gaussian-GaussView. (http://www.gaussian.com/) GaussView is a relatively
new product (summer 1997 release) with the ability to visualize results from the
Gaussian program including molecular orbitals, electron density surfaces,
electrostatic potential surfaces, and animation of the normal modes corresponding
to vibrational frequencies. GaussView is also helpful for setting up Gaussian jobs.

MOLDEN. (http://www.caos.kun.nl/~schaft/molden/molden.html)
G.Schaftenaar, CAOS/CAMM Center, the Netherlands. Molden is a package for
displaying molecular density from the ab initio packages GAMESS-UK ,
GAMESS-US and GAUSSIAN and the semiempirical packages Mopac and
Ampac. Molden reads all the required information from the GAMESS /
GAUSSIAN output file.

PCMODEL. (http://serenasoft.com/) Serena Software, Dr. Kevin Gilbert,
P.O. Box 3076, Bloomington, IN 47402-3076, U.S.A. Tel. 812-333-0823, 812-
855-1302/9415, fax 812-332-0877, e-mail gilbert@indiana.edu. PCMODEL
calculates heats of formation along with steric energies, handles conjugated
systems as well as non-conjugated, and handles transition metal complexes. The
thermochemical estimates are derived from the MMX parameterization, a
molecular mechanics force field (see MM3 below). PCMODEL also provides a
simple interface to the Gaussian program.

MM3. (http://www.osc.edu/ccl/qcpe/QCPE/ent/special_packages/mm396.html)
QCPE, Creative Arts Building 181, Indiana University, Bloomington, IN 47405,
U.S.A. Tel. 812-855-4784, fax 812-855-5539, e-mail qcpe@ucs.indiana.edu.
MM3 is a molecular mechanics program with a well refined force field that is
optimized to give good geometries and to account for the internal energy of a
molecule. The heat of formation is computed from the steric energy derived from
the force field, empirically derived bond enthalpies, a partition function
contribution, and a SCF pi-bond resonance stabilization contribution if appropriate.

SPARTAN. (http://www.wavefun.com/) Wavefunction, Inc., Dr. Warren J.
Hehre, 18401 Von Karman Avenue, Suite 370, Irvine, CA 92715, U.S.A. Tel.

714-955-2120, fax 714-955-2118, e-mail support@wavefun.com. Spartan is a very good program for visualization of molecular structures and properties and has a forcefield, semiempirical and *ab initio* computational engines with a modest set of capabilities. Spartan provides an interface to other quantum mechanics programs including Gaussian as well as the molecular mechanics program MM3. Available for UNIX, PC, and Macintosh computers.

The molecular coordinate files generated with the above applications are commonly edited in a subsequent step to ensure the proper input keywords and designation of the molecular charge and multiplicity. The amount of editing differs among the available quantum chemistry programs and careful attention must be paid to the specific input requirements of each since the expense of the computation is now going to increase rapidly. The quantum chemistry application of choice at the Dow Chemical Company is currently the Gaussian 94 program from Gaussian, Incorporated. This program is unique in the arena of quantum thermochemistry because it offers keywords that reduce the composite set of computations in a G2 or CBS methodology down to a single input option. Thus the "G2" keyword results in the full series of calculations needed for the G2 estimate. The G2 component energies are tabulated at the end of the output which can be brought into a spreadsheet for final conversion to a heat of formation. Prior to the introduction of these keywords we had developed our own program to automate the creation of input files, job submission scripts, and the collection of output results for the G2 type of calculation.

Listed below are some general software recommendations that are useful for doing computations leading to various types of thermochemical estimates from quantum mechanical results.

ACES II. Dr. Rodney J. Bartlett, Quantum Theory Project, 362 Williamson Hall, University of Florida, Gainesville, FL 32611-2085, U.S.A. Tel. 904-392-1597, fax. 904-392-8722, e-mail aces2@qtp.ufl.edu. The ACES II program is a general *ab initio* molecular orbital program where special attention was given to taking advantage of symmetry toward reducing computational expense.

CADPAC. Lynxvale WCIU Programs, Dr. Roger Amos, 20 Trumpington Street, Cambridge CB2 1QA, England, U.K. Tel.44-223-336384, e-mail cadpac@theory.chemistry.cambridge.ac.uk. CADPAC is a general *ab initio* molecular orbital program where recent emphasis has been placed on methods in density functional theory.

GAMESS. (http://www.msg.ameslab.gov/GAMESS/GAMESS.html) Dr. Michael Schmidt, Department of Chemistry, Iowa State University, Ames, IA 50011, U.S.A. Tel. 515-294-9796, fax 515-294-5204, e-mail mike@si.fi.ameslab.gov; theresa@si.fi.ameslab.gov. The GAMESS program is a general *ab initio* molecular orbital program with an especially nice MCSCF module.

Gaussian. (http://www.gaussian.com/) Gaussian, Inc., Dr. Michael Frisch, Carnegie Office Park, Building 6, Pittsburg, PA 15106, U.S.A. Tel. 412-279-6700, fax 412-279-2118, e-mail info@gaussian.com. The Gaussian program is a general *ab initio* molecular orbital program offering many built-in basis sets and theoretical methods.

MOLCAS. (http://garm.teokem.lu.se/MOLCAS/index.html) Mats Olsson, Department of Theoretical Chemistry, University of Lund, Chemical Center, P.O.B. 124, S-22100 Lund, Sweden. MOLCAS is a suite of ab initio programs emphasizing the multiconfigurational approach. Molcas contains codes for general and effective multiconfigurational SCF calculations including gradient techniques. Especially useful is the CASPT2 technique which combines a multideterminant description of the wave function with second order perturbation theory. The result is a very accurate description of relative energies for potential energy surfaces including reaction barriers and reaction energies.

MOLPRO. (http://www.tc.bham.ac.uk/molpro) Prof. Peter J. Knowles; School of Chemistry, University of Birmingham, Edgbaston, Birmingham, B15 2TT, UK. Tel. +44-121-414-7472, Fax: +44-121-414-7471, e-mail: molpro-request@tc.bham.ac.uk. MOLPRO is a complete system of ab initio programs for molecular electronic structure calculations, written and maintained by H. J. Werner and P. J. Knowles, and containing contributions from a number of other authors. As distinct from other commonly used quantum chemistry packages, the emphasis is on highly accurate computations for small molecules, with extensive treatment of the electron correlation problem through the multiconfiguration-reference CI, coupled cluster and associated methods.

The following semiempirical quantum mechanical applications also provide estimates of thermochemical properties of molecules. There are numerous caveats to the reliable application of these methods, as with any. We have found these programs or their various incarnations very useful for modeling at the Dow Chemical Company.

AMPAC. (http://www.semichem.com/ampac.html) 7204 Mullen, Shawnee, KS 66216. e-mail: sales@semichem.com. AMPAC is Semichem's flagship product and is a fully-featured semiempirical quantum mechanical program. It also includes a graphical user interface (GUI) that builds molecules and offers full visualization of results. PCMODEL is the recommended GUI on personal computers. The AMPAC GUI on UNIX systems provides an interface to Gaussian.

AMSOL. (http://pollux.chem.umn.edu/~amsol/) QCPE, Creative Arts Building 181, Indiana University, Bloomington, IN 47405, U.S.A. Tel. 812-855-4784, fax 812-855-5539, e-mail qcpe@ucs.indiana.edu. The AMSOL program is able to provide semiempirical estimates of the thermodynamics of solvation and in particular the free energy of solvation.

MOPAC. QCPE, Creative Arts Building 181, Indiana University, Bloomington, IN 47405, U.S.A. Tel. 812-855-4784, fax 812-855-5539, e-mail qcpe@ucs.indiana.edu MOPAC is a generally recommended semiempirical molecular orbital program. Many of the above mentioned *ab initio* programs have also implemented some of the methods available here.

Published Literature Sources for Thermodynamic Information

The listing below is based primarily on a reference volume collection we have maintained at Dow Chemical Thermal Laboratory in Midland Michigan for the last 20 years and tends to focus on data collections published in monograph form. Many of the references given are now out of print but fortunately still available from many libraries. The following several references have been sufficient for the bulk of the reaction heat prediction problems encountered at the Dow Chemical Company: 86PED/NAY (now superseded by 94PED), 82WAG/EVA, 70COX/PIL, and 76BEN.

Note that there are two main types of thermodynamic data collections. One manner of presenting the data is by a detailed table of temperature dependent quantities (enthalpy, entropy, and related quantities) usually presented in increments of 100 K. This is the so-called "JANAF" format. The other type of data presentation (more common) is at a single temperature, usually 298 K (or sometimes at 0 K and 298 K). In the descriptions below, unless otherwise noted, the data are in the latter format. Also, unless noted, data are presented for species in both the condensed (solid and liquid) and the gas phase. In some instances, there may be more recent editions of monographs than those listed.

61LAN/BOR. K. Schafer and E. Lax, *Caloric Variables of State*; Landolt-Börnstein, Sixth Edition, Vol. II, Part 4, Springer-Verlag: Berlin, 1961. An older but occasionally useful source of standard thermodynamic data for organic and inorganic species. The data are typically presented as reported in the original citation with no critical review.

69STU/WES. D. R. Stull, E. F. Westrum, Jr. and G. C. Sinke, *The Chemical Thermodynamics of Organic Compounds*; Wiley: New York, 1969. Republished (with corrections) by Krieger: Malabar, Florida, 1989. A classic compendium of thermochemical data for over 4000 organic compounds. Most of the data from this book may be found in the more recent references by Pedley, et al., below. Many tables of data are also included in the JANAF format (ca. 700 species). Also includes tabular data for the elements and some inorganic species.

70COX/PIL. J. D. Cox and G. Pilcher, *Thermochemistry of Organic and Organometallic Compounds*; Academic: London, 1970. A large collection of thermochemical data for organic species and a useful reference for data on oganometallic species. This book is unfortunately out of print but available from many libraries. The organic data is now largely superseded by 94PED.

Additionally, this book has the advantage of presenting the original experimental calorimetric data (including transition heats) with critical review. These data often allow the user to help sort out discrepancies in published data.

70KAR/KAR. M.Kh. Karapet'yants and M.L. Karapet'yants, *Thermodynamic Constants of Inorganic and Organic Compounds*; translated by J. Schmorak, Humphrey Science: Ann Arbor (1970). A large compilation of data as gleaned from the literature. The values are presented as originally published with no recommendations. All data are at 298K. Contains a large amount of data on inorganic and organometallic species.

76BEN. S. W. Benson, *Thermochemical Kinetics*; Second Edition, Wiley Interscience: New York, 1976. The classic text book on kinetics which contains an excellent discussion of Benson's method. Tables in the appendices contain detailed listings of gaseous second order group thermochemical values. The reader is also encouraged to review the pertinent literature cited in the chapter in this book by Benson and Cohen.

77PED/RYL. J. B. Pedley and J. Rylance, *Sussex-NPL Computer Analyzed Thermochemical Data: Organic and Organometallic Compounds*; University of Sussex: Sussex, UK, 1977. Largely superseded, first by 86PED/NAY, then by 94PED (see separate discussions above for each of these publications) except for organometallics.

82WAG/EVA. D. D. Wagman, W. H. Evans, V. B. Parker, R.H. Schumm, I. Halow, S.M. Bailey, K.L. Churney, and R.L. Nuttall "The NBS Tables of Chemical Thermodynamic Properties," *J. Phys. Chem. Ref. Data* **1982**, *11*, Suppl. 2. Errata, *JPCRD*, **1990**, *19*, 1042. A comprehensive thermochemical data compendium for mainly inorganic species and organic species (up to and including species containing two carbon atoms). A "must" for anyone needing accurate thermochemical data. Tabulated thermodynamic data include: enthalpies of formation at 0 K and 298.15 K, $\Delta_f G°_{298.15}$, $H°_{298.15} - H°_0$, $S°_{298.15}$, and $Cp_{298.15}$. The latter data are typically not available for all species. As discussed in the introduction to this section, the authors' have found this volume to be invaluable as a source of critically evaluated thermodynamic data.

84DOM/EVA. E. S. Domalski, W. H. Evans, and E. D. Hearing, "Heat Capacities and Entropies of Organic Compounds in the Condensed Phase," *J. Phys. Chem. Ref. Data*, **1984**, *13*, Suppl. 1. A tremendous source for the subject data. We also find it useful for the identification of polymorphic phases and transition temperatures for organic species. Newer editions were published in 1990 (90DOM/HEA below) and in 1996 (96DOM/HEA below).

85CHA/DAV. M. W. Chase, C. A. Davies, J. R. Downey, D. J. Frurip, R. A. McDonald, and A. N. Syverud, "JANAF Thermochemical Tables, Third Edition," *J. Phys. Chem. Ref. Data*, **1985**, *14*, Suppl. 1. The data in this compilation are presented in temperature dependent, table form and are also available electronically. Although most of the species in this work are inorganic, there are some simple organic species. Data for each table has been extensively reviewed and critically evaluated. A more detailed description of this database may be found in the Electronic Database section of this appendix.

85MAJ/SVO. V. Majer and V. Svoboda, *Enthalpies of Vaporization of Organic Compounds*; Blackwell: Oxford, 1985. This is an extensive compilation of calorimetric data for vaporization heats. The original data and data corrected to 298 K are presented.

86PED/NAY. J. B. Pedley, R. D. Naylor and S. P. Kirby, *Thermochemical Data of Organic Compounds*; Second Edition, Chapman and Hall: London, 1986. Although largely supplanted by Pedley's latest volume (see 94PED below), this work does also contain all of the experimental reaction heat data used to extract the enthalpy data. These source experimental data are sometimes useful. Enthalpy of formation data (at 298 K only) are presented for species in all three phases if available. Pedley's empirical group contribution method for the estimation of thermochemical data is also presented in this volume (a more detailed discussion of this method may be found in the description of 94PED below).

87CHI. J. S. Chickos, "Heats of Sublimation," in *Molecular Structure and Energetics, Physical Measurements* (Volume 2), J. F. Liebman and A. Greenberg (editors), VCH: Weinheim, Germany (and Deerfield Beach, Florida), 1987, Chapter 3, pp. 67-150. Experimental sublimation heat data for several hundred (mainly) organic compounds. The reader is also encouraged to refer to citations and recommendations in the chapter by Chickos, Acree, and Liebman in this Volume.

87REI/PRA. R. C. Reid, J. M. Prausnitz, and B. E. Poling, *The Properties of Gases and Liquids*; 4th ed., McGraw-Hill, New York, 1987. A standard reference book for the prediction of a multitude of physical properties. Two complete chapters are devoted to the prediction of thermodynamic properties. A good description of Benson's method is included and it also describes other group contribution methods. Many examples are given. Note: the reader is advised to be extremely careful in using the list of Benson groups in this volume. We have determined that this list has been combined from the values in different Benson group compilations and are incompatible. We recommend the self-consistent lists in either Benson's original book (76BEN) or the values published by Cohen (N. Cohen, *J. Phys. Chem. Ref. Data* **1996**, *25*, 1411).

89BAR. I. Barin, *Thermochemical Data for Pure Substances*; VCH: Weinheim, Germany, 1989. Tabular data for mainly inorganic species in all three phases.

89BRA/IMM. J. Brandrup and E. H. Immergut, *Polymer Handbook*; 3rd ed., John Wiley: New York, 1989. A comprehensive section contains polymerization enthalpies (experimental) for various monomers.

89GUR/VEY. L.V. Gurvich, I.V. Veyts, C.B. Alcock, *Thermodynamic Properties of Individual Substances* Hemisphere: New York, 1989. A large compilation of mainly inorganic data presented in the JANAF format.

90BAR/BEA. J. A. Martinho Simões and J. L. Beauchamp, *Chem. Rev.* **1990**, *90*, 629-688. A lengthy review of organometallic thermochemistry with several hundred references.

90DOM/HEA. E. S. Domalski and E. D. Hearing, "Heat Capacities and Entropies of Organic Compounds in the Condensed Phase - Vol. II", *J. Phys. Chem. Ref. Data* **1990**, *19*, 881. This is the second volume in the series starting with 84DOM/EVA (see description above).

92TRC/HYD. *TRC Thermodynamic Tables, Hydrocarbons*, and **92TRC/NON** *TRC Thermodynamic Tables, Non-Hydrocarbons*; Thermodynamics Research Center, Texas A&M University, College Station. Tabular thermodynamic and transport data, published as loose leaf sheets, for hydrocarbons and non-hydrocarbons. The data are for species in the gas phase (mainly) and arranged by families.

94FRE/MAR. M. Frenkel, K.N. Marsh, R.C. Wilhoit, G.J. Kabo, and G.N. Roganov, *Thermodynamics of Organic Compounds in the Gas State*; Volumes I and II, TRC Data Series, College Station, Texas, U.S.A., 1994. A large collection of tabular data presented in the "JANAF" format. This work is valuable because it contains a large introductory section on the treatment techniques for thermodynamic data.

94PED. J. B. Pedley, *Thermochemical Data and Structures of Organic Compounds - Vol. I*; TRC Data Series: College Station, Texas, 1994. An update to 86PED, this volume contains enthalpy of formation data for ca. 3000 compounds. In addition, the volume describes newly fitted parameters for Pedley's method which allows the prediction of enthalpies of formation for the gaseous and liquid states. This prediction scheme is, in effect, a third order group contribution method which takes into account effects two atoms removed from the central atom. In this book, the data are arranged by families of compounds and the predicted values are tabulated with the experimental data. A highly recommended source of thermodynamic data for organic species.

96DOM/HEA. E. S. Domalski and E. D. Hearing, "Heat Capacities and Entropies of Organic Compounds in the Condensed Phase - Vol. III", *J. Phys. Chem. Ref. Data* **1996**, *25*, 1-525. The most up to date publication for the subject matter. See description under 84DOM/HEA above.

Other Published Literature of Note:

PATAI. S. Patai (and occasional co-editors), from the series *The Chemistry of Functional Groups*; Wiley: ca. 1964 - current date. A large and very comprehensive series of books devoted to a single or several functional groups (examples: Carboxylic Acids and Esters 1969; Carbon Nitrogen Double Bonds, 1970 ; Carbon - Carbon Triple Bond, 1978). There is always a chapter on thermochemistry and many times these chapters are the only available source of collected data. A recent chapter on sulfonic acids and derivatives is an excellent example of the detailed and comprehensive nature of these chapters (*4*).

Electronic Databases for Thermochemistry

The following databases are available in electronic format and in many cases are available from on-line searching services such as STN (Scientific and Technical Information Network: http://www.cas.org/stn.html). Costs for searching this and other databases should be obtained from the individual companies. Much of the material below was obtained directly from the subject web sites.

DIPPR - PURE COMPONENT DATA COMPILATION V11.3.
(http://www2.hrz.th-darmstadt.de/ze/online/stn-info/ONLINE/DBSS/dipprss.html) Technical Database Services, Inc. Tel: (212) 245-0044. DIPPR Version 10.2 from TDS Numerica provides pure component data for 26 constant and 13 temperature-dependent properties of over 1,500 industrially important chemicals. The DIPPR PC database and its software interface were developed by TDS to meet the technical specifications of the Design Institute of Physical Property Data (DIPPR) of the American Institute of Chemical Engineers (AIChE), the sponsors of the compilation. The PC version of DIPPR is available through Numerica , the collection of online information services and user-hosted products distributed by Technical Database Services.
      The DIPPR Project 801 Data Compilation contains evaluated data from both experimental sources and estimations developed by the project staff in the Department of Chemical Engineering at the Pennsylvania State University. DIPPR data are analyzed according to the laws of physical chemistry and show uniformity and consistency of homologous series by chemical groups.

GMELIN. (http://info.cas.org/ONLINE/DBSS/gmelinss.html) Gmelin Online is a property database of inorganic and organometallic compounds, available on STN. It contains a wide range of chemical and physical properties including thermodynamic data. The data come from the Gmelin Handbook of Inorganic and Organometallic Chemistry (all volumes published up to 1975) and from the primary literature (1983 to date; literature years 1976 to 1982 are currently being added). As of late 1995, the database contained nearly 1,000,000 substances, among them about 450,000 coordination compounds.

FACT. (http://www.crct.polymtl.ca/FACT/factwhat.htm) F*A*C*T (Facility for the Analysis of Chemical Thermodynamics) is a fully integrated thermochemical database which couples software with self-consistent critically assessed thermodynamic data. F*A*C*T is mainly an inorganic and metallurgical tool which contains 12 program modules for performing various types of calculations in chemical thermodynamics. The product is available for use on-line or as a PC version. The interested reader is encouraged to visit the web site above for more detailed information.

NIST JANAF Thermochemical Tables Database.
(http://www.nist.gov/srd/janaf.htm) Standard Reference Data Program, National Institute of Standards and Technology, Bldg. 820, Rm. 113, Gaithersburg, MD 20899-0001. Tel. (301) 975-2208, fax (301) 926-0416, e-mail srdata@nist.gov. Price: $200. The NIST JANAF Thermochemical Tables provide a compilation of critically evaluated thermodynamic properties of approximately 1800 substances over a wide range of temperatures. Recommended temperature-dependent values are provided for inorganic substances and for organic substances containing only one or two carbon atoms. These tables cover the thermodynamic properties with single-phase and multi-phase tables for the crystal, liquid, and ideal gas.

The properties tabulated are heat capacity, entropy, Gibbs energy function, enthalpy, enthalpy of formation, Gibbs energy of formation, and logarithm of the equilibrium constant for formation of each compound from the elements in their standard reference states. Transition data are included. This database is consistent with the Third Edition of the JANAF Thermochemical Tables, published as Supplement No. 1 to Vol. 14 of the Journal of Physical and Chemical Reference Data. The database is currently available online through STN.

NIST Chemical Thermodynamics Database.
(http://www.nist.gov/srd/therch.htm) Standard Reference Data Program, National Institute of Standards and Technology, Bldg. 820, Rm. 113, Gaithersburg, MD 20899-0001. Tel. (301) 975-2208, fax (301) 926-0416, e-mail srdata@nist.gov. Price: $200. This database contains recommended values for selected thermodynamic properties of more than 15,000 inorganic substances and organic substances containing only one or two carbon atoms. These properties include the following: standard state properties at 298.15 K and 1 bar; enthalpy of formation from the elements in their standard state; gibbs energy of formation for the elements in their standard state; enthalpy $H°(298.15 \text{ K}) - H°(0 \text{ K})$; heat capacity at constant pressure; entropy. The data files are also available online through STN.

IVTANTHERMO-PC. (http://www.nist.gov/srd/spec5.htm) Special Database 5, Standard Reference Data Program, National Institute of Standards and Technology, Bldg. 820, Rm. 113, Gaithersburg, MD 20899-0001. Tel. (301) 975-2208, fax (301) 926-0416, e-mail srdata@nist.gov. Price: $950. See also the published volume 89GUR/VEY in the Published Literature Section above.

IVTANTHERMO is a computerized system providing information on the thermodynamic properties of about 2,300 substances (containing 85 elements and the electron) in the standard state over a wide temperature range. It was developed by scientists at the Institute of High Temperatures in Moscow. The software permits the calculation of thermodynamic parameters of chemical reactions and the composition of chemical systems. All recommended values are cited with a reliability assessment.

This database is capable of thermodynamic analysis of: new high-temperature processes, including combustion processes; the optimization of chemical processes, including synthesis of refractory materials and microelectronic materials; stability of materials at high temperatures and in various media; chemical processes occurring in power-generating facilities, including nuclear plants; the optimization of raw materials; use and waste management; the emissions of incinerators and industrial exhaust gases into the atmosphere. The software provides a choice of formats, temperature scales, and energy units. The database contains the following information on each substance: substance name and chemical formula; accuracy of thermodynamic properties; isobaric heat capacity; entropy; change of enthalpy; Gibbs energy function; equation(s) fitting tabulated values of Gibbs energy function; enthalpy of formation; equilibrium constant. The software provides a choice of formats, temperature scales, and energy units.

The methodology of the evaluation process of IVTANTHERMO is described in Volume 1 of the hard copy series Thermodynamic Properties of Individual Substances which is available from CRC Press, Inc. The data contained in this database will be included in a series of publications (5 volumes) available from CRC Press, Inc. Volume 3 was published in February 1994; Volume 4 is in preparation (English version).

Beilstein. (http://www2.hrz.th-darmstadt.de/ze/online/stn-info/ONLINE/CATALOG/beilstein.html) Beilstein Institute for Organic Chemistry. Beilstein contains organic chemical structures, preparation and reaction information, and numeric property information (including thermodynamic properties). The source for BEILSTEIN is the Beilstein Handbook of Organic Chemistry (main volume and supplements 1-5) covering literature from 1779-1979. Also included are data from 84 leading journals in organic chemistry for the period 1980-1991. Records (6.1 million) are searchable by structure, and all text fields are searchable.

CODATA Key Values for Thermodynamics.
(http://www.nrc.ca/programs/codata/databases/key1.html) The Committee on Data for Science and Technology (CODATA) has conducted a project to establish internationally agreed values for the thermodynamic properties of key chemical substances. This table presents the final results of the project. Use of these recommended, internally consistent values is encouraged in the analysis of thermodynamic measurements, data reduction, and preparation of other thermodynamic tables. The table includes the standard enthalpy of formation at 298.15 K, the entropy at 298.15 K, and the quantity $H°(298.15 \text{ K}) - H°(0)$.

Web Sites with Thermochemical Themes

Other than those listed above, there are a number of web sites of note with thermochemical themes. Prof. Mansoori from the University of Illinois Chicago (UIC) Thermodynamics Research Laboratory has consolidated an index of many of these links into two web sites. Because there is always the potential for any of these sites to disappear without notice, we only give the web addresses for the UIC sites and a few others of note below. In a personal communication with Prof. Mansoori, we were assured that his compiled site will be supported as a long-term project. Readers are encouraged to assist in this valuable effort by informing Prof. Mansoori of any other sites which fit the theme.

The two sites at the UIC Thermodynamics Research Laboratory are *Thermodynamic and Related Data Sites* (http://www.uic.edu/~mansoori/Thermodynamic.Data.and.Property_html) and *Classical, Quantum & Statistical Mechanics & Thermodynamics Educational* Sites (http://www.uic.edu/~mansoori/Thermodynamics.Educational.Sites_html)

A partial listing of links and selected URL addresses from the *Thermodynamic and Related Data Sites* web page is given below:

- CODATA http://www.cisti.nrc.ca/programs/codata/
- F*A*C*T (Facility for the Analysis of Chemical Thermodynamics) / C.W. Bale http://www.crct.polymtl.ca/fact/fact
- Fundamental Constants / NIST http://physics.nist.gov/funcon.html htm
- NPL/UK Metallurgy. Thermochem. Sec. Thermodynamic Database
- Molar SS Thermodynamic Properties of Infinitely Dilute Aqueous Solutes / J.A. Plambeck
- Molar SS Thermodynamic Properties of Pure Substances / J.A. Plambeck
- NIST WebBook: A gateway to the data collection of the National Institute of Standards and Technology http://webbook.nist.gov/
- Thermodynamics databases for inorganic and metallurgical systems / SGTE
- Thermodynamic and heat transfer properties of THERMINOLS / Monsanto
- Thermodynamics and Thermochemistry Data Program / NIST http://www.nist.gov/srd/thermo.htm
- Thermodynamic and Thermophysical Data to Download / K&K Assoc.
- Thermodynamic Properties of Minerals Databases / T.H. Brown
- ThermoDex: An Index of Selected Thermodynamic Data Handbooks
- ASME Thermodynamic Software to Download
- Boiling Point Online Prediction / ACD Labs
- ChemFinder substructure-searchable small-molecule database / CambrigeSoft
- Mathtrek Chemical Equilibrium Software Systems / W. Smith
- MINEQL+: A Chemical Equilibrium Modeling System
- Mixcalc: Gas mixture thermodynamic property calculation / F. Mejia
- Property Prediction of Inorganic Compounds / ChemSage
- Pure Mineral Solids Thermodynamic Properties / M.S. Ghiorso
- Silicate Liquids Thermodynamic Properties / M.S. Ghiorso
- Simulation Of Phase Equilibria Download / Shell

- SteamTab / ChemicaLogic
- Thermo-Calc: For Thermodynamic Calculations of Inorganic Systems / RIT, Sweden http://www.met.kth.se/mse/tc-leaf.html
- ThermoChemical Calculator / D.G. Goodwin
- Thermodynamic Diagrams Page / Quest Consultants
- Thermodynamic Properties Page / Quest Consultants
- Thermodynamic Properties Page with final pressure other than one atmosphere / Quest Consultants
- Vapor Pressure Online Prediction / ACD Labs
- Water Thermodynamic Properties / M.S. Ghiorso

A partial listing of links in the *Classical, Quantum & Statistical Mechanics & Thermodynamics Educational Sites* web page:

- About Temperature and Intro. to Thermodynamics / UNIDATA
- Properties of Heat and Matter / UC Berkeley
- First Year Engineering Thermodynamics / Loughborough U
- Basics of thermodynamic laws / WPI
- The 1st, 2nd and 3rd Laws of Thermodynamics / UC Berkeley
- Entropy on the World Wide Web / Washington U
- Fundamentals of Thermodynamics / U of Pittsburgh
- Basic Thermodynamic Cycles and Components / Virginia Tech
- CyclePad: To construct and analyze thermodynamic cycles / Northwestern U
- Introduction to Thermodynamics Courseware Module / Tuskegee U
- Thermodynamics Demonstrations/U North Carolina
- An Introduction to Applied Thermodynamics / Michigan State & U of Akron
- Equation of State / Shell
- Phase Equilibria / Shell
- Global phase diagrams / U of Cologne
- Review of Basic Chemical Equilibrium / U of Richmond
- A Brief Review of Elementary Quantum Chemistry / UC Berkeley
- Visual Quantum Mechanics / Kansas State
- Quantum Mechanics, Statistical Mechanics, Thermodynamics / Brown U
- Thermal and Statistical Physics plus Mathcad / Cleveland State
- Statistical Mechanics: 1. Michigan State ; 2. UC San Diego ; 3. U of Chicago
- Molecular Monte Carlo Method / Cooper Union
- Quantum Mechanics / UC San Diego
- Ergodic Theroy / U of East Anglia
- Metallurgical: 1.Thermodynamics/Tulane U; 2.Phase Diagram Basics/Virginia Tech; 3. The Phase Diagram Web/Georgia Tech
- Polymers Thermodynamics and Statistical Mechanics / UTK
- Heavy Organics (Asphaltene/Bitumen, Resin, Wax, etc.) Deposition from Petroleum Fluids / UIC
- Thesaurus / NASA
- Encyclopedia of Thermodynamics / U of Basel
- Searching For Chemical And Physical Properties Of Substances / Indiana U

NIST WebBook. (http://webbook.nist.gov) The NIST WebBook will provide access to the full array of data compiled and distributed by NIST under the Standard Reference Data Program.

The current edition, the Chemistry WebBook, contains thermochemical data for over 5000 organic and small inorganic compounds, phase transition enthalpies and temperatures, vapor pressure, heat of reaction data for over 3000 reactions, IR spectra for over 5000 compounds, mass spectra for over 8000 compounds, ion energetics data for over 14000 compounds, and thermophysical property data for 13 fluids. You may search for data on specific compounds in the WebBook based on name, chemical formula, CAS registry number, molecular weight, ionization energy or proton affinity.

Literature Cited

1. K.B. Lipkowitz and D.B. Boyd, "Reviews in Computational Chemistry 7", Appendix entitled "Compendium of Software for Molecular Modeling" by D.B. Boyd, VCH Publishing, 1996.
2. Curtiss, L. A.; Raghavachari, K.; Trucks, G. W.; Pople, J. A. *J. Chem. Phys.* **1991**, *94*, 7221-7230.
3. Foresman, J. B., Frisch Æ. "Exploring Chemistry with Electronic Structure Methods: A Guide to Using *Gaussian*", Gaussian, Inc. 4415 Fifth Ave. Pittsburgh, PA ISBN 0-9636769-0-3, p. 116.
4. J. F. Liebman, "Thermochemistry of Sulfonic Acids and Their Derivatives," in The Chemistry of Sulfonic Acids, Esters, and Their Derivatives, S. Patai and Z. Rappaport (editors), John Wiley and Sons Ltd. (1991), Chapter 8, pp. 283-321.

Appendix B

Essential Statistical Thermodynamics

Karl K. Irikura

Physical and Chemical Properties Division, National Institute of Standards and Technology, Gaithersburg, MD 20899

Some computational methods, particularly *ab initio* techniques, produce detailed molecular information but no thermodynamic information directly. Further calculations are needed to generate familiar, ideal-gas quantities such as the standard molar entropy ($S°$), heat capacity ($C_p°$), and enthalpy change [$H°(T)$-$H°(0)$]. This Appendix details the necessary procedures, including worked examples. Thermochemical calculations can be extended to transition states of chemical reactions. Procedures are provided for converting such information into rate constants. Tables are also provided for unit conversions and physical constants.

Statistical thermodynamics calculations are necessary to compute properties as functions of temperature. In some computations, such as *ab initio* electronic calculations of molecular energy, the raw results do not even correspond to properties at absolute zero temperature and must always be corrected. All the corrections are based upon molecular spectroscopy, with temperature-dependence implicit in the molecular partition function, Q. The partition function is used not only for theoretical predictions, but also to generate most published thermochemical tables. Many data compilations include descriptions of calculational procedures (*1-3*).

Corrections Unique to *Ab Initio* Predictions

By convention, energies from *ab initio* calculations are reported in hartrees, the atomic unit of energy (1 hartree = 2625.5 kJ/mol = 627.51 kcal/mol = 219474.6 cm^{-1}) (*4*). These energies are negative, with the defined zero of energy being the fully-dissociated limit (free electrons and bare nuclei). *Ab initio* models also invoke the approximation that the atomic nuclei are stationary, with the electrons swarming about them. This is a good approximation because nuclei are much heavier than electrons. Consequently, the resulting energies are for a hypothetical, non-vibrating molecule. Although oscillators may be at rest in classical mechanics, real (quantum-mechanical) oscillators are always in motion. The small residual motion at absolute zero temperature is the *zero-point vibrational energy*, abbreviated ZPVE or ZPE. For a simple harmonic oscillator, the ZPE equals one-half the vibrational frequency. Although all real molecular vibrations are at least slightly anharmonic, they are usually approximated as harmonic. Thus, the molecule's ZPE may be taken as one-half the sum of the vibrational frequencies. In equation 1, N is the number of atoms in the molecule and the v_i are the fundamental

$$ZPE = \frac{1}{2} \sum_{i=1}^{3N-6} v_i \tag{1}$$

vibrational frequencies. There are $3N-6$ vibrations in a non-linear molecule and $3N-5$ in a linear molecule; equation 1 is for the more common non-linear case. The ZPE must be added to the raw *ab initio* energy to obtain an energy corresponding to absolute zero temperature, $T = 0$ K.

In practice, the ZPE correction is slightly complicated by the observation that *ab initio* vibrational frequencies are often in error by +5% to +10%. To compensate for this error, the computed frequencies are usually multiplied by empirical scaling factors. The most recent recommendations are those of Scott and Radom (5). For example, they suggest scaling HF/6-31G* frequencies by 0.8953 to predict vibrational spectra (i.e., fundamental frequencies), by 0.9135 for the computation of ZPEs, by 0.8905 to predict enthalpy differences $H°(298.15) - H°(0)$, and by 0.8978 to predict $S°(298.15)$. The methods for computing these quantities are described below. Common abbreviations and acronyms of the *ab initio* literature are defined in the glossary (Appendix D) of this book. In this Appendix, the degree sign (°) that indicates ideality and standard pressure (1 bar) is omitted except where the thermal electron convention for ions is being emphasized (see below).

Enthalpies of formation depend upon the thermodynamic conventions for reference states of the elements. Since this information is not intrinsic to an isolated molecule, an *ab initio* reaction energy (i.e., energies for at least two molecules) must be combined with experimental data to compute an enthalpy of formation, $\Delta_f H°$.

Example: $\Delta_f H°_0$ of hydrogen fluoride. There are many levels of approximation in *ab initio* theory; several are described in the chapters of this book. For the present example, we choose the CCSD(T)/aug-cc-pVTZ//B3LYP/6-31G(d) level. The notation indicates that (1) molecular geometries are calculated at the density-functional B3LYP level using the 6-31G(d) basis set and (2) molecular electronic energies are calculated at the high CCSD(T) level of theory using the rather large aug-cc-pVTZ basis set. To compute an enthalpy of formation for HF, we must also choose a balanced chemical reaction for which to calculate an energy. We choose arbitrarily the reaction shown in equation 2. Note that the ideal-gas energy and enthalpy are equal at 0 K, since $H = E + PV = E + nRT = E$.

$$H_2 + F_2 \rightleftharpoons 2\,HF \tag{2}$$

The optimized B3LYP/6-31G(d) bond lengths are 0.743, 1.404, and 0.934 Å for H_2, F_2, and HF respectively, in reasonably good agreement with the experimentally derived values $r_e = 0.741$, 1.412, and 0.917 Å respectively (6). The calculated (harmonic) B3LYP/6-31G(d) vibrational frequencies are 4451, 1064, and 3978 cm^{-1} for H_2, F_2, and HF respectively, in modest agreement with the experimentally derived harmonic frequencies $\omega_e = 4401$, 917, and 4138 cm^{-1} respectively (6). Since 1 cm^{-1} equals only 0.01196 kJ/mol, small errors in ZPE will not cause significant errors in the final enthalpy of formation. Scaling the calculated frequencies by 0.9806 (5) and substituting them into equation 1 yields ZPE = 2182, 522, and 1950 cm^{-1} = 0.009943, 0.002377, and 0.008887 hartree for H_2, F_2, and HF respectively. At these optimized geometries, the CCSD(T)/aug-cc-pVTZ energies are E_e = -1.172636, -199.313519, and -100.349402 hartree for H_2, F_2, and HF respectively. Adding the ZPEs thus leads to enthalpies (or energies) at $T = 0$ K of E_0 = -1.162693, -199.311142, and -100.340515 hartree for H_2, F_2, and HF respectively. The calculated enthalpy change is then $\Delta_r H°_0$(reaction 2) = -0.207194 hartree = -544.0 kJ/mol. Using the experimental (defined, in these cases) enthalpies of formation for H_2 and F_2 of 0 and 0 kJ/mol (1), we obtain $\Delta_f H°_0$(HF) = -272.0 kJ/mol. This is in good agreement with the experimental value of -272.5 ± 0.8 kJ/mol (1). All the *ab initio* calculations for this example were done on a personal computer.

General Relationships of Statistical Thermodynamics

In the present context, statistical thermodynamics is meant to include the methods used to convert molecular energy levels into macroscopic properties, especially enthalpies, entropies, and heat capacities. Molecular energy levels arise from molecular translation (i.e., motion through space), rotation, vibration, and electronic excitation. This information constitutes the spectroscopy of the molecule of interest and can be obtained experimentally or from calculations.

Partition Function. The molecular energy levels ϵ_i are used to compute the *molecular partition function*, usually denoted by the symbol Q, as shown in equation 3. The sum

$$Q(T) = \sum_i \exp(-\epsilon_i / kT) \tag{3}$$

extends over all energy levels. (Sometimes this sum is written only over all unique energy levels, in which case a level degeneracy g_i must be included in the sum.) However, for very high temperatures at which the molecule becomes unstable, the extent of the sum may be ambiguous. Tabulated thermochemical data must be used with caution under such conditions; the values (1) may depend strongly upon the high-energy cutoff procedure adopted and (2) may deviate implicitly from the ideal-gas model.

One typically chooses the lowest energy level to be the zero of energy, so that no levels lie at negative energies. From equation 3 it follows that the largest contributions to Q are from the lowest energy levels. Conversely, levels that lie far above kT (207 cm^{-1} at room temperature) have only a minor effect on Q and its derivative thermodynamic quantities.

Thermodynamic Functions. Given the partition function, the usual molar thermodynamic functions can be calculated based upon the following general equations. Equation 4 is for the entropy, equation 5 for the heat capacity at constant volume, equation 6 for the heat capacity at constant pressure, and equation 7 for the enthalpy

$$S = Nk\left[\frac{\partial}{\partial T}(T\ln Q) - \ln N + 1\right] \tag{4}$$

$$C_v = NkT\frac{\partial^2}{\partial T^2}(T\ln Q) \tag{5}$$

$$C_p = C_v + R \tag{6}$$

$$H(T) - H(0) = \int_0^T C_p\, dT = \frac{RT^2}{Q}\frac{\partial Q}{\partial T} + RT \tag{7}$$

difference relative to absolute zero temperature. N is Avogadro's number (6.022137×10^{23}), k is the Boltzmann constant (1.38066×10^{23} J/K), and the ideal-gas constant $R \equiv Nk$ (*4*). The last two terms inside the brackets in equation 4 arise from the indistinguishability of identical molecules, which requires a factor of $(1/N!)$ in the partition function for the ensemble. Expressions 4-7 may more easily be evaluated using equations 8-11 for the various derivatives.

$$\frac{\partial}{\partial T}(T\ln Q) = \ln Q + \frac{T}{Q}\frac{\partial Q}{\partial T} \tag{8}$$

$$\frac{\partial^2}{\partial T^2}(T\ln Q) = \frac{2}{Q}\frac{\partial Q}{\partial T} + \frac{T}{Q}\frac{\partial^2 Q}{\partial T^2} - \frac{T}{Q^2}\left(\frac{\partial Q}{\partial T}\right)^2 \tag{9}$$

$$\frac{\partial Q}{\partial T} = \frac{1}{kT^2}\sum_i \epsilon_i \exp(-\epsilon_i/kT) \tag{10}$$

$$\frac{\partial^2 Q}{\partial T^2} = \frac{-2}{T}\frac{\partial Q}{\partial T} + \frac{1}{k^2T^4}\sum_i \epsilon_i^2 \exp(-\epsilon_i/kT) \tag{11}$$

Practical Calculations

A complete set of molecular energy levels is almost never available. To simplify the problem, one usually adopts a model in which translation, rotation, vibration, and electronic excitation are uncoupled. In other words, one makes the approximation that the different types of motion are unaffected by each other and do not mix together. This leads to a separability of Q into four factors that correspond to separate partition functions for translation, rotation, vibration, and electronic excitation. This is shown in equation 12, where the explicit dependence upon temperature has been dropped for

$$Q = Q_{trans}Q_{rot}Q_{vib}Q_{elec} \tag{12}$$

simplicity. When electronically excited states are considered, one often assumes that the translational, rotational, and vibrational spectra of the excited state are the same as those of the ground electronic state. This is crude, but is convenient when no other information is available. Moreover, if the excited state lies far above kT, the final results will not be sensitive to such details.

Different energy units are used conventionally in the fields of molecular spectroscopy, quantum chemistry, and thermochemistry. To provide some feeling for magnitudes, the values of the thermal energy kT, at "room temperature" (298.15 K) and at 1000 K, are listed in Table I in several units. In this Appendix, all units are of the SI (*Système International*: kg, m, s, Pa, K) unless otherwise indicated.

Table I. Thermal energy (kT) at two temperatures, expressed in various units

Unit	"Room temperature"	1000 K
kelvin (K)	298.15	1000
wavenumber (cm^{-1})	207.2	695.0
Hertz (s^{-1})	6.212×10^{12}	2.084×10^{13}
kJ/mol	2.479	8.314
kcal/mol	0.592	1.987
electron volt (eV)	0.0257	0.0862
hartree (atomic unit)	0.000944	0.003167

Translational Partition Function. Rigorously, Q_{trans} must be calculated from a sum over all the translational energy levels that are available to a molecule confined to a cubic box of volume $V = RT/p$ (molar volume of an ideal gas at temperature T and pressure p). This is seldom done. Instead, the sum is approximated as an integral to obtain equations 13-16. This approximation is good as long as $m^{3/2}T^{5/2}p^{-1} \gg h^3(2\pi)^{-3/2}k^{-5/2}$ (3). At the standard pressure $p = 1$ bar $= 10^5$ Pa $= 0.986923$ atm, this condition is met for sufficiently heavy molecules, m (in amu) $\gg 11.4\ T^{-5/3}$, and for sufficiently high temperatures, $T \gg 4.31\ m^{-3/5}$ (m expressed in amu). Fortunately, this covers the conditions of common chemical interest. For an atomic ideal gas, there is no vibrational or rotational motion.

$$Q_{trans} = (2\pi mkT)^{3/2}h^{-3}\ V \tag{13}$$

$$S_{trans} = R[(3/2)\ln(2\pi m/h^2) + (5/2)\ln(kT) - \ln(p) + 5/2] \tag{14}$$

$$C_{p,trans} = (5/2)\,R \tag{15}$$

$$[H(T)-H(0)]_{trans} = (5/2)\,RT \tag{16}$$

As an example, we can calculate the standard entropy for neon ideal gas at T = 298.15 K. The atomic mass is converted to SI units using the equivalence N_A amu = 0.001 kg, where N_A = 6.022 × 10²³ mol⁻¹ is the Avogadro constant. Thus for ²⁰Ne (m = 19.992 amu), m = 3.320 × 10⁻²⁶ kg. The values of the physical constants are h = 6.626 × 10⁻³⁴ J s, k = 1.381 × 10⁻²³ J K⁻¹, and $R = kN_A$ = 8.3145 J mol⁻¹ K⁻¹ (*7,4*). The standard pressure is p = 10⁵ Pa. Substituting these values into equation 14 yields $S_{298.15}$(²⁰Ne) = 146.21 J mol⁻¹ K⁻¹. For ²¹Ne (m = 20.994 amu), $S_{298.15}$(²¹Ne) = 146.82 J mol⁻¹ K⁻¹, and for ²²Ne (m = 21.991 amu), $S_{298.15}$(²²Ne) = 147.40 J mol⁻¹ K⁻¹. Averaging these values using the natural abundances of 90.48%, 0.27%, and 9.25%, respectively (*7,8*), we find $S_{298.15}$ = 146.32 J mol⁻¹ K⁻¹ for the naturally occurring isotopic distribution. This agrees well with the accepted value of 146.33 J mol⁻¹ K⁻¹ (*9*).

Rotational Partition Function. The free rotation of a rigid molecule is also quantized (the angular momentum and its projection are integer multiples of $h/2\pi$), so the rotational energy is restricted to certain discrete levels. Rotational spectra are characterized by the constants A, B, and C, where $A \equiv h/(8\pi^2 I_A)$ and likewise for B and C. The quantities $I_{A,B,C}$ are the principal moments of inertia of the molecule, with the convention $I_A \le I_B \le I_C$ (or $A \ge B \ge C$). Many computer programs, including *ab initio* packages, report the rotational constants when provided with a molecular geometry. The moments can also be calculated manually as the eigenvalues of the inertial tensor, which has elements like $I_{xy} = -\sum m_i x_i y_i$ and $I_{xx} = +\sum m_i (y_i^2 + z_i^2)$, where the index i runs over all atoms in the molecule and the coordinate origin is at the center of mass. Linear molecules ($I_A = 0$) are described by a single rotational constant, B, and a single moment of inertia, I. Details may be found in textbooks of molecular spectroscopy.

Fortunately, at high enough temperatures ($kT \gg hA$), the sum can be replaced by an integral as it is for translation. In the general case, the rotational partition function is given by equation 17. For linear molecules, equation 18 should be used instead. In these

$$Q_{rot} = \frac{8\pi^2}{\sigma h^3}(2\pi kT)^{3/2}(I_A I_B I_C)^{1/2} = (kT/h)^{3/2}(ABC)^{-1/2}\pi^{1/2}\sigma^{-1} \tag{17}$$

$$Q_{rot}^{linear} = \frac{8\pi^2 I k T}{\sigma h^2} = kT/(\sigma hB) \tag{18}$$

and subsequent equations, the symbol σ denotes the "rotational symmetry number" or "external symmetry number" for the molecule. This is the number of unique orientations of the rigid molecule that only interchange identical atoms. It preserves parity restrictions on the interchange of identical nuclei when summation is replaced by integration. Identifying the correct symmetry number is a common point of difficulty; it is discussed further below.

For the typical case (equation 17), the thermodynamic functions are given by equations 19-21. For linear molecules (equation 18), equations 22-24 are used instead.

$$S_{rot} = R[\ln(8\pi^2/\sigma) + (3/2)\ln(2\pi kT/h^2) + (1/2)\ln(I_A I_B I_C) + 3/2]$$
$$= R[(3/2)\ln(kT/h) - (1/2)\ln(ABC/\pi) - \ln(\sigma) + 3/2] \tag{19}$$

$$C_{p,rot} = (3/2)R \tag{20}$$

$$[H(T) - H(0)]_{rot} = (3/2)RT \tag{21}$$

$$S_{rot}^{linear} = R[\ln(8\pi^2 IkT/\sigma h^2) + 1] = R[\ln(kT/\sigma hB) + 1] \tag{22}$$

$$C_{p,rot}^{linear} = R \tag{23}$$

$$[H(T) - H(0)]_{rot}^{linear} = RT \tag{24}$$

External Symmetry Number. Some computer programs, such as many *ab initio* packages, determine the molecular symmetry and external symmetry number (σ) automatically. If such a program is unavailable, σ may be determined by hand. With practice, this becomes very fast.

If you are familiar enough with group theory to identify the molecule's point group (*10*), then σ can be determined from Table II (*11*). Without identifying the point group, one can count manually the number of orientations of the rigid molecule that interchange only identical atoms.

Table II. Symmetry numbers corresponding to symmetry point groups

Group	σ	Group	σ	Group	σ	Group	σ
$C_1, C_i, C_s, C_{\infty v}$	1	$D_{\infty h}$	2	T, T_d	12	O_h	24
I_h	60	S_n	$n/2$	C_n, C_{nv}, C_{nh}	n	D_n, D_{nh}, D_{nd}	$2n$

For example, the benzene molecule (C_6H_6) belongs to the D_{6h} point group. From Table II, $\sigma = 12$. Alternatively, one can draw the molecule as a hexagon with numbered vertices. Rotating the drawing by $n \times 60°$, where n runs from 0 to 5, generates six different orientations that are distinguished only by the artificial numbering of the vertices. Each of these six orientations can be flipped over to generate another orientation, for a total of 12 unique orientations, $\sigma = 12$.

Another example is methyl chloride, CH_3Cl. This belongs to the C_{3v} point group, so $\sigma = 3$. Alternatively, one can artificially number the hydrogen atoms and see that there are three unique orientations, related by rotations of $n \times 120°$ ($n = 0\text{-}2$) around the C-Cl bond axis.

Chlorobenzene, C_6H_5Cl, belongs to the C_{2v} point group, so $\sigma = 2$. Alternatively, one can again number the hydrogen atoms and see that there are two unique orientations, related by rotations of $n \times 180°$ ($n = 0$-1) around the C-Cl bond axis. In contrast, toluene ($C_6H_5CH_3$) belongs to the C_s point group, so $\sigma = 1$. There are no ways to rotate or flip the molecule *rigidly* that will leave it unchanged. Allowing the methyl group to rotate leads to an *internal* symmetry number which is discussed below, following the section on internal rotation.

Vibrational Partition Function. To complete the simple rigid-rotator/harmonic oscillator (RRHO) model, one must consider the molecular vibrations. As indicated in the discussion of ZPE (equation 1), a molecule that contains N atoms has $3N$-6 vibrational frequencies ($3N$-5 for linear molecules). The partition function is given in equation 25, where the product runs over all vibrational frequencies ν_i. The corresponding thermodynamic functions are given by equations 26-28.

$$Q_{vib} = \prod_i \left(1 - e^{-h\nu_i/kT}\right)^{-1} \tag{25}$$

$$S_{vib} = -R\sum_i \ln\left(1 - e^{-h\nu_i/kT}\right) + R\sum_i \frac{h\nu_i}{kT} \frac{e^{-h\nu_i/kT}}{\left(1 - e^{-h\nu_i/kT}\right)} \tag{26}$$

$$C_{p,\,vib} = R\sum_i \left(\frac{h\nu_i}{kT}\right)^2 \frac{e^{-h\nu_i/kT}}{\left(1 - e^{-h\nu_i/kT}\right)^2} \tag{27}$$

$$[H(T) - H(0)]_{vib} = RT\sum_i \left(\frac{h\nu_i}{kT}\right) \frac{e^{-h\nu_i/kT}}{\left(1 - e^{-h\nu_i/kT}\right)} \tag{28}$$

Example: Hydrogen Fluoride. Earlier we used the results of *ab initio* calculations to obtain a value for $\Delta_f H°_0(HF)$. The other equations above permit us to compute *ab initio* thermodynamic functions, which will provide an enthalpy of formation at the more useful temperature of 298.15 K. Results are summarized in Table III. For simplicity, we will neglect the naturally occurring heavy isotopes of hydrogen. The molecular weight of $^1H^{19}F$ is 20.006 amu. Using equation 14, as done above for neon, leads to $S_{trans} = 146.22$ J mol^{-1} K^{-1}. HF is a linear molecule, so we use equation 22 to calculate S_{rot}. The *ab initio* calculation reports a rotational constant $B = 605.64$ GHz $= 6.0564 \times 10^{11}$ s^{-1} based upon the calculated B3LYP/6-31G(d) equilibrium geometry and the most common isotopes. This molecule belongs to the $C_{\infty v}$ point group ($\sigma = 1$); there are no identical nuclei that can be interchanged by any rotation. Hence $S_{rot} = 27.67$ J mol^{-1} K^{-1}. For the vibrational contribution, we scale the B3LYP/6-31G(d) frequency of 3987 cm^{-1} by 1.0015 as suggested for entropies (5) to obtain $\nu = 3993$ cm^{-1}. This is multiplied by the speed of light, $c = 2.998 \times 10^{10}$ cm s^{-1} (7,4), to convert wavenumbers to SI frequency units, $\nu = 1.197 \times 10^{14}$ s^{-1}. Thus $h\nu/kT = 19.27$ and equation 26 yields $S_{vib} = 7.22 \times 10^{-7}$ J mol^{-1} K^{-1}. The total entropy is $S_{298.15} = S_{trans} + S_{rot} + S_{vib} = 173.89$ J mol^{-1} K^{-1}, in good agreement with the accepted value of 173.78 J mol^{-1} K^{-1} (9).

For enthalpy and heat capacity, the B3LYP/6-31G(d) frequency is scaled by 0.9989 (5) to obtain $\nu = 3983$ cm^{-1}. The heat capacity C_p(HF) is calculated using equations 15, 23, and 27, leading to $C_p = C_{p,\,trans} + C_{p,\,rot} + C_{p,\,vib} = (5/2)R + R + 1.38 \times 10^{-5}$ J mol^{-1} K^{-1} = 29.10 J mol^{-1} K^{-1}. This compares well with the accepted value of 29.14 J mol^{-1} K^{-1} (1). Finally, the enthalpy difference can be computed using equations 16, 24, and 28 to be $[H(298.15)\text{-}H(0)] = (5/2)\,RT + RT + 2.14 \times 10^{-4}$ J mol^{-1} = 8.68 kJ mol^{-1}. This can be used to compute $\Delta_f H^{\circ}_{\,298.15}$(HF) = $\Delta_f H^{\circ}_{\,0}$(HF) + $[H(298.15)\text{-}H(0)]_{\text{HF}}$ - $[H(298.15)\text{-}H(0)]_{\text{elements}}$. Taking the *ab initio* value $\Delta_f H^{\circ}_{\,0}$(HF) = -272.0 kJ/mol from above, the calculated enthalpy difference of 8.68 kJ/mol for HF, and the accepted enthalpy differences of (8.47)/2 and (8.83)/2 kJ/mol for ½H$_2$ and ½F$_2$ (9), we obtain $\Delta_f H^{\circ}_{\,298.15}$(HF) = -272.0 kJ/mol, in agreement with the accepted value of -273.3 ± 0.7 kJ mol^{-1} (9) [-272.5 ± 0.8 kJ/mol is listed in older ref (1)].

Table III. Results for Hydrogen Fluoride Example

Contribution	S, J/(mol·K)	C_p, J/(mol·K)	[H(298.15)-H(0)], kJ/mol
Translation	146.22	20.79	6.20
Rotation	27.67	8.31	2.48
Vibration	7×10^{-7}	1×10^{-5}	2×10^{-4}
Total	*173.89*	*29.10*	*8.68*

Electronic Partition Function. Although they may not have low-lying electronic excited states, some molecules have degenerate electronic ground states. Free radicals are a common example. They may have unpaired electrons in their electronic ground states and a net electron spin of $S = n_{unpaired}/2$, where $n_{unpaired}$ is the number of unpaired electrons. (Beware not to confuse the spin quantum number S with the entropy.) The multiplicity, or degeneracy g, of such a state is $g = (2S+1)$. Using degeneracy numbers is equivalent to an explicit count of all states, including degenerate ones. Thus, $Q_{elec} = g$ is a constant and only affects the entropy: $S_{elec} = R\ln(g)$ and $C_{p,\,elec} = [H(T)\text{-}H(0)]_{elec} = 0$. Since most free radicals have only a single unpaired electron, the usual effect is to increase the entropy by $R\ln(2)$. In addition to spin degeneracies, some states have spatial degeneracies. This situation is most common for diatomic molecules. Linear molecules with a spatial symmetry other than Σ (e.g., Π or Δ) have a spatial degeneracy of 2. For example, the OH radical has a $^2\Pi$ ground state, so its degeneracy is $g = 2$ (spin) × 2 (spatial) = 4. If there are both spin and spatial degeneracies, spin-orbit coupling lifts the degeneracy, often significantly. In the example of OH, the 4-fold degenerate ground state is split into two doubly-degenerate levels separated by 139.2 cm^{-1} (6). In such a case the low-lying excited states should be included in the calculation of thermodynamic quantities. The partition function is given by equation 29, where ϵ_i and g_i are the excitation energies (spectroscopic T_0) and degeneracies of the excited states, g_0 and $\epsilon_0 \equiv 0$ are for the ground state, and the sum runs over all the electronic states being considered, including the ground state. The contributions to the thermal functions are given by equations 30-32. This treatment assumes, rather crudely, that the rotations and vibrations are unaffected by electronic excitation.

$$Q_{elec} = \sum g_i \exp(-\epsilon_i/kT) \tag{29}$$

$$S_{elec} = R\ln\left(\sum g_i e^{-\epsilon_i/kT}\right) + R\frac{\sum g_i(\epsilon_i/kT)e^{-\epsilon_i/kT}}{\sum g_i e^{-\epsilon_i/kT}} \tag{30}$$

$$C_{p,\,elec} = R\sum g_i(\epsilon_i/kT)^2 e^{-\epsilon_i/kT} + R\left(\frac{\sum g_i(\epsilon_i/kT)e^{-\epsilon_i/kT}}{\sum g_i e^{-\epsilon_i/kT}}\right)^2 \tag{31}$$

$$[H(T)-H(0)]_{elec} = RT\frac{\sum g_i(\epsilon_i/kT)e^{-\epsilon_i/kT}}{\sum g_i e^{-\epsilon_i/kT}} \tag{32}$$

Example: Entropy of Methyl Radical. For simplicity, we again neglect minor isotopes. Results are summarized in Table IV. The molecular weight of CH_3 is then $m = 15.023$ amu, so $S_{trans} = 142.65$ J mol^{-1} K^{-1}. This is a flat, triangular molecule that belongs to the D_{3h} point group, $\sigma = 6$. The experimental bond length is $r_e = 1.0767$ Å $= 1.0767 \times 10^{-10}$ m (*12-14*). The moments of inertia can be evaluated using the symmetry of this oblate top, or more generally by diagonalizing the inertial tensor. We place the molecule in the yz plane with one hydrogen atom on the z axis. The center of mass coincides with the carbon atom. The cartesian coordinates then lead to an inertial tensor with components $I_{xx} = 3m_H r_e^2$, $I_{xy} = I_{yx} = 0$, $I_{xz} = I_{zx} = 0$, $I_{yy} = (3/2)m_H r_e^2$, $I_{yz} = I_{zy} = 0$, and $I_{zz} = (3/2)m_H r_e^2 = 2.910 \times 10^{-47}$ kg m^2. This is already diagonal, with eigenvalues $I_A = I_B = 2.910 \times 10^{-47}$ kg m^2 and $I_C = 5.820$ kg m^2 so that $S_{rot} = 43.50$ J mol^{-1} K^{-1} (equation 19). The observed vibrational frequencies of CH_3 are 3004.4, 606.5, 3160.8, and 1396 cm^{-1} for ν_1, ν_2, ν_3, and ν_4 respectively (*15*). Since ν_3 and ν_4 are both doubly degenerate (e' symmetry), they are counted twice and we have the correct number of vibrations, $3N-6 = 6$. Converting to SI units leads to $S_{vib} = 6.51 \times 10^{-5} + 1.84 + 6.42 \times 10^{-5} + 0.15 = 1.99$ J mol^{-1} K^{-1}. In this case, this is a radical with one unpaired electron, $S = \frac{1}{2}$ (electronic ground state is $^2A_2''$), so the degeneracy $g = 2$ and $S_{elec} = R\ln(2)$. Adding the four contributions to the entropy gives $S_{298.15} = 193.9$ J mol^{-1} K^{-1}, in agreement with the literature value of 194.2 ± 1.3 J mol^{-1} K^{-1} (*1*).

Example: Entropy of Hydroxyl Radical. For simplicity, we again neglect minor isotopes. Results are summarized in Table IV. The molecular weight of OH is then $m = 17.003$ amu, so $S_{trans} = 144.19$ J mol^{-1} K^{-1}. Again using the simple RRHO model, the observed bond length is $r_e = 0.96966$ Å (*6*) and the symmetry number $\sigma = 1$ ($C_{\infty v}$ point group). Hence $I = 1.480 \times 10^{-47}$ kg m^2 and $S_{rot} = 28.22$ J mol^{-1} K^{-1}. Using the observed vibrational fundamental for the vibrational frequency, $v = \omega_0 = 3568$ cm^{-1} (6) leads to S_{vib} = 5.04×10^{-6} J mol^{-1} K^{-1}. If spin-orbit splitting is ignored, $Q_{elec} = 4$ as explained above, so $S_{elec} = 11.53$ J mol^{-1} K^{-1}. Combining these four contributions yields $S = 183.9$ J mol^{-1} K^{-1}. If instead the spin-orbit splitting is included, so that $g_0 = g_1 = 2$ and $\epsilon_1 = 139.2$ cm^{-1} = 2.765×10^{-21} J, then equation 30 yields $S_{elec} = 11.08$ J mol^{-1} K^{-1} and so the total $S =$ 183.5 J mol^{-1} K^{-1}. The literature value is 183.71 ± 0.04 J mol^{-1} K^{-1} (1).

Table IV. Methyl and Hydroxyl Examples, S°_{298}, J/(mol·K)

Contribution	Methyl	Hydroxyl[a]	Hydroxyl[b]
Translation	142.65	144.19	144.19
Rotation	43.50	28.22	28.22
Vibration	1.99	5×10^{-6}	5×10^{-6}
Electronic	5.76	11.53	11.08
Total	193.9	183.9	183.5

[a]Spin-orbit splitting ignored. [b]Spin-orbit splitting included.

Internal Rotation. This refers to torsional motion, most commonly involving methyl groups. There are three ways to treat such a rotor, depending upon its barrier to rotation. The free and hindered rotor models require that an internal symmetry number, σ_{int}, be included. σ_{int} equals the number of minima (or maxima) in the torsional potential energy curve. The harmonic oscillator model does not require σ_{int} because it ignores all but one of the energy minima. For intermediate barrier heights (hindered rotor), the appropriateness of an internal symmetry number may be confusing. In such cases, avoid over- or under-counting states by ensuring that the limiting case of infinite barriers (harmonic oscillator model, no σ_{int}) moves smoothly into the limiting case of zero barrier (free rotor model, σ_{int} needed) as the barrier height decreases. Note that the vibrational frequency corresponding to the torsion must be deleted if the torsion is treated as a free or hindered rotation.

Free Rotor. If the barrier to rotation is much less than kT, then the rotor may be considered freely rotating. For a symmetric rotor such as a methyl group, the partition function is given by equation 33, where I_{int} is the reduced moment of inertia for the internal rotation and is given by equation 34 (3). Asymmetric rotors can be treated using

$$Q_{free\ rotor} = (\sigma_{int} h)^{-1} \left(8\pi^3 I_{int} kT\right)^{1/2} \tag{33}$$

$$I_{int} = I_{top} - I_{top}^2 \left(\frac{\alpha^2}{I_A} + \frac{\beta^2}{I_B} + \frac{\gamma^2}{I_C} \right) \tag{34}$$

an appropriate formula for the reduced moment (3). In equation 34, I_{top} is the moment of inertia of the rotating fragment about the axis of internal rotation. This is expressed as $I_{top} = \sum m_i r_i^2$, where the m_i are atomic masses, r_i is the distance of atom i from the axis of internal rotation, and the sum runs over all atoms in the rotating fragment. The quantities α, β, and γ are the cosines of the angles formed between the internal rotation axis and the principal axes of the overall molecule that correspond to I_A, I_B, and I_C, respectively. Contributions to the thermodynamic functions are given in equations 35-37.

$$S_{free\ rotor} = R[(1/2)\ln(8\pi^2 I_{int} kT) - \ln(\sigma_{int} h) + (1/2)] \qquad (35)$$

$$C_{p,\ free\ rotor} = (1/2)R \qquad (36)$$

$$[H(T) - H(0)]_{free\ rotor} = (1/2)RT \qquad (37)$$

Harmonic Oscillator. If the barrier to internal rotation is much greater than kT, one can consider the torsion to be a non-rotating, harmonic oscillator. Treatment is the same as for other vibrations.

Hindered Rotor. This is the common, intermediate case, when the torsional barrier V is comparable to kT. If the torsional potential is assumed to have the simple form $U(\phi) = V(1 - \cos\sigma_{int}\phi)/2$, then the tables of Pitzer and Gwinn are usually used to compute the contribution of the hindered rotor to the thermodynamic functions (16,17). Their tables are in terms of the dimensionless variables x and y, where $x = V/(kT)$ and $y = \sigma_{int} h(8\pi^3 I_{int} kT)^{-1/2}$ and I_{int} is defined as for a free rotor (see above).

Example: Entropy of Ethane at $T = 184$ K [adapted from ref (17)]. Results are summarized in Table V. Ignoring minor isotopes as before, for C_2H_6 we have $m = 30.047$ amu, so $S_{trans} = 141.26$ J mol^{-1} K^{-1}. The experimental geometry is staggered (D_{3d} point group, $\sigma = 6$) and defined by $r_{CC} = 1.535$ Å, $r_{CH} = 1.094$ Å, and $\theta_{CCH} = 111.2°$ (12-14). If we choose coordinates so that the origin is at the center of mass, the carbon atoms lie on the z-axis, and the yz plane is a reflection plane of symmetry, then the elements of the inertial tensor are, in (amu Å^2), $I_{xx} = I_{yy} = 25.46$, $I_{xy} = I_{yx} = I_{xz} = I_{zx} = I_{yz} = I_{zy} = 0$, and $I_{zz} = 6.291$ amu Å^2. This is already diagonal, with eigenvalues $I_A = 1.045 \times 10^{-46}$ kg m^2 and $I_B = I_C = 4.228 \times 10^{-46}$ kg m^2 so that $S_{rot} = 62.17$ J mol^{-1} K^{-1} (equation 19). The observed vibrational frequencies (18) are 2954, 1388, 995, 289 (torsion), 2896, and 1379 cm^{-1} (non-degenerate), and 2969, 1468, 1190, 2985, 1469, and 822 cm^{-1} (doubly degenerate), for a total of $3N-6 = 18$ vibrational modes and $S_{vib} = 3.36$ J mol^{-1} K^{-1} (0.25 J mol^{-1} K^{-1} excluding the torsional mode). The total entropy in the RRHO model is thus $S°_{184} = 206.8$ J mol^{-1} K^{-1}, which is below the experimental value $S = 207.7 \pm 0.6$ J mol^{-1} K^{-1} (17).

If we consider the torsion to be a free, unhindered rotor, then we require the corresponding reduced moment of inertia. In this case, in the rotor axis is aligned with the A axis of the molecule, so that $\alpha = 1$ and $\beta = \gamma = 0$ in equation 34. This gives $I_{int} = I_{top} - I_{top}^2/I_A$. The symmetry of this molecule requires that $I_A = 2I_{top}$, because the moment of inertia of the whole molecule around the A axis (viz., the C-C bond axis) is twice that of a single methyl group. Thus $I_{int} = 2.613 \times 10^{-47}$ kg m^2. The internal symmetry number is $\sigma_{int} = 3$, since there are three equivalent values of the torsion angle (0°, 120°, and 240°). Equation 35 yields $S_{free\ rotor} = 5.33$ J mol^{-1} K^{-1}, for a total entropy of $S°_{184} = 209.0$ J mol^{-1} K^{-1} in the free-rotor model, which is higher than the experimental value.

    To apply the hindered-rotor model we need a value for the torsional barrier height. This can be estimated from the observed torsional vibrational frequency v (in s^{-1}) as $V \approx 8\pi^2 I_{int} v^2/\sigma_{int}^2 = 1.720 \times 10^{-20}$ J (or 10.4 kJ/mol). Thus the parameters are $x = 6.77$ and $y = 0.490$. Interpolating within the standard tables (16,17), $S_{hindered\ rotor} = 3.99$ J mol^{-1} K^{-1}, so that the total entropy is $S = 207.7$ J mol^{-1} K^{-1}, in agreement with the experimental value.

Table V. Ethane Example, $S°_{184}$ (J mol^{-1} K^{-1})

Contribution	Harmonic Rotor	Free Rotor	Hindered Rotor
Translation	141.26	141.26	141.26
Rotation	62.17	62.17	62.17
Vibration	0.25	0.25	0.25
Torsion	3.11	5.33	3.99
Total	206.8	209.0	207.7

Charged Molecules: Two Conventions. The balanced chemical equation describing an ionization process involves at least one free electron. There are two major conventions for the thermodynamic properties of the electron. Most compilations of thermochemical data adopt the *thermal electron convention*. In this convention, the free electron is treated as a chemical element, so that its ideal-gas enthalpy of formation is zero at all temperatures. In contrast, most of the literature in mass spectrometry and ion chemistry adopts the *ion convention*, sometimes also called the *stationary electron convention*. In this convention, the enthalpy content of the electron is ignored. At absolute zero temperature there is no difference between the two conventions, but in general enthalpies of formation under the two conventions are related by equation 38, where q is the

$$\Delta_f H°_{thermal\ electron\ convention} = \Delta_f H_{ion\ convention} + (5/2)qRT \qquad (38)$$

(signed) charge on the ion in question (±1 in most cases). A thorough discussion is provided in the introduction to the *GIANT* tables (19). When reporting the thermochemistry of ions, it is important always to indicate which convention is being used. Especially beware not to combine enthalpies of formation that were derived using different conventions

Chemical Kinetics

The equilibrium constant for a reaction is $K_{eq} = \exp(-\Delta G/RT)$, where $\Delta G = \Delta H - T\Delta S$ and the differences are between the reactants and products, e.g., $\Delta S = S_{products} - S_{reactants}$. Simple transition-state theory for chemical kinetics assumes that the reaction rate is limited by formation of a transient *transition state*, which is the point of maximum energy along the path from reactants to products. The transition state is considered to be in quasi-equilibrium with the reactants. If differences between reactants and the transition state are denoted with a double dagger, e.g., $\Delta S^{\ddagger} = S_{TS} - S_{reactants}$, then the rate constant (denoted r here to avoid confusion with the Boltzmann constant) is given by equation 39.

$$r(T) = \frac{kT}{h}\exp\left(-\frac{\Delta G^{\ddagger}}{RT}\right)$$ (39)

As for stable species, $\Delta G^{\ddagger} = \Delta H^{\ddagger} - T\Delta S^{\ddagger}$. Thus rate constants can be calculated easily from the "thermochemistry" for transition states. In such calculations, the imaginary vibrational frequency is ignored, so that there are only $3N$-7 molecular vibrations in the transition structure ($3N$-6 if linear). If all internal and external symmetry numbers are included in the rotational partition functions, then any reaction path degeneracy will usually be included automatically. Occasionally, however, stereochemical factors are also needed (*20*).

Experimental, temperature-dependent rate constants are often presented as an *Arrhenius plot* of $r(T)$ vs. $1/T$. This is motivated by the observation that such plots are nearly linear, $r(T) = A \exp(-E_a/RT)$, where the pre-exponential factor A is usually called simply the *A-factor* and E_a is the phenomenological *activation energy*. It is often desirable to report A and E_a in computational studies, for comparison with the values derived from experimental data. They may be determined using equations 40 and 41,

$$E_a = RT^2\frac{\partial}{\partial T}(\ln k) = \Delta H^{\ddagger} + MRT$$ (40)

$$A = r(T)\exp(+E_a/RT)$$ (41)

where M is the molecularity of the reaction (e.g., $M = 1$ for a unimolecular and $M = 2$ for a bimolecular reaction). The derived A and E_a are weakly temperature-dependent. This is consistent with experimental results, which are often fitted using the three-parameter modified Arrhenius expression $r(T) = A' T^n \exp(-E_a'/RT)$. This functional form leads to a better fit than the ordinary Arrhenius expression, but the parameters may have little physical interpretation. If the RRHO approximation is accepted, then the three parameters are given by equations 42-44, where the y_i in equation 42 are the reduced vibrational frequencies, $y_i = h\nu/kT$. However, in practice it is often best to determine the parameters A', n, and E_a' by fitting calculated rate constants to the modified Arrhenius expression.

$$n = 1 + \sum_{TS} \frac{y_i^\ddagger e^{-y_i^\ddagger}\left(1 + e^{-y_i^\ddagger}\right)}{\left(1 - e^{y_i^\ddagger}\right)^2} - \sum_{reactants} \frac{y_i e^{-y_i}\left(1 + e^{-y_i}\right)}{\left(1 - e^{y_i}\right)^2} \qquad (42)$$

$$E_a' = RT^2 \frac{\partial}{\partial T}(\ln k) - nRT = \Delta H^\ddagger + (M - n)RT \qquad (43)$$

$$A' = r(T)\, T^{-n} \exp(+E_a'/RT) \qquad (44)$$

Ab initio energies are now precise enough that it is becoming common to use kinetic theories more sophisticated than simple transition-state theory. When the reaction coordinate is dominated by motion of a hydrogen atom, corrections for quantum-mechanical tunneling are often made (*21*). The simplest is the Wigner correction, which requires only the imaginary vibrational frequency $\nu^\ddagger i$ associated with the reaction coordinate. To apply this correction, the calculated rate is multiplied by F_{tunnel} (equation 45). Better results may be obtained by fitting the energy profile of the reaction to an

$$F_{tunnel} = 1 + \frac{1}{24}\left(h\nu^\ddagger/kT\right)^2 \qquad (45)$$

Eckart potential function (see the chapter by Petersson in this book). The location of the transition state can also be refined. Variational transition-state theory defines the transition state as the maximum in the free energy along the reaction path, instead of the maximum along the vibrationless potential energy curve. Such a definition is essential for reactions such as simple bond cleavage, which usually has no barrier in excess of the bond energy.

Units and Constants

In actual calculations, many practical difficulties involve incompatible units. In addition to the standard units of the SI, many others are in use, usually for historical reasons. Conversion factors among selected units are provided in Table VI. For convenience, the values of commonly-used constants are collected in Table VII. Detailed information is available on-line at http://www.physics.nist.gov/PhysRefData/contents.html#SI.

Table VI. Unit Conversions

Quantity	Unit	Conversion[a]	SI Unit
energy	hartree (atomic unit)	2 625.500	kJ/mol
energy	cal	4.184	J
energy	cm^{-1} (wavenumber)	0.011 962 66	kJ/mol
energy	eV	96.485 31	kJ/mol
energy	K (temperature)	$8.314 511 \times 10^{-3}$	kJ/mol
distance	Å	10^{-10}	m
distance	bohr (atomic unit)	$5.291 772 \times 10^{-11}$	m
mass	amu *or* u	$1.660 540 \times 10^{-27}$	kg
pressure	bar	10^{5}	Pa
pressure	atm	101 325	Pa
pressure	Torr *or* mm-Hg	133.322 37	Pa
pressure (density)	cm^{-3} (at 298.15 K; ideal gas)	$4.166 43 \times 10^{-15}$	Pa
pressure (density)	cm^{-3} (arb. temp.; ideal gas)	$10^{6} kT$	Pa
pressure (density)	M^{-1} *or* mol/L (ideal gas)	$10^{3} RT$	Pa
dipole moment	atomic unit	$8.478 358 \times 10^{-30}$	C·m
dipole moment	D (debye)	$3.335 641 \times 10^{-30}$	C·m

[a]Multiply the quantity expressed in the units of column 2 by the conversion factor in column 3 to obtain the quantity expressed in units of column 4 (SI units).

Table VII. Physical Constants

Quantity	Value
k	$1.380 66 \times 10^{-23}$ J K^{-1}
N_A	$6.022 137 \times 10^{23}$ mol^{-1}
$R = kN_A$	$8.314 510$ J mol^{-1} K^{-1}
h	$6.626 076 \times 10^{-34}$ J s
c	$299 792 458$ m s^{-1}

Literature Cited

1. Chase, M. W., Jr.; Davies, C. A.; Downey, J. R., Jr.; Frurip, D. J.; McDonald, R. A.; Syverud, A. N. *J. Phys. Chem. Ref. Data* **1985**, *14*, Suppl. 1.
2. *Thermodynamic Properties of Individual Substances*; 4th ed.; Gurvich, L. V.; Veyts, I. V.; Alcock, C. B., Eds.; Hemisphere: New York, 1989.
3. Frenkel, M.; Marsh, K. N.; Wilhoit, R. C.; Kabo, G. J.; Roganov, G. N. *Thermodynamics of Organic Compounds in the Gas State*; Thermodynamics Research Center: College Station, Texas, 1994; Vol. I.
4. Cohen, E. R.; Taylor, B. N. *The 1986 CODATA Recommended Values of the Fundamental Physical Constants*; http://physics.nist.gov/PhysRefData/codata86/article.html.
5. Scott, A. P.; Radom, L. *J. Phys. Chem.* **1996**, *100*, 16502.
6. Huber, K. P.; Herzberg, G. *Molecular Spectra and Molecular Structure: IV. Constants of Diatomic Molecules*; van Nostrand Reinhold: New York, 1979.
7. *CRC Handbook of Chemistry and Physics*; 76th ed.; Lide, D. R.; Frederikse, H. P. R., Eds.; CRC Press: Boca Raton, 1995.
8. Winter, M. *WebElements*; http://www.shef.ac.uk/~chem/web-elements/.
9. Cox, J. D.; Wagman, D. D.; Medvedev, V. A. *CODATA Key Values for Thermodynamics*; Hemisphere: New York, 1989.
10. Cotton, F. A. *Chemical Applications of Group Theory*; 2nd ed.; Wiley--Interscience: New York, 1971.
11. Herzberg, G. *Molecular Spectra and Molecular Structure: II. Infrared and Raman Spectra of Polyatomic Molecules*; van Nostrand: New York, 1945.
12. *Structure Data of Free Polyatomic Molecules*; Kuchitsu, K., Ed.; *Landolt-Börnstein New Series*; Madelung, O., Ed.; Springer: Berlin, 1992; Vol. II:21.
13. *Structure Data of Free Polyatomic Molecules*; Hellwege, K.-H.; Hellwege, A. M., Eds.; *Landolt-Börnstein New Series*; Madelung, O., Ed.; Springer: Berlin, 1987; Vol. II:15.
14. *Structure Data of Free Polyatomic Molecules*; Hellwege, K.-H.; Hellwege, A. M., Eds.; *Landolt-Börnstein New Series*; Hellwege, K.-H., Ed.; Springer: Berlin, 1976; Vol. II:7.
15. Jacox, M. E. *Vibrational & Electronic Energy Levels of Small Polyatomic Transient Molecules (VEEL): Version 4.0*; NIST-SRD 26 (electronic database); Natl. Inst. of Standards and Technology: Gaithersburg, MD, 1995.
16. Pitzer, K. S.; Gwinn, W. D. *J. Chem. Phys.* **1942**, *10*, 428.
17. Janz, G. J. *Thermodynamic Properties of Organic Compounds*; Academic: New York, 1967.
18. Shimanouchi, T. *Tables of Molecular Vibrational Frequencies, Consolidated Volume I*; NSRDS-NBS 39; U.S. GPO: Washington, D.C., 1972.
19. Lias, S. G.; Bartmess, J. E.; Liebman, J. F.; Holmes, J. L.; Levin, R. D.; Mallard, W. G. *J. Phys. Chem. Ref. Data* **1988**, *17*, Suppl. 1.
20. Gilbert, R. G.; Smith, S. C. *Theory of Unimolecular and Recombination Reactions*; Blackwell Scientific: Oxford, 1990.
21. Bell, R. P. *The Tunnel Effect in Chemistry*; Chapman and Hall: London, 1980.

Appendix C

Worked Examples

Karl K. Irikura[1] and David J. Frurip[2]

[1]Physical and Chemical Properties Division, National Institute of Standards and Technology, Gaithersburg, MD 20899
[2]Analytical Sciences, Reactive Chemical/Thermal Analysis/Physical Properties Discipline, The Dow Chemical Company, Building 1897F, Midland, MI 48667

Many methods for computing enthalpies of formation are described in the main chapters of this book. Several of these techniques require arithmetic computations beyond those that are performed automatically by widely-disseminated software. This Appendix provides worked examples of these manual calculations.

Group additivity (GA) is usually the method of choice for thermochemical predictions of organic compounds containing only the more common functional groups. Although GA software is available (see Appendix A), it is not widely distributed and GA calculations are still often done by hand. Similarly, for many sophisticated *ab initio* methods, commercial software (see Appendix A) executes the quantum calculations but may not produce enthalpies of formation. In such cases, the output of the quantum code must be processed further to reach the quantity of interest. The chapters of this book that describe techniques generally also provide enough information to implement them, but this Appendix provides worked examples to aid in learning the methods. Examples are provided for GA ("group additivity;" see the chapter by Benson and Cohen), BAC-MP4 ("bond-additivity-corrected MP4;" chapter by Zachariah and Melius), G2 and G2(MP2) (Curtiss and Raghavachari), G2M(CC) (*1*), PCI-X ("parametrized configuration interaction;" Blomberg and Siegbahn), and "scaling all correlation" (SAC) in the MP2-SAC and CCD-SAC variants (*2-5*). The very precise methods described in the chapter by Martin, which are very expensive and require expert treatment of auxiliary quantities such as ZPEs, are beyond the scope of this tutorial.

For the *ab initio*-based methods, this Appendix provides examples of determining the composite molecular energy. To convert this to an enthalpy of formation, a reaction energy must be calculated using the composite energies. This reaction energy is then combined with reference thermochemical data to yield a prediction for the species of interest. See Appendix B and the overview chapter for a fuller discussion and an example, and for the mechanics of computing thermochemistry at temperatures above $T = 0$ K.

Group Additivity

There are many empirical estimation schemes; several are discussed in the overview chapter of this book. Reference (6) also provides brief summaries of a number of methods. Pedley has recently developed a new computerized method (7). Among the empirical methods, however, the one developed by Benson and coworkers is the most popular and is therefore illustrated below.

The first step in using GA is to identify all the groups in the molecule of interest. Their group additivity values (GAVs) are then summed to obtain an estimate for $\Delta_f H°_{298}$, the enthalpy of formation of the molecule at $T = 298.15$ K. Since there are very many possible groups, many GAVs may be needed. Moreover, additional corrections may be required to include effects such as ring strain and 1,4 (*gauche*) steric interactions. In addition to the chapter by Benson and Cohen in this book, refer to Benson's book (8) for a full description of GA and to the recent evaluation by Cohen for updated parameter values for groups involving carbon, hydrogen, and oxygen (9). Among the software packages implementing GA, the *NIST Structures & Properties* (10) and *CHETAH* (11) programs are the most popular (see Appendix A).

Within a molecule, there is one group for each polyvalent atom, with the exception that the carbonyl (>C=O) and ketenyl (>C=C=O) moieties each constitute a single divalent group. Atom types may include information about the chemical environment of the atom. For example, common carbon atom types are sp^3-hybridized (C, tetravalent as in alkanes), sp^2 (C_d, as in alkenes, considered divalent), aromatic sp^2 (C_b, as in benzene, considered monovalent), and sp (C_t, as in alkynes, considered monovalent).

Example: $\Delta_f H°_{298}$ **of 2-Amino-4-hydroxypentane.** The molecule is shown at right. There are seven polyvalent atoms and seven groups. Table I lists the notations and GAVs for the groups in the molecule, first the carbon groups from left to right in the molecule and then the amino and hydroxyl groups. In this example, there are no corrections such as ring strain or other steric effects. The sum of the GAVs equals the estimation $\Delta_f H°_{298} = -295.3$ kJ/mol.

Example: $\Delta_f H°_{298}$ **of Acrylic Acid.** The molecule is shown at right and the groups and GAVs are included in Table I. Note that the sp^2-hybridized carbon atoms are categorized either as C_d or as the special carbonyl "atom" CO. The CO "atom" is divalent as would be expected. C_d is also considered divalent because it is implicit that it is double-bonded to another C_d atom. The estimate obtained from GA, $\Delta_f H°_{298} = -329.9$ kJ/mol, is in acceptable agreement with the experimental value of -322.2 kJ/mol (10).

Example: $\Delta_f H°_{298}$ **of Chlorocyclopropane.** The groups and GAVs are listed in Table II. Table II differs from Table I by including the multiplicity of repeated groups, rather than repeating them explicitly as done for the first example above. In addition to the GAVs, a large ring-strain correction (RSC) is needed for the cyclopropyl ring. The predicted enthalpy of formation is $\Delta_f H°_{298}$ = 12.4 kJ/mol.

Table I. Group Additivity Estimations of $\Delta_f H°_{298}$ (kJ/mol)

Group	GAV^a	Group	GAV^a
2-amino-4-hydroxypentane		**Acrylic Acid**	
C-(C)(H)$_3$	-42.2	C$_d$-(H)$_2$	26.2
C-(C)$_2$(H)(N)	-21.8	C$_d$-(CO)(H)	20.9
C-(C)$_2$(H)$_2$	-20.6	CO-(C$_d$)(O)	-133.9
C-(C)$_2$(H)(O)	-30.1	O-(CO)(H)	-243.1
C-(C)(H)$_3$	-42.2	*Total*	*-329.9*
N-(C)(H)$_2$	20.1		
O-(C)(H)	-158.5		
Total	*-295.3*		

^{a}GAVs from ref. (*11*).

Table II. More Group Additivity Estimations of $\Delta_f H°_{298}$ (kJ/mol)

Group	Number	GAV^a	Group	Number	GAV^a
Chlorocyclopropane			**Chlorobenzene**		
C-(C)$_2$(H)$_2$	2	2 × -20.6	C$_b$-(H)	5	5 × 13.8
C-(C)$_2$(H)(Cl)	1	-61.9	C$_b$-(Cl)	1	-15.9
cyclopropane RSC	1	115.5	*Total*		*53.1*
Total		*12.4*			

^{a}GAVs from ref. (*11*).

Example: $\Delta_f H°_{298}$ **of Chlorobenzene.** The groups and GAVs are included in Table II. In this case, all the sp^2-hybridized carbon atoms are of the aromatic C$_b$ type. This is considered monovalent because it is implicitly part of an aromatic ring, with three bonds to other carbon atoms in the ring. There is no RSC for the aromatic ring. The resulting estimate is $\Delta_f H°_{298}$ = 53.1 kJ/mol, in good agreement with the experimental value, 54.5 kJ/mol (*10*).

G2 and Variants

G2 and G2(MP2). As described in the chapter by Curtiss and Raghavachari, these composite methods involve calculating the molecular vibrational frequencies at the HF/6-31G(d) level and the geometry at the MP2/6-31G(d) level with all electrons active. Freezing the core in the MP2 geometry optimization would be less expensive than using "full" MP2 and give results that differ negligibly. However, at the time the methods were developed, the Gaussian program was unable to do efficient geometry optimizations for frozen-core MP2. Thus, the G2 prescriptions specify "full" MP2-level geometries.

The electronic energy is an estimation of what would be obtained at the QCISD(T)/6-311+G(3df,2p) level. The G2 and G2(MP2) methods differ in how this estimation is made. Both methods will be illustrated within each example below. Note that in the most recent version of the Gaussian program (*12,13*), the G2 and G2(MP2) composite energies can be obtained automatically, using a single keyword. We choose to illustrate the methods here because they are used widely and can be determined using *ab initio* programs other than Gaussian94. To illustrate the affordability and accessibility of the methods, all the calculations for the following two examples were done on a personal computer (180-MHz PPro, 64 Mb RAM, 1.5 Gb temporary disk space) (*13*); the allyl chloride calculations (four "heavy" atoms) required about three days (wallclock). On an IBM RS/6000-560 workstation (*13*), the allyl chloride calculations required less than two days (wallclock).

Example: $\Delta_f H°_0$ of Methanol. These results are summarized in Table III. In most *ab initio* programs, the presence of symmetry speeds the calculations, so it is advantageous to specify any symmetry that may be present. CH_3OH has a plane of symmetry, so it belongs to the C_s point group. Methanol has 38 basis functions when the 6-31G(d) basis set (also denoted 6-31G*) is used with the usual choice of six d-functions per atom (cartesian d-functions). Hartree-Fock geometry optimization yields E = -115.035418 hartree and selected geometric parameters r(C-O) = 1.400 Å, r(O-H) = 0.946 Å, θ(H-O-C) = 109.4°, and r(C-H_{unique}) = 1.081 Å, where H_{unique} is *anti* to the hydroxyl hydrogen atom.

At the HF/6-31G* level, the harmonic vibrational frequencies are, in cm^{-1}, 4117, 3306, 3186, 1663, 1637, 1508, 1189, and 1165 (a' symmetry) and 3232, 1652, 1290, and 349 (a" symmetry). The vibrational zero-point energy (see overview chapter) is half the sum of the frequencies, or 12147 cm^{-1}. But in G2 and G2(MP2) theories the vibrational frequencies are scaled by 0.8929, so ZPE = 10846 cm^{-1} = 0.049417 hartree.

At the MP2/6-31G(d) level with all electrons correlated, the optimized geometry is characterized by energy E = -115.353295 hartree and selected geometric parameters r(C-O) = 1.423 Å, r(O-H) = 0.970 Å, θ(H-O-C) = 107.4°, and r(C-H_{unique}) = 1.090 Å. This geometry is used for the subsequent correlated, frozen-core energy calculations collected in Table III. The "higher-level correction," denoted HLC, is given (in hartrees) by HLC = $-0.00019n_\alpha -0.00481n_\beta$, where n_α and n_β are the numbers of valence up- and down-spin electrons, respectively. The total electron count is just the sum of the atomic numbers, or 18 for CH_3OH. The core comprises the 1s orbitals on C and on O, for 4 core electrons. Of the 14 valence electrons, 7 are up- and 7 down-spin, since the system is a

Table III. G2 and G2(MP2) Energy Components (hartree)

Calculation	CH_3OH	C_3H_5Cl	C (triplet)	O (triplet)	H	Cl
MP2/6-311G(d,p)	-115.436205	-576.596917	-37.745023	-74.918145	-0.499810	-459.585137
MP4/6-311G(d,p)	-115.468475	-576.662975	-37.764302	-74.933327	-0.499810	-459.602633
QCISD(T)/6-311G(d,p)	-115.468766	-576.664518	-37.766680	-74.934022	-0.499810	-459.603286
MP2/6-311+G(d,p)	-115.444855	-576.601903	-37.745874	-74.921781	-0.499810	-459.586161
MP4/6-311+G(d,p)	-115.477333	-576.668288	-37.765200	-74.937240	-0.499810	-459.603772
MP2/6-311G(2df,p)	-115.492731	-576.704972	-37.755185	-74.946524	-0.499810	-459.628998
MP4/6-311G(2df,p)	-115.527874	-576.781471	-37.774830	-74.964781	-0.499810	-459.656289
MP2/6-311+G(3df,2p)	-115.513660	-576.725095	-37.756851	-74.952421	-0.499810	-459.633378
Δ(+)	-0.008858	-0.005313	-0.000898	-0.003913	0	-0.001139
Δ(2df)	-0.059399	-0.118496	-0.010528	-0.031454	0	-0.053656
Δ(QCI)	-0.000291	-0.001543	-0.002378	-0.000695	0	-0.000653
ZPE	0.049414	0.068882	0	0	0	0
HLC	-0.035000	-0.060000	-0.005380	-0.010380	-0.000190	-0.015190
Δ(G2)	-0.012279	-0.015137	-0.000815	-0.002261	0	-0.003356
Δ(MP2)	-0.077455	-0.128178	-0.011828	-0.034276	0	-0.048241
E(G2)	-115.534888	-576.794582	-37.784301	-74.982030	-.500000	-459.676627
E(G2MP2)	-115.531807	-576.783814	-37.783888	-74.978678	-.500000	-459.666717
$\Delta_f H°_0$(G2; rxn 1 or 3)	-195.7[a]	12.0[a]				
$\Delta_f H°_0$(G2MP2; rxn 1 or 3)	-197.5[a]	11.0[a]				
$\Delta_f H°_0$ (exptl, kJ/mol)	-190.1[a,b]	11.7[a,c]	711.2±0.5[a,d]	246.8±0.1[a,d]	216.035±0.006[a,d]	119.621±0.006[a,d]

[a]kJ/mol. [b]Ref (18). [c]Derived from refs (25,18,17). [d]Ref (17).

spin singlet, so $n_\alpha = n_\beta = 7$. In Table III, note that for the atoms $n_\alpha > n_\beta$ because they are open-shell. Specifically, the (n_α, n_β) pairs are (3, 1) for triplet C, (4, 2) for triplet O, (1, 0) for doublet H, and (4, 3) for doublet Cl.

For the G2 energy, we calculate energy differences relative to the reference MP4/6-311G(d,p) energy. Thus, $\Delta(+)$ refers to the MP2/6-311+G(d,p) energy (68 basis functions), Δ(2df) refers to the MP4/6-311G(2df,p) energy (84 basis functions), and Δ(QCI) refers to the QCISD(T)/6-311G(d,p) energy. In addition, the quantity Δ(G2) is computed from MP2 energies, Δ(G2) = E[MP2/6-311+G(3df,2p)] + E[MP2/6-311G(d,p)] - E[MP2/6-311G(2df,p)] - E[MP2/6-311+G(d,p)]. (For methanol, the 6-311+G(3df,2p) basis set involves 114 basis functions.) These corrections, plus the ZPE and HLC, are added to the MP4/6-311G(d,p) energy to obtain the G2 energy, E(G2) = -115.534888 hartree.

The G2(MP2) energy is slightly simpler to compute. The difference Δ(MP2) = E[MP2/6-311+G(3df,2p)] - E[MP2/6-311G(d,p)], plus the ZPE and HLC, is added to the QCISD(T)/6-311G(d,p) energy to obtain the G2(MP2) energy, E(G2MP2) = -115.531807 hartree.

Although isogyric and isodesmic reactions usually yield better thermochemical results, it is common to combine G2 or G2(MP2) atomization energies with the experimental enthalpies of formation of the monatomic gases. For methanol, we must first obtain the G2 and G2(MP2) energies for atomic C (spin triplet), H (spin doublet), and O (spin triplet). These energies can be obtained from the literature (14,15) but are included in Table III as additional examples. The atomization energy of CH_3OH is the energy change (equal to the enthalpy change at $T = 0$ K) for reaction 1, and is calculated

$$CH_3OH \rightarrow C(^3P) + O(^3P) + 4\ H(^2S) \qquad (1)$$

$$\Delta_f H^\circ_0(CH_3OH) = \Delta_f H^\circ_0(C) + \Delta_f H^\circ_0(O) + 4\ \Delta_f H^\circ_0(H) - E_{at} \qquad (2)$$

to be 0.768557 hartree = 2017.8 kJ/mol at the G2 level and 0.769241 hartree = 2019.6 kJ/mol at the G2(MP2) level. The corresponding enthalpy of formation of methanol is given by equation 2. Accepting the experimental enthalpies of formation of the atoms in Table III leads to a prediction $\Delta_f H^\circ_0(CH_3OH)$ = -195.7 kJ/mol from the G2 energies and -197.5 kJ/mol from the G2(MP2) energies. Since thermochemistry from these methods is typically reliable to about 10 kJ/mol, the agreement with the experimental value of -190.1 kJ/mol (Table III) is good.

Example: $\Delta_f H^\circ_0$ of 3-Chloropropene (Allyl Chloride). These results are summarized in Table III. We choose the most stable *gauche* conformation of the chloromethyl group, in which a hydrogen atom eclipses the C=C double bond. This has no symmetry and belongs to the C_1 point group. Allyl chloride has 74 basis functions in the 6-31G(d) basis set. Hartree-Fock geometry optimization yields E = -575.971804 hartree and selected geometric parameters r(C=C) = 1.317 Å, r(C-C) = 1.496 Å, r(C-Cl) = 1.805 Å, θ(C=C-C) = 123.5°, and θ(C-C-Cl) = 111.4°.

At the HF/6-31G* level, the harmonic vibrational frequencies are, in cm^{-1}, 3418, 3364, 3347, 3334, 3285, 1879, 1636, 1588, 1451, 1424, 1346, 1221, 1137, 1093, 1022,

992, 818, 651, 435, 310, and 113. The vibrational zero-point energy (see overview chapter) is half the sum of the frequencies, or 16931 cm^{-1}. After scaling by 0.8929, the ZPE = 15118 cm^{-1} = 0.068882 hartree.

At the MP2/6-31G(d) level with all electrons correlated, the optimized geometry is characterized by energy E = -576.507301 hartree and selected geometric parameters $r(C=C)$ = 1.337 Å, $r(C-C)$ = 1.490 Å, $r(C-Cl)$ = 1.796 Å, $\theta(C=C-C)$ = 123.0°, and $\theta(C-C-Cl)$ = 111.2°. The total electron count for C_3H_5Cl is 40. The core comprises $1s^2$ on each C and $1s^2 2s^2 2p^6$ on Cl, for 16 core electrons. That leaves 24 valence electrons, so for this spin singlet, $n_\alpha = n_\beta = 12$.

The atomization reaction in this case is shown in equation 3. The calculations yield E_{at} = 1.265052 hartree = 3321.4 kJ/mol at the G2 level and E_{at} = 1.265433 hartree = 3322.4 kJ/mol at the G2(MP2) level. The corresponding enthalpies of formation of allyl chloride are given by equation 4. Accepting the experimental enthalpies of

$$CH_2=CH-CH_2Cl \rightarrow 3\,C(^3P) + Cl(^2S) + 5\,H(^2S) \qquad (3)$$

$$\Delta_f H°_0(C_3H_5Cl) = 3\,\Delta_f H°_0(C) + \Delta_f H°_0(Cl) + 5\,\Delta_f H°_0(H) - E_{at} \qquad (4)$$

formation of the atoms in Table III leads to a prediction $\Delta_f H°_0(CH_2CHCH_2Cl)$ = 12.0 kJ/mol from the G2 energies and 11.0 kJ/mol from the G2(MP2) energies, in excellent agreement with the experimental value of 11.7 kJ/mol. We note that Colegrove and Thompson have recently published comprehensive calculations also supporting this experimental value (*16*).

G2M(CC). This is one of the more successful of the many variations of G2 theory (see the chapter by Curtiss and Raghavachari for a summary of the current variations). Although the G2M(RCC) is preferred, especially for cases with spin contamination, it requires restricted open-shell ROCCSD(T) calculations, which are not yet available in the popular Gaussian program (*13*). We therefore choose here to illustrate the G2M(CC) instead, which is also somewhat less expensive than G2M(RCC). The cost of G2M(CC) is slightly greater than that of unmodified G2 theory. The performance of this method is discussed in the literature (*1*), and also briefly in the chapter by Martin.

In the G2M(CC) method (*1*), the molecular geometry and vibrational frequencies are calculated at the B3LYP/6-311G(d,p) level. Vibrational frequencies are not scaled. CCSD(T) is used instead of QCISD(T). For open-shell systems, spin-projected MP4 energies (PMP4) are used instead of the usual unprojected energies, but unprojected MP2 energies (UMP2) are still used. For singlets, UCCSD(T) is used instead of the usual closed-shell calculation if there is an RHF→UHF instability in the reference HF wavefunction. Finally, the empirical "higher-level correction" is re-fitted to obtain HLC = $-0.00019 n_\alpha - 0.00578 n_\beta$. Results for methanol, allyl chloride, and the corresponding free atoms are summarized in Table IV.

We choose methanol and allyl chloride for the G2M(RCC) examples to allow a close comparison among the three G2 variants presented here. All the calculations for

Table IV. G2M(CC) Energy Components (hartree)

Calculation	CH_3OH	C_3H_5Cl	C (triplet)	O (triplet)	H	Cl
UMP2/6-311G(d,p)	-115.436311	-576.596169	-37.745023	-74.918145	-0.499810	-459.585137
PMP4/6-311G(d,p)	-115.468574	-576.662450	-37.764797	-74.934002	-0.499810	-459.603084
CCSD(T)/6-311G(d,p)	-115.468567	-576.663621	-37.766669	-74.933997	-0.499810	-459.603250
UMP2/6-311+G(d,p)	-115.444922	-576.601095	-37.745874	-74.921781	-0.499810	-459.586161
PMP4/6-311+G(d,p)	-115.477391	-576.667708	-37.765808	-74.938247	-0.499810	-459.604242
UMP2/6-311G(2df,p)	-115.492805	-576.704493	-37.755185	-74.946524	-0.499810	-459.628998
PMP4/6-311G(2df,p)	-115.527946	-576.781214	-37.775441	-74.965544	-0.499810	-459.657136
UMP2/6-311+G(3df,2p)	-115.513733	-576.724370	-37.756851	-74.952421	-0.499810	-459.633378
$\Delta(+)$	-0.008817	-0.005258	-0.001011	-0.004245	0	-0.001158
$\Delta(2df)$	-0.059372	-0.118764	-0.010644	-0.031542	0	-0.054052
$\Delta(CC)$	0.000007	-0.001171	-0.001872	0.000005	0	-0.000166
ZPE	0.051097	0.071324	0	0	0	0
HLC	-0.041790	-0.071640	-0.006350	-0.012320	-0.000190	-0.018100
Δ'	-0.012317	-0.014951	-0.000815	-0.002261	0	-0.003356
E[G2(CC)]	-115.539766	-576.802910	-37.785489	-74.984365	-.500000	-459.679916
$\Delta_f H^\circ{}_0$(G2CC; rxn 1 or 3)	-199.3[a]	8.1[a]				
$\Delta_f H^\circ{}_0$ (exptl, kJ/mol)	-190.1[a,b]	11.7[a,c]	711.2±0.5[a,d]	246.8±0.1[a,d]	216.035±0.006[a,d]	119.621±0.006[a,d]

[a]kJ/mol. [b]Ref (18). [c]Derived from refs (25,18,17). [d]Ref (17).

these two examples were done on an IBM RS/6000-560 workstation (*13*); the allyl chloride calculations required about three days (wallclock).

Example: $\Delta_f H°$ **of Methanol.** The B3LYP/6-311G(d,p) geometry optimization yields E(B3LYP) = -115.757394 hartree and r(C-O) = 1.422 Å, r(O-H) = 0.962 Å, θ(H-O-C) = 108.3°, and r(C-H$_{unique}$) = 1.091 Å, where H$_{unique}$ is *anti* to the hydroxyl hydrogen atom. The harmonic vibrational frequencies are, in cm^{-1}, 3839, 3101, 2972, 1508, 1489, 1380, 1086, and 1053 (a' symmetry) and 3013, 1492, 1168, and 328 (a″ symmetry). The vibrational zero-point energy (see overview chapter) is half the sum of the frequencies, or ZPE = 11214.6 cm^{-1} = 0.051097 hartree.

The various corrections to the base energy, $E_{base} \equiv E$[PMP4/6-311G(d,p)], are Δ(+) $\equiv E$[PMP4/6-311+G(d,p)] - E_{base}, Δ(2df) $\equiv E$[PMP4/6-311G(2df,p)] = E_{base}, Δ(CC) $\equiv E$[CCSD(T)/6-311G(d,p)] - E_{base}, Δ' $\equiv E$[UMP2/6-311+G(3df,2p)] + E[UMP2/6-311G(d,p)] - E[UMP2/6-311G(2df,p)] - E[UMP2/6-311+G(d,p)], where "UMP2" indicates that unprojected energies are used for open-shell systems. For closed-shell systems such as this, the usual RHF-based results are used. The RHF/6-311G(d,p) wavefunction for methanol is found to be stable. Since $n_\alpha = n_\beta = 7$, we have HLC = -0.041790 hartree. The energies and results, E[G2(CC)], are collected in Table IV.

We again use the atomization reaction for computational convenience. The sum of the atomic energies is -114.769854 hartree, so that E_{at} = 0.769912 hartree = 2021.4 kJ/mol. Subtracting this from the sum of the atomic enthalpies of formation gives $\Delta_f H°_0$ = 1822.1 - 2021.4 = -199.3 kJ/mol, in fair agreement with the experimental value $\Delta_f H°_0$ = -190.1 kJ/mol (equation 2).

Example: $\Delta_f H°_0$ **of 3-Chloropropene (Allyl Chloride).** These results are summarized in Table IV. We choose the *gauche* conformation of the chloromethyl group as above (C_1 point group). B3LYP/6-311G(d,p) geometry optimization yields E(B3LYP) = -577.565050 hartree and selected geometric parameters r(C=C) = 1.329 Å, r(C-C) = 1.489 Å, r(C-Cl) = 1.837 Å, θ(C=C-C) = 123.5°, and θ(C-C-Cl) = 111.3°. The harmonic vibrational frequencies are, in cm^{-1}, 3219, 3158, 3141, 3132, 3084, 1705, 1486, 1449, 1327, 1289, 1227, 1119, 1029, 966, 950, 908, 732, 590, 405, 283, and 110. Half the sum of the frequencies is ZPE = 15654 cm^{-1} = 0.071324 hartree. The RHF/6-311G(d,p) wavefunction is stable. Since $n_\alpha = n_\beta = 12$, we have HLC = -0.071640 hartree. The energies and results, E[G2(CC)] = -576.802910 hartree, are collected in Table IV.

We again use the atomization reaction for computational convenience. The sum of the atomic energies is -575.536383 hartree, so that E_{at} = 1.266527 hartree = 3325.3 kJ/mol. Subtracting this from the sum of the atomic enthalpies of formation (3333.4 kJ/mol) gives $\Delta_f H°_0$ = 8.1 kJ/mol, in good agreement with the experimental value $\Delta_f H°_0$ = 11.7 kJ/mol (equation 2).

BAC-MP4

The BAC-MP4 method involves calculating the molecular geometry and vibrational frequencies at the HF/6-31G(d) level, the electronic energy at the MP4/6-31G(d,p) level, and then applying empirical corrections ("bond-additivity corrections") to the energy.

The parametrization scheme has the effect of mimicking an isodesmic reaction. The method and parameters are summarized in the chapter by Zachariah and Melius (which we will refer to below as ZM). All the calculations for the following two examples were done on a personal computer.

BAC-MP4 Energies of Atoms. Enthalpies of formation are obtained from BAC-MP4 calculations by means of the atomization reactions (see the G2 examples above). This requires that BAC-MP4 energies for the isolated atoms be computed. The atoms relevant for the two examples below are C, H, N, O, and Cl. The (unrestricted) MP4/6-31G(d,p) energies for the atoms in their ground states are collected in Table V The unrestricted and spin-projected MP3 energies and the corresponding corrections for spin contamination, $E_{spin}(UHF) = E(UMP3) - E(PMP3)$ (equation 5 in ZM) are also included in Table V. The $E_{spin}(UHF)$ are subtracted from the MP4 energies to obtain BAC-MP4 energies for the atoms. Despite published statements to the contrary, it is these BAC-MP4 atomic energies that are combined with molecular energies to obtain BAC-MP4 atomization energies. Thus, the molecules and atoms are treated consistently.

Table V. BAC-MP4 Energies of Selected Atoms (hartree)

	C (triplet)	H (doublet)	N (quartet)	O (triplet)	Cl (doublet)
MP4	-37.750434	-0.498233	-54.473256	-74.895973	-459.569835
MP3	-37.746364	-0.498233	-54.470652	-74.893218	-459.567105
PMP3	-37.746731	-0.498233	-54.470993	-74.893871	-459.567544
$E_{spin}(UHF)$	0.000368	0.000000	0.000341	0.000653	0.000439
BAC-MP4	-37.750802	-0.498233	-54.473597	-74.896626	-459.570274
$\Delta_f H^\circ_0{}^a$	711.2±0.5	216.035±0.006	470.8±0.1	246.8±0.1	119.621±0.006

^{a}Experimental values in kJ/mol from ref. (*17*)

Example: $\Delta_f H^\circ_0$ of Nitromethane. The first step is to determine the lowest-energy conformer of the molecule. We consider two conformations of CH_3NO_2, both belonging to the C_s point group. The *syn* conformation, with one hydrogen eclipsing one oxygen, has an energy of -243.661983 at the HF/6-31G(d) level. The other C_s conformer has a plane of symmetry that bisects the O-N-O angle and passes through one hydrogen atom. Also at the HF/6-31G(d) level, this has an energy of -243.661992, very slightly (0.02 kJ/mol) more stable than the *syn* conformer. We choose the conformer with the lower-energy equilibrium (r_e) structure, although in this case it makes little difference. Its geometry includes r(C-N) = 1.478 Å, r(N-O) = 1.192 Å, r(C-H$_{unique}$) = 1.080 Å, r(C-H$_{not\ unique}$) = 1.076 Å, and θ(O-N-O) = 125.8°. Also at the HF/6-31G(d) level, the vibrational frequencies are, in cm^{-1}, 3369, 3272, 1688, 1620, 1571, 1271, 1052, 744, and 690 (a$'$ symmetry) and 3404, 1880, 1607, 1233, 530, and 27 (a$''$ symmetry). One-half the sum of these frequencies is ZPE = 11979 cm^{-1} = 0.054581 hartree. Although the vibrational frequencies are scaled by 0.8929 for determining the vibrational partition function, they are left unscaled when calculating the ZPE. The very low frequency corresponds to the torsional motion. The ZPE is not much affected by the treatment of hindered rotors since they contribute little to it. In contrast, low-frequency motions dominate the entropy and heat capacity. See Appendix B in this book for further details.

At the HF/6-31G(d) geometry, the MP4/6-31G(d,p) energy is $E(MP4) = -244.390002$ hartree. Adding the ZPE and subtracting the result from the sum of the atomic MP4 energies (first row of Table V) yields an atomization energy $E_{at} = 0.825086$ hartree = 2166.3 kJ/mol. Subtracting E_{at} from the sum of the atomic enthalpies of formation (last row of Table V) yields $\Delta_f H°_0(CH_3NO_2) = 157.4$ kJ/mol. This is a straight MP4 prediction, without any bond-additive corrections, and disagrees wildly with the experimental value $\Delta_f H°_0 = -60.8$ kJ/mol (*18*). The number that is commonly reported as "uncorrected MP4" actually includes E_{spin} for the atoms, as mentioned above; this corresponds to $\Delta_f H°_0 = 162.7$ kJ/mol. The empirical BAC for the molecule is subtracted from this value as shown below.

Nitromethane is a singlet, closed-shell molecule but has an RHF→UHF instability, i.e., there is a UHF wavefunction of lower energy than the RHF energy. Using the 6-31G(d,p) basis set, $E(UHF) = -243.669688$ hartree $< E(RHF) = -243.666887$ hartree. The singlet ground state is spin-contaminated with some triplet character to the extent that $\langle S^2 \rangle = 0.306$ (computed using the 6-31G(d,p) basis). For comparison, a pure spin singlet ($S = 0$) has $\langle S^2 \rangle = S(S+1) = 0$, a pure doublet ($S = \frac{1}{2}$) has $\langle S^2 \rangle = S(S+1) = 0.75$, and a pure triplet ($S = 1$) has $\langle S^2 \rangle = S(S+1) = 2$. The appropriate correction is given by equation 6 in ZM, $E_{spin}(S^2) = (10 \text{ kcal/mol})\langle S^2 \rangle = 3.06$ kcal/mol = 12.8 kJ/mol.

The bond corrections involve considerably more accounting. For the central bond in the bonded sequence of atoms A_k-A_i-A_j-A_l, they are given by equations 1-4 in ZM. Equations 5-7 below are an alternative form of the same equations. The values of the

$$E_{BAC}(A_i - A_j) = f_{ij} \prod_k (1 - h_{ij} h_{ik}) \prod_l (1 - h_{ji} h_{jl}) \tag{5}$$

$$h_{ij} = B_j \exp[-\alpha_{ij}(r_{ij} - 1.4 \text{ Å})] \tag{6}$$

$$f_{ij} = A_{ij} \exp(-\alpha_{ij} r_{ij}) \tag{7}$$

parameters are listed in Table I of ZM. Note that $A_{ij} = A_{ji}$ is listed in ZM in kcal/mol (1 kcal = 4.184 kJ), that $\alpha_{ij} = \alpha_{ji}$ is listed as a_{ij}, that $r_{ij} = r_{ji}$ is the distance between atoms A_i and A_j expressed in Å (1 Å = 10^{-10} m), and that $h_{ij} \neq h_{ji}$ unless $B_i = B_j$. The intermediate quantities in these calculations are illustrated in Tables VI and VII. In Table VI, only a single instance of a hydrogen neighbor (i.e., $A_k = H$) is considered, since hydrogen neighbors never affect the results ($B_H = 0$). We find that the total bond correction is $E_{BAC} = 206.1$ kJ/mol. Subtracting the corrections $E_{spin}(S^2)$ and E_{BAC} from the uncorrected enthalpy of formation yields $\Delta_f H°_0(CH_3NO_2) = 162.7 - 12.8 - 206.1 = -56.2$ kJ/mol, in good agreement with the experimental value of -60.8 kJ/mol.

Example: $\Delta_f H°_0$ of Chloroacetic Acid. Again we seek the lowest-energy conformation. We consider the six conformations described approximately by $\phi_1 = 0$, 120°, and 180° and by $\phi_2 = 0$, 180°, where ϕ_1 is the dihedral angle Cl-C-C=O and ϕ_2

Table VI. Components of BAC-MP4 Bond Corrections[a] for Nitromethane and Chloroacetic Acid

k	i	j	l	α_{ik}	r_{ik}	α_{ij}	r_{ij}	α_{jl}	r_{jl}	B_k	B_i	B_j	B_l	h_{ik}	h_{ij}	h_{ji}	h_{jl}	A_{ij}	f_{ij}
								Nitromethane, CH_3NO_2											
N	C	H_1		2.8	1.478	2.0	1.080			0.2	0.31	0		0.161	0	0.59		161.5	18.63
N	C	H_2		2.8	1.478	2.0	1.076			0.2	0.31	0		0.161	0	0.59		161.5	18.77
H_1	C	N	O	2.0	1.080	2.8	1.478	2.1	1.192	0	0.31	0.2		0	0.161	0.249	0.348	1934.3	30.85
C	N	O		2.8	1.478	2.1	1.192			0.31	0.2	0.225		0.249	0.348	0.310		945.6	77.37
O	N	O		2.1	1.192					0.225				0.348	0.348				
								Chloroacetic Acid, $ClCH_2COOH$											
C_1	C_2	Cl		3.8	1.513	2.0	1.766			0.31	0.31	0.42		0.202	0.202	0.149		1273.2	37.24
Cl	C_2	H		2.0	1.766	2.0	1.080			0.42	0.31	0		0.202	0	0.588		161.5	18.63
C_1	C_2	H		3.8	1.513					0.31	0.31			0.202	0				
Cl	C_2	C_1	O	2.0	1.766	3.8	1.513	2.14	1.180	0.42	0.31	0.31	0.225	0.202	0.202	0.202	0.360	6042.1	19.24
C_2	C_1	O	O_H	3.8	1.513	2.14	1.180	2.14	1.330	0.31	0.31	0.225	0.225	0.202	0.360	0.496	0.261	734.7	58.81
O_H	C_1	O		2.14	1.330					0.225				0.261	0.360				
C_2	C_1	O_H		3.8	1.513	2.14	1.330			0.31	0.31	0.225		0.202	0.261	0.360		734.7	42.66
O	C_1	O_H		2.14	1.180					0.225				0.360	0.261				
C_1	O_H	H		2.14	1.330	2.0	0.953			0.31	0.225	0		0.360	0	0.550		303.1	45.06

[a] A_{ij}, B_n, and f_{ij} in kJ/mol, α_{nm} in Å⁻¹, and r_{nm} in Å. H_1 and H_2 refer to the unique and non-unique hydrogen atoms, respectively. C_1 and C_2 refer to the carboxylic and chloromethyl carbons, respectively. O_H refers to the hydroxyl oxygen atom.

Table VII. BAC-MP4 Bond Corrections^a for Nitromethane and Chloroacetic Acid

Bond	f_{ij}	$\Pi_k(1-h_{ij}h_{ik})\,\Pi_l(1-h_{ji}h_{jl})$	$E_{BAC}(A_i\text{-}A_j)$
C-H$_1$	18.63	1	18.63
C-H$_2$	18.77	1	18.77
C-H$_2$	18.77	1	18.77
C-N	30.85	(1)(1)(1)·(0.913)(0.913)	25.72
N-O	77.37	(0.913)(0.879)	62.11
N-O	77.37	(0.249)(0.348)	62.11
Total for Nitromethane			*206.11*
C-Cl	37.24	0.959	35.72
C-H	18.63	1	18.63
C-H	18.63	1	18.63
C$_2$-C$_1$	19.24	(0.959)(1)(1)·(0.927)(0.947)	16.21
C=O	58.81	(0.927)(0.906)	49.41
C-O$_H$	42.66	(0.947)(0.906)	36.61
O-H	45.06	1	45.06
Total for Chloroacetic Acid			*220.27*

$^a f_{ij}$ and E_{BAC} in kJ/mol. H$_1$ and H$_2$ refer to the unique and the non-unique hydrogen atoms, respectively. O$_H$ refers to the hydroxyl oxygen atom.

is the dihedral angle H-O-C=O. After geometry optimization at the HF/6-31G(d) level, the most stable structure is that of C_s symmetry with $\phi_1 = \phi_2 = 0$. Among the geometric parameters are the distances r(C-Cl) = 1.766 Å, r(C-H) = 1.080 Å, r(C-C) = 1.513 Å, r(C=O) = 1.180 Å, r(C-O$_H$) = 1.330 Å, and r(O-H) = 0.953 Å and the bond angles θ(Cl-C-C) = 113.3°, θ(C-C=O) = 127.4°, and θ(C-C-O$_H$) = 108.9°. Also at the HF/6-31G(d) level, the vibrational frequencies are, in cm^{-1}, 4050, 3289, 2068, 1610, 1536, 1427, 1291, 981, 864, 675, 433, and 234 (a$'$ symmetry) and 3346, 1331, 1048, 698, 546, and 68 (a$''$ symmetry). One-half the sum of these frequencies is ZPE = 12747 cm^{-1} = 0.058079 hartree.

At the HF/6-31G(d) geometry, the MP4/6-31G(d,p) energy is E(MP4) = -687.519376 hartree. Adding the ZPE and subtracting the result from the sum of the atomic MP4 energies (first row of Table V) yields an atomization energy E_{at} = 1.103950 hartree = 2898.4 kJ/mol. Subtracting E_{at} from the sum of the atomic enthalpies of formation (last row of Table V) yields $\Delta_f H°_0$(ClCH$_2$CO$_2$H) = -214.7 kJ/mol. This is a straight MP4 prediction, without any BAC-type corrections. Using the BAC-MP4 atomic energies yields -208.2 kJ/mol, again in terrible disagreement with the experimental value $\Delta_f H°_0$ = -427.3 kJ/mol (*18*).

Chloroacetic acid is a closed-shell singlet molecule with a stable HF/6-31G(d) wavefunction, so $E_{spin}(S^2) = 0$. The bond corrections, done using equations 5-7, are summarized in Tables VI and VII. In this case the total correction is E_{BAC} = 220.3 kJ/mol, so the corrected, BAC-MP4 enthalpy of formation is $\Delta_f H°_0$ = -208.2 - 220.3 = -428.5 kJ/mol, in excellent agreement with the experimental value of -427.3 kJ/mol.

PCI-X

In the PCI-X ("parametrized configuration interaction with parameter X") family of methods, discussed in this book in the chapter by Blomberg and Siegbahn, a single parameter (X) is used to scale the contribution of electron correlation to the thermochemical quantity of interest (*19,20*). A optional correction for basis set incompleteness, at the Hartree-Fock (HF) level, may be included for additional accuracy, especially in molecules with multiple bonding between first-row elements. When calculating atomization energies, it is recommended (*19*) that each atomic energy be calculated for a low-lying electronic state with a single valence s-electron (s^1 state), followed by a correction based upon the experimental excitation energy (*21*). Molecular geometries and vibrational frequencies are typically at the HF level using a modest basis set.

If the quantity of interest (e.g., a bond strength or reaction energy) is denoted E, then the corrected value is determined from equation 8. This is identical to a single-parameter SAC method (see the next section). If the basis set correction is desired, then equation 9 is used instead. The value of X is determined by fitting to a data set, and

$$E_{PCI-X}{}' = E_{HF} + (E_{correlated} - E_{HF})/(X/100) \qquad (8)$$

$$E_{PCI-X} = E_{PCI-X}{}' + (E_{HF\ limit} - E_{HF}) \qquad (9)$$

depends upon the choice of correlated *ab initio* theory and upon the choice of basis set. For the basis set correction, the value at the HF limit is approximated using a large basis set with six s-, five sets of p-, three sets of d- and two sets of f-basis functions (6s, 5p, 3d, 2f) on first-row elements (Li-F) and (4s, 3p, 2d) on hydrogen (*19*). For convenience in the examples below, we use instead the more standard and somewhat less expensive aug-cc-pVTZ basis set (*22*), which is (5s, 4p, 3d, 2f) on first-row atoms and (4s, 3p, 2d) on hydrogen.

The PCI-X methods were developed for molecules that contain transition metals. This makes them unique among the methods described in this book. In the examples below, we choose simple organic compounds instead, to allow closer comparisons with the other methods illustrated in this Appendix. All the calculations for the following two examples were done on a personal computer, although the molecular HF/aug-cc-pVTZ calculation, which is most expensive step, required about 2.5 days (wallclock) for each molecule.

Example: $\Delta_f H°_0$ of 2-Butanone. We choose the conformation of methyl ethyl ketone (MEK) with the methyl groups *anti* and with a hydrogen atom on each methyl group *anti* to the central C-C bond (C_s point group). We choose to optimize the molecular geometry at the HF/6-31G(d) level (91 basis functions), which yields $r(C=O) = 1.193$ Å, $r(C_1-C_2)$ = 1.515 Å, $r(C_2-C_3) = 1.519$ Å, $r(C_3-C_4) = 1.524$ Å, $\theta(C_1-C_2-C_3) = 116.5°$, and $\theta(C_2-C_3-C_4) = 113.6°$. Also at the HF/6-31G(d) level, the vibrational frequencies are, in cm^{-1}, 3322, 3283, 3220, 3210, 3192, 2016, 1644, 1614, 1607, 1571, 1553, 1524, 1302, 1208, 1072, 1034, 827, 636, 428, and 267 (a′ symmetry) and 3299, 3265, 3219, 1640, 1623, 1409, 1251, 1049, 826, 516, 234, 111, and 31 (a″ symmetry). One-half their sum is 26501 cm^{-1}, so scaling by 0.9135 (*23*) yields ZPE = 24209 cm^{-1} = 0.110303 hartree.

At the HF/6-31G(d) geometry, we calculate the HF and frozen-core CCSD energies using the cc-pVDZ basis set for the heavy atoms. For hydrogen, for convenience we use the D95 basis set supplied with the Gaussian program, supplemented with a p-polarization function with $\zeta_p = 0.727$. This basis set is quite close to that for which the "CCSD/VDZ" value $X = 76.8$ was optimized (*19*). With this basis (110 basis functions) we obtain $E(HF) = -231.024798$ and $E(CCSD) = -231.820542$ hartree. As mentioned above, the HF limit is estimated using the aug-cc-pVTZ basis set (414 basis functions) to give $E(HF\text{-limit}) = -231.086013$. The ZPE is then added to these energies; the results are summarized in Table VIII.

For computational convenience, we choose to evaluate the enthalpy of formation of MEK by way of its atomization energy, so that $E \equiv E_{at}$ in equation 8. Atomic energies and the resulting atomization energies are included in Table VIII. For the carbon atom, the excited state ^5S (s^1p^3) is used instead of the ground state ^3P (s^2p^2); its experimental excitation energy 33735.2 cm^{-1} = 403.56 kJ/mol (*21*). Experimental enthalpies of formation, $\Delta_f H°_0$, are also included in Table VIII. For the C atom, the value listed includes the excitation energy.

From Table VIII, we have that $E_{HF} = 4669.1$ kJ/mol and $E_{correlated} = 6183.0$ kJ/mol. The uncorrected $E_{correlated}$ value is in error by -7.0%. Taking $X = 76.8$ (*19*), equation 8 yields $E_{PCI-X'} = 6640.3$ kJ/mol. This corrected value is in error by only -0.2%. Subtracting $E_{PCI-X'}$ from the sum of the atomic enthalpies of formation gives $\Delta_f H°_0$(MEK)

= -206.0 kJ/mol without the HF-limit correction, in agreement with the experimental value $\Delta_f H°_0$ = -217.1 kJ/mol (*18*).

Including the HF-limit correction by means of equation 9, with $E_{HF\ limit}$ = 4665.4 kJ/mol, yields E_{PCI-X} = 6636.6 kJ/mol. This corrected value is still in error by -0.2%. Subtracting this from the sum of the atomic enthalpies of formation leads to $\Delta_f H°_0$ = -202.3 kJ/mol, which is 15 kJ/mol higher than the experimental value. Thus, in this case the HF-limit correction makes the prediction slightly worse.

Table VIII. Intermediate Quantities for PCI-X Calculations

Species	HF[a]	CCSD[a]	"HF-limit"[a]	exptl $\Delta_f H°_0$[b]
2-butanone	-230.914494	-231.710239	-230.975709	-217.1
cyclobutanol	-230.867560	-231.668679	-230.929952	-169.
C (^{5}S)	-37.590713	-37.616240	-37.596797	1114.8
H (^{2}S)	-0.497637	-0.497637	-0.499821	216.035
O (^{3}P)	-74.792166	-74.909201	-74.812982	246.8
(4C+8H+O)	-229.136116	-229.355259	-229.198741	6434.3
E_{at}(2-butanone)	1.778379 (4669.1)[c]	2.354980 (6183.0)[c]	1.776968 (4665.4)[c]	6651.4
E_{at}(cyclobutanol)	1.731444 (4545.9)[c]	2.313420 (6073.9)[c]	1.731211 (4545.3)[c]	6603.

[a]In hartree, including ZPE. "HF-limit" is actually HF/aug-cc-pVTZ. [b]In kJ/mol, from refs (*17*) and (*18*). [c]In kJ/mol.

Example: $\Delta_f H°_0$ of Cyclobutanol. The molecule is illustrated at right. We considered four conformations: with the hydroxyl hydrogen *gauche* or *anti* to the neighboring hydrogen, and with the hydroxyl group equatorial or axial. Geometry optimization at the HF/6-31G(d) level indicates that the most stable conformer (although by only 0.5 kJ/mol) is the one of C_s symmetry with the OH equatorial and with the hydrogens *anti* (i.e., ϕ(H-O-C-H) =  180°). At the HF/6-31G(d) level we obtain $\phi(C_1\text{-}C_2\text{-}C_3\text{-}C_4)$ = 20.3°, $r(C_1\text{-}C_2)$ = 1.540 Å, $r(C_1\text{-O})$ = 1.391 Å, r(O-H) = 0.947 Å, $\theta(C_4\text{-}C_1\text{-}C_2)$ = 88.7°, and θ(H-O-C_1) = 109.5°. The vibrational frequencies are, in cm^{-1}, 4098, 3295, 3281, 3260, 3237, 3214, 1667, 1633, 1582, 1458, 1408, 1330, 1263, 1186, 1052, 982, 820, 663, 492, and 190 (a' symmetry) and 3274, 3208, 1625, 1444, 1393, 1377, 1312, 1143, 1026, 992, 855, 421, and 323 (a'' symmetry). One-half their sum is 27254 cm^{-1}, so scaling by 0.9135 (*23*) yields ZPE = 24897 cm^{-1} = 0.113438 hartree.

Using the same basis as for MEK, we obtain $E(\text{HF}) = -230.980998$ and $E(\text{CCSD})$ $= -231.782117$ hartree. The HF limit is again estimated using the aug-cc-pVTZ basis set (414 basis functions) as $E(\text{HF-limit}) = -231.043390$. The ZPE is added to these energies and the results summarized in Table VIII. As above, we choose to evaluate the enthalpy of formation of cyclobutanol by way of its atomization energy, so that $E \equiv E_{at}$ in equations 8 and 9. Atomic energies and the resulting atomization energies are included in Table VIII.

From Table VIII, we have that $E_{\text{HF}} = 4545.9$ kJ/mol, $E_{\text{correlated}} = 6073.9$ kJ/mol, and $E_{\text{HF limit}} = 4545.3$ kJ/mol. The uncorrected $E_{\text{correlated}}$ value is in error by -8.0%. Taking $X = 76.8$ (*19*), equation 8 yields $E_{\text{PCI-X'}} = 4545.9 + (6073.9 - 4545.9)/0.768 = 6535.5$ kJ/mol (no HF-limit correction). This is only 1.0% lower than the literature value. Subtracting this from the sum of the atomic enthalpies of formation gives $\Delta_f H°_0(\text{cyclobutanol}) = 6434.1 - 6535.5 = -101.2$ kJ/mol, which is 68 kJ/mol higher than the literature value $\Delta_f H°_0 = -169$ kJ/mol (*18*). We include the HF-limit correction using equation 9, which yields $E_{\text{PCI-X}} = 6534.8$ kJ/mol, little different. Subtracting $E_{\text{PCI-X}}$ from the sum of the atomic enthalpies of formation gives $\Delta_f H°_0(\text{cyclobutanol}) = -100.6$ kJ/mol, still 68 kJ/mol above the literature value.

Atomization energies are usually a poor choice of reaction, as evinced by the widely differing HF and CCSD results (see the overview in this book). A much better

$$\text{cyclobutanol } (C_4H_8O) \rightarrow 2\text{-butanone } (C_4H_8O) \qquad (10)$$

reaction scheme is the isomerization reaction 10. For this reaction, E represents the enthalpy change at $T = 0$ K. Including ZPE, $E_{\text{HF}} = -0.046935$ hartree $= -123.2$ kJ/mol and $E_{\text{correlated}} = -0.041560$ hartree $= -109.1$ kJ/mol. The literature values (Table VIII) provide $E_{\text{lit}} = -48$ kJ/mol, so the uncorrected $E_{\text{correlated}}$ value is too negative by 61 kJ/mol. Taking $X = 76.8$ (*19*), equation 8 yields $E_{\text{PCI-X'}} = -104.9$ kJ/mol (no HF-limit correction), still too low. Accepting the experimental enthalpy of formation for MEK, this corresponds to $\Delta_f H°_0(\text{cyclobutanol}) = -112.2$ kJ/mol, which is 47 kJ/mol higher than the literature value in Table VIII. Including ZPE, $E_{\text{HF limit}} = -0.045757$ hartree $= -120.1$ kJ/mol, so applying equation 9 gives $E_{\text{PCI-X}} = -101.8$ kJ/mol (with the HF-limit correction) or $\Delta_f H°_0 = -115.3$ kJ/mol, still 54 kJ/mol above the literature value.

The poor agreement with the literature value for cyclobutanol might suggest that the PCI-X procedure works poorly. However, the literature value cited in reference (*18*) is not a measurement but an estimate. Room-temperature ($T = 298.15$ K) estimates of $\Delta_f H°_{298}(\text{cyclobutanol})$ in the literature are -194 kJ/mol (*18*) and -134 kJ/mol (*24*), and group additivity (GA) provides -141 kJ/mol (*10*). Thus, the value from reference (*18*) is 53-60 kJ/mol lower than alternative values. Since the PCI-X value for $\Delta_f H°_0$ based upon reaction 10 is 54 kJ/mol higher than that from reference (*10*) (Table VIII), the calculation strongly supports the higher values (*24,10*).

MP2-SAC and CCD-SAC

These are two of a family of methods (*2-5*) that are very similar to the PCI-X methods. CCD-SAC is nearly the same as PCI-X, except that the parameter is called F instead of

X and has the value 0.709 for the cc-pVDZ basis set (5). In MP2-SAC, the main difference from PCI-X is that the single parameter X in PCI-X is replaced by the three-parameter expression $F_0 + F_1 x_H + F_2 x_H^2$, where x_H is the fraction of atoms in the system

$$E_{SAC} = E_{HF} + \frac{E_{correlated} - E_{HF}}{F_0 + F_1 x_H + F_2 x_H^2}$$ (11)

that are hydrogen atoms (equation 11). For MP2/cc-pVDZ calculations, the values of the parameters are 0.918, -0.143, and -0.176 for F_0, F_1, and F_2, respectively (5). The molecular geometry in the SAC methods is calculated at the same level of theory as the energy. All the calculations for the following examples were done on a personal computer, although the CCD calculations required 35 and 27 hours (wallclock) for norbornadiene and 3-methyl pyrrole, respectively.

Example: $\Delta_f H°_0$ **of Norbornadiene.** The molecule, also called bicyclo[2.2.1]hepta-2,5-diene, is shown at right (C_{2v} point group). Its empirical formula is C_7H_8, so $x_H = 8/15 = 0.533$. We choose to compute an initial geometry and associated vibrational frequencies at the HF/6-31G(d) level. The vibrational frequencies are, in cm^{-1}, 3415, 3292, 3237, 1822, 1654, 1396, 1235, 1015, 968, 846, 836, and 463 (a$_1$ symmetry), 3385, 1449, 1396, 1237, 1055, 986, 775, and 486 (a$_2$ symmetry), 3412, 3287, 1780, 1381, 1193, 1139, 987, 764, and 557 (b$_1$ symmetry), and 3387, 3289, 1483, 1409, 1315, 1082, 1028, 965, 871, and 600 (b$_2$ symmetry). One half their sum, scaled by 0.9135 (23), provides ZPE = 30439 cm^{-1} = 0.126692 hartree. The geometry is then reoptimized at the frozen-core MP2/cc-pVDZ level to obtain $r(C_2=C_3) = 1.355$ Å, $r(C_1-C_2) = 1.541$ Å, $r(C_1-C_7) = 1.559$ Å, $\theta(C_1-C_2-C_3) = 106.9°$, and $\theta(C_1-C_7-C_4) = 106.9°$. The molecular energies are -269.670812 and -270.618272 hartree at the HF/cc-pVDZ and MP2/cc-pVDZ levels, respectively, at the MP2 geometry. Atomic and molecular energies (including ZPE) and experimental enthalpies of formation are collected in Table IX.

The MP2 energies, including ZPE, lead to an atomization energy of 6115.3 kJ/mol. Subtracting this from the sum of the atomic enthalpies of formation gives $\Delta_f H°_0$ = 591.4 kJ/mol, far above the experimental value (Table IX). Using the values for the F_i and for x_H given above, equation 11 yields $E_{MP2-SAC}$ = 4579.1 + (6115.3 - 4579.1)/(0.792) = 6519.5 kJ/mol. This corresponds to $\Delta_f H°_0$ = 187.2 kJ/mol, still 83 kJ/mol lower than the experimental value $\Delta_f H°_0$ = 270.4 kJ/mol.

At the CCD-SAC level, the geometry should be reoptimized at the CCD/cc-pVDZ level (5). However, such a geometry optimization is very expensive, so for convenience we continue to use the MP2/cc-pVDZ geometry instead. Thus, the methods used here would be better denoted "CCD-SAC//MP2" instead of "CCD-SAC." We expect very minor differences between results from the two procedures.

The CCD/cc-pVDZ energy is -270.666416 hartree; the value in Table IX includes the ZPE. The CCD atomization energy is 5861.1 kJ/mol, corresponding to $\Delta_f H°_0$ = 845.5 kJ/mol, well above the experimental value. Using $F = 0.709$ (5) gives $E_{CCD-SAC}$ =

4579.1 + (5861.1 - 4579.1)/(0.709) = 6387.3 kJ/mol, or $\Delta_f H°_0$ = 319.3 kJ/mol. This is 49 kJ/mol higher than the experimental value $\Delta_f H°_0$ = 270.4 kJ/mol.

Example: $\Delta_f H°_0$ of 3-Methyl Pyrrole. The molecule is shown at right. Its empirical formula is C_5H_7N, so x_H = 7/13 = 0.538. At the HF/6-31G(d) level, the most stable conformation of the methyl group has a hydrogen eclipsing the $C_2=C_3$ double bond (C_s point group). The HF/6-31G(d) vibrational frequencies are, in cm^{-1}, 3927, 3448, 3437, 3414, 3265, 3196, 1783, 1670, 1646, 1608, 1572, 1535, 1404, 1385, 1247, 1185, 1155, 1094, 1038, 974, 683, and 345 (a′ symmetry) and 3244, 1629, 1171, 974, 885, 808, 694, 682, 455, 280, and 123 (a″ symmetry). One half their sum, scaled by 0.9135 (*23*), provides ZPE = 32730 cm^{-1} = 0.108123 hartree. The geometry at the frozen-core MP2/cc-pVDZ level includes $r(\text{N-}C_2)$ = 1.377 Å, $r(C_2\text{-}C_3)$ = 1.397 Å, $r(C_3\text{-}C_4)$ = 1.430 Å, $r(C_3\text{-}CH_3)$ = 1.504 Å, and $\theta(C_2\text{-N-}C_5)$ = 110.1°. The molecular energies are -247.865813 and -248.709878 hartree at the HF/cc-pVDZ and MP2/cc-pVDZ levels, respectively, at the MP2 geometry. Atomic and molecular data are included in Table IX.

Table IX. Intermediate Quantities for SAC Calculations

Species	HF^a	$MP2^a$	CCD^a	$exptl\ \Delta_f H°_0{}^b$
norbornadiene	-269.544120	-270.491580	-270.539724	270.4
3-methyl pyrrole	-247.757691	-248.601756	-248.638141	92.7
C (^{3}P)	-37.686544	-37.738310	-37.759015	711.2
H (^{2}S)	-0.499278	-0.499278	-0.499278	216.035
N (^{4}S)	-54.391115	-54.461855	-54.477728	470.8
(7C+8H)	-267.800038	-268.162395	-268.307335	6706.7
(5C+7H+N)	-246.318786	-246.648352	-246.767754	5539.0
E_{at}(norbornadiene)	1.744082 $(4579.1)^c$	2.329185 $(6115.3)^c$	2.232389 $(5861.1)^c$	6436.3
E_{at}(3-methyl pyrrole)	1.438905 $(3777.8)^c$	1.953404 $(5128.7)^c$	1.870388 $(4910.7)^c$	5446.3
C_2H_4	-77.989025	-78.265269	-78.293326	61.0
CH_4	-40.154906	-40.316552	-40.339589	-66.9
NH_3	-56.161170	-56.348698	-56.364294	-38.9
H_2	-1.119034	-1.145537	-1.153827	0

[a]In hartree, at the MP2/cc-pVDZ geometry, including scaled HF/6-31G* ZPE. [b]In kJ/mol, from refs (*17*) and (*18*). [c]In kJ/mol.

The MP2 energies, including ZPE, lead to an atomization energy of 1.953404 hartree = 5128.7 kJ/mol. Subtracting this from the sum of the atomic enthalpies of formation gives $\Delta_f H°_0$ = 410.4 kJ/mol, far above the experimental value (Table IX). Using the values for the F_i and for x_H given above, equation 11 yields E_{SAC} = 3777.8 + (5128.7 -3777.8)/(0.790) = 5487.8 kJ/mol. This corresponds to $\Delta_f H°_0$ = 51.2 kJ/mol, much closer to the experimental value $\Delta_f H°_0$ = 92.7 kJ/mol but still 42 kJ/mol too low.

As above, we depart from the CCD-SAC prescription (5) and use the MP2/cc-pVDZ geometry instead of the CCD/cc-pVDZ geometry, for economy. The CCD//MP2/cc-pVDZ energy is -248.746264 hartree; the value in Table IX includes the ZPE. The CCD atomization energy is 1.870388 hartree = 4910.7 kJ/mol, corresponding to $\Delta_f H°_0$ = 628.3 kJ/mol, far from the experimental value. Using F = 0.709 as above gives $E_{CCD-SAC}$ = 3777.8 + (4910.7 -3777.8)/(0.709) = 5375.7 kJ/mol, or $\Delta_f H°_0$ = 163.4 kJ/mol. This is closer but still 71 kJ/mol above the experimental value $\Delta_f H°_0$ = 92.7 kJ/mol.

Example: Avoiding Atomization Energies. In the examples above, $\Delta_f H°_0$ values from MP2-SAC disagree with experimental values by -83 and -42 kJ/mol for norbornadiene and 3-methyl pyrrole, respectively. The corresponding errors for CCD-SAC, albeit not quite at the right geometries, are +49 and +71 kJ/mol. These are based upon the atomization reactions, which are neither isogyric or isodesmic (see overview chapter and Appendix D to this book). Better results are expected from reactions 12 and 13, which are isogyric. Additional calculations on ethylene, methane, ammonia, and hydrogen are required; intermediate results, determined as above, are included in Table IX.

$$\text{norbornadiene } (C_7H_8) \ + \ 6H_2 \ \rightarrow \ 2\,C_2H_4 + 3\,CH_4 \qquad (12)$$

$$3\text{-methyl pyrrole } (C_5H_7N) \ + \ 4H_2 \ \rightarrow \ 2\,C_2H_4 + CH_4 + NH_3 \qquad (13)$$

Table X. Reaction Energies (kJ/mol) for Atomization and for Isogyric Partial Hydrogenation

Energy	Norbornadiene		3-Methyl Pyrrole	
	atomization	reaction 12	atomization	reaction 13
uncorr. HF	4579.1	-484.3	3777.8	-158.3
uncorr. MP2	6115.3	-303.0	5128.7	-31.2
uncorr. CCD	5861.1	-374.7	4910.7	-97.4
MP2-SAC	6519.5	-436.3	5487.8	-142.8
CCD-SAC	6387.3	-329.8	5375.7	-72.3
experiment[a]	6436.3	-349.1	5446.3	-76.0

[a] From refs (17) and (18).

Reaction energies are calculated using individually SAC-corrected molecular energies. The reaction energies listed in Table X, along with reaction energies calculated using uncorrected energies (i.e., electronic energies plus ZPE). For norbornadiene, the partial hydrogenation reaction energy (equation 12 and Table X) leads to a prediction $\Delta_f H^{\circ}_0 = 357.6$ kJ/mol at the MP2-SAC level, 87 kJ/mol higher than the experimental value $\Delta_f H^{\circ}_0 = 270.4$ kJ/mol. At the CCD-SAC level, the corresponding value is $\Delta_f H^{\circ}_0 = 251.1$ kJ/mol, in acceptable agreement with experiment. For 3-methyl pyrrole, using the isogyric reaction 13 leads to $\Delta_f H^{\circ}_0 = 159.0$ at the MP2-SAC level, 66 kJ/mol above the experimental value. The CCD-SAC prediction is $\Delta_f H^{\circ}_0 = 88.5$ kJ/mol, in good agreement with the experimental value $\Delta_f H^{\circ}_0 = 92.7$ kJ/mol. These examples illustrate that better results are indeed obtained using the isogyric reactions 12 and 13 than using atomization reactions.

Summary

Computational thermochemistry includes many powerful techniques, but is not yet a mature field. New methods are continually being developed. They may offer more precision, more robustness (i.e., tolerance for poor choices of reaction), greater range of chemical applicability, lower computational expense, or other improvements. However, new methods seldom offer user-friendly software to automate the calculations. If such software is produced at all, years may elapse before it becomes generally available. Thus, users must usually apply new methods by hand or develop their own software. We hope that the worked examples in this Appendix are helpful.

Acknowledgments

KKI thanks Drs. Michael Zachariah and Donald Burgess for instruction in the BAC-MP4 method and its implementation.

Literature Cited

1.   Mebel, A. M.; Morokuma, K.; Lin, M. C. *J. Chem. Phys.* **1995**, *103*, 7414.
2.   Gordon, M. S.; Truhlar, D. G. *J. Am. Chem. Soc.* **1986**, *108*, 5412.
3.   Gordon, M. S.; Truhlar, D. G. *Int. J. Quantum Chem.* **1987**, *31*, 81.
4.   Gordon, M. S.; Nguyen, K. A.; Truhlar, D. G. *J. Phys. Chem.* **1989**, *93*, 7356.
5.   Rossi, I.; Truhlar, D. G. *Chem. Phys. Lett.* **1995**, *234*, 64.
6.   Reid, R. C.; Prausnitz, J. M.; Poling, B. E. *The Properties of Gases and Liquids*; 4th ed.; McGraw-Hill: New York, 1987.
7.   Pedley, J. B. *Thermochemical Data and Structures of Organic Compounds*; Thermodynamics Research Center: College Station, Texas, 1994; Vol. 1.
8.   Benson, S. W. *Thermochemical Kinetics*; 2nd ed.; Wiley: New York, 1976.
9.   Cohen, N. *J. Phys. Chem. Ref. Data* **1996**, *25*, 1411.
10.   Lias, S. G.; Liebman, J. F.; Levin, R. D.; Kafafi, S. A. "Structures and Properties, vers. 2.02," NIST Standard Reference Database 25, National Institute of Standards and Technology, 1994.

11.   Downey, J. R.; Frurip, D. J.; LaBarge, M. S.; Syverud, A. N.; Grant, N. K.;
      Marks, M. D.; Harrison, B. K.; Seaton, W. H.; Treweek, D. N.; Selover, T. B.
      "CHETAH 7.0, The ASTM Computer Program for Chemical Thermodynamic
      and Energy Release Evaluation (NIST Special Database 16)," ASTM Data Series
      DS 51B, 1994.

12.   *Gaussian 94*. Frisch, M. J.; Trucks, G. W.; Schlegel, H. B.; Gill, P. M. W.;
      Johnson, B. G.; Robb, M. A.; Cheeseman, J. R.; Keith, T.; Petersson, G. A.;
      Montgomery, J. A.; Raghavachari, K.; Al-Laham, M. A.; Zakrzewski, V. G.;
      Ortiz, J. V.; Foresman, J. B.; Cioslowski, J.; Stefanov, B. B.; Nanayakkara, A.;
      Challacombe, M.; Peng, C. Y.; Ayala, P. Y.; Chen, W.; Wong, M. W.; Andres,
      J. L.; Replogle, E. S.; Gomperts, R.; Martin, R. L.; Fox, D. J.; Binkley, J. S.;
      Defrees, D. J.; Baker, J.; Stewart, J. P.; Head-Gordon, M.; Gonzalez, C.; Pople,
      J. A. Gaussian, Inc., Pittsburgh, PA, 1995.

13.   Certain commercial materials and equipment are identified in this paper in order
      to specify procedures completely. In no case does such identification imply
      recommendation or endorsement by the National Institute of Standards and
      Technology, nor does it imply that the material or equipment identified is
      necessarily the best available for the purpose.

14.   Curtiss, L. A.; Raghavachari, K.; Trucks, G. W.; Pople, J. A. *J. Chem. Phys.*
      1991, *94*, 7221.

15.   Curtiss, L. A.; Raghavachari, K.; Pople, J. A. *J. Chem. Phys.* **1993**, *98*, 1293.

16.   Colegrove, B. T.; Thompson, T. B. *J. Chem. Phys.* **1997**, *106*, 1480.

17.   Chase, M. W., Jr.; Davies, C. A.; Downey, J. R., Jr.; Frurip, D. J.; McDonald, R.
      A.; Syverud, A. N. *J. Phys. Chem. Ref. Data* **1985**, *14*, Suppl. 1.

18.   Frenkel, M.; Marsh, K. N.; Wilhoit, R. C.; Kabo, G. J.; Roganov, G. N.
      Thermodynamics of Organic Compounds in the Gas State; Thermodynamics
      Research Center: College Station, Texas, 1994; Vol. I.

19.   Siegbahn, P. E. M.; Svensson, M.; Boussard, P. J. E. *J. Chem. Phys.* **1995**, *102*,
      5377.

20.   Siegbahn, P. E. M.; Blomberg, M. R. A.; Svensson, M. *Chem. Phys. Lett.* **1994**,
      223, 35.

21.   Moore, C. E. *Atomic Energy Levels*; NSRDS-NBS 35, reprint of NBS Circular
      467; U.S. Govt. Printing Office: Washington, D.C., 1971.

22.   Dunning, T. H., Jr. *J. Chem. Phys.* **1989**, *90*, 1007.

23.   Scott, A. P.; Radom, L. *J. Phys. Chem.* **1996**, *100*, 16502.

24.   Lias, S. G.; Bartmess, J. E.; Liebman, J. F.; Holmes, J. L.; Levin, R. D.; Mallard,
      W. G. *J. Phys. Chem. Ref. Data* **1988**, *17*, Suppl. 1.

25.   Stull, D. R.; Westrum, E. F., Jr.; Sinke, G. C. *The Chemical Thermodynamics of
      Organic Compounds*; Krieger: Malabar, Florida, 1987.

Appendix D

Glossary of Common Terms and Abbreviations in Quantum Chemistry

Karl K. Irikura

Physical and Chemical Properties Division, National Institute of Standards and Technology, Gaithersburg, MD 20899

The literature of quantum chemistry, covering both molecular orbital theory and density functional theory, is cluttered with abbreviations, acronyms, and jargon. Some of the more common terminology is explained in this glossary.

An extensive list of acronyms has been published recently (*1*), and a major reference work is currently in preparation. The list below is adapted from one that was developed for a brief course within the Physical and Chemical Properties Division at NIST.

Glossary

3-21G	A VDZ basis set (see). See also 6-31G.
3-21G*	The asterisk indicates that a set of polarizing d-functions (6D) is included to supplement the 3-21G basis, but only on second-row and heavier atoms (beyond neon). Also denoted 3-21G(*) or 3-21G(d).
5D	Indicates that five functions are used in each d-set.
6D	Indicates that six (cartesian) functions are used in each d-set. This includes the s-like combination $(x^2 + y^2 + z^2)$.
6-31G	A VDZ basis set (see) from the Pople school. Very popular, often used with a set of heavy-atom polarization functions (see 6-31G*). The "6" indicates that each core basis function is built using six primitives (see). The "3" indicates that the inner valence basis functions are each built using three primitives. The "1" indicates that the outer valence basis functions are each built using a single uncontracted primitive. The "G" stands for "Gaussian", indicating the type of primitive function. (Recently-developed basis sets don't

include the "G" or its equivalent since they are essentially all based upon Gaussian functions.)

6-311G	A popular VTZ basis set (see) similar to the small 6-31G set. Usually supplemented with polarization functions. Built like 6-31G but with a third layer of valence functions composed of a single, uncontracted primitive set. Some workers consider this basis to be less flexible than a "real" triple-zeta basis.
6-311G*	The asterisk indicates that a set of polarization d-functions (5D) has been added to heavy atoms to supplement the 6-311G basis; also denoted 6-311G(d).
6-311G**	The second asterisk indicates that a set of polarization p-functions has been added to hydrogen; also denoted 6-311G(d,p).
6-311+G(3df,2p)	In addition to the 6-311G basis, the "+" indicates that diffuse s- and p-functions are added to heavy atoms, the "3df" indicates that three sets of polarization d-functions and one set of polarization f-functions are added to heavy atoms, and the "2p" indicates that two sets of polarization p-functions are added to hydrogen.
6-31G*	A polarized VDZ basis set (see) from the Pople school. Maybe the most popular basis set in use today. The single asterisk indicates that a set of polarizing d-functions (6D) is included on "heavy" atoms (beyond helium). Also denoted 6-31G(d).
6-31G**	A polarized VDZ basis set (see) from the Pople school. A set of polarizing d-functions (6D) is included on "heavy" atoms and a set of p-functions on hydrogen. Also denoted 6-31G(d,p).
6-31+G*	Augmented 6-31G* basis (see); the single "+" indicates that a set of diffuse (see) s-functions and a set of diffuse p-functions has been added to each heavy atom. Also denoted 6-31+G(d).
6-31++G*	Augmented 6-31+G* basis (see); the second "+" indicates that a set of diffuse s-functions has been added to each hydrogen atom. Also denoted 6-31++G(d).
α electron	An electron with spin up ($\uparrow$).
ACES II	An *ab initio* software package that emphasizes coupled-cluster methods (2,3).
ACPF	Averaged coupled-pair functional. Pretty high level of multi-reference electron correlation, requires skill to use.
AIM	Atoms-in-molecules. An analysis method based upon the shape of the total electron density; used to define bonds, atoms, etc. (4). Atomic charges computed using this theory are often quite different from those from other analyses (e.g., from Mulliken populations). Such charges are probably the most justifiable theoretically, but meet some resistance because the values obtained may be quite different from those from older theories (5).
AM1	Austin model 1. One of the most popular semi-empirical MO theories.
ANO	Atomic natural orbital. Very large basis sets derived from correlated atomic calculations. More expensive to use than the corresponding

	correlation consistent basis sets (e.g., cc-pVTZ) but not often significantly more accurate.
aug-cc-pVDZ	Augmented cc-pVDZ basis (diffuse functions added).
basis set	The set of mathematical functions (basis functions) used for the expansion of the molecular orbitals. Gaussian functions predominate heavily, but occasional papers use the old "Slater" orbitals or functions, which are exponentials.
β electron	An electron with spin down ($\downarrow$).
B3LYP	DFT using Becke exchange functional and Lee-Yang-Parr correlation functional, as well as Hartree-Fock exchange. A hybrid method; the parameters were optimized for thermochemistry, but using a different functional and numerical (basis set-free) code (*6,7*). Very popular.
BAC-MP4	Bond-additivity-corrected MP4. A method in which an MP4 energy is corrected using empirical parameters that depend upon the atoms in the molecule, bond distances, and nearest neighbors. Method developed by Carl Melius (Sandia); not well-described in the open literature.
BD	Brueckner doubles. Very similar to CCSD and QCISD. Although BD only involves doubles, the Brueckner orbitals are those for which the singles contribution is zero. So BSD would be the same thing.
BFGS	An optimization method often used in geometry optimization (four people with initials B, F, G, S).
BLYP	DFT using the Becke exchange functional and the Lee-Yang-Parr correlation functional.
bohr	One atomic unit of distance, equal to 0.5292 Å.
bond function	Special basis functions that are centered on a bond midpoint (for example) rather that on an atomic nucleus. Not commonly used.
BP	Becke-Perdew. A non-local DFT method employing the Becke exchange and Perdew correlation functionals.
BSSE	Basis-set superposition error. An insidious artifact traceable to the fact that one can seldom afford to use a really big basis set. It causes an extra decrease in energy (i.e. more negative energy, greater stability) when two systems (atoms or molecules) are brought together. The energy of one fragment is lowered because its orbitals can use the basis functions on the other fragment, even if the actual electrons and nuclei on the other fragment are not included in the calculation. With a complete basis set, there is no BSSE because the other fragment's basis functions are superfluous. BSSE is usually ignored in thermochemical calculations, except for studies of weak, non-bonded interactions. See "counterpoise."
CADPAC	Another *ab initio* package; acronym stands for "Cambridge analytical derivatives package."
cartesian coordinates	The positions in space of the atoms in a molecule listed as triples (x, y, z).

CASPT2	Complete active space, second-order perturbation theory. This is one formulation of MP2 theory using a CASSCF reference instead of a HF reference. A high-level multireference theory.
CASSCF	Complete active space self-consistent field. A type of MCSCF calculation (see) in which the configurations are chosen to be all those obtainable (i.e., full CI) using a specified number of electrons and a specified set of orbitals. The set of orbitals is called the "active space," and the specified electrons are called "active." In many cases, it requires experience and skill to select the active space correctly.
CBS	Complete basis set. Indicates that some method of basis set extrapolation was applied in an attempt to determine the result that would have been obtained using an infinitely large basis set. The two major extrapolation methods are (1) repeating the calculation with increasingly large basis sets and making an empirical extrapolation, and (2) using analytical formulas that are correct to second-order. See the chapters by Martin and by Petersson in this book.
cc-pVDZ	Correlation-consistent polarized valence double-zeta basis set. The smallest in a series of "correlation consistent" basis sets developed by Dunning and coworkers for high-level calculations. It has been observed that properties computed using successively larger basis sets of this series appear to converge exponentially, presumably to the corresponding CBS values.
cc-pVTZ	Correlation-consistent polarized valence triple-zeta basis set. See cc-pVDZ.
cc-pVQZ	Correlation-consistent polarized valence quadruple-zeta basis set. See cc-pVDZ.
cc-pV5Z	Correlation-consistent polarized valence quintuple-zeta basis set. See cc-pVDZ.
CCD	Coupled-cluster, doubles. A theory of electron correlation that is complete to infinite order but only for a subset of possible excitations (doubles, for CCD). See "CI."
CCSD	Coupled-cluster, singles and doubles. (See CCD.)
CCSD(T)	Coupled-cluster, singles and doubles with approximate triples. (See CCD.) Triples contributions are determined perturbatively. CCSD(T) is the cheapest of the usual approximations to full CCSDT, but appears to be the best. The most popular high-level (i.e., lots of electron correlation) method. Size-consistent. Very expensive.
CCSDT	Coupled-cluster, singles and doubles and triples. (See CCD.) Extra-high level of electron correlation: incredibly expensive, rarely used.
CEPA	Coupled electron pair approximation. An approximate coupled-cluster-type method. Pretty high level.
cGTO	Contracted gaussian-type orbital. The usual basis function; it's a linear combination of gaussian functions with the linear coefficients fixed, then multiplied by an angular function. So one set of p-

functions contains three cGTO's (p_x, p_y, and p_z), i.e, three basis functions. See "primitives."

CI Configuration interaction. A theory of electron correlation. A large set of Hartree-Fock-type configurations (Slater determinants) is used as a many-electron basis set. The coefficient of each configuration is determined variationally so as to minimize the total energy of this wavefunction. Recovers the "dynamic" electron correlation important in bonding. Reliability and expense depend upon the size of the CI, e.g., CISD was popular before coupled-cluster methods caught on. Ordinary, truncated CI (CIS, CISD, etc.) is not size-consistent, so determining bond energies requires "supermolecule" calculations.

CID Configuration interaction, doubles. A CI (see) that only includes those determinants that correspond to double excitations from the reference (which is usually Hartree-Fock).

CIS Configuration interaction, singles. (See CID.) The simplest method for calculating electronically excited states; limited to singly-excited states. Contains no electron correlation and has no effect on the ground state (Hartree-Fock) energy.

CISD Configuration interaction, singles and doubles. A CI (see) that only includes those determinants that correspond to single or double excitations from the reference (which is usually Hartree-Fock). Declining popularity.

CNDO A semi-empirical method ("complete neglect of differential overlap").

contraction Refers to the particular choice of scheme for generating the linear combinations of gaussian functions that constitute a contracted basis set. (See cGTO.) A "generally-contracted" basis set is one in which each primitive is used in many basis functions. A "segmented" basis set, in contrast, is one in which each primitive is used in only one (or maybe two) contracted function.

correlation effects The effect upon the quantity of interest attributable to the inclusion of dynamic electron correlation.

correlation energy The difference between the Hartree-Fock energy and the FCI energy for a given basis set. Most of this energy is attributable to the correlation among the positions of electrons of opposite spin, caused by their coulombic repulsion.

counterpoise The most common, but still controversial, correction for BSSE (see). The BSSE is approximated as the energy difference between (1) an isolated fragment and (2) the fragment accompanied by the basis functions, but not the atoms, of its companion fragment(s).

coupled cluster In diagrammatic perturbation theory, an excited configuration that is "coupled" to the reference configuration.

Davidson correction A correction sometimes made to some types of correlated calculations (esp. CISD and MR-CISD) to estimate higher-order

	contributions to the energy and correct approximately for size-inconsistency.
DCI	See CID. In very recent papers, "DCI" may mean "direct" (see) CI.
DFT	Density-functional theory. *Ab initio* method not based upon a wavefunction. Instead, the energy is computed as a functional of the electron density. Sometimes called Kohn-Sham theory. The correct functional has not yet been found, but many approximations are in use.
diffuse	Diffuse basis functions are typically of low angular momentum (unlike polarization functions) but with much smaller exponents, so that they spread more thinly over space. Usually essential for calculations involving negative ions.
dihedral angle	The angle between two intersecting planes. In a molecule with atoms A-B-C-D, the dihedral angle ϕ(A-B-C-D) is the angle between the planes defined by (A, B, C) and by (B, C, D). By convention, the angle is positive for a right-handed rotation from the first plane to the second, i.e., for a right-handed twist along the sequence A-B-C-D.
DIIS	Direct inversion in the iterative subspace. An extrapolation procedure used to accelerate the convergence of SCF calculations.
direct	Indicates that integrals are not pre-computed and then written to disk, but instead are computed as needed. Done to save I/O time during a calculation.
dynamic correlation	
	All the correlation energy or correlation effect (see) that is not considered "nondynamic" (see) or "static."
DZ	Double-zeta. A basis set for which there are twice as many basis functions are minimally necessary (see "MBS"). "Zeta" (Greek letter ζ) is the usual name for the exponent that characterizes a Gaussian function.
DZP	Double-zeta with polarization. DZ (see) with polarization basis functions added. A polarization set generally has an angular momentum one unit higher than the highest valence function. So a polarization set on carbon is a set of d-functions.
ECP	Effective core potential. The core electrons have been replaced by an effective potential. Saves computational expense. May sacrifice some accuracy, but can include some relativistic effects for heavy elements (see RECP).
ESP	Electrostatic potential. The electrical potential due to the nuclei and electrons in the molecule, as experienced by a test charge.
exchange energy	Also called "exchange correlation energy." The energy associated with the correlation among the positions of electrons of like spin. This is included in Hartree-Fock calculations.
expensive	Requiring large resources: cpu time, memory, and/or disk space.
FCI	Full configuration interaction. A CI (see) that includes **all** possible determinants. FCI is the best wavefunction (and provides the lowest

	varitaional energy) obtainable using a given basis set. Almost never affordable.
frozen	Indicates that some orbitals were not included in the treatment. Usually used as "frozen-core," to indicate that the core orbitals were left uncorrelated in a correlated calculation. Sometimes (esp. in some DFT programs) it means that core orbitals were fixed as taken from calculations on isolated atoms.
G1	Gaussian-1. A composite method for computing thermochemistry, involving extrapolations and a few parameters (*8*). Effectively superseded by G2 and G2(MP2).
G2	Gaussian-2. A composite method for computing thermochemistry, involving extrapolations and a few parameters (*9*). Very popular but quite expensive; practical for up to seven "heavy" atoms (i.e., non-hydrogen) on a Cray supercomputer.
G2(MP2)	Gaussian-2, second-order variant. A composite method for computing thermochemistry, involving extrapolations and a few parameters (*10*). Less-expensive alternative to G2, of comparable accuracy.
GAMESS	General Atomic and Molecular Electronic Structure System. A free *ab initio* software package that emphasizes multi-reference calculations (*11*). There is also a British program with the same name, distinguished as GAMESS-UK (vs. GAMESS-US).
Gaussian	The leader in the *ab initio* software industry is Gaussian Inc., whose most recent program is "Gaussian 94" (*12*). The name refers to the common use of gaussian functions as basis functions in quantum chemistry.
generally contracted	
	See "contraction."
GGA	See "nonlocal."
ghost function	A basis function that is not accompanied by an atomic nucleus, usually for counterpoise corrections for BSSE.
GIAO	Gauge-independent atomic orbitals. A method specially designed for calculation of NMR shifts. Currently used with methods such as HF, MP2, or DFT. Research codes can handle CCSD.
GTO	Gaussian-type orbital. Basis function consisting of a Gaussian function, i.e., $\exp(-\zeta r^2)$, multiplied by an angular function. If the angular function is "cartesian", there are six d-functions, ten f-functions, etc. (6d, 10f). If the angular function is spherical, there will the usual number of functions (5d, 7f).
GVB	Generalized valence bond. A limited type of MCSCF, in which excitations are taken within an electron pair but not between orbitals in different pairs. Dissociation-consistent. If restricted to doubles, is called "perfect pairing" (GVB-PP). If includes both singles and doubles, is called "restricted configuration interaction" (GVB-RCI).
hartree	One atomic unit of energy, equal to 2625.5 kJ/mol, 627.5 kcal/mol, 27.211 eV, and 219474.6 cm^{-1}.

heavy atom	An atom beyond helium, i.e., $Z > 2$.
Hartree-Fock	Simplest and least expensive *ab initio* wavefunction. Involves only a single Slater determinant (a single electron configuration). Orbitals that contain electrons are "occupied," those that are vacant are called "virtual."
HF	Hartree-Fock (see). In the Gaussian programs, HF denotes RHF for closed-shell molecules and UHF for open-shell.
HFS	Hartree-Fock Slater. An older DFT method involving only local, Slater-style exchange. Often synonymous with the Xα method.
HONDO	An *ab initio* program package; genealogically related to GAMESS.
HOMO	Highest occupied molecular orbital. The energy of this orbital approximates the ionization energy of the molecule (Koopmans' theorem).
HOMO-1	The second-highest occupied molecular orbital.
instability	A wavefunction is expressed as a long list of parameters (basis-set expansion) that are adjusted to minimize the total energy. Sometimes the global minimum is not obtained; the local minimum that *is* obtained may be unstable with respect to various perturbations or liberalization of constraints. Such a wavefunction is said to be unstable. One of the more common instabilities is an RHF→UHF instability, which indicates that the UHF solution (different α and β orbitals) is of lower energy than the RHF solution (identical α and β orbitals) for a closed-shell system. This may be encountered, for example, when a bond is stretched.
internal coordinates	Bond lengths, bond angles, and dihedral (torsional) angles; sometimes called "natural coordinates."
IRC	Intrinsic reaction coordinate. An optimized reaction path that is followed downhill, starting from a transition state, to approximate the course (mechanism) of an elementary reaction step.
isodesmic	Refers to a chemical reaction that conserves types of chemical bond. Due to better cancellation of systematic errors, energy changes computed using such reactions are expected to be more accurate than those computed using reactions that do not conserve bond types. Example: $CH_3CH_2F + CH_4 \rightarrow CH_3CH_3 + CH_3F$ for computing the C-F bond strength in fluoroethane.
isogyric	Refers to a chemical reaction that conserves net spin. Due to better cancellation of systematic errors, energy changes computed using such reactions are expected to be more accurate than those computed using reactions that do not conserve spin. Example: $CH_3CH_2F + H^{\cdot} \rightarrow CH_3CH^{\cdot}_2 + HF$ for computing the C-F bond strength in fluoroethane.
Koopmans	His theorem for approximate ionization energies (see HOMO) and poor electron affinities (see LUMO).
LDA	see LSDA.

local	In DFT, a functional that depends only upon the value of the density, $f[\rho]$. This is the simplest and least expensive type of functional.
LSDA	Local spin-density approximation. A DFT method involving only local functionals (i.e., no dependence upon the gradient of the electron density). In the Gaussian programs "LSDA" is equivalent to the "SVWN" keyword.
LST	Linear synchronous transit. An interpolative method used to guess a transition state structure given the structures of the products and of the reactants.
LUMO	Lowest unoccupied molecular orbital. The energy of this orbital is sometimes used to approximate the electron affinity of the molecule, but this usually works badly.
MBPT	Many-body perturbation theory. Synonymous with MP (Møller-Plesset) perturbation theory.
MBS	Minimal basis set. Only enough basis functions are supplied to put all the electrons somewhere; the number of basis functions is equal to the number of orbitals. The most common of these is "STO-3G" (see). Qualitative results at best.
MCPF	Modified coupled-pair functional. Pretty high-level theory.
MCSCF	Multi-configuration self-consistent field. More than one configuration (Hartree-Fock-type determinant) is used to describe the wavefunction. Both the coefficients of the configurations and the orbital coefficients are optimized. This is a limited type of CI (configuration interaction), with the added feature of orbital optimization. See CASSCF.
MNDO	A semi-empirical method ("minimal neglect of differential overlap").
MO	Molecular orbital.
MOLCAS	An *ab initio* software package that emphasizes electron correlation, esp. CASPT2 (*13*).
MOLPRO	An *ab initio* software package that emphasizes electron correlation, esp. very large MRCI (*14*).
MOPAC	The most popular package for semiempirical MO calculations (*15*).
MP2	Second-order Møller-Plesset perturbation theory. Standard Rayleigh-Schrödinger perturbation theory taken to second order. The least-expensive traditional method for including electron correlation. For open-shell cases with a UHF reference, MP2 is sometimes denoted "UMP2."
MP2=fc	Frozen-core MP2 calculation in Gaussian-style notation. (See "Frozen" and "MP2.")
MP2=fu	MP2=full.
MP2=full	MP2 calculation in which the core orbitals are active and **not** frozen.
MP4	Fourth-order MBPT. (See MP2.) At the heart of the BAC-MP4 method.
MRCI	Multi-reference configuration interaction. CI (see) using more that one reference determinant, instead of the usual single Hartree-Fock reference. Among multi-reference theories, MR-CISD (singles and

doubles CI) is popular and high-level (but not dissociation consistent).

Mulliken population

A procedure for assigning net atomic charges within a molecule. It includes an arbitrary choice involving overlap populations, and more seriously is very sensitive (values varying by more than 100%) to the choice of basis set. Still used mostly for convenience, since it has no cost and is included in all *ab initio* program packages. Superseded by NPA and AIM methods of population analysis.

NO   Natural orbital (see).

NBO   Natural bond order. See NPA.

natural orbital   The natural orbitals are those for which the first-order density matrix is diagonal; each will contain some non-integer number of electrons between 0 and 2. Usually discussed in the context of a correlated calculation. RHF calculations give MOs that are also NOs. The NOs are the orbitals for which the CI expansion converges fastest.

nondynamic correlation

Also called "static" correlation. The part of the correlation that is ascribed to the "multireference" nature of the problem at hand, i.e., to the qualitative failure of Hartree-Fock theory to describe the system. The best-known stable molecule with important nondynamic correlation is singlet methylene, CH_2 ($\tilde{a}$ $^1A_1$), for which two configurations are important: $(a_1)^2(b_1)^0$ and $(a_1)^0(b_1)^2$. In many cases, the distinction between nondynamic and dynamic correlation is rather arbitrary. When nondynamic correlation is important, single-reference theories may be unreliable.

nonlocal   In DFT, indicates that a functional of the density gradient (i.e., $f[\nabla\rho]$) is included in addition to a local functional. The most popular NL exchange functional is that by Becke. Popular NL correlation functionals are those by Lee/Yang/Parr and by Perdew. A functional that includes nonlocal terms is sometimes called "gradient-corrected" or a "GGA," which stands for "generalized-gradient approximation."

NPA   Natural population analysis. Considered better than Mulliken populations for assigning atomic charges; results are fairly independent of the basis set. Theory based upon chemical concepts of bonds, lone pairs, etc.

orbital   Usually an eigenfunction of a one-electron hamiltonian, e.g., from Hartree-Fock theory. A spin orbital has an explicit spin (α or β) and a spatial orbital does not. Orbitals are probably the most useful concept from quantum chemistry: one can think of an atom or molecule as having a set of orbitals that are filled with electrons (occupied) or vacant (unoccupied or "virtual").

PES   Potential energy surface. The $3N$-6 (or $3N$-5, for linear molecules) dimensional function that indicates how the molecule's energy depends upon its geometry. (Not to be confused with experimental photoelectron spectroscopy.)

PM3	A semi-empirical method.
PMP2	Spin-projected MP2 energy. Analog of PUHF (see), but for UMP2 energy instead of UHF energy. Likewise, the PMP3 and PMP4 energies are the UMP3 and UMP4 analogs.
polarized	A "polarized" basis set includes functions that are of higher angular momentum than is minimally required. For example, carbon atoms have 1s, 2s, and 2p orbitals, so a polarized basis set would also include at least a set of d-functions. The added functions are often called "polarization functions." Polarization functions help to account for the fact that atoms within molecules are not spherical.
primitives	Also called "primitive functions." The individual gaussian functions that are summed to produce a contracted basis function (cGTO). So a set of p-functions is three basis functions, but may be many primitives ($3n$, where there are n primitives in the cGTO).
pseudopotential	ECP (see).
PUHF	Spin-projected UHF energy. An approximation intended to provide the energy that would result from a UHF calculation if it did not suffer spin-contamination. The PUHF energy is usually lower than the UHF energy because the contributions of higher-multiplicity states, which usually have high energies, have been (approximately) subtracted.
QCI	Quadratic configuration interaction. A CI method to which terms have been added to confer size-consistency. May also be considered to be an approximation to coupled-cluster theory.
QCISD	Quadratic configuration interaction, singles and doubles. (See QCI.)
QCISD(T)	QCISD (see) with a correction for triples. This method is at the core of the popular G2 theory. Usually gives similar results as CCSD(T), of which it is a truncation.
RECP	Relativistic effective core potential. The core electrons have been replaced by an effective potential that is based upon relativistic quantum calculations of the free atoms. Saves cost because of fewer explicit electrons and also includes some relativistic effects, especially the contraction of core s- and p-orbitals.
redundant internal coordinates	
	Internal coordinates (see) that overdetermine the molecular geometry, i.e., are more numerous than $3N$-6 for non-linear molecules or $3N$-5 for linear molecules.
reference	As in "single-reference" or "multi-reference," refers to the number of configurations (or really Slater determinants) in the 0th-order description of the wavefunction. Most methods that don't begin with "MR," "MC," or "CAS" are single-reference methods.
RHF	Spin-restricted Hartree-Fock. Closed-shell singlet with two electrons in each occupied orbital.
ROHF	Spin-restricted open-shell Hartree-Fock. For open-shell molecules. Except for the odd electron(s), there are two electrons in each occuped orbital. (See UHF.)

SCF	Self-consistent field. The orbitals (i.e., the coefficients of the atomic basis functions in each molecular orbital) are adjusted until they are optimal in the mean electric field that they imply. Implicit for Hartree-Fock calculations. Sometimes the term "SCF" is used interchangeably with "HF," but it also applies to most DFT calculations and to all MCSCF calculations.
SCRF	Self-consistent reaction field. A continuum method for treating solvation. The simplest formulation involves placing the molecule in a spherical hole in a polarizable (dielectric) continuum. The molecule polarizes the solvent, which in turn affects the electron distribution in the molecule; this is iterated to self-consistency.
SDCI	See CISD.
scaling	Multiplying calculated results by an empirical fudge factor in the hope of getting a more accurate prediction. Very often done for vibrational frequencies computed at the HF/6-31G* level, for which the accepted scaling factor is 0.893 (16).
segmented	See "contraction."
size-consistent	Describes a calculation that gives the same energy for two atoms (or molecular fragments) separated by a large distance as is obtained from summing the energies for the atoms (or molecular fragments) computed separately. So for a size-consistent method, the bond energy in N_2 is $D_e = 2E(N) - E(N_2)$. For a method that is not size-consistent, a "supermolecule" calculation with a big distance (e.g., 100 Å) is required: $D_e = E(N......N) - E(N_2)$.
SOMO	Singly-occupied molecular orbital (for radicals).
spin-contamination	See UHF.
spin density	The amount of excess α (over β) spin; useful for identifying the location of unpaired electrons in radicals and for interpreting ESR experiments.
spin-polarized	See UHF.
spin-unrestricted	See UHF.
split-valence	Refers to a basis set that is more than minimal (see MBS) for the valence orbitals, i.e., at least VDZ. 3-21G is one example of a split-valence basis. 6-311G might be called a triple-split-valence basis.
static correlation	See "nondynamic correlation."
STO	Slater-type orbital. Basis function with an exponential radial function, i.e., $\exp(-\zeta r)$. Also used to denote a fit to such a function using other functions, such as gaussians. For example, STO-3G is an MBS that uses 3 gaussians to fit an exponential. Exponentials are probably better basis functions than gaussians, but are so much more difficult computationally that they were abandoned by most people a long time ago.
STO-3G	The most popular MBS (see MBS and STO).
supermolecule	A system composed of two or more atoms or molecules separated by large distances. See "size-consistent."

SVWN	Slater exchange functional, Vosko-Wilk-Nusair correlation functional. A local DFT method.
T_1 diagnostic	An indication of how far the canonical HF orbitals differ from the Brueckner orbitals. It has been used as an indicator of multi-reference character and therefore of the reliability of coupled cluster calculations (*17,18*), although this usage has been challenged.
TDA	Tamm-Dancoff approximation. Synonymous with CIS (see).
TZ	Triple-zeta. See "DZ."
TZ2P	Triple-zeta with two sets of polarization functions. See "DZP."
TZP	Triple-zeta with polarization. See "DZP."
UHF	Spin-unrestricted Hartree-Fock. For open-shell molecules. There are separate orbitals for spin-up (alpha) and for spin-down (beta) electrons. UHF wavefunctions are usually not eigenfunctions of spin, and are often contaminated by states of higher spin multiplicity (which usually raises the energy; see PUHF).
UMP2	MP2 theory using a UHF reference. Likewise UMP3, UMP4, UCCSD, etc.
unrestricted	See UHF.
unstable	See instability.
virtual	An unoccupied orbital.
VDZ	Valence double-zeta. A minimal basis is used to describe core electrons, but the valence electrons have twice the minimum number of functions (see "DZ").
$X\alpha$	X-alpha. A venerable, local DFT method in which the coefficient alpha is taken as an adjustable parameter, usually 0.7.
Z-matrix	A common format for specifying molecular geometry in terms of internal coordinates (see).

Literature Cited

1. Brown, R. D.; Boggs, J. E.; Hilderbrandt, R.; Lim, K.; Mills, I. M.; Nikitin, E.; Palmer, M. H. *Pure Appl. Chem.* **1996**, *68*, 387.
2. (a) ACES II, an ab initio program system authored by J. F. Stanton, J. Gauss, J. D. Watts, W. J. Lauderdale, and R. J. Bartlett. The package also contains modified versions of the MOLECULE Gaussian integral program of J. Almlöf and P. R. Taylor, the ABACUS integral derivative program of T. U. Helgaker, H. J. A. Jense, P. Jorgensen, and P. R. Taylor, and the PROPS property integral package of P. R. Taylor. (b) http://www.qtp.ufl.edu/Aces2/
3. Stanton, J. F.; Gauss, J.; Watts, J. D.; Lauderdale, W. J.; Bartlett, R. J. *Int. J. Quantum Chem.* **1992**, *S26*, 879.
4. Bader, R. F. W. *Atoms in Molecules: A Quantum Theory*; Clarendon: Oxford, 1990.
5. Wiberg, K. B.; Rablen, P. R. *J. Comput. Chem.* **1993**, *14*, 1504.
6. Becke, A. D. *J. Chem. Phys.* **1993**, *98*, 5648.
7. Stephens, P. J.; Devlin, F. J.; Chabalowski, C. F.; Frisch, M. J. *J. Phys. Chem.* **1994**, *98*, 11623.

8. Pople, J. A.; Head-Gordon, M.; Fox, D. J.; Raghavachari, K.; Curtiss, L. A. *J. Chem. Phys.* **1989**, *90*, 5622.
9. Curtiss, L. A.; Raghavachari, K.; Trucks, G. W.; Pople, J. A. *J. Chem. Phys.* **1991**, *94*, 7221.
10. Curtiss, L. A.; Raghavachari, K.; Pople, J. A. *J. Chem. Phys.* **1993**, *98*, 1293.
11. (a) Schmidt, M. W.; Baldridge, K. K.; Boatz, J. A.; Elbert, S. T.; Gordon, M. S.; Jensen, J. H.; Koseki, S.; Matsunaga, N.; Nguyen, K. A.; Su, S. J.; Windus, T. L.; Dupuis, M.; Montgomery, J. A. *J. Comput. Chem.* **1993**, *14*, 1347. (b) http://www.msg.ameslab.gov/GAMESS/GAMESS.html
12. (a) *Gaussian 94*. Frisch, M. J.; Trucks, G. W.; Schlegel, H. B.; Gill, P. M. W.; Johnson, B. G.; Robb, M. A.; Cheeseman, J. R.; Keith, T.; Petersson, G. A.; Montgomery, J. A.; Raghavachari, K.; Al-Laham, M. A.; Zakrzewski, V. G.; Ortiz, J. V.; Foresman, J. B.; Cioslowski, J.; Stefanov, B. B.; Nanayakkara, A.; Challacombe, M.; Peng, C. Y.; Ayala, P. Y.; Chen, W.; Wong, M. W.; Andres, J. L.; Replogle, E. S.; Gomperts, R.; Martin, R. L.; Fox, D. J.; Binkley, J. S.; Defrees, D. J.; Baker, J.; Stewart, J. P.; Head-Gordon, M.; Gonzalez, C.; Pople, J. A. Gaussian, Inc., Pittsburgh, PA, 1995. (b) http://www.gaussian.com/
13. (a) *MOLCAS version 3*. Andersson, K.; Blomberg, M. R. A.; Fülscher, M. P.; Karlström, G.; Kellö, V.; Lindh, R.; Malmqvist, P.-Å.; Noga, J.; Olsen, J.; Roos, B. O.; Sadlej, A. J.; Siegbahn, P. E. M.; Urban, M.; Widmark, P.-O. University of Lund, Sweden, 1994. (b) http://garm.teokem.lu.se/MOLCAS/index.html
14. http://www.tc.bham.ac.uk/molpro/
15. http://www.iti2.net/jstewart/
16. Scott, A. P.; Radom, L. *J. Phys. Chem.* **1996**, *100*, 16502.
17. Lee, T. J.; Taylor, P. R. *Int. J. Quantum Chem.: Quantum Chem. Symp.* **1989**, *23*, 199.
18. Jayatilaka, D.; Lee, T. J. *J. Chem. Phys.* **1993**, *98*, 9734.

Subject Index

458